Proceedings of the Institution of Mechanical Engineers

Bulk 2000:
Bulk Materials Handling—Towards the Year 2000

International Conference

29–31 October 1991
Kensington Conference Centre, London

Sponsored by
Bulk Materials Handling Committee of the
Process Industries Division of the
Institution of Mechanical Engineers

In association with
British Materials Handling Board
Institution of Chemical Engineers
Mechanical Handling Engineers' Association
Solids Handling and Processing Association
Institution of Engineers, Australia
Verein Deutscher Ingenieure
American Society of Mechanical Engineers
Japanese Society of Mechanical Engineers

IMechE 1991–12

Published for IMechE by
Mechanical Engineering Publications Limited

First Published 1991

ISBN 0 85298 766 8

A CIP catalogue record for this book is available from the British Library.

Printed by Waveney Print Services Ltd, Beccles, Suffolk

Contents

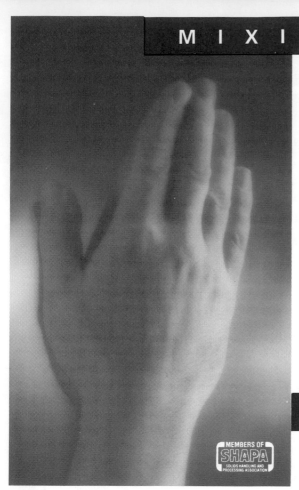

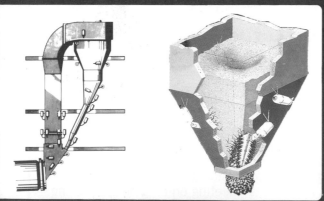

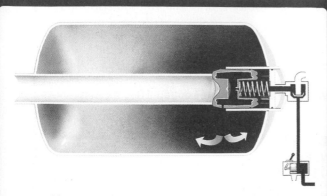

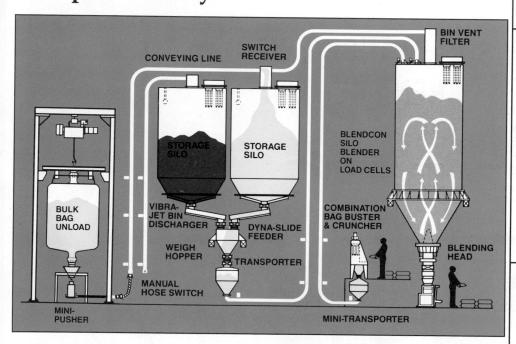

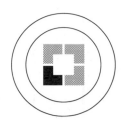

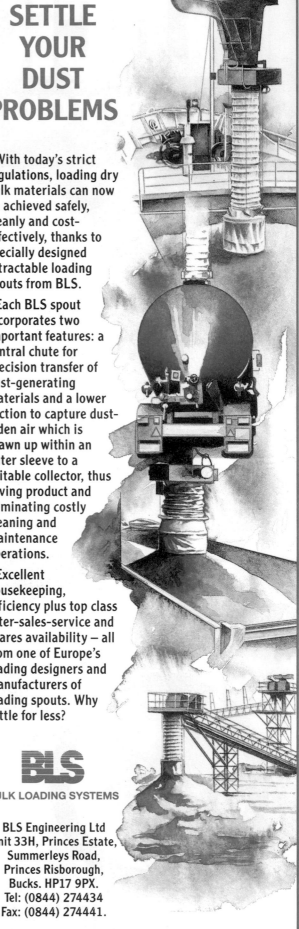

C418/035

Strength criteria for steel silo structures

J M ROTTER, MA, PhD, CPEng, FIEAust
Department of Civil Engineering, University of Edinburgh, Scotland

SYNOPSIS Knowledge of silo loadings, silo structural behaviour and silo strength has advanced very considerably over the last decade. Some areas of knowledge have now been very extensively researched and further work is probably not urgent. However, many other aspects of silo structural behaviour and design have never been addressed, and some of these have been the cause of recent catastrophic failures. These failures cannot simply be attributed to the use of low design pressures, because the causes of failure were often quite different from the events imagined by the designer. Many failures would not have been prevented by increased codified loadings. The many failures indicate that a number of issues still need to be addressed in silo structural design.

This paper reviews the current state of knowledge applicable to the design of steel silo structures for strength. A brief review is given of the critical features and common failure modes found in modern steel silos. For space reasons, complete design recommendations are not presented, but extensive references are given to recent studies in which more detailed information may be found. A short summary of current knowledge of the pressures on silo walls is followed by a review of each principal mode of structural failure.

1 INTRODUCTION

Many new large circular steel silos have been built in recent years. Some are flat-bottomed and supported directly on the ground, whilst others are elevated with conical hoppers (Fig. 1). This paper describes the commoner aspects of steel silo strengths. The term silo is used here for all bulk solids storage structures: silos, hoppers, bunkers, bins, tanks or containment structures.

Steel silos have several advantages over concrete silos. They are much lighter structures, quick to erect and dismantle, deforming readily and reversibly when subject to unsymmetrical loads, and placing smaller loads on their foundations. Steel silos are increasingly being built as light squat structures (H/D < 1) supported directly on the ground.

This paper deals with the strengths of cylindrical steel silos under symmetrical filling and discharge conditions, eccentric filling, eccentric discharge, wind when empty, and earthquakes, and it outlines conditions for the failure of conical hoppers. Column-supported elevated steel silos are also mentioned briefly, as several special and difficult features occur in these structures.

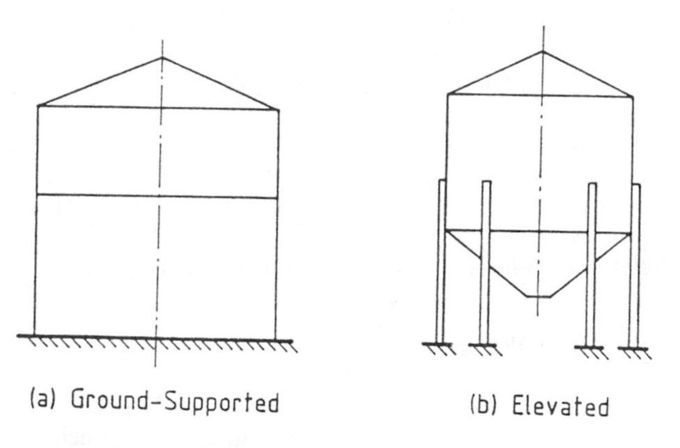

Fig. 1 Elevated and ground-supported silos

2. PRESSURES ON SILO WALLS

2.1 Symmetrical Filling and Discharge

The walls of silos are subjected to both normal pressures and frictional shears or tractions which vary all over the wall. Normal pressures give rise to circumferential (or hoop) tensions, and frictional tractions cause cumulative axial (or vertical) compressive stresses in the silo wall. However, the real loading pattern is much more complex, and other load cases give rise to different stress patterns in the shell.

Janssen's [1] theory is probably the simplest silo pressure theory for vertical walls. It is widely accepted that Janssen values are close to the wall pressures on first filling. Much larger pressures can occur during discharge. These are termed overpressures or flow pressures, and are often explained in terms of a 'switch' from an active stress field after filling to a passive stress field during discharge [2-4]. Several theories attempt to predict flow pressures. The large differences between these theories are reflected in the considerable differences between national silo design standards.

Many designers are confused by the wide differences between the design pressures given in different national codes, but the resulting steel silo designs are generally similar. This is because normal pressures do not control the design of many cylindrical walls, and the initial-filling condition is generally critical for discharge hoppers [5].

Almost all silo pressure theories assume a perfect silo geometry with homogeneous isotropic stored solid. The pressures are assumed constant at a given height. A cylindrical silo can carry symmetrical pressures well. By contrast, experiments on full scale silos [6] show that unsymmetrically-distributed patches of high local pressure occur on the wall during flow [7]. These have been shown [8] to be potentially very serious in steel silos. Unfortunately, no known silo pressure theory treats randomly-occurring unsymmetrical pressures (patch pressures) and existing experiments are too limited to define these patches reliably.

2.2 Eccentric Filling and Cleanout

Large squat silos built on the ground are sometimes eccentrically filled (Fig. 2), as this allows for faster rates of filling. These silos are also often discharged from a central slot across the diameter beneath the base, or by using front-end loaders (Fig. 2). All three conditions can lead to heaps of solid being piled up against one side of a circular wall, leading to unsymmetrical loading. These pressures have not yet been investigated, but Gaylord and Gaylord [9] and Rotter [10] have suggested design pressures. Verification of these proposals is required.

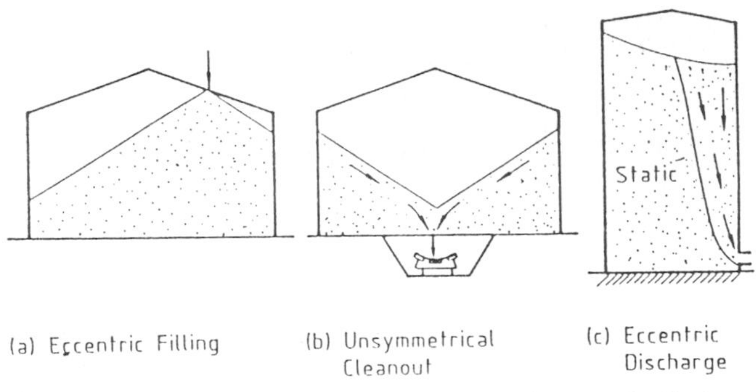

Fig.2 Eccentric filling, eccentric cleanout and eccentric discharge

2.3 Eccentric Discharge

The discharge of solids from a point eccentric to the vertical centreline of a silo is often desired by silo operators (Fig. 2). The use of multiple outlets at the base may be needed, or an outlet at the wall may be required to reduce mechanical handling costs. Many failures have arisen from the use of eccentric discharge, in both concrete and steel silos. Failures are often accidental, but in steel silos they are usually catastrophic, leading to total collapse.

The pressures arising from eccentric discharge have been investigated in a many experimental studies [6,7,11,12], and some theories advanced for the design of silos under eccentric discharge [13-15]. However, these experiments pose great difficulties of interpretation. The pattern of solids flow often was not or could not be observed, so it is difficult to develop matching theories of pressure for structural and design predictions. The pattern of pressures changes with time during the discharge [12], and the most detrimental instant or pattern is not easily determined. Many investigations have reported only the highest or the mean pressures observed at different points in the structure, but even the simplest analysis of steel silo buckling failures shows that this is not the critical information. Most studies of eccentric discharge have been concerned with bending moments in the silo wall, but bending moments do not even give a qualitative explanation of the observed failure modes of thin-walled steel silos [16]. A discussion of failure under eccentric discharge is given below.

3. MODES OF FAILURE IN SILOS

Silos are subject to many different loading conditions, so that many different modes of failure are possible. However, the stresses in the wall generally lead to one of only a few modes. These are:

For the cylindrical wall:

- bursting
- buckling under axial (vertical) compressive stresses
- buckling under circumferential compressive stresses
- buckling under membrane shear stresses

For the conical hopper:

- plastic collapse in the hopper body
- rupture of the hopper/ring junction
- buckling of the ring support
- plastic collapse of the ring support

These modes are discussed briefly below. More extensive but less recent advice on steel silo design is given in Trahair et al [17], Gaylord and Gaylord [9] and Rotter [18,19].

4. CYLINDRICAL WALLS UNDER BULK SOLIDS LOADING

4.1 Bursting of the Cylindrical Wall

The literature on over-pressures in silos suggests that very high internal or 'switch' pressures are to be expected on limited zones of the wall. It is commonly recommended that the entire wall be designed to sustain the envelope of highest pressures. However, very few bursting failures are seen in steel silos in service, and even these are generally attributed to poor detailing rather than excessive pressures. This mode is relatively easy to design against, though practical designs are not often governed by bursting.

4.2 Buckling under Symmetrical Axial Compression

The commonest mode of failure of silos in service is buckling under axial or vertical compressive stresses. Under symmetrical filling loads this is usually the controlling load case for much of the wall. Other load cases induce higher axial stresses over limited regions on the wall. In particular, eccentric discharge [15], eccentric filling [10], unsymmetrical patches of high pressure during discharge [8], earthquake loading on squat silos [20] and column-support forces in elevated silos [21,22] are all potential causes of buckling failure.

In traditional design [9,17], a single empirical strength relation was used to define the probable buckling strength. However, the strength is extremely sensitive to initial wall imperfections [23], which in turn depend on the silo fabrication. Unfortunately, most design rules are empirically derived from laboratory tests, which have little relevance to silos in service. Recent work has therefore been on the buckling of cylinders with practical imperfections [24,25].

Stored solids pressurise the cylinder and increase the buckling strength. Useful advice may be found in ECCS [26] and Rotter and Teng [25]. The bulk solid has a finite shear strength, so it restrains the silo wall against buckling. Information on the strengthening effect of granular solids may be found in Rotter [27] and Rotter and Zhang [28]. Other failure modes which have recently been studied include "elephant's foot" buckling at a boundary condition [29] and the buckling of lap jointed cylinders [30].

Many common loading cases lead to high local axial compressive stresses. Very few studies have explored buckling under local high stresses, the first really useful study being Peter [31]. The only work on imperfection-sensitivity under local high stresses appears to be Teng and Rotter [32]. The only known design suggestions of general applicability are those of Rotter [15]. Much further work in this area is needed.

4.3 Buckling under Eccentric Filling

Under eccentric filling or eccentric clean-out of squat silos (Fig. 2), the non-uniform normal wall pressures cause much higher compressive stresses than the frictional drag. The wall is also under non-uniform membrane shear. This problem has been studied theoretically by Rotter [10] and Gaylord and Gaylord [9]. Experiments on eccentric filling show that two buckle types occur: an 'elephant's foot' buckle at the base, and a membrane shear buckle. The buckling strengths of cylinders under this varying membrane shear were studied by Jumikis and Rotter [33], who produced a simple design rule.

4.4 Buckling under Earthquake Loading

The quasi-static response of squat silos under earthquake loading has a close similarity to the response under eccentric filling, and failures by buckling under axial compression and in membrane shear are to be expected [20]. Elevated silos may be treated in the same manner as elevated tanks [17].

4.5 Buckling under Eccentric Discharge

As noted above, most studies of eccentric discharge have been concerned with reinforced concrete silos, in which failure is by circumferential bending in an inadequately reinforced wall. Thin-walled steel silos respond quite differently. Experiments on the buckling failure of thin-walled silos under eccentric discharge have been described by Rotter et al [16]. In all, approximately 150 experiments were undertaken in this series, using different silo geometries and discharge eccentricities. Observations of the flow patterns in the funnel-flowing stored solid were made. Many silo geometries can successfully support eccentric discharge without distress and the buckling modes are demonstrably elastic. These observations indicate that the simple theory of Jenike [13], based on bending moments causing yielding, cannot be right.

The only published rational theory for the pressures and wall stresses under eccentric discharge, together with an appropriate criterion for buckling strength assessment, appears to be that of Rotter [15].

4.6 Buckling above Column Supports

In elevated silos supported on columns, high local vertical forces occur at the columns. Several methods (Fig. 3) are used by designers to avoid buckling problems above the column: many columns, engaged columns, columns extending to the roof, stiffening rings, and deep skirts below the transition are all common. Each technique may be appropriate to a different class and size of silo, and each has been successfully used.

Information on the stresses in column-supported silos is fairly scarce, and very little is known about buckling failures above column supports. The only known study [32] shows that the mean stress above the column termination is critical, and that imperfections near this point can affect the strength markedly.

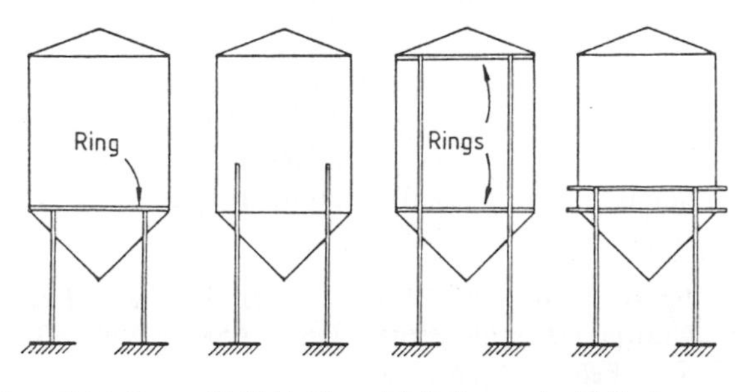

Fig. 3 Alternative column support arrangements

5. CYLINDRICAL WALLS UNDER WIND

Both silos and tanks for liquid storage are susceptible to buckling failure under external pressure when empty. External pressure arises from severe windstorms, and rapid cooling or solids withdrawal with inadequate venting. Both conditions are more severe in squat ground-supported structures, because the wall is thin and the diameter large.

The reference buckling pressure is generally the classical uniform external critical pressure [23]. However, many silos which are most at risk are light-gauge stiffened structures, so buckling predictions for eccentrically stiffened orthotropic shells are often used [26]. Several different buckling heights must be checked, as the critical mode cannot normally be deduced by inspection [19]. Procedures for orthotropic (corrugated) sheeting, smearing of stiffeners, determination of an equivalent uniform pressure and working stress design pressures were given by Trahair et al [17] and greatly improved by Blackler [34].

Silo structures are also susceptible to being torn from their foundations in windstorms. A simple design assuming the wind load to be carried by the silo as a cantilever in global bending is at great risk. The problem has been described by Gaylord and Gaylord [9].

6. CONICAL HOPPERS AND TRANSITION RINGS

6.1 The Hopper Body

The most important loading condition for the steel hopper is often initial filling [5,35], with the minimum value for the cylinder wall friction coefficient to maximise the total hopper load. Stresses in the hopper body follow the predictions of shell membrane theory closely, but the most highly stressed point varies according to the depths of the hopper and cylinder [5]. Welded hoppers may fail in a plastic collapse mechanism, but bolted hoppers often rupture down a meridional seam. In large hoppers with small cylinders, these potential failures are initiated at quite different locations.

The first thorough study of the plastic collapse of welded hoppers was Teng and Rotter [36], which included multi-strake hoppers of changing thickness. Under large deformations, higher hopper loads can be carried, and failure is by rupture at the top of the hopper [37]. This matches most field observations. The top of the hopper is a critical field joint, which supports most of the weight of the stored solid, so it must be carefully fabricated. Detailed design recommendations for light gauge bolted hoppers have been given by Rotter [35].

6.2 The Continuously-Supported Transition Ring

The hopper carries its load chiefly by meridional (sloping) tension. The tension at the hopper top may be from global equilibrium (Fig. 4a). The vertical component of this tension is resisted by the support (Fig. 4b), but the radial component must be resisted by a ring. Consequently, the ring carries a high circumferential compressive stress. If there is no ring, the hopper, cylinder and skirt will deform to form an effective ring.

Failure modes of the ring involve the hopper, cylinder and skirt and involve either plastic collapse or buckling. Outlines of the potential failure modes and design rules were given by Rotter [38,35]. The hopper prevents in-plane ring buckling, and out-of-plane buckling is quite local. Cross-section stiffeners do not strengthen the structure much.

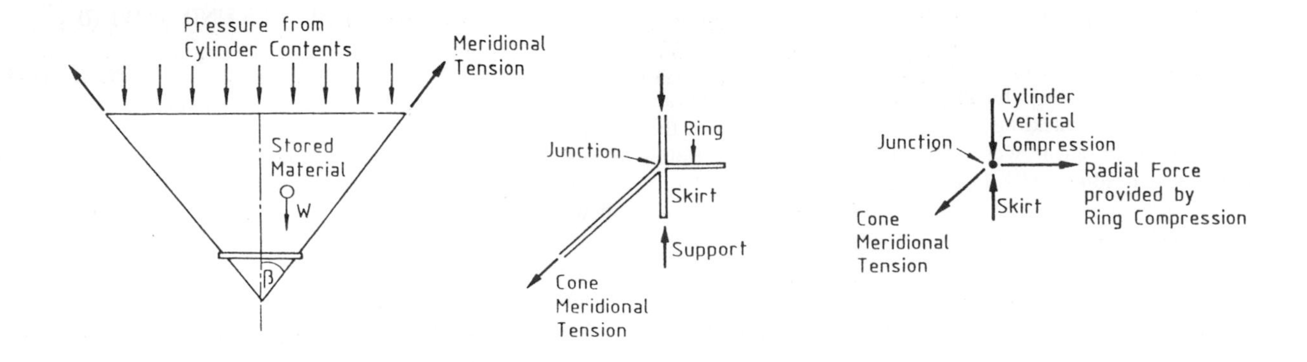

Fig. 4 Hopper and transition ring equilibrium

6.3 Rings in a Column-Supported Silo

Larger stresses occur in column-supported silos, where the ring is a bowgirder beam flange spanning between column supports. In larger silos, a beam section like that of Fig. 3d is often used, but sometimes the columns are extended to the eaves (Fig. 3c). Lighter silos are often built with terminating engaged columns (Fig. 3b), but occasionally the upper ring is placed below the silo eaves (Fig. 3c). These four structural types act differently and are not simple to analyse. Design recomendations for some configurations are given by Gaylord and Gaylord [9] and Rotter [38].

All the above structures must be designed to resist both buckling and plastic collapse of the transition ring. A recent study of ring buckling under the non-uniform stresses of a column-supported structure [39] showed that design to the highest compressive stress provides a conservative criterion for buckling, but it is sometimes very conservative.

7. SUMMARY

A brief review has been presented of current knowledge of the behaviour, strength and design of circular steel silos. A brief description has been given of each of the principal strength considerations for these large and complex shell structures. Extensive references to more detailed sources have been given.

8. ACKNOWLEDGEMENTS

This paper describes part of a research programme into the loading, behaviour, analysis and design of silo and tank structures at the Universities of Sydney and Edinburgh. Support for this program from the Australian Research Grants Scheme, these two universities and cooperating commercial organisations is gratefully acknowledged.

9. REFERENCES

1) Janssen, H.A. "Versuche uber Getreidedruck in Silozellen", Zeitschrift des Vereines Deutscher Ingenieure, 39, No 35, 1895, pp 1045-1049.

2) Jenike, A.W. and Johanson, J.R. " Bin Loads", J. Struct. Divn., ASCE, 94, ST4, 1968, 1011-41.

3) Nanninga, N. "Gibt die ubliche Berechtungsart der Drucke auf die Wande und den Boden von Silobauten sichere Ergebnisse?" Der Ingenieur, 68 (44), 1956, 190-194.

4) Walters, J.K. "A Theoretical Analysis of Stresses in Silos with Vertical Walls", Chem Eng Sci, 28, 1973, 13-21.

5) Rotter, J.M. "On the Significance of Switch Pressures at the Transition in Elevated Steel Bins", 2nd Int Conf Bulk Matls Stor Handl & Tptn, IE Aust., 1986, 82-88.

6) Hartlen, J. et al. "The Wall Pressure in Large Grain Silos", Swedish Council for Building Res, Stockholm, D2:1984.

7) Ooi, J.Y., Rotter, J.M. and Pham, L. "Systematic and Random Features of Measured Pressures on Full-Scale Silo Walls", Engrg Structures, Vol. 12, No. 2, April 1990, pp 74-87.

8) Rotter, J.M., Pham, L. and Nielsen, J. "On the Specification of Loads for the Structural Design of Bins and Silos", Proc., 2nd Int Conf on Bulk Matls Storage Handling & Transportn, IEAust., Wollongong, July 1986, pp 241-247.

9) Gaylord, E.H. and Gaylord, C.N. Design of Steel Bins for Storage of Bulk Solids, Prentice Hall, 1984.

10) Rotter, J.M. "Structural Effects of Eccentric Loading in Shallow Steel Bins", Proc., 2nd Int Conf on the Design of Silos for Strength and Flow, Stratford upon Avon, Nov. 1983, pp 446-463.

11) Hampe, E. and Kaminski, M. "Der Einfluss exzentrischer Entleerung auf die Druckverhaltnisse in Silos", Bautechnik, 61, 3, 1984, pp 73-82 and 4, pp 136-42.

12) Gale, B.R., Hoadley, P.J. and Schmidt, L.C. "Aspects of Eccentric Discharge of Granular Material from a Circular Silo", Proc., 2nd Int Conf on Bulk Matls Storage Handling & Transportn, IEAust., Wollongong, July 1986, pp 258-263.

13) Jenike, A.W. "Denting of Circular Bins with Eccentric Drawpoints", Jnl Struct. Div., ASCE, Vol. 93, No. ST1, Feb 1967, pp 27-35.

14) Wood, J.G.M. "The Analysis of Silo Structures Subject to Eccentric Discharge", Proc., 2nd Int. Conf. on Design of Silos for Strength and Flow, Stratford-upon-Avon, 1983, pp 132-144.

15) Rotter, J.M. "The Analysis of Steel Bins Subject to Eccentric Discharge", Proc., 2nd Int Conf. on Bulk Materials Storage Handling and Transportation, IEAust., Wollongong, July 1986, pp 264-271.

16) Rotter, J.M., Jumikis, P.T., Fleming, S.P. and Porter, S.J. "Experiments on the Buckling of Thin-Walled Model Silo Structures", J. Constructional Steel Res., 13, 4, 1989, pp 271-299.

17) Trahair, N.S., Abel, A., Ansourian, P., Irvine, H.M. and Rotter, J.M. Structural Design of Steel Bins for Bulk Solids, Australian Institute of Steel Construction, Sydney, 1983.

18) Rotter, J.M. (Ed) "Design of Steel Bins for the Storage of Bulk Solids", University of Sydney, 1985.

19) Rotter, J.M. "Recent Research on the Strength of Circular Steel Silos", Proc., International Conference: 'Silos- Forschung und Praxis' Universitat Karlsruhe, October 1988, pp 265-286.

20) Rotter, J.M. and Hull, T.S. "Wall Loads in Squat Steel Silos during Earthquakes", Engrg Structs, 11, 3, July 1989, pp 139-147.

21) Rotter, J.M. "Analysis of Ringbeams in Column-Supported Bins", Eighth Austral. Conf. on Mechanics of Structs and Matls, Univ. of Newcastle, Aust., Aug 1982.

22) Ory, H. and Reimerdes, H.G. "Stresses in and Stability of Thin Walled Shells under Non-ideal Load Distribution", Proc. Int. Colloq. Stability of Plate and Shell Structs, Gent, Belgium, 1987, ECCS, 555-561.

23) Yamaki, N. "Elastic Stability of Circular Cylindrical Shells", North Holland, Elsevier Applied Science Publishers, Amsterdam, 1984.

24) Rotter, J.M and Seide, P. "On the Design of Unstiffened Cylindrical Shells Subject to Axial Load and Internal Pressure", Proc., Int. Colloq. Stability of Plate and Shell Structs, Ghent, 1987, 539-548.

25) Rotter, J.M. and Teng, J.G. "Elastic Stability of Cylindrical Shells with Weld Depressions", J. Struct. Engrg, ASCE, 115, 5, May 1989, pp 1244-1263.

26) ECCS European Recommendations for Steel Construction: Buckling of Shells, 4th edition, 1988, European Convention for Constructional Steelwork, Brussels.

27) Rotter, J.M. "Buckling of Ground-Supported Cylindrical Steel Bins under Vertical Compressive Wall Loads", Proc., Metal Structures Conference, IEAust., Melbourne, May 1985, pp 112-127.

28) Rotter, J.M. and Zhang, Q. "Elastic Buckling of Imperfect Cylinders containing Granular Solids", J. Struct. Engrg, ASCE, 116, 8, August 1990, pp 2253-2271.

29) Rotter, J.M. "Local Inelastic Collapse of Pressurised Thin Cylindrical Steel Shells under Axial Compression", J. Struct. Engrg, ASCE, 116, 7, July 1990, pp 1955-1970.

30) Rotter, J.M. and Teng, J.G. "Elastic Stability of Lap-Jointed Cylinders", J. Struct Engrg, ASCE, 115, 3, March 1989, pp 683-697.

31) Peter, J. "Zur Stabilitat von Kreiscylinderschalen unter ungleichmassig verteilten axialen Randbelastungen", Dissertation, Univ Hannover, 1974.

32) Teng, J.G. and Rotter, J.M. "A Study of Buckling in Column-Supported Cylinders", Proc., IUTAM, Symp. on Contact Loading and Local Effects in Thin-Walled Plated and Shell Structures, Prague, August 1990, Preliminary Report 1990, pp. 39-48.

33) Jumikis, P.T. and Rotter, J.M. "Buckling of Cylindrical Shells under Non-Uniform Torsion", Proc., 10th Austral. Conf. on Mechs of Structures & Materials, Adelaide, Aug. 1986, pp 211-216.

34) Blackler, M.J. "Stability of Silos and Tanks under Internal and External Pressure", PhD Thesis, University of Sydney, 1986.

35) Rotter, J.M. "Structural Design of Light Gauge Silo Hoppers", J. Struct. Engrg, ASCE, 116, 7, July 1990, pp 1907-1922.

36) Teng, J.G. and Rotter, J.M. "Plastic Collapse of Restrained Steel Silo Hoppers", J. Construct. Steel Res., 14, 2, 1989, pp 139-158.

37) Teng, J.G. and Rotter, J.M. "The Collapse Behaviour and Strength of Steel Silo Transition Junctions" in two parts, J. Struct. Engrg, ASCE, 1991, (accepted for publication).

38) Rotter, J.M. "Analysis and Design of Ringbeams", in Design of Steel Bins for the Storage of Bulk Solids, edited by J.M. Rotter, University of Sydney, March 1985, pp 164-183.

39) Teng J.G. and Rotter, J.M. "Buckling of Rings in Column-Supported Bins and Tanks", Thin Walled Structs, 7, 3 and 4, 1989, pp 257-280.

C418/048

Silo-quaking – a pulsating load problem during discharge in bins and silos

A W ROBERTS, BE, PhD, ASTC, FIEAust, MIMechE, FTS and O J SCOTT, BE, ME
Department of Mechanical Engineering, University of Newcastle, New South Wales, Australia
S J WICHE, BE
Tunra Bulk Solids Research Associates, University of Newcastle, Australia

SYNOPSIS This paper examines the phenomenon of pulsating loads in bins. Factors affecting the flow behaviour such as bin flow patterns and flow properties of bulk solids, notably wall friction are reviewed with a view to identifying the conditions under which flow modes may change and pulsating loads may be induced. Mechanisms under which pulsating loads may occur in tall mass-flow and funnel-flow bins and silos and in squat funnel-flow and expanded-flow bins are investigated. Two case study examples of coal bins exhibiting severe pulsating loads are presented. Studies using a pilot scale mass-flow silo with wheat are described and results depicting the conditions under which load pulsations occur are given.

1 INTRODUCTION

Over the past three decades much progress has been made in the theory and practice of bulk solids handling. Reliable test procedures for determining the strength and flow properties of bulk solids have been developed and analytical methods have been established to aid the design of bulk solids storage and discharge equipment. There has been wide acceptance by industry of these tests and design procedures and, as a result, there are numerous examples in various countries of the world of modern industrial bulk solids handling installations which are performing efficiently and reliably.

Yet, as is often the case, the solution of one problem which leads to an improvement in plant performance exposes other problems which require further research and development. This applies particularly to gravity flow in storage bins and silos where the application of known theories for reliable discharge, such as by mass-flow, can give rise to dynamic or pulsating flow effects. These effects are normally imperceptible as far as bin discharge is concerned having no detrimental effect on the plant operation. However, the pulsating flow can have a significant influence on the loads acting on bin walls.

Self excited, pulsating flows in bins and silos are known to occur in a number of industrial installations. The phenomenon is often described as 'silo quaking' and gives rise to significant shock loads on the bin or silo structure. Despite the common knowledge that such loadings may exist, there is limited information to indicate precisely what is actually happening during gravity discharge. A much greater concentration of research on this subject is required. In the meantime, there is a need to identify the conditions under which cyclical shock loads or 'silo quaking' is likely to occur so that designers of bins and silos can be better informed. To achieve this goal, the following course of action is suggested;

(i) Assemble as much information as possible from case study examples from actual industrial installations where the 'silo quaking' phenomenon is known to exist.

(ii) Examine current bulk solid testing procedures to ensure that the parameters measured, such as wall friction, accurately reflect the true conditions experienced in practice.

(iii) Examine current theories on bin flow and wall loadings to highlight any shortcomings and identify areas where further refinement is necessary. In this respect, the present state-of-knowledge relates primarily to steady state conditions, either static or steady flow. A better understanding of flow transients is required.

This paper reviews the foregoing objectives. The material presented is based on the research and industrial consulting activities of the bulk solids research group of the University of Newcastle, Australia. The paper includes test results on a pilot scale model silo; this particular area of research is part of collaborative work being conducted jointly with the Technical University of Denmark.

2 BASIC CONCEPTS

It is generally known that the phenomenon of 'silo quaking' may be due to two primary causes which may occur either independently or in an interactive mode:

(i) Variations of the wall friction in the hopper which causes the flow to oscillate between mass-flow and funnel-flow. The limits for mass and funnel-flow need to be defined.

(ii) Variations in the degree of dilation in the different flow zones within the hopper during gravity discharge.

Some basic concepts which relate to the silo quaking problem are now reviewed.

2.1 Mass-Flow and Funnel-Flow Limits for Symmetrical Hoppers

(a) Established Theory Due to Jenike

The mass-flow and funnel-flow limits have been defined by Jenike on the assumption that a radial stress field exists in the hopper [1]. These limits are well known and have been used extensively and successively in bin design. The limits for axi-symmetry or conical hoppers and plane symmetry depend on the hopper half angle α and effective angle of internal friction δ. Once the wall friction angle ϕ and effective angle of internal friction have been determined by laboratory tests, the hopper half-angle may be determined. In functional form

$$\alpha = f(\phi, \delta) \tag{1}$$

Jenike showed that the bounds for mass-flow in conical hoppers are well defined. Since flow in a hopper is based on the radial stress field theory, no account is taken of the influence of the cylinder on the flow pattern developed, particularly in the region of the transition. For plane flow hoppers, the bounds between mass and funnel-flow are much less severe than for conical hoppers. In plane flow hoppers, much larger hopper half-angles are possible which means that the discharging bulk solid will undergo a significant change in direction as it moves from the cylinder to the hopper. In view of this, Jenike has recommended the application of design limits. An empirical relationship for these limits has been established [2].

(b) Modification to Limits - More Recent Research

More recent research has shown that the mass-flow and funnel-flow limits require further explanation and refinement. For instance, Jenike [3] published a new theory to improve the prediction of funnel-flow; this led to new limits for funnel-flow which give rise to larger values of the hopper half-angle. In a very comprehensive study of flow in silos, Benink [4] has identified three flow regimes, mass-flow, funnel-flow and an intermediate flow as illustrated in Figure 2. Whereas the radial stress theory ignores the surcharge head, Benink has shown that the surcharge head has a significant influence on the flow pattern generated.

2.2 Wall Friction

(a) Effect on Flow Pattern

Wall friction has a major influence on the behaviour of bulk solids in bins. The subject of wall friction has been discussed in several papers [5-8]. It has been shown that friction depends on the interaction between the relevant properties of the bulk solid and lining surface, with external factors such as the loading condition and environmental parameters such as temperature and moisture having a significant influence.

The interaction between surface roughness and particle size is one set of parameters that effects the magnitude of the friction generated. For instance, tests conducted using coal in contact with two samples of Stainless Steel type 304 with 2B finish, one a 3mm thick plate and the other a 5mm thick plate, showed an appreciable difference in the measured Wall Yield Loci [6]. For the range of normal pressures consistent with those experienced in a hopper, the 5mm thick plate yielded friction angles in the order of 3° larger than that of the 3mm plate. Even though both samples were within the accepted tolerance band for a 2B finish, the roughness of the 5mm plate was observably higher than that of the 3mm plate; the two plates had been manufactured by different rolling processes.

Of the various external factors, vibrations are one of the most significant in terms of the influence in lowering the magnitude of wall friction [9-10]. In particular, high frequency vibrations in the order of 100 Hertz and higher, and low amplitude can reduce the wall friction angle by several degrees.

During continued operation, the surface condition of bins and silos can change leading to a change in the wall friction. For example, a surface may become smoother during operation leading to a reduction in wall friction. On the other hand, surface corrosion due to bulk solid and liner contact can lead to corrosion and surface pitting, resulting in a substantial increase in wall friction angle.

Any of the foregoing influences can change the flow pattern in a bin from funnel-flow to mass-flow and vice versa. In many cases, this can be a contributing factor in the creation of pulsating flow. Furthermore, it is well known that a reduction in wall friction will lead to an increase in the magnitude of bin wall pressures.

(b) Measurement of Wall Friction

It is common to measure wall friction using the direct shear cell apparatus [1,2]. In some respects this can lead to conservative values of the measured friction angles, which may be satisfactory for determining bin geometry, but may lead to problems due to variations in bin flow patterns. Tests conducted using the linear action, friction/wear apparatus [8] resemble more closely the friction values experienced in practice. This machine allows the rate of rubbing velocity to be varied to match those expected to occur in practice.

By way of example, comparative tests were conducted using the Jenike Shear Tester and the Linear Friction/Wear Tester to determine the Wall Yield Loci for Bauxite in contact with Bisalloy 500. The Linear Friction/Wear Tester measured a lower Wall Yield Locus than the Jenike Tester, with differences in friction angle of approximately 1.5^{o}; although this is quite a small difference, in other cases a greater difference has been measured. In general, the Linear Friction/Wear Tester permits more accurate measurement of the kinetic friction occurring during flow in hoppers and chutes.

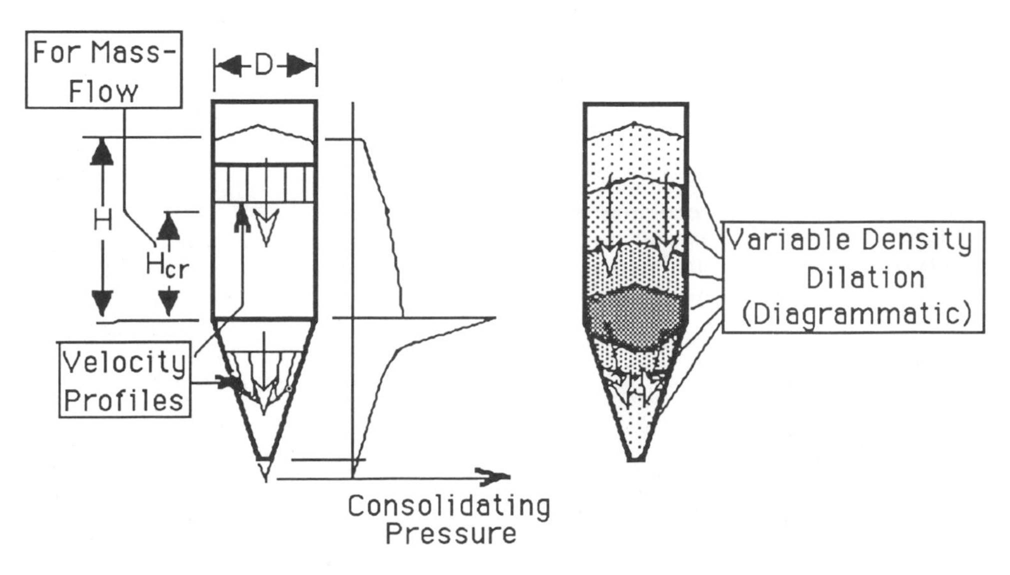

(a) Velocity Profiles and Pressure Distribution (b) Variable Density and Dilation

Figure 1. Mass-Flow Bin

As discussed in Section 4.2, the measurement of normal and shear stresses in silo walls using load cells permits the determination of the wall friction angle under operating conditions.

2.3 Flow Patterns and Dilation of Bulk Solids during Flow

The discussion that follows provides a qualitative view of the 'silo quaking' problem as it relates to mass-flow, funnel-flow and expanded-flow bins.

In the mass flow bin of Figure 1(a), there is a minimum level H_{cr} which is required to enforce mass-flow in the hopper. As reported by Thomson, this height ranges from approximately 0.75 D to 1.0 D [11]. This critical height has been studied in some detail by Benink [4] who derived a fundamental relationship for H_{cr} in terms of the various bulk solid and hopper geometrical parameters. As the material flows, it dilates leading to variations in density from the static condition. This is depicted pictorially in Figure 1(b). With $H > H_{cr}$, the flow in the cylinder is uniform or 'plug-like' over the cross-section, with flow along the walls. In the region of the transition, the flow starts to converge due to the influence of the hopper and the velocity profile is no longer uniform. The velocity profile is further developed in the hopper as shown . As the flow pressures generate in the hopper the further dilation of the bulk solid occurs. As a result of the dilation, it is possible that the vertical supporting pressures decrease slightly reducing the support given to the plug of bulk solid in the cylinder. This causes the plug to drop momentarily giving rise to a load pulse. The cycle is then repeated.

A similar action to that described above may occur in tall funnel-flow bins or silos where the effective transition intersects the wall in the lower region of the silo. As a result, there is flow along the walls of a substantial mass of bulk solid above the effective transition.

During funnel-flow in bins of squat proportions, where there is no flow along the walls, as depicted in Figure 2, dilation of the bulk solid occurs as it expands in the flow channel. As a result some reduction in the radial support given to the stationary material may occur. If the hopper is fairly steeply sloped, say [$\theta \geq \delta$)], then the stationary mass may slip momentarily causing the pressure in the flow channel to increase as a result of the 'squeezing' action. The cycle then repeats.

Expanded-flow bins are commonly used to store bulk solids in large tonnages. Such bins have the advantage of providing large storage capacity while maintaining low overall height. As illustrated in Figure 3, the upper section of the bin operates under funnel-flow, while the flow is expanded through the mass-flow hopper. For complete discharge, the bin diameter at the transition should be at least equal to the critical rathole diameter corresponding to the consolidation pressure at this level. Pulsating loads can occur in such bins, particularly if the slope angle θ of the transition is too steep. Owing to segregation on filling, larger size particles are more likely to be located adjacent to the sloping surface at the lower end of the funnel-flow section. Such particles tend to roll as well as slide, aggravating the load slipping problem and giving rise to load pulsations.

3 SILO QUAKING PROBLEM - TWO CASE STUDY EXAMPLES

Field observations confirm the foregoing description of the mechanisms of 'silo quaking'. To illustrate this, two case study examples of pulsating flow in coal bins in Australia are presented.

3.1 Single Outlet Mass-Flow Bin

Figure 4 illustrates a symmetrical coal bin which, while giving totally reliable discharge, exhibited 'silo quaking'. As shown, the bin is of squat proportions with H/D = 1.0. Discharge was at the rate of approximately 350 t/h and when the bin was substantially full, severe load pulses at a period of approximately 3 sec. were observed. As indicated, the hopper was lined with stainless steel only part of the height. The rationale for this is that due to segregation on filling, larger coal particles were in contact with the mild steel section of the hopper; these larger particles were substantially 'free flowing' and flow along the walls could occur. Since the finer particles congregate in the central region of the bin, the stainless steel lining is necessary for mass-flow. It is evident that the mild steel section of the hopper was at a critical slope angle and the flow could trip in and out of the mass-flow zone. In effect the load pulsations in this case are similar to those illustrated in Figures 2 and 3.

3.2 Multi-Outlet Coal Bin

Figure 5 shows a 6000 tonne coal bin constructed of reinforced concrete. The bin is 21m diameter and 24 m tall above the hoppers; it is supported by a number of columns. The bin has seven outlets, six around an outer pitch circle and one located centrally. The hopper geometries provide for reliable flow permitting complete discharge of the bin contents. Coal was discharged by means of seven vibratory feeders onto a centrally located conveyor belt. When the bin was full or near full, severe shock loads were observed at approximately 3 second intervals during discharge. The discharge rate from each feeder was in the order of 300 t/h. When the level in the bin had dropped to approximately half the height, the shock loads had diminished significantly. With all the outlets operating, the effective transition was well down towards the bottom of the bin walls and the critical head H_m was of the same order as the bin diameter and greater than

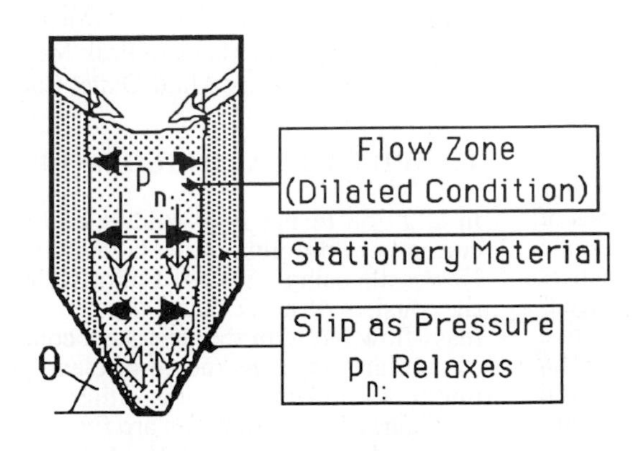

Figure 2 . Funnel Flow Bin

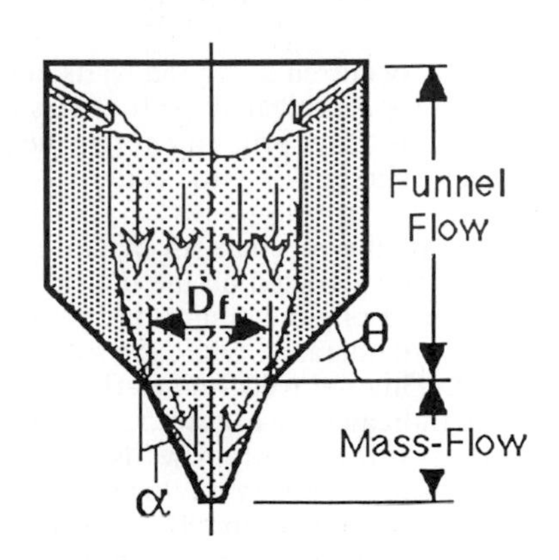

Figure 3. Expanded-Flow Bin

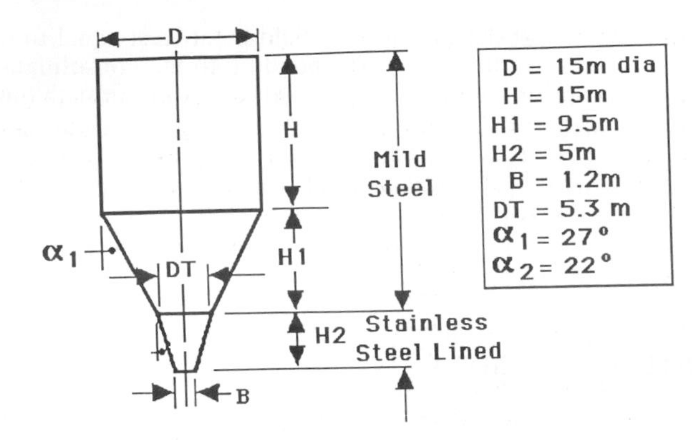

Figure 4. Mass-Flow Coal Bin Subject to 'Silo Quaking'

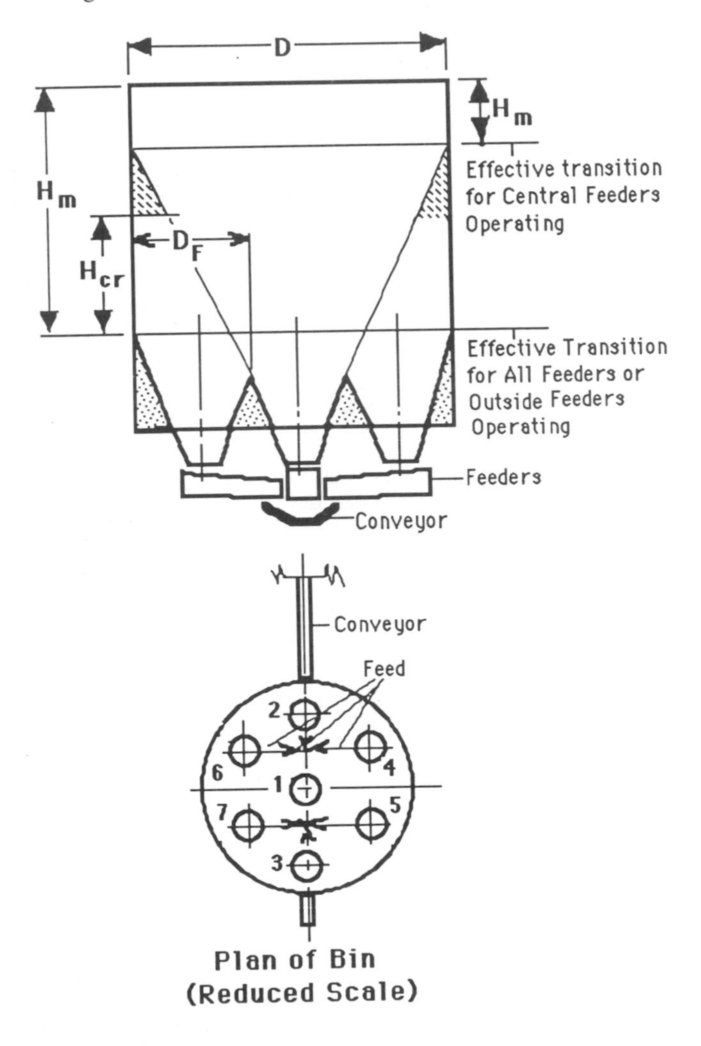

Figure 5. 6000 tonne Multi-Outlet Coal Bin

D_F. Substantial flow occurred along the walls, and since the reclaim hoppers were at a critical slope for mass and funnel-flow as determined by flow property tests, the conditions were right for severe 'silo quaking' to occur.

Confirmation of the mechanism of silo quaking was obtained in field trials conducted on the bin. In one series of tests the three feeders along the centre line parallel with the reclaim conveyor were operated, while the four outer feeders were not operated. This induced funnel-flow in a wedged-shaped pattern as indicated in Figure 5, with the effective transition occurring well up the bin walls, that is $H_m < H_{cr}$ ($=D_F$) or $H_m << D$. The same was true when only the central feeder (Fdr. 1) was operated; in this case the stationary material in the bin formed a conical shape. Under these conditions, the motion down the walls was greatly restricted and, as a result, the load pulsations were barely perceptible.

In a second set of trials, the three central feeders were left stationary, while the four outer feeders were operated. This gave rise to the triangular prism shaped dead region in the central region, with substantial mass-flow along the walls. The loads pulsations were just as severe in this case as was the case with all feeders operating. Dynamic strain measurements were made using strain gauges mounted on selected support columns. A sample of the dynamic strain record for the case when the four outer feeders were operating is shown in Figure 6.

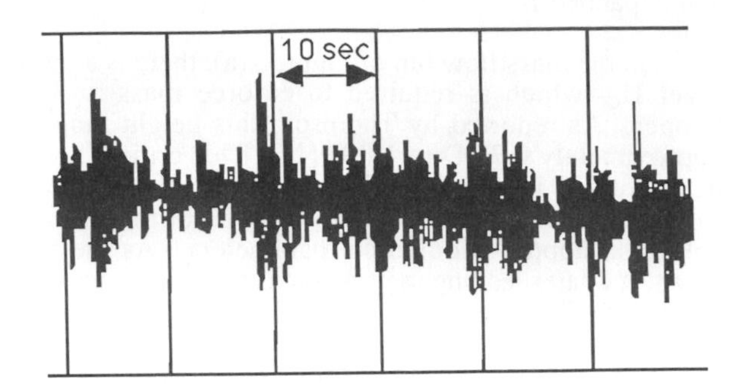

Figure 6. Dynamic Strain Measurements for Four Outer Feeders Operating. 6000 tonne Coal Bin

Figure 7 shows the variation of dynamic strain amplitudes as a function of bin level when different combinations of feeders are used. When either the central (Fdr. 1) or the three centre line feeders (Fdrs. 1, 2 and 3) are operating, the dynamic strains were quite low and virtually unaffected by the level of coal in the bin. However, when either the four outer feeders (Fdrs. 4, 5, 6 and 7) are used or all seven feeders are used the dynamic strains are quite high particularly when the bin is full. This clearly demonstrates that in order to avoid shock loading, the effective transition must be high up in the bin so as to limit the height of coal moving against the wall.; that is $H_m \leq D_F$ (Figure 5).

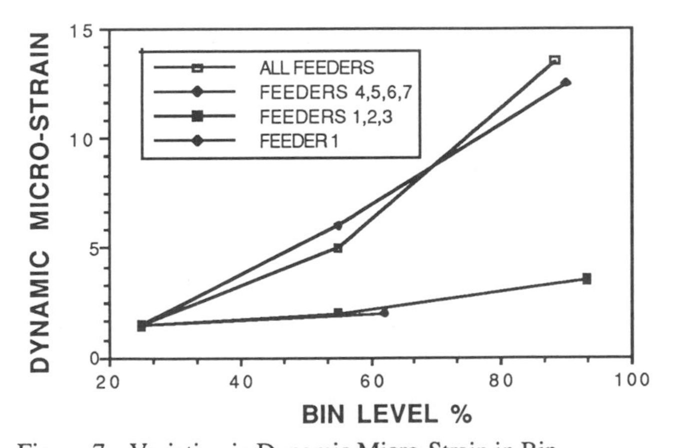

Figure 7. Variation in Dynamic Micro-Strain in Bin Column. (Peak to Peak Measurement) 6000 tonne Multi-Outlet Coal Bin

4. PILOT SCALE SILO MODEL

In addition to the field studies outlined above, research is currently in progress in the laboratories of The University of Newcastle using a series of pilot scale silo test rigs. One test rig consists of a 1.2 metre diameter by 3.5 metre tall steel, mass-flow bin with stainless steel conical hopper and another of similar size, constructed of clear 'perspex', but operating under funnel-flow with either single or multi-outlet configurations. Both bins are fitted with wall pressure cells designed by Professor V. Askegaard of the Technical University of Denmark. Studies are aimed at examining the presence of flow instabilities through the recording of dynamic load pulsations for various bulk materials both non-cohesive, such as wheat, and cohesive under varying discharge conditions and surcharge heads. The two bins, together with a third holding bin form a complete test facility which also includes a reclaim conveyor, bucket elevator and distribution conveyor.

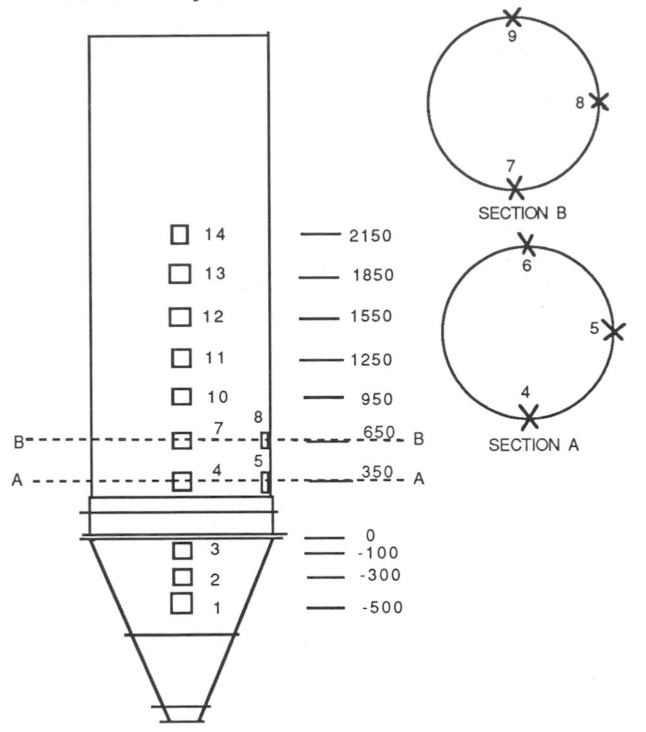

Figure 8. Mass-Flow Test Bin

By way of illustration, some test results with wheat using the mass-flow steel bin are presented. A schematic arrangement of this bin is shown in Figure 8. Fourteen load cells, located as shown, are capable of measuring both normal pressure and wall shear stress. For the purpose of the present discussion on load pulses, the loads acting in the cylinder are examined.

4.1 Effect of Surcharge Head on Wall Pressures and Shear Stress

A series of tests were conducted for different stored heads in which the normal pressures and shear stresses at several locations were recorded during filling, during undisturbed storage and during emptying. Referring to Figure 8, the locations at which the measurements are reported are 5, 7, 10, 12, 13 and 14. Three head heights of the wheat in the cylinder were examined:

H	=	3.44 m	2.41m	1.1m
H/D	=	2.9	2.0	0.9

(a) Pressures During and After Filling

By way of example Figure 9, shows sample records for location 14 for the ratio H/D = 2.9. A dynamic 'slip stick' effect is depicted in both the normal pressure and shear records during filling; this effect continues on after the bin is filled, the effect being most pronounced in the shear stress record where the pulses are most evident. It is quite clear that during undisturbed storage, the stored mass approaches, asymptotically, its critical state consolidation condition by a pulse type settling action. The phenomena depicted in Figure 9 was observed at all locations, with the pulses at each location occurring at the same time intervals. Records of the undisturbed settling were taken over prolonged time periods. The same effect was observed for the three H/D ratios examined. As the settling time increased, the pulse period also increased in an exponential manner as illustrated in Figure 10. This shows that the settling characteristics for the two cases H/D = 2.9 and H/D = 2.0 are virtually the same.

(b) Pressures During Emptying

Figure 11 shows for the case of H/D = 2.9, the normal wall pressures and shear stresses at locations 5 and 14. At location 5, the dynamic wall pressure and shear stress have each taken approximately one minute to reach their maximum values. At this location, the influence of the hopper transition is felt and the long rise time is no doubt associated with the time taken to establish the arched stress field in the hopper. The influence of a pulsating flow effect is depicted in the records for location 5, as for all other locations, even though the amplitude of the pulsations in the normal pressure and shear stress at location 5 is small compared with the average values. The amplitude increased relative to the average values at higher locations in the bin. For instance, as Figure 11(b) shows, at location 14 the pulsations are quite pronounced, the period of the pulsations averaging approximately 4 seconds.

(c) Effect of Surcharge Head

Similar dynamic effects were experienced for the case of H/D = 2.0, but when the head was reduced to H/D = 0.9, there was no evidence of any pulsations during discharge. This confirms the conclusion that self excited pulsating flows in bins which give rise to the 'silo quaking' phenomenon are associated with surcharge heads greater than a minimum value of, say, H/D = 1.0.

4.2 Wall Friction

The normal and shear stress records permit the wall friction to be determined. Values determined from the test results range from 12° to 14°, which are significantly lower than the friction angle of 18° measured by the Jenike Shear Tester. The latter value represents static friction, whereas the values determined from the silo tests represnt kinetic friction which accounts for the lower values. It is also be possible that some

Figure 9. Normal Pressure and Shear Stress Records for Location 14 of Test Bin

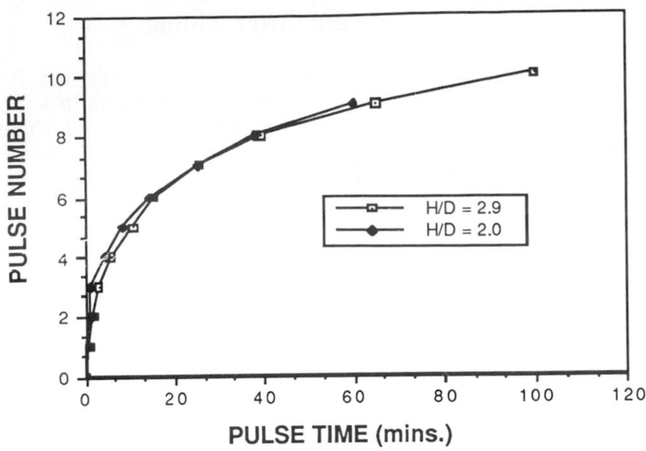

Figure 10. Load Pulsations during Undisturbed Storage of Wheat in Test Bin.

rolling as well as sliding of the grain particles may have occurred which would account for the lower angles kinetic friction. These results highlight the importance of accurately determining wall friction.

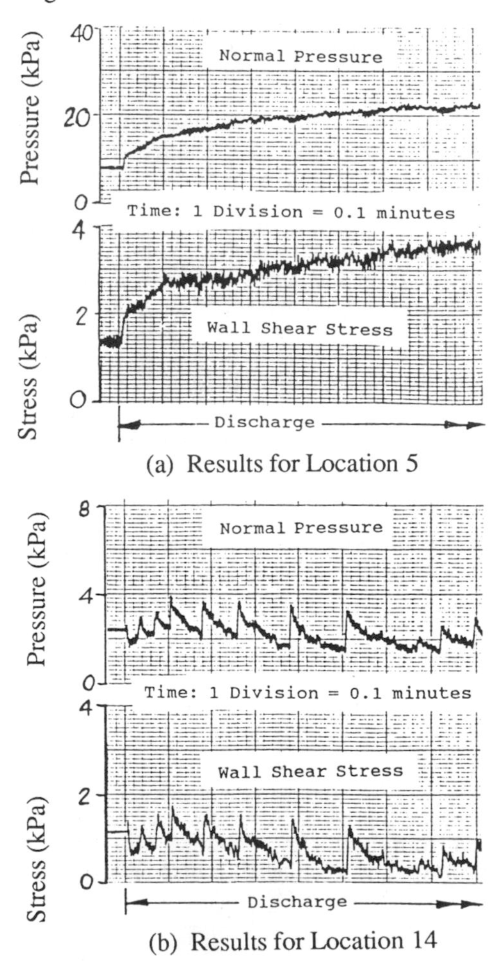

Figure 11. Pressures and Shear Stresses During Emptying H/D = 2.9

5. CONCLUDING REMARKS

The paper has demonstrated the possible causes of 'silo quaking' in tall mass-flow and expanded-flow bins and in squat funnel-flow and expanded-flow bins. The case study examples have demonstrated the theories put forward and highlighted the problems existing in multi-outlet bins. The experimental work using the pilot scale mass-flow bin is providing evidence to support the theory although further work needs to be undertaken.

6 REFERENCES

1. Jenike, A.W. "Gravity Flow of Bulk Solids". Bul. 123, The Univ. of Utah, Engn Exp. Station, USA 1964.

2. Arnold, P.C., McLean, A.G. and Roberts, A.W. "Bulk Solids: Storage, Flow and Handling". The University of Newcastle Research Associates (TUNRA), Australia, 1982.

3. Jenike, A.W. "A Theory of Flow of Particulate Solids in Converging and Diverging Channels Based on a Conical Yield Function". Powder Tech., Vol.50. (pp. 229-236).

4. Benink, E.J. "Flow and Stress Analysis of Cohesionless Bulk Materials in Silos Related to Codes". Doctoral Thesis, The University of Twente, Enschede, The Netherlands. 1989.

5. Ooms, M. and Roberts, A.W. "Significant Influence of Wall Friction in the Gravity Flow of Bulk Solids". Intl. Jnl. of Bulk Solids Handling, Vol. 5, No. 6, 1985 (pp.1271-1277)

6. Roberts, A.W., Ooms, M. and Scott, O.J. "Surface Friction and Wear in the Storage, Gravity Flow and Handling of Bulk Solids". Proc. Conf. 'War on Wear', Wear in the Mining and Mineral Extraction Industry, Instn. of Mech. Engnrs, Nottingham U.K., 1984. (pp.123-134).

7. Roberts, A.W. "Friction, Adhesion and Wear in Bulk Materials Handling". Proc., AntiWear 88, The Royal Soc. London. 1988. Inst. of Metals, I.Mech. E.

8. Roberts, A.W., Ooms, M. and Wiche, S.J. "Concepts of Boundary Friction, Adhesion and Wear in Bulk Solids Handling Operations". Intl. Jnl. of Bulk Solids Handling, Vol.10, No. 2, May 1988.

9. Roberts, A.W."Vibrations of Powders and Bulk Solids". Chapter 6, Handbook on Powder Science & Technology. (1984) Van Nostrand.

10. Roberts, A.W., Ooms, M. and Scott, O.J. "Influence of Vibrations on the Strength and Boundary Friction Characteristics of Bulk Solids and the Effect on Bin Design". Intl. Jnl. of Bulk Solids Handling, Vol.6, No.1. 1986. (pp.161-169).

11. Thomson F.M. "Storage of Particulate Solids". Chapter 9, Handbook on Powder Science & Technology. (1984) Van Nostrand.

C418/016

Techniques to assist in the in-plant handling of slurries, pastes and cakes

N I HEYWOOD, BSc(Eng), MSc, PhD, CEng, FIChemE, ACGI, DIC,
Marine Pollution and Bulk Materials Division, Warren Spring Laboratory,
Stevenage, Hertfordshire

SYNOPSIS Solid-liquid mixtures such as slurries and pastes (essentially pumpable materials), and solid-liquid-air compressible mixtures such as some filter or centrifuge cakes, can be notoriously difficult to store or convey in processing plants. However, over the last few years, technical developments have occurred to assist in improving the handling of 'wet solids'. This has partly been brought about through a greater understanding of the importance of identifying and measuring the bulk properties of 'wet solids' and the use of these property data to help specify the design and operating conditions of handling equipment. This paper briefly reviews some of the more significant advances and classifies the improved techniques according to the nature of the 'wet solids' and the type of handling operation. The paper is based on a detailed review written in the Wet Solids Handling Project at Warren Spring Laboratory.

1 INTRODUCTION

The design of equipment, and the specification of equipment operating conditions, for the handling of liquids and dry powders has come a long way in the last 50 years. Plant availability is consequently high for such materials, often between 90 to near 100%. The art of system design for the handling of 'wet solids' has not kept pace with these developments (1) although progress is being made (2-5). This is partly the result of perhaps disproportionately large research attention devoted to 'wet solids' unit operations such as filtration, centrifugation, crystallisation and drying compared with lesser effort aimed at improving the links between unit operations. Typically at the design stage attention is focused much more on the specification of the operating conditions of the processing stages rather than the storage/conveying requirements.

However, various techniques to assist in both the storage and conveying of high concentration slurries, pastes, cakes, wet powders and other 'wet solids' have been developed. In some cases, these techniques are more the result of progressive improvements in understanding the underlying bulk properties of 'wet solids' and the way in which they are responsible for determining their handling behaviour. These techniques relate to:

(a) the modifications of one or more of the bulk properties of wet solids to enable 'conventional' equipment and operating conditions to be used, either at all or else more effectively;

(b) the development of handling equipment designed specifically with the handling problems of 'wet solids' in mind;

(c) the careful specification of operating conditions of either 'conventional' or novel equipment.

In developing design methods for storage and conveying systems for wet solids, it is essential that if these methods are to be reliable and therefore used by design engineers, they must take into account the relevant flow properties of the material of interest. Measurements of these flow properties must be considered as a crucial starting point for any system design or equipment selection procedure (6,7).

2 FLOW OR DEFORMATION PROPERTIES OF BULK SOLIDS

2.1 Flow properties relevant to slurry and paste handling

Slurries can often be conveniently categorised as 'non-settling' or 'settling', although sometimes the dividing line can be indistinct and appropriate consideration must be made of the residence time of the slurry in process equipment in order to try to classify the slurry type for pipeline design (8-10). 'Non-settling' slurries (to which the term paste often refers) can be treated for design purposes as a pseudohomogeneous mixture. There will be no significant phase separation or settling under gravity during the storage or conveying operation. 'Settling' slurries will not remain a pseudohomogeneous mixture under gravity or other body forces and consequently parameters such as particle size and size distribution, the density difference between the solid and liquid phases, and liquid viscosity are important.

An important flow property of 'non-settling' slurries is its shear behaviour. This is summarised by its flow curve (a plot of shear stress versus shear rate) which gives the slurry's viscosity at different shear rate conditions. This can be measured using a variety of viscometers/rheometers (11). Slurries can exhibit Newtonian (constant viscosity) or non-Newtonian (either shear-thinning or shear-thickening) behaviour. Shear-thinning slurries (viscosity decreases with increasing shear rate) are generally much more commonly met in industry than shear-thickening slurries (viscosity increasing with increasing shear rate); the latter are normally concentrated, non-flocculated slurries and can create severe pumping problems. In addition, pastes often appear to exhibit a 'yield stress', the shear stress required in order to initiate and sustain flow at a practical rate. This can often be a difficult quantity to measure. The choice of measurement method and the criterion adopted for yield should be related to the application for the data.

A yield stress is often exhibited by slurries exhibiting time-dependent flow property; the most commonly met example being thixotropy. A thixotropic slurry will by definition reduce in viscosity if its internal structure is broken down by shear (12,13). This structure breakdown is only temporary for a classical thixotropic material and, if the slurry is left undisturbed after being sheared, its viscosity will eventually recover its original value. With some slurries this recovery is incomplete or occasionally does not occur at all, so it is important first to characterise the slurry shear property if some time is likely to elapse following the shearing process. Thixotropic behaviour is regularly confused with pseudoplastic, or shear-thinning, behaviour. Thixotropy is a time-dependent property (viscosity decreases with time when a constant shear rate is applied), while pseudoplasticity is a time-independent property (viscosity decreases as shear rate is increased). Slurries can, of course, exhibit both properties and often do.

Pastes can also adhere to equipment surfaces because of the attractive van der Waal's forces between the atoms which comprise the particle and equipment surfaces and because of the suction forces generated through the surface tension of the liquid phase. This can cause caking problems in handling equipment (14,15). Using linings on surfaces or changing the paste moisture content or physical form can alleviate this problem (see Section 4).

2.2 Flow properties relevant to wet, granular solids handling

As with 'dry' bulk solids, the two most important properties are the internal shear strength (typically measured using a Jenike cell or an annular ring shear cell) and wall friction on chosen equipment surfaces. However, because of the presence of moisture these materials will exhibit viscous as well as granular shear property (3) and the viscous component will tend to become progressively more important as the moisture content increases.

Viscous behaviour reflects the relationship between shear strength (or stress) and shear rate, and it is assumed that this relationship for slurries is essentially independent of the applied normal stress (and hence the resulting bulk density). Granular behaviour on the other hand relates to the dependence of shear strength on normal stress and for 'dry' materials this relationship is independent of shear rate. With 'wet solids' no viscometer or shear cell is capable of characterising shear properties so instrument development is called for.

As is the case with saturated pastes, unsaturated wet, granular material will tend to adhere to equipment surfaces causing cake formation. Some success has been achieved in identifying the conditions of moisture content and stresses under which cake will form on a surface using an annular ring shear cell (16) and these stresses can be related to those occurring on walls of process equipment such as a hopper (Section 5.2).

3 MODIFICATION OF FLOW PROPERTIES OF SLURRIES AND PASTES

Two main problems occur in the handling of slurries and pastes. Firstly the viscosity of a 'non-settling' slurry or paste is potentially so high that (a) it cannot readily be discharged from a storage vessel or (b) a pump cannot be primed. Secondly, the slurry may contain large particles which in the time frame of a storage or conveying operation will settle out under gravity onto the bottom of a tank or pipeline. Thus continuous or intermittent agitation is required in slurry tanks and pipeflow velocities need to be sufficiently large to prevent solids deposition and pipe blockage.

3.1 Viscosity reduction

If it is possible to make small adjustments to the operating conditions of some upstream processing stages which are responsible for determining slurry properties, then there are several options open to reduce slurry or paste viscosity. The most important parameter is the solids concentration. It is useful to remember that although small changes in solids concentration can have a relatively small effect on viscosity levels at the lower concentration end, a large effect is possible at high solids concentrations; a few per cent increase in solids concentration at high concentrations can easily double or triple slurry viscosity.

Slurries containing a significant proportion of particles below about 2 μm may often be in a flocculated state. This has the effect of immobilising the carrier liquid which would otherwise be available to assist the shear of the slurry. The result is often high viscosities coupled with non-Newtonian, shear-thinning behaviour. Dramatic reduction in these viscosities is possible through the addition of an appropriate dispersant at the correct concentration (17,18). The dispersant creates a charge barrier between the particles which prevents them from flocculating. This charge barrier can be characterised by measuring the zeta potential and a direct relationship between slurry viscosity and zeta potential has been demonstrated (18).

For a dispersed or deflocculated slurry, the particle size distribution can have an additional significant effect on viscosity level (19,20). To minimise viscosity the widest practical particle size distribution should be strived for, particularly when the overall solids concentration exceeds around 30 to 40% by volume. The same applies if the prime requirement is to maximise the solids content of a slurry while maintaining its viscosity below a certain critical level.

3.2 Stabilised slurries

Stabilised slurries are formulated so that the suspending medium is capable of supporting coarse particles in the slurry for a defined time period and often indefinitely. Normally, if the suspending medium is water, or some other low viscosity liquid, the coarse particles would readily settle under gravity. To counteract the influence of gravity in storage vessels or in transfer pipelines, the suspending medium comprises a homogeneous mixture of fine, colloidal particles and water (21).

These particles are normally in a flocculated state imparting a shear-thinning flow property to the slurry. This provides a relatively high viscosity at the relevant low shear rates defined by the settling process so enabling the coarse particles to be supported, but often much lower viscosities at the higher shear rates occurring at the wall of a pipeline facilitating pipe flow at reasonable energy consumptions (see Section 6.1).

4 MODIFICATIONS TO BULK PROPERTIES OF WET, GRANULAR SOLIDS

4.1 Size reduction

Wet materials emerging from various types of filter or centrifuge (4) tend to have one of two main physical forms: brittle cakes which appear dry on their surface (typical of compression filters), or pliable wetter pastes (typical of rotary vacuum filters). In both cases, they can be in a form which is inconvenient to convey readily without some reduction in their size and attempts are therefore made to create more uniformly-sized material.

Adequate size reduction of cakes is sometimes achieved simply by allowing them to fall either directly onto a conveyor belt or through a horizontal mesh of bars before falling onto the conveyor. The conveyor is typically used to transfer the material to temporary intermediate storage or to a dryer. When this approach gives insufficient size reduction, it can be followed by a crusher or lump-breaker both of which rely upon the brittle nature of the material to function effectively. Commercial size reducers typically employ a series of fixed hammers working between fixed anvils and perforated screens. Other home-made designs involve two horizontal, contrarotating cylinders each having a number of spikes set normally to the cylinder surface and distributed over the surface.

More pliable pastes are generally not amenable to crushing or lump-breaking techniques. This is because these pastes are not sufficiently brittle and because they tend to adhere to equipment surfaces with a resultant build-up and eventual blockage of size-reducing equipment. Instead there are a number of ways of preforming pastes. Preforming is often necessary to facilitate paste feeding to dryers and to improve dryer efficiency. Partial surface drying of the paste can be combined with preforming by using a grooved rotating drum which is normally steam-heated. A second plain roller forces the paste into the grooves and a doctor knife subsequently scours the grooves to release the preformed material onto a suitable conveyor.

Other preformers extrude the paste through a cylindrical mesh or a bank of extrusion dies. The former type uses a roller to force the paste through the mesh while the latter requires the paste to be pressurised using a suitable pump. In either case, the paste can sometimes be formulated such that suitable lengths of extrudate are created through the continuous rupture of the extruded material as a result of its own weight after extrusion. If this is not possible a rotating knife is necessary.

4.2 Size enlargement

This refers to the process by which small particles are gathered into larger, permanent masses in which the original particles can still be identified. The process can result in bulk solids which are less cohesive or even free-flowing compared with the original material. It can also prevent cake or lump formation during various processing and handling stages and can densify the material for more convenient storage and shipment.

Various processes are available to achieve size enlargement: granulation, agglomeration, pelletisation/spheronisation and briquetting. Wet granulation is widely used and achieved using various processes at ambient or near-ambient pressures such as pan or fluid-bed granulation. Pelletisation tends to produce highly densified masses formed under pressure which because of their dense character and uniform size tend to be relatively easy to handle. Spheronisation involves extrusion of a wet mass to produce short and dense pellets which are then transferred to a spheroniser, a short vertical cylinder with a spinning disc in its base, where the pellets are made nearly spherical and of uniform size. Briquetting is also carried out under pressure using either a roll-type machine or a reciprocating press.

4.3 Shear strength reduction

This is sometimes possible through mechanical working of an initially brittle cake which might emerge, for instance, from a compression filter. The required work input may be achieved through subsequent processing or handling where the cake is sheared e.g.

conveying using a screw conveyor or preforming by extrusion. It is unclear why the cake strength should reduce but it is likely that the initial high strength is caused by the high surface friction between particles which are locked together by capillary forces. Shearing the cake redistributes the moisture more uniformly and reduces the effect of the capillary forces.

One disadvantage of working a cake is that the moisture which was previously confined primarily to the cake centre is now brought to the surface. The result is a lower strength but more sticky material which will tend to adhere more readily to equipment surfaces. However, working a cake may facilitate the pumping of a cake which would not otherwise be possible.

Another way of reducing the strength of some pastes or cakes is achieved by reducing the surface tension of the liquid phase through the addition of aqueous surfactants. These are sometimes referred to as 'flow improvers' (22) and can be applied using simple spraying equipment. Dose rates might be typically 2 to 10 litres per tonne for wet coal. The surfactants appear to work in a number of ways: by reducing the forces created by liquid bridging between particles, by promoting liquid drainage under gravity, and by lubricating the sliding action between particle surfaces.

The application of vibration, either deliberate or unintentional, is known to be very effective in reducing both the shear strength (23) and the wall friction (24) of some wet materials. It has been argued (25) that one of the reasons why we still do not have a good understanding of the underlying mechanisms responsible for strength reduction under vibration is that past experimentation and thinking has focused on too long a time-scale and that instantaneous changes in material properties over fractions of a second should be investigated.

4.4 Adhesion/stickiness reduction

The tendency of pastes and cakes with a significant amount of surface moisture to adhere to equipment surfaces can be reduced under some circumstances. Surface moisture can be removed on a small-scale by blowing ambient or warm air through a travelling bed of material. This may then facilitate subsequent transport using a screw, belt or vibratory conveyor (see Section 7).

Another way of removing moisture, this time from the bulk of the material, is by recycling a proportion of the material which has already been passed through a dryer and combining it with the wet feed. This may also be necessary because some dryer types (e.g. pneumatic or fluid-bed) are not capable of handling the original wet material.

Cooling equipment surfaces to just above freezing is another useful approach which may also reduce wall friction. In the food industry, stainless steel surfaces are often cooled to between 1 and 5°C, primarily for hygiene reasons, but adhesion is also minimised. Because of the high cost of cooling surfaces, this approach is probably limited to low tonnage, higher value materials.

5 DESIGN FOR STORAGE OF PASTES AND CAKES

In most solids handling plant designs, attempts are generally made to minimise the occasions when the intermediate or final storage of pastes or cakes are necessary. Discharge under gravity of wet materials from storage vessels is generally very difficult. However, if the moisture level is low and shear and wall friction tests indicate that the internal shear strength and wall friction property can be accommodated in the established Jenike design approach, then it may be possible to specify mass flow discharge under gravity through conventional conical or wedge-shaped bottom hoppers.

The Jenike design approach takes into account neither viscous shear nor adhesion effects arising from the presence of moisture and consequently caking can occur on the inside wall of the hopper, so preventing mass flow conditions. A method is being developed to predict this situation at the design stage (Section 5.2), but, for obviously sticky and highly cohesive wet solids, various types of integral mechanical discharge aid are used in flat-bottomed silos.

5.1 Storage vessel design with integral discharge facility

A variety of commercial designs is available. The Saxlund sliding frame discharger is located at the bottom of a flat-bottomed vessel and reciprocates back and forth to push the material to the discharge opening where a screw conveys it away. Another Saxlund design can be used in rectangular silos in which the floor is ribbed to assist in the funnelling of material to the outlet where it falls down a chute to a discharging screw conveyor. A main advantage claimed for this design is the small amount of preparatory civil work required.

The Bowerhill/Parcey planetary extractor uses a single screw extending across the radius of the silo at the silo flat-bottomed base. This screw conveys to the central outlet by circling slowly around the base thus discharging the material on a first-in/first-out basis. When high discharge rates are required, two screws located across opposite radii are employed. The screw is driven centrally and this has the disadvantage of necessitating internally mounted motors and drives which can present maintenance problems.

A similar geometrical design is the Mitchell Storall Feeder, but here the screw rotates in a fixed position and the silo base revolves slowly to maintain a steady feed of material to the screw. With some designs the discharge is through the centre of the base whilst others extend the screw across a diameter and the discharge is at the silo side. Driving bars are fitted at the base to reduce possible slippage of the material. Sometimes the silo shell has divergent walls to aid the flow of difficult materials.

The Euro-Mammoth silo (26) uses screw discharge but the screws are located at the surface of the material. They are used during the filling stage to distribute material evenly across the silo to its walls, and their direction of rotation is reversed during emptying so that material is fed to the central discharge position. A possible disadvantage of this system is the first-in last-out mode of operation. Other silo designs include the Louise Centrex System and the Big Atlas Silo System (27).

5.2 Discharge from hoppers

The Jenike design procedure for specifying the angle of the conical or wedge-shaped base and the outlet dimensions of a hopper for the mass flow of 'dry' bulk solids is well-known. However, the extent to which this procedure can be used with confidence for wet materials has yet to be fully established. As already mentioned, one of the main problems is the potential for wet materials to cake on the hopper walls. Many hopper walls are still constructed from mild steel which can rust in the presence of wet solids and which can absorb the moisture.

Thus, even if the hopper is designed for mass flow discharge, this surface deterioration can cause either core flow or no flow instead. Various wall lining materials are possible (28) which reduce wall friction and can also minimise wear through abrasion. These include aluminium oxide sinter, fused cast basalt, glass, glass fibre, rubber (29), stainless steel (30), and ultra-high molecular weight polyethylene (UHMW-PE) (31,32).

For any given liner type, it has been demonstrated (16) that the propensity for a wet material to cake on a hopper wall can be established using a simple laboratory test. If the shearing surface of an annular ring shear cell tester is fitted with a sample of the liner material, the conditions under which caking will occur can be determined. These conditions have been summarised as 'caking envelopes', which indicate when caking is and is not expected to occur as a function of moisture content and the shear properties of the material (the major consolidating pressure and the confined yield strength).

These 'caking envelopes' are time-consuming to develop but, in principle, a number of possible lining materials could be evaluated using this technique to see which one gives the least potential for caking under the proposed storage conditions. Predictions of hopper wall stresses combined with caking envelope data have shown (16) that caking can typically occur at the top of the cylindrical hopper section and at the base of the conical section.

5.3 Discharge from drums and cans

Wet sticky pastes and cakes stored in drums and cans which need to be up-ended for the material to be discharged can cause particular problems unless some simple operating rules are followed (33). These problems arise because the material sticks to the bottom and walls of the container, phase separation under gravity occurs causing liquid containing small amounts of solid to form on the top of compacted solids, and caking can result on a paste surface if substantial evaporation occurs. All these difficulties contribute to producing what is often an even more inhomogeneous material compared with the original wet solids prior to loading the drum or can.

Discharge from drums or cans is assisted if any container emptying system fully inverts the drum through 180° and not the 150° which is frequently adopted in commercial equipment designs. However, full inversion will not guarantee discharge. Sometimes very little material emerges because of the partial vacuum created as the slug of paste tries to leave the drum or, even if most of the material does flow out, wall deposits remain.

One solution is to use flexible plastic liners (e.g. reinforced PVC) which are tailored to suit the internal shape of the container and which completely cover the inner walls of circular or rectangular cross-section containers. The material either discharges leaving the liner in place, or if it adheres to the liner, it will draw the liner out of the container with it. Securing the liner to the outer lip will allow the liner to reverse and be peeled off by the discharging material, provided there are a number of holes in the liner to allow air to flow in behind the liner, so preventing the formation of a partial vacuum (33).

5.4 Removal of sediment from storage vessels

Resuspension of settled coarse or fine particles is a common industrial problem. In the former case, the sediment is fairly free-flowing, particularly when there are few fines present and tends to present less of a problem than when the majority of the particles are fine (i.e. less than 50 µm) and are partially or fully deflocculated or dispersed.

The Marconaflo system (34,35) is designed to retrieve solids which have been allowed to settle. Resuspension is accomplished by introducing high pressure liquid through rotating jets located in the bottom of the retaining vessel. The system was first used on ships carrying iron ore concentrates in slurry form, but it has also been applied to in-plant handling (34), including coal supply and storage for power plants, nickel laterites and storage of concentrates.

For compacted sediments, a number of direct pumping systems are available. For instance, the Hazleton twin-volute submersible centrifugal pump employs an agitator on the end of the pump shaft. This agitator is positioned outside the pump casing and rotates at the pump impeller speed of around 1400 rpm. The agitator mobilises the sediment not by cutting into it but by creating vapour bubbles in the sediment liquid. It is believed that these bubbles subsequently collapse causing pressure waves to travel through the sediment whose strength is reduced by the dissipation of the wave energy. The commercial design is covered by UK patent 2 131 486 B (17 Dec 1986).

Vacuum sludge pumps have been available for some time but although high suction is often achieved, comparatively low throughputs are possible with some designs. The 'BREVAC' design (Building Research Establishment, Garston Moor, UK) can give both high vacuum and high throughputs. The latter is necessary when operating with an air bleed to the suction pipe which is necessary for particularly viscous sludges so that the sludge is broken up in the pipe (36).

6 PIPELINE DESIGN FOR PASTES AND SLURRIES

The formulation of stabilised slurries (21), as already described in Section 3.2 and methods for reducing the frictional head loss for the flow of either a 'non-settling' or a 'settling' slurry in a pipe (37) can provide a number of potential benefits. Clearly head loss reduction will lead to power savings but it may also facilitate the priming of a preferred pump for the duty. The latter implies that a 'non-pumpable' slurry may be modified to a 'pumpable' slurry. Power savings can, in addition, mean that either a different type of pump can be used compared with that originally envisaged or a smaller pump size of the same type can be employed. Either situation can lead to a capital as well as operating cost saving.

6.1 Stabilised slurries

Slurries stabilised or semi-stabilised by the presence of fine particles which create a non-Newtonian, shear-thinning medium (see Section 3.2) can be transported by pipe at reduced velocities because the critical velocity for the coarse particles to settle out is correspondingly reduced. However, stabilised slurries are normally conveyed in the laminar flow regime because this tends to reduce both energy requirements and pipe wall wear.

Appropriate slurry formulation can lead to substantial energy reductions. It is reported (38) that silica sand transported at 20% by volume of solids in a carrier slurry containing 5% by volume fine clay resulted in an excess head loss compared to the transport of water alone of only one third that expected for the sand alone. British Petroleum have carried out extensive tests (39) on coarse coal stabilised in a fine coal slurry medium which was pumped at high concentration through several kilometres of 300 mm pipeline with one third of the coal particles being sub-200 µm. CSIRO in Australia has transported coal with a top-size of 20 mm through 152 mm pipe using a fine coal carrier.

Coarse material can also be conveyed using fibres in water as the carrier (40). The pipeline flow behaviour is complex. Frictional pressure loss is often low and under some circumstances drag reduction can occur, that is the pressure loss is less than for water flowing at the same mean flow velocity through the same pipe size.

6.2 Head loss reduction for 'non-settling' slurries

Methods for reducing the frictional head loss for the pipe flow of 'non-settling' slurries are summarised in Table 1, which provides references for the underlying mechanisms and advice on the specification of the design parameters. All the methods are effective for laminar flow of the slurry only. Thus the flow curve for the material has to be measured and an appropriate Reynolds number defined and estimated to determine whether the flow is laminar. In addition, the table shows that in many cases the laminar flow property must be non-Newtonian and shear-thinning (i.e. the viscosity must decrease with increasing shear rate). The vast majority of industrially-important non-Newtonian slurries are in fact shear-thinning.

Methods A to E require changes to the slurry formulation (see Section 3.1), while other methods F to H focus on modifications to the pipeline operating conditions and leave the slurry formulation unchanged.

6.3 Head loss reduction for 'settling' slurries

Methods for reducing the frictional head loss for the pipeflow of 'settling' slurries are summarised in Table 2, with references included for further information on the reduction mechanism and design parameter specification. All five approaches are effective in the turbulent regime; laminar flow of the slurry would be inappropriate because this would allow the particles to settle out and ultimately to block the pipe.

Methods A and B in Table 2 require changes to the slurry formulation which may not be permitted. Methods C to E alter the pipeline operating conditions by allowing a lower minimum mean flow velocity in the pipe to ensure all particles are in suspension. This reduced velocity provides lower frictional pressure loss and lower pipe wall abrasion rates. In addition, for friable particles attrition rates can be reduced.

7 CONVEYOR DESIGN FOR PASTES AND CAKES

Many fully saturated pastes and some unsaturated cakes can be pumped successfully with specialist designs of reciprocating piston pumps and their associated valve designs (49-51). Pump discharge pressure requirements can be high and transfer distances short but provided it can be primed, a pump can often function if there is adequate moisture present. Unfortunately there is currently no small-scale method for predicting whether a given pump can be primed successfully with an unsaturated cake. Nor is there a method to predict the resulting pressure gradient generated in the discharge pipe. Only full-scale trials can determine these design parameters.

If these trials indicate that even a specialist piston pump is inappropriate, there are various conventional 'dry' bulk solids conveyors which can be employed. A selection of these, together with points to consider for wet materials, is given below.

7.1 Screw and ribbon conveyors

Screw conveyors are generally most suited to 'dry' bulk solids. However, if attention is paid to the design of the screw (for instance, an appropriate pitch-to-diameter ratio and the screw diameter) and the associated operating conditions (per cent loading and screw rotational speed), a wide range of wet materials, typically filter and centrifuge cakes can be transported.

There are various design options to the main principle of screw conveyor operation, such as helical ribbon (with or without a central shaft), interrupted flight and intermeshing, contrarotating conveyors. The latter are used to prime piston pumps (50,51). Because screw conveyors operate on the basis of material sliding on the screw flights but moving axially with respect to the trough axis, parallel, equi-spaced flat bars secured to the inner trough wall can increase the friction between the material and the trough wall so encouraging forward transport of the material. These trough bars can also support the screw in long conveyors where intermediate hanger bearings may act as sites for material build-up.

The various screw designs are attempts to address the main problem of progressive caking which occurs primarily on the screw flight surfaces and on the central shaft. In addition, material deposited in the conveyor trough can become compacted causing screw flight wear, bending of the conveyor shaft and a steep rise in the power requirement. The latter can cause the conveyor motor to trip out.

Caking also leads to reduced conveyor efficiency as conveyor capacity is being removed owing to the presence of the cake. Thus it is important to develop a method which can predict the onset of caking at the design/specification stage. Warren Spring Laboratory is currently applying the ideas of Matchett et al (16) (see Section 6.2) to this caking problem within the industry-sponsored Wet Solids Handling Project.

7.2 Vibratory conveyors

These can be used for wet materials provided attention is paid to the main design variables of vibration intensity, bed depth and trough material. Poorly selected levels of frequency and amplitude of vibration can create simultaneous phase separation of the liquid and solid components of the wet material and a sharp decrease in its internal strength. This phenomenon is generally referred to as 'liquefaction' (52) and one result is a tendency for the material to adhere to the conveyor trough.

The time required for a wet particle or wet agglomerate to deform and break away from the trough following each impact in the vibration cycle may constitute a significant proportion of this cycle. One approach for wet materials is therefore to decrease the frequency of vibration while increasing the amplitude. This gives an overall increase in vibration intensity which is proportional to the vibration amplitude multiplied by the frequency squared. The intensity determines whether the acceleration imparted to the particles is sufficient to cause them to leave the surface of the trough and be conveyed by the hopping mechanism.

As bed depth is increased so the transport velocity is adversely affected by the material's internal friction and hence wet materials show a greater reduction in velocity than 'dry' materials. A high transport velocity coupled with a low bed depth should be maintained for wet materials to reduce the likelihood of adhesion to the trough.

Adhesive forces can also be minimised by using hydrophobic surfaces or coatings to the trough. Stainless steel and ultra-high molecular weight polyethylene are often the first choice, with PTFE ('Teflon' or 'Fluon') useful if the solids are non-abrasive. For abrasive solids, natural (and to a lesser extent synthetic) rubber is preferred. The ability of rubber to flex can also help prevent build-up of sticky materials.

7.3 Belt conveyors

The main advantage of using belt conveyors for wet materials over other types of conveyor is that the only relevant bulk property of the solids is the adhesion characteristic. Thus, increases in internal shear strength brought about by the presence of moisture are unimportant. However, one of the biggest problems is belt cleaning and hence the avoidance of spillage and wasted material (53). Effective cleaning is essential if a weighing section to the belt is incorporated.

Much effort has been devoted to the development of a cleaning technique which is universally satisfactory for all types of materials conveyed. This is probably an unrealistic approach and the many devices commercially available probably have their own ranges of application. Fixed scrapers are the simplest method but cause more wear than other types, need continual adjustment and cause problems with mechanical joints to belt sections. Revolving brushes can be successful if they do not clog (or if they are cleaned frequently if they do clog) and the cost of the electric motor can be saved by driving the cleaner from the tail end pulley motor which must be uprated by a kW or so.

The most successful cleaning devices employ two spiral assemblies of scrapers or brushes placed at two separate locations (53,54). High velocity water jets are often used if other methods have failed. Belt thumpers have been used to dislodge sticky sludge (54) but this is not normally recommended unless the belt is cleated or has a rivetted ridge on it. Whichever cleaning method is found to work best, the importance of regular maintenance of any device cannot be overstressed.

7.4 Chutes

Transfer chutes are common devices which are in essence very simple but which can cause difficulties when conveying wet materials. Ideally chutes should be vertical and have as large a diameter as possible commensurate with the feed and receiving equipment. The practice of placing some types of valves (e.g. butterfly valve) somewhere along the length of the chute and of using rubber bellows to limit the deleterious effects of vibration and temperature gradients should be discouraged as both act as sites for cake build-up.

Sometimes a vessel is charged with paste or cake using a gravity chute where the material originates from an upended drum, can (see Section 5.3) or from a filter. A grid placed at the entrance to the chute acts as both a safety measure and creates a hold-up of the material to prevent large quantities of it dropping into the chute at a time and causing blockage. By using a simple vibratory grid (vibrating in the vertical plane), the material enters the chute at a more controlled rate (33). The vibration will often liquefy the material which will assist flow.

Some paste types (particularly those showing thixotropic property and Bingham plastic property) require alternative vibration forms to facilitate complete discharge from a container. An oscillating, circular motion was found to be necessary for one material type (33). It is believed that this motion causes the rapid shearing of the material in contact with the container wall. After a while the paste adjacent to the wall can no longer support the bulk of the material which then discharges as a complete slug.

Consideration should be given to chute linings to minimise material build-up and eventual blockage. Lining material options are similar to those listed for hoppers in Section 5.2. Alternatively the entire chute can be made from a low friction or flexible material. Both glass fibre chutes and chutes made from terylene cloth, flexing as a result of vibration from a feed centrifuge, have been found to be very effective under some circumstances (1).

ACKNOWLEDGEMENT

This paper is an abbreviated version of a state-of-the-art review (SAR) written in the Wet Solids Handling Project, a research and information consortium funded jointly by 25 industrial member companies and the Department of Trade and Industry. The Project was launched in 1984 and will run until at least 1994. New members can obtain copies of the full review and 30 other SAR's, design/ selection guides and research reports issued in the Project.

REFERENCES

(1) HEYWOOD, N.I., CARLETON, A.J. and BRANSBY, P.L. Problems in handling and processing wet solids. Chem. Engr., Dec 1982, 465-471.

(2) HEYWOOD, N.I. and MCDONAGH, M.A. A review of the present state of the art of selecting and using wet solids handling equipment. "Selection and Use of Wet Solids Handling Equipment", Manchester, UK, June 1985, (I. Chem. E., N.W. Branch).

(3) HEYWOOD, N.I. and BRANSBY, P.L. Needs for research on the handling of wet, particulate solids. CHISA 84 Conference, Prague, Sept 1984.

(4) CARLETON, A.J. and HEYWOOD, N.I. Can you handle sticky cakes? Filtration and Separation, Sept/Oct 1983, 357-360.

(5) BROWN, N.P. and HEYWOOD, N.I. (eds). Slurry Systems: A Design Handbook, 1991 (Elsevier Applied Science Publishers).

(6) HEYWOOD, N.I. The importance of small-scale testwork in the design and selection of bulk materials handling systems. "Getting Better Value for Money in Bulk Handling Plants", London, Feb 1989 (I. Mech. E.).

(7) HEYWOOD, N.I. and KIRBY, J.M. Laboratory measurement techniques for the bulk properties of wet granular solids (in relation to the design and operation of materials handling equipment). Warren Spring Laboratory Report LR 543 (MH), 1985, Stevenage, UK.

(8) HEYWOOD, N.I. Pipeline design for non-Newtonian fluids. I. Chem. E. Symposium Series No. 60, Feb 1980, pp 33-52.

(9) HEYWOOD, N.I. and CHENG, D.C.-H. Comparison of methods for predicting head loss in turbulent pipe flow of non-Newtonian fluids. Trans. Inst. Meas. Cont., 1984, 6 (1), 33-45.

(10) CHEREMISINOFF, N.P. (ed). Encyclopedia of Fluid Mechanics. Vol 5: Slurry Flow Technology. Gulf Publishing Corporation, 1986.

(11) HEYWOOD, N.I. Selecting a viscometer. Chem. Engr., June 1985, pp 16-23.

(12) MOORE, K. Paste preparation for dryers within ICI Organics Division. North West Branch I. Chem. E. Papers, 1984, No. 1.

(13) WANT, F.M., ZOBOROWSKI, M.E. and LAUTERS, C.L. Alcoa of Australia Ltd, Karinama refinery pilot plant for red mud disposal. Proc. Hydrotransport 9, Rome, Italy, 1984, paper F1, pp 237-250 (BHRA Fluid Engng, Cranfield, UK).

(14) TAYLOR, P.D., MATCHETT, A.J. and PEACE, J. Wall effects in the transport and storage of water-contaminated bulk solids. I. Chem. E. Annual Research Meeting, Univ Nottingham, April 1987.

(15) TAYLOR, P.D., MATCHETT, A.J. and PEACE, J. A site investigation of cake formation in materials handling equipment. Chem. Eng. Sci., 1987, 42 (4), 921-922.

(16) MATCHETT, A.J., TAYLOR, P.D. and PEACE, J. Adhesion in materials handling equipment. "Solutions to Industrial Dust, Mess and Spillage Problems", London, April 1989 (I. Mech. E).

(17) HEYWOOD, N.I. and RICHARDSON, J.F. Rheological behaviour of flocculated and dispersed kaolin suspensions. J. Rheol., 1978, 22 (6), 599-613.

(18) HORSLEY, R.R. and REIZES, J.A. The effect of zeta potential on the head loss gradient for slurry pipelines with varying slurry concentrations. Proc. Hydrotransport 7, Sendai, Japan, 1980, pp 163-172 (BHRA Fluid Engng, Cranfield, UK).

(19) FARRIS, R.J. Prediction of the viscosity of multimodal suspensions from unimodal viscosity data. Trans. Soc. Rheol., 1968, 12, 281-301.

(20) SADLER, L.Y. and SIM, K.G. Minimise solid-liquid mixture viscosity by optimising particle size distribution. Chem. Eng. Prog, March 1991, pp 68-71.

(21) DUCKWORTH, R.A. and ADDIE, G.R. Application of a non-Newtonian carrier to transport coarse coal refuse. Proc. 12th Slurry Transport Association Conference, April 1987, New Orleans, Louisiana, USA, pp 21-26.

(22) KESTNER, M.O. Improving the flow properties of wet coal. Proc. 10th Annual Powder and Bulk Solids Conference, Rosemount, Illinois, USA, May 1985, pp 753-761.

(23) ROESSLER, M.L. Application of vibrating equipment for storage and handling of coal filter cake and refuse. 4th Kentucky Coal Refuse Disposal and Utilisation Seminar, Univ Kentucky, Lexington, June 1978, pp 57-63 (Inst. Mining and Minerals Research).

(24) AKIYAMA, T., NAITO, T. and KANO, T. Vibrated beds of wet particles. Powder Technol., 1986, 45, 215-222.

(25) MATCHETT, A.J. Vibration in materials handling systems. Proc. of Int. Symp. on Materials Handling in Pyrometallurgy, Hamilton, Canada, Aug 1990, pp 75-79 (Metallurgical Society of CIM, Pergamon Press).

(26) RADEMACHER, F.J.C. Advantages of the Euro-Mammoth silo for the covered storage of bulk solids. Bulk Solids Handling, 1984, 3 (3), 519-.

(27) AIHARA, S. and SAIDA, Y. The Big Atlas silo system. Bulk Solids Handling, 1985, 5 (3), 627-632.

(28) NUTTALL, R.J. The selection of abrasion resistant lining materials. Bulk Solids Handling, 1985, 5 (5), 1053-1055.

(29) EPHITHITE, H.J. Rubber lining: the soft option against abrasion. Bulk Solids Handling, 1985, 5 (5), 1035-1039.

(30) PEACE, J. Material flow improvement in hoppers, bunkers and silos using low friction lining materials. Concrete Society Conference, Newcastle, UK, Dec 1984, pp 3-15.

(31) PRATKO, M.J. UHMW polymer: a successful approach to lining bulk handling equipment. Bulk Solids Handling, 1983, 3 (1), 197-200.

(32) STEPPLING, K. and HOSSFELD, R.J. Ultra-high molecular weight polyethylene abrasion resistant liners facilitate solids flow in hoppers. Bulk Solids Handling, 1985, 5 (5), 1049-1052.

(33) HARRINGTON, P.S. Filter cake handling in the chemical industry. Filtration and Separation, March/April 1984, 92-94.

(34) GALLOWAY, R.M., RATHBURN, D.R. and KIRSHENBAUM, N.W. In-plant materials handling using the Marconaflo system concept. Proc. Hydrotransport 2, Coventry, UK, 1972 (BHRA Fluid Engng, Cranfield, UK).

(35) SIMS, W.N., BEEBE, R.R., ANDERSON, A.K. and BHASIN, A.K. Evaluation of basic slurry properties as design criteria for the Marconaflo system. Bulk Solids Handling, 1981, 1 (3), 455-462.

(36) HEYWOOD, N.I. and RICHARDSON, J.F. Head loss reduction by gas injection for highly shear-thinning suspensions in horizontal pipelines. Proc. Hydrotransport 5, Hanover, Germany, May 1978, Paper C1 (BHRA Fluid Engng, Cranfield, UK).

(37) HEYWOOD, N.I. A review of techniques for reducing energy consumption in slurry pipelining. Proc. Hydrotransport 10, Innsbruck, Austria, Oct 1986, paper K3, pp 319-332 (BHRA Fluid Engng, Cranfield, UK).

(38) HISAMIJSU, N., SHOJI, Y. and KOSUGI, S. Effect of added fine particles on flow properties of settling slurries. Proc. Hydrotransport 5, Hanover, Germany, May 1978, paper D3, (BHRA Fluid Engng, Cranfield, UK).

(39) BROOKES, D.A. and SNOEK, P.E. Stabflow slurry development. Proc. Hydrotransport 10, Innsbruck, Austria, Oct 1986, paper C3 (BHRA Fluid Engng, Cranfield, UK).

(40) DUFFY, G.G. A new suspending medium for the pipeline transport of coarse, high density and dense-phase particles and capsules. Proc. Hydrotransport 9, Rome, Italy, Oct 1984, paper E2, pp 227-234 (BHRA Fluid Engng, Cranfield, UK).

(41) ROUND, G.F., HAMEED, A. and LATTO, B. Pulsing flows of clay suspensions. Proc. 6th Slurry Transport Association Conference, Las Vegas, Nevada, USA, March 1981, pp 121-30.

(42) DEYSARKAR, A.K. and TURNER, G.A. Flow of paste in a vibrated tube. J. Rheol., 1981, 25 (1), 41-54.

(43) KOKINI, J.L., DERVISOGLU, M. and KILLOPS, R. Facilitating the transport of very viscous suspensions. J. Rheol., 1986, 30, S61-S74.

(44) ZAKIN, J.L., POREH, M., BROSH, A. and WARSHAVSKY, M. Chem. Eng. Prog. Symp. Ser. No. 11, 1971, 67, 85-89.

(45) GOLDA, J. The effect of polymeric additives on the hydraulic transport of coal in pipes. Proc. 9th Slurry Transport Association Conference, Lake Tahoe, Nevada, USA, 1984, pp 55-62.

(46) ROUND, G.F. Pulsating flows - applications to the pipeline flow of solid-liquid suspensions. Proc. Hydrotransport 3, Golden, Colorado, USA, May 1974, paper B3 (BHRA Fluid Engng).

(47) SHRIEK, W., SMITH, L.G., HAAS, D.B. and HUSBAND, W.H.W. The potential of helically-ribbed pipes for solids transport. CIM Bulletin, 1974, 67, No. 750, 94-91.

(48) SAUERMANN, H.B. Hydraulic transport of coarse solids in circular and segmented pipes. Proc. 7th Slurry Transport Association Conference, Lake Tahoe, Nevada, USA, 1982, PP 51-60.

(49) FEHN, B. Two-cylinder piston pumps with hydraulic drive: technical characteristics and application options. Proc. Hydrotransport 9, Rome, Italy, 1984, paper H3 (BHRA Fluid Engng, Cranfield, UK).

(50) CARROLL, E. Pumps for stiff and aggressive wet solids. Int. Conf. on Positive Displacement Pumps, Oct 1986, paper B1 (BHRA Fluid Engng, Cranfield, UK).

(51) KRONENBERG, J. New Putzmeister developments. Bulk Solids Handling, 1988, 8 (3), 47-53.

(52) KIRBY, J.M. Liquefaction of cargoes - a literature survey. Warren Spring Laboratory Report LR 388 (MP), 1981, Stevenage, UK.

(53) STAHURA, R.P. The price of DURT - what fugitive materials cost conveyors and how improved systems can save money. Proc. Int. Symp. on Materials Handling in Pyrometallurgy, Hamilton, Canada, Aug 1990, pp 217-226 (Metallurgical Society of CIM, Pergamon Press).

(54) DAVIS, T.D. Design of belt conveyors for dewatered sewage sludge and compost. Proc. 10th Annual Powder and Bulk Solids Conference, Rosemount, Illinois, USA, May 1985, pp 180-192.

Table 1 Summary of techniques to reduce head loss for pipeline flow of 'non-settling' slurries in the laminar regime

Technique	Requirement for Suspension Flow Property	Essential Mechanism
A. Increase in particle size	Newtonian or non-Newtonian	Reduction in Brownian motion contribution to viscosity and/or reduction in increase in effective volume fraction of solids arising from reduced importance of adsorbed layer effect.
B. Broader particle size distribution (19,20)	Newtonian or non-Newtonian	Wider particle size distribution at fixed solids concentration leads to lower slurry viscosity. (Alternatively higher solids loading is achievable at a fixed suspension viscosity and hence fixed head loss.)
C. Reduce particle angularity	Newtonian or non-Newtonian	Reduction in viscosity arising from reduced surface friction contacts between particles at high solids concentrations.
D. Addition of soluble ionic compounds as dispersants (17,18)	Usually non-Newtonian, shear-thinning	Changes pH and/or zeta potential on fine particle surfaces, causing deflocculation and hence reduces suspension viscosity.
E. Periodic injection of water (or other suspending medium) along pipeline length (43)	Newtonian or non-Newtonian but probably latter with high yield stress	Water injection gives local reduced solids concentration at pipe wall and hence lower head loss.
F. Air injection into flowing slurry (36)	Must be non-Newtonian, shear-thinning with or without yield stress	Presence of air reduces wetted pipe wall area on which slurry shear stress acts so reducing drag and head loss. Increase in slurry velocity at fixed slurry flow-rate tends to increase head loss marginally owing to shear-thinning behaviour.
G. Oscillation of slurry flow or applied pressure gradient (41)	Non-Newtonian, shear-thinning. Viscoelastic property may enhance effect	Probable creation of annulus adjacent to pipe wall which is depleted in particles and hence lower wall shear stress occurs.
H. Pipe vibration (42)	Non-Newtonian, shear-thinning	Possible creation of slip layer near to wall depleted in solids. May also be thixotropic structure breakdown effect occurring.

Table 2 Summary of techniques to reduce head loss for pipeline flow of 'settling' slurries in the turbulent flow regime

Technique	Essential Mechanism
A. Addition of surfactants (44)	Probable reduction in inter-particle friction.
B. Addition of soluble, long-chain polymers (45)	Toms' effect occurs in turbulently-flowing suspending medium. Mechanism still unclear but probably dampening of turbulent eddies.
C. Oscillation of slurry flowrate (46)	Reduction in deposit velocity arising from more effective particle suspension in oscillating flow.
D. Use of helical ribs attached to inner pipe wall (47)	Ribs reduce deposit velocity by continuously picking up particles tending to settle. Lower velocity means lower head loss.
E. Use of segmented pipe (48)	Mechanism unclear but deposit velocity reduced and hence pipeline can be operated at lower slurry flowrates and hence lower head losses are incurred.

C418/017

Ash handling on large power stations in South Africa – the material, the equipment, the changes

A G LORENZATO, BSc(Eng), PrEng
ESKOM, Johannesburg, South Africa

SYNOPSIS With the recent changes in technologies used by Eskom for the removal and disposal of ash from its 6 x 600 to 665 MW power stations in South Africa, generic problems have become evident. The paper discusses some of the problems experienced with these ash handling systems. This paper also introduces a plant which for the purpose of transporting ash to the disposal site, uses the concept of mixing coarse ash with fly ash and waste water, to produce a slurry with a concentration of 50 % by mass. The plant is to be commissioned in July 1991, on a 6 x 600 MW power station.

1 INTRODUCTION

The quantity of ash produced from a 600 to 665 MW boiler-generator set is largely dependent on the quality of coal burnt at the specific power station and varies in Eskom from 50 T/hr to 145 T/hr. Given the typical arrangement of 6 x 600 to 665 MW boiler-generator sets on each of the larger power stations, this ash quantity equates to between 300 T/hour to 876 T/hour, when the power station is at full load.

Coarse ash is removed from the bottom of the boiler where it has been quenched in a 'bath' of water. Fly ash which is collected dry at between 120 °C and 140 °C is removed from the multiple hoppers under the electrostatic precipitators.

The ash is dumped at a disposal site located either above ground or over an opencast mine. Sales of ash are minimal and the bulk of the ash is transported to the disposal sites located approximately 2 to 4 km from the power station terrace.

2 TYPICAL PARTICLE SIZE DISTRIBUTIONS

The coarse ash and fly ash have distinctly different characteristics in terms of size and shape. Their typical differences in particle size distribution are shown in Figure 1, however, 'coarser' and 'finer' particle size distributions will occur, for example:

(a) Between different power stations (Figure 2).

(b) Between fly ash samples taken from hoppers under different fields in the electrostatic precipitator.

(c) With changes in pulverised fuel supply, and/or combustion conditions.

(d) With particle degradation during transport.

A typical relative density for fly ash is approximately 2.2 and for coarse ash approximately 2.38.

In terms of total ash production, the percentage split between fly ash production and coarse ash production varies between 95:5 and 80:20 (fly ash : coarse ash).

3 DESCRIPTION OF ASH TRANSPORTATION SYSTEMS WITHIN ESKOM

The majority of the ash handling systems utilised by Eskom can be generally categorised into 'wet' and 'dry' type systems [1]. See Figures 3 and 4.

The basic distinction between 'dry' and 'wet' ash handling systems is derived from the characteristics of the product conveyed to the disposal site. (These basic differences are listed in items 3.1 and 3.2 below.) Typically, 'dry' ash is transported on conveyor belts and 'wet' ash is pumped by centrifugal pumps.

3.1 **'Dry' Ash Transport - Normal Operation**

Moisture contents is defined by:

$$\frac{\text{mass of water}}{\text{mass of water + mass of ash}} \times 100$$

(a) Moisture content of fly ash
= 9 % to 21 % (by mass)

(b) Moisture content of coarse ash
= 42 % to 56 % (by mass)

These ranges in moisture contents result principally from the different ash characteristics on power stations, with a narrower band being applicable for any specific plant.

3.2 **'Wet' Ash Transport - Normal Operation**

Slurry concentration is defined by:

$$\frac{\text{mass of ash}}{\text{mass of ash + mass of water}} \times 100$$

(a) Slurry concentration with fly ash
= 12 % to 40 % (by mass)

(b) Slurry concentration with coarse ash
= 4 % to 25 % (by mass)

As a result of the system design, periods of pumping clear water and dilute slurries are regular occurrences. Although relatively 'dense' slurries can be achieved for short periods, slurry concentration over the complete ashing cycle tends towards the 'dilute' slurry concentrations.

Fly ash and coarse ash are transported at different times from the same pumping system. This is largely due to system design constraints, but also a requirement for building the walls of the ash dam.

Resulting from the extensive design, operating and maintenance experience on 'wet' ash handling systems (installed on all Eskom power stations completed before 1985), the problems [1] associated with this type of plant are well understood and are largely inherent to the type of equipment and concept utilised. In contrast to the 'wet' ash handling plant, the problem areas associated with 'dry' ash handling plant are, at this stage of investigation, a combination of

(a) design, erection, commissioning, operating and maintenance problems, and

(b) inherent problems particular to the type of equipment and concepts utilised.

This is due to the relatively short period of operation of the 'dry' ash handling plants. This type of plant is installed on four 6 x 665 MW power stations which are presently in various stages of commissioning. The longest running unit was first commissioned in 1985.

Figures 3 and 4 show water ejectors and chain conveyors respectively for handling fly ash from beneath the electrostatic precipitators. Although not shown, pneumatic conveying systems are also used by Eskom.

4 TYPICAL PROBLEMS ON 'DRY' ASH HANDLING PLANTS

4.1 Equipment Problems

A large percentage of the problems have resulted from 'standard' materials handling equipment and system designs being applied without adequate consideration for

(a) the ash properties,

(b) the change in the ash properties with varying conditions, e.g. moisture content,

(c) the general ash handling environment, e.g. dust.

Examples of the properties of ash which affect the system designs, include:

(a) The 'flowability' of different particle size distributions of fly ash when hot and/or aerated.

(b) The relative reduction or lack of 'flowability' of fly ash upon the introduction of moisture, either from external sources or as a result of its temperature dropping below the dew point of the flue gas.

(c) The change in the consistency of fly ash when mixed with different quantities of water, i.e. as the moisture content is increased, it transforms from a dusty powder to:
- a moist 'cake-like' mixture,
- a paste,
- a slurry.

At each of these possible consistencies, the behaviour of the ash is unique.

(d) Moist ash hangs up/bonds to surfaces and to itself. Irregularities on the contact surfaces propagate the 'bonding' effect. The extent of this 'bonding' is also dependent on any dynamics which may exist at the point of contact, e.g. 'flexing' of the surface, changes in angle of impingement, etc.

(e) Vapour and airborne fines are produced from the process of mixing hot fly ash with ambient temperature water. The rising 'sticky' mixture creates a build-up of solids wherever it comes into contact with a surface.

(f) Dry fly ash is extremely dusty.

(g) Where fly ash has been in contact with sufficient water to have formed a slurry, the bulk of the fly ash will harden once it has 'dried out', but a dusty surface layer will remain.

(h) The fineness of the coarse ash influences its dewatering properties, the coarser material dewatering easier than the finer material (finer particle size distributions than those shown in Figure 2 do occur).

(i) The finer portion of dewatered coarse ash is 'sticky'.

(j) Coarse ash and fly ash are abrasive products.

The abovementioned ash properties have transition zones where the materials' behaviour in terms of, for example, 'bonding', 'flowability' or 'dewatering', can change significantly with even minor changes in the ash/water or ash/air ratios and/or ash particle size distribution.

Having identified some of the basic ash properties which influence ash handling plant designs, the following are some examples of typical areas where 'standard' designs have been applied without taking adequate cognisance of these material properties:

(a) The design of chutes at transfer points between belt conveyors.

(b) Mechanical belt cleaning devices.

(c) Sealing arrangements on bearings, shafts, etc.

(d) Control of fly ash flow through chain conveyors.

(e) Inclined chain conveyors used on fly ash (run-back of material).

(f) Primary control and instrumentation devices, e.g. silo (fly ash) level measurement and control, blocked chute detection, belt and chain conveyor protection.

(g) Apron conveyors on dewatered coarse ash applications.

(h) Materials used in the construction of ash conditioner (mixer) blades.

(i) 'Steep' inclined troughed belt conveyor used on dewatered coarse ash (run-back of material).

(j) Cleaning facilities for spillages.

(k) Containment of dust and vapour.

Some of the consequences of such problem areas are:

(a) Blockages in transfer chutes resulting in a system shutdown.

(b) Spillages causing excessive cleaning costs and damage to equipment.

(c) Rapid wear rates and equipment failures.

(d) Loss of control and protection on system operation.

(e) Unpleasant working environment.

(f) Negative attitudes of personnel operating, maintaining and cleaning the plant.

(g) Unpredictable/erratic plant performance (deviations from performance setpoints).

For many of the problems on the ash plants, various solutions with varying degrees of success are available. It is not the intention of this paper to cover this extensive and relatively detailed topic. However, some examples are given of areas in the design of 'dry' ash handling plant in Eskom which have the potential for further development, in order to avoid them from becoming considered inherent problems on 'dry' ash handling plants. Examples are:

(a) Development of low maintenance, effective mechanical belt cleaning device/s.

(b) Development of low maintenance, reliable primary control and instrumentation devices.

4.2 Skills Problems

Associated with the shift in technology from 'wet' ash handling plants to 'dry' ash handling plants (Figures 3 and 4), was a relative increase in complexity concerning operating and maintenance requirements. In general, this shift in technology was not accompanied initially by a suitable increase in the level of skills of operating and maintenance staff. This resulted in typical plant performance shortfalls associated with poor operating and maintenance practices.

This problem can be avoided, principally by ensuring that there is a 'match' at the design stage between

(a) the available/achievable skills of operating and maintenance staff, and

(b) the level of technology inherent in the new ash plant.

5 A NEW TYPE 'WET' ASH HANDLING SYSTEM IN ESKOM

5.1 Background

Matla Power Station is an existing 6 x 600 MW power station which produces up to 580 T/hr of ash. The existing ash handling plant is a 'wet' type ash handling plant, principally as shown in Figure 3.

As the result of an agreement to cease utilisation of the existing ash disposal site by January 1991, numerous alternative ash disposal sites and ash handling systems were investigated (in accordance with Eskom's procedures for selecting an ash handling system [1]).

From this investigation, the system described below was selected and is presently under construction with a target date for commissioning of July 1991.

5.2 Basic Concept for the New 'Wet' Ash Handling System

Coarse Ash Transportation

Part of the existing equipment on the 'wet' ash handling system at Matla Power Station will remain in service, namely the system for removing coarse ash from the bottom of the boiler and the associated water supply system for this purpose. (See Figure 5.)

Coarse ash accumulates in the water-filled hoppers below each of the boilers and is removed typically once every 8 hours. Removal is via a system of hydraulically operated doors on the hoppers and high pressure water jets inside the hoppers and along sluiceways. Varying degrees of manual intervention are required to break up and remove large fused lumps of coarse ash which block the discharge system at the hoppers.

Oversize lumps in the ash slurry from the sluiceways are crushed to approximately 10 - 25 mm before passing into the ash sump for further transport by centrifugal pumps. The slurry which is pumped has a relatively uncontrolled slurry concentration, with the characteristics described in Section 3.2 (b).

The present system for handling coarse ash is now modified by diverting each of the existing pipelines (which are transporting the 'dilute' slurry of coarse ash) into dewatering bins. (See Figure 6.)

When a bin is full of coarse ash, the feed to the bin is stopped and the water is drained from the ash. Draining continues until the coarse ash is of a suitable quality that its rate of discharge from the bin can be controlled. Its moisture content must also be suitable for transportation on belt conveyors.

The water which overflows the bin while it is being filled with a dilute slurry of coarse ash, as well as the water which is later drained from the coarse ash, is discharged into tanks where settlement takes place. The settled ash in these tanks is continuously removed and pumped into 'sludge' tanks containing sludge and effluent water. There it forms part of the overall water supply used to create the 'dense' ash slurries. The 'clear' water which overflows from the settling tanks is pumped to an existing head tank. From the head tank, this 'clear' water is re-circulated on terrace for further 'dilute' slurry transport of coarse ash from the boiler hoppers to the dewatering bins.

Fly Ash Transportation

The existing water ejector type ash removal system under the hoppers of the electrostatic precipitators will be removed and replaced by a pneumatic conveying system.

The first stage of the pneumatic conveying system will transport the fly ash from the hoppers to an intermediate silo, located adjacent to each of the electrostatic precipitators. From each of the six silos, a second stage of pneumatic conveying will transport the fly ash to one of three centralised silos, located at the edge of the power station terrace.

Mixing Plant

The silos for fly ash storage, together with the bins for coarse ash storage, are located in close proximity to one another for the purpose of combining their contents into a mixing plant.

Fly ash is discharged from the silo, at a controlled rate, into a paddle-type mixer. Here it is mixed with waters from

(a) the station's water treatment plant and

(b) 'excess' water which is re-circulated from the disposal site after decanting from the slurry.

The paddle-type mixer is linked to a tank ('BBA tank'), through which the slurry of fly ash must pass to enter the first stage of the pumping system.

With this system operational, coarse ash is discharged from the dewatering bin, at a controlled rate, into the BBA tank. At this point, the coarse ash will mix with the fly ash slurry passing through the tank, prior to being pumped to the ash disposal site.

The control for the feed of dry fly ash is by means of a flow control valve, interlocked with a mass measuring device. This control equipment is situated in the air activated gravity conveyors which feed the fly ash from the silos to the mixers.

The control for the feed of dewatered coarse ash is by means of a beltfeeder located below the bins and interlocked with a mass measuring device. Belt conveyors will transport the metered quantities of coarse ash to the mixing tank (BBA tank).

The entire mixing process will be controlled by a process logic controller (PLC).

Ash Disposal

One variable speed pump and four fixed speed centrifugal pumps form the series pumping arrangement, to deliver the slurries to a selected discharge point on the ash dam. Flowmeters, pressure switches and density meters will be used to control the conditions within the pipeline.

The disposal area is an opencast coal mine, which has largely been rehabilitated after mining operations were completed in 1985. The slurry is delivered by means of a ringfeed system of pipelines to a selected discharge point on the disposal area. The nearest discharge point is approximately 3,5 km from the pumping plant and the furthest approximately 8 km.

The excess water from the slurry transport, as well as stormwater, is accumulated in the open 'cut' of the opencast mine. This water is returned to the power station, for further ash transport, by means of barge-mounted vertical spindle pumps in series with a booster pumpstation.

5.3 Objectives of the New 'Wet' Ash Handling System

The principal objective with this plant is to achieve a continuously CONTROLLED slurry delivered to the disposal site. The basic features of the slurry which are to be controlled include:

i) Slurry concentration (50 % by mass).

ii) Ratio of fly ash to coarse ash (2 parts fly ash to 1 part coarse ash, by mass).

Some other operating criteria for the system are:

(a) Coarse ash must only be pumped as a mixture with fly ash (see item (ii) above), but in a backlog condition for coarse ash, it must be possible to pump coarse ash alone to a selected discharge point.

(b) Only slurries containing fly ash and no coarse ash must be pumped to the periphery of the ash dam, for building the walls of the dam.

(c) Sludge and effluents from the water treatment plants must be disposed of by the ash handling plant. However, during the pumping of fly ash to the wall of the ash dam (see item b above), effluent must not be mixed into this slurry. (The effluent affects the bonding properties of the fly ash.)

(d) The number of stop/starts of any slurry pumpset/pipeline to the disposal site must be minimal, i.e. 'continuous' operation with a controlled slurry is the objective.

Tests were conducted for Eskom by research institutes, to obtain information on settling velocities and friction gradients for slurries in pipelines. Tests included the behaviour of slurries in inclined sections of pipeline after shutdown and re-start of a pumping system. These tests were conducted in closed loop pumping systems at the research centres, for various slurry concentrations and various ratios of fly ash to coarse ash, using ash samples from Matla Power Station.

5.4 Comments on the New 'Wet' Ash Handling Plant

Concerning the water aspects of slurries, there are numerous advantages to pumping the controlled 'dense' slurries as opposed to the 'dilute' slurries of the traditional 'wet' ash handling systems in Eskom. Some of these advantages are:

(a) Water saving.

(b) Reduced power costs, where the water must be returned for re-use.

(c) Easier management of the water balance.

(d) Easier ash dam construction.

(e) Reduced risk of environmental pollution through seepage.

When pumping slurries with coarse ash alone, it is not possible to achieve the slurry concentrations equivalent to those of 'dense' fly ash slurries. By producing slurries of coarse ash mixed with fly ash, the resulting concentrations for the overall ash production, pumped by centrifugal pumps to the disposal site, remain 'dense' slurries (50 % concentration by mass), i.e. no 'dilution' required for pumping the coarse portion of ash.

Although no tests were conducted to establish the effects on the wear of pumps, pipelines, valves, etc., it is expected that by mixing coarse ash into the fly ash, the wear effects of the coarse ash will be reduced.

However, in converting to the new 'wet' ash handling plant, some of the benefits of the traditional 'wet' ash handling systems in Eskom are lost, namely:

(a) The relative simplicity of the equipment, operation and control and instrumentation.

(b) The relative independence of the ash handling plants for each of the six boiler units on common plant.

(c) The familiarity which the operators and maintenance staff now have for the operation and maintenance of the existing system.

In addition, since the new 'wet' ash handling system incorporates various areas of plant which are untested, both in Eskom and by the contractor, there are inherent risks with such an extensive change to the system.

6 CONCLUSION

The properties of ash and particularly fly ash, make it possible to handle and transport this material by numerous methods. Some implications are:

(a) This can be a disadvantage since it can lead to the arbitrary use of materials handling equipment, where inadequate attention is given to the detail design features required to accommodate, without problems, the particular ash characteristics and their variations in a particular application.

(b) The advantage is that there is extensive potential for the further development of both new equipment and the modification/improvement of existing equipment, to handle and transport ash.

(c) The level of technologies can vary widely between different ash handling systems, particular in the area of automation (control and instrumentation). It is important that from conceptual design stage,the designer and the client understand the level of operating and maintenance skills required to be used on the plant.

(d) The most cost effective system for handling ash at a particular power station will be unique in terms of either concept, plant type, layout or detail design. Therefore each power station site requires its own evaluation process for the numerous ash handling plant options available.

(e) Any ash handling plant will normally be made up of a series of different types of materials handling plants, each of which require certain ash properties in order to convey the material, e.g. 'wet' or 'dry' ash. Since compatibility is required between these systems, it is important that the system be finalised as a 'whole' from source to disposal and not as separate subsystems.

(f) The client and the designer of a new ash handling plant should be fully aware at the time of system selection process, of the limitations of the particular plant type, i.e. in terms of the range of ash properties which it can successfully handle and the consequences if ash properties deviate beyond these limits.

Due to the close proximity of the ash handling plant with other power plant equipment, spillages from the ash plant are 'not welcome', since besides the effects on the working environment, dust blows can ultimately also have a detrimental effect on nearby equipment. Therefore, in addition to features on ash plants which prevent/contain spillages, facilities must be provided for cost effective cleaning, since spillages can occur suddenly and in large quantities.

REFERENCE

[1] LORENZATO, AG and BLACKBEARD, PJ. Collection, Storage and Transport of Ash at Large Power Stations in South Africa, International Symposium on Ash - A Valuable Resource, Pretoria, Republic of South Africa, 2-6 February 1987, Paper IV - Session 7, Council for Scientific and Industrial Research

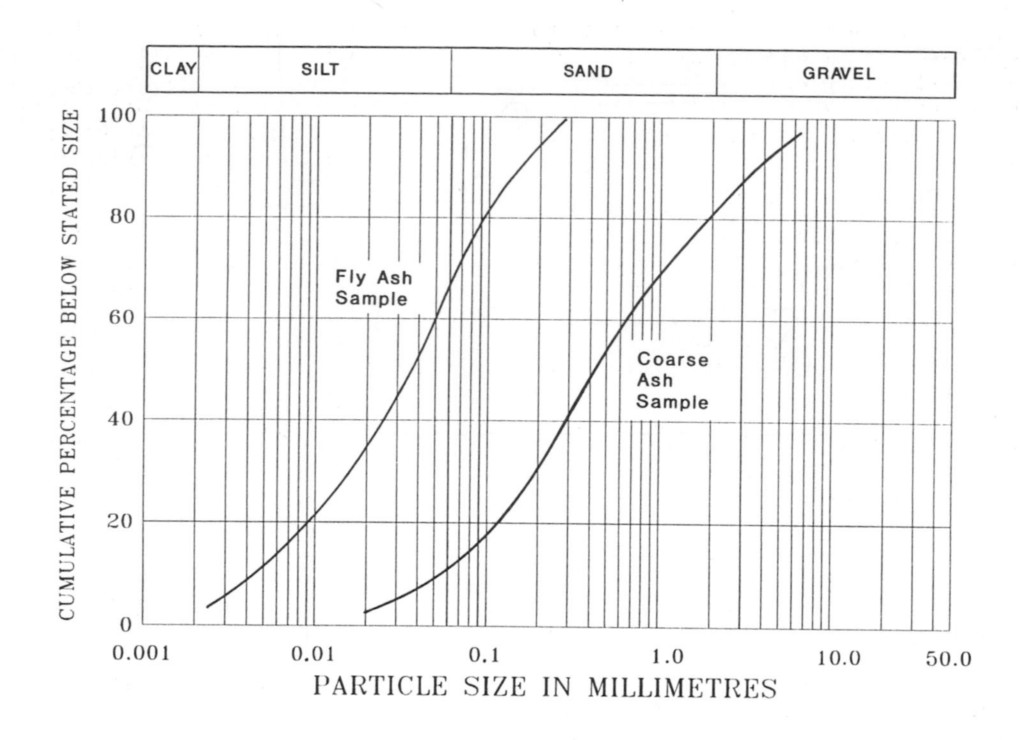

Fig. 1: TYPICAL DIFFERENCE IN PARTICLE SIZE DISTRIBUTION BETWEEN FLY ASH AND COARSE ASH

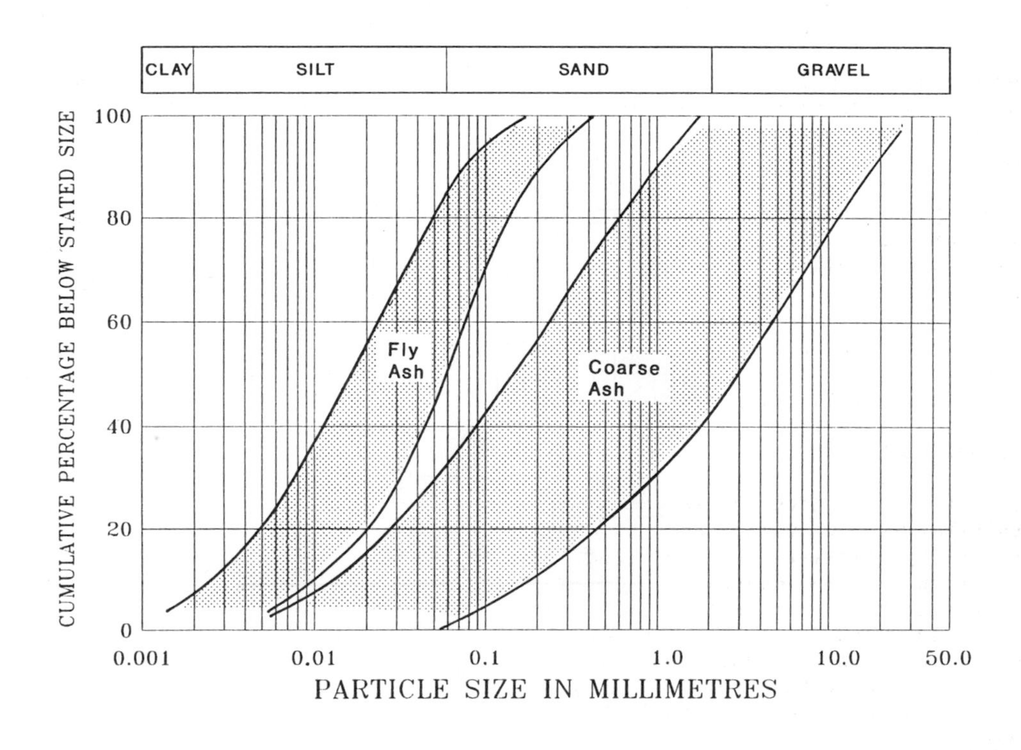

Fig. 2: ENVELOPE OF ASH PARTICLE SIZE DISTRIBUTIONS FOR DIFFERENT POWER STATIONS

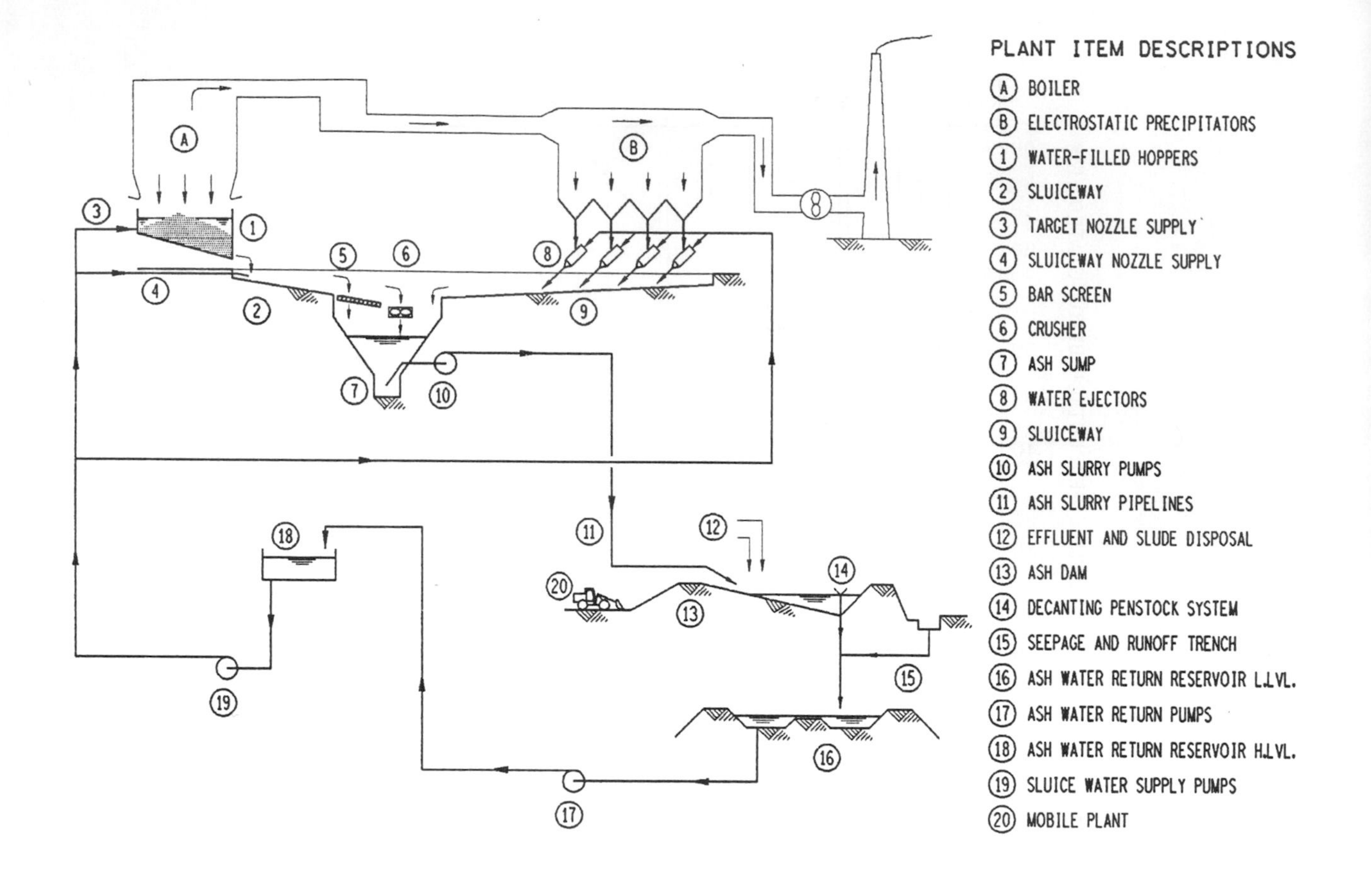
PLANT ITEM DESCRIPTIONS
A BOILER
B ELECTROSTATIC PRECIPITATORS
1 WATER-FILLED HOPPERS
2 SLUICEWAY
3 TARGET NOZZLE SUPPLY
4 SLUICEWAY NOZZLE SUPPLY
5 BAR SCREEN
6 CRUSHER
7 ASH SUMP
8 WATER EJECTORS
9 SLUICEWAY
10 ASH SLURRY PUMPS
11 ASH SLURRY PIPELINES
12 EFFLUENT AND SLUDE DISPOSAL
13 ASH DAM
14 DECANTING PENSTOCK SYSTEM
15 SEEPAGE AND RUNOFF TRENCH
16 ASH WATER RETURN RESERVOIR L.LVL.
17 ASH WATER RETURN PUMPS
18 ASH WATER RETURN RESERVOIR H.LVL.
19 SLUICE WATER SUPPLY PUMPS
20 MOBILE PLANT

Fig. 3.0: BASIC FLOWSHEET FOR 'WET' ASH HANDLING SYSTEM

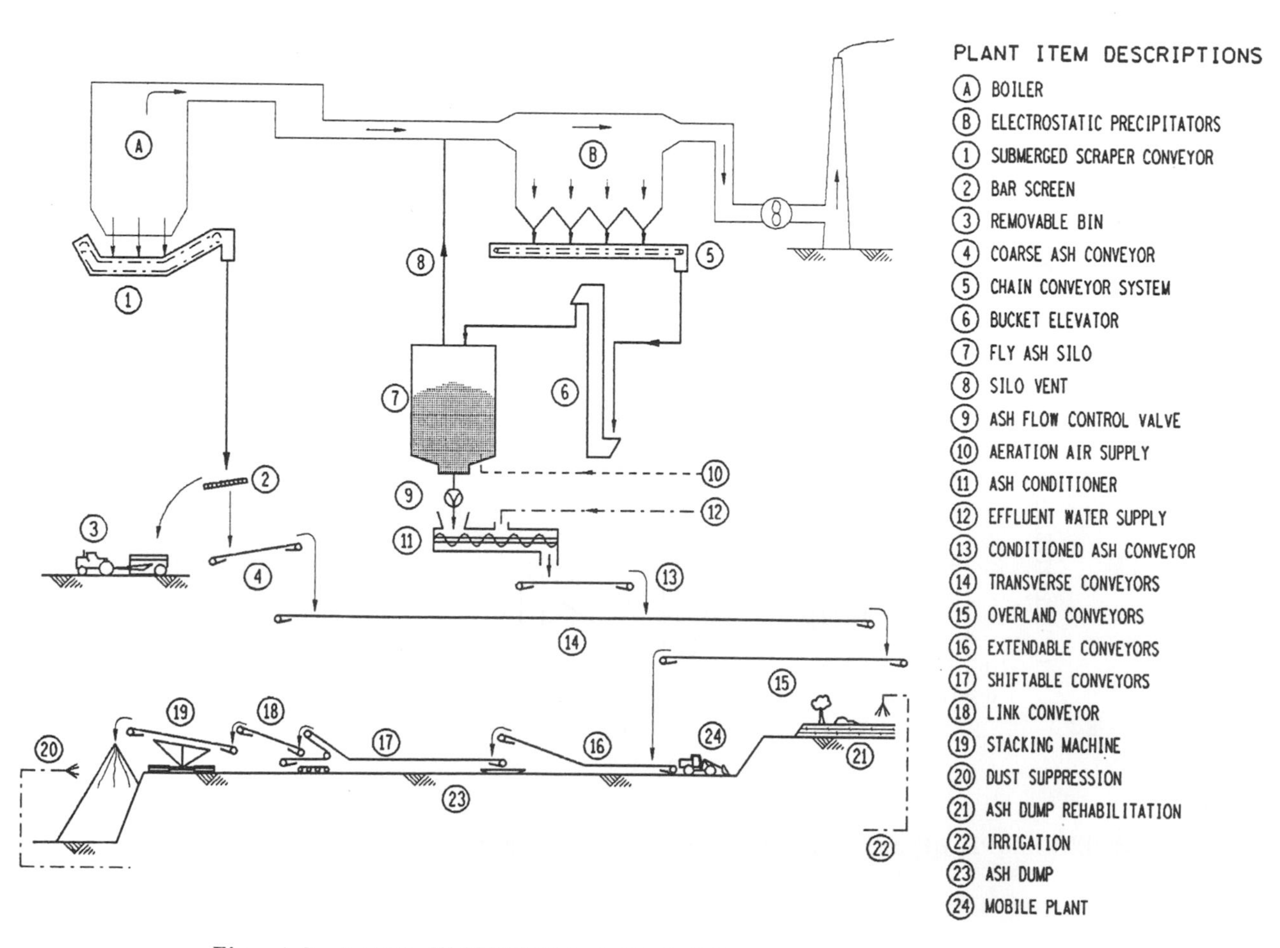
PLANT ITEM DESCRIPTIONS
A BOILER
B ELECTROSTATIC PRECIPITATORS
1 SUBMERGED SCRAPER CONVEYOR
2 BAR SCREEN
3 REMOVABLE BIN
4 COARSE ASH CONVEYOR
5 CHAIN CONVEYOR SYSTEM
6 BUCKET ELEVATOR
7 FLY ASH SILO
8 SILO VENT
9 ASH FLOW CONTROL VALVE
10 AERATION AIR SUPPLY
11 ASH CONDITIONER
12 EFFLUENT WATER SUPPLY
13 CONDITIONED ASH CONVEYOR
14 TRANSVERSE CONVEYORS
15 OVERLAND CONVEYORS
16 EXTENDABLE CONVEYORS
17 SHIFTABLE CONVEYORS
18 LINK CONVEYOR
19 STACKING MACHINE
20 DUST SUPPRESSION
21 ASH DUMP REHABILITATION
22 IRRIGATION
23 ASH DUMP
24 MOBILE PLANT

Fig. 4.0: BASIC FLOWSHEET FOR 'DRY' ASH HANDLING SYSTEM

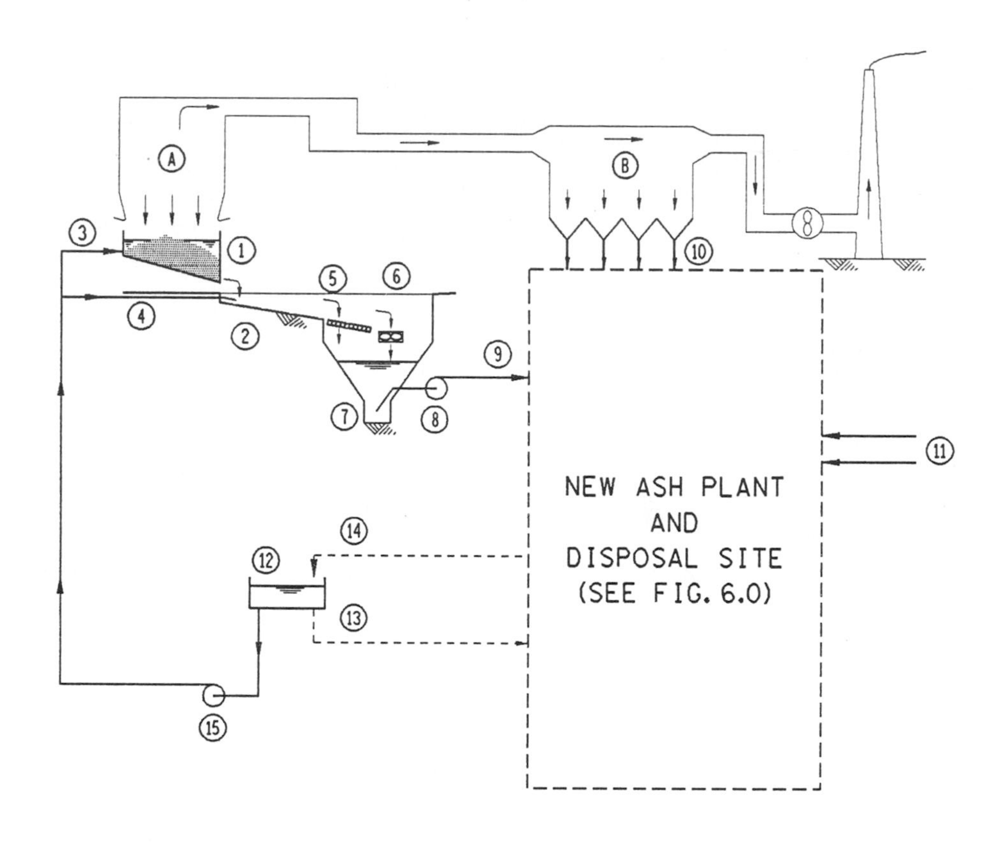

- (A) BOILER
- (B) ELECTROSTATIC PRECIPITATORS
- (1) WATER-FILLED HOPPERS
- (2) SLUICEWAY
- (3) TARGET NOZZLE SUPPLY
- (4) SLUICEWAY NOZZLE SUPPLY
- (5) BAR SCREEN
- (6) CRUSHER
- (7) ASH SUMP
- (8) ASH SLURRY PUMPS
- (9) ASH SLURRY PIPELINES
- (10) FLY ASH FEED FROM HOPPERS
- (11) EFFLUENT AND SLUDGE LINES
- (12) ASH WATER RETURN RESERVOIR H.LVL.
- (13) NEW ASH WATER RETURN CONNECTION - SUPPLY
- (14) NEW ASH WATER RETURN CONNECTION
- (15) SLUICE PUMPS

Fig. 5.0: BASIC FLOWSHEET SHOWING PRINCIPLE TIE-IN INTERFACES BETWEEN EXISTING AND NEW ASH PLANT AT MATLA POWER STATION

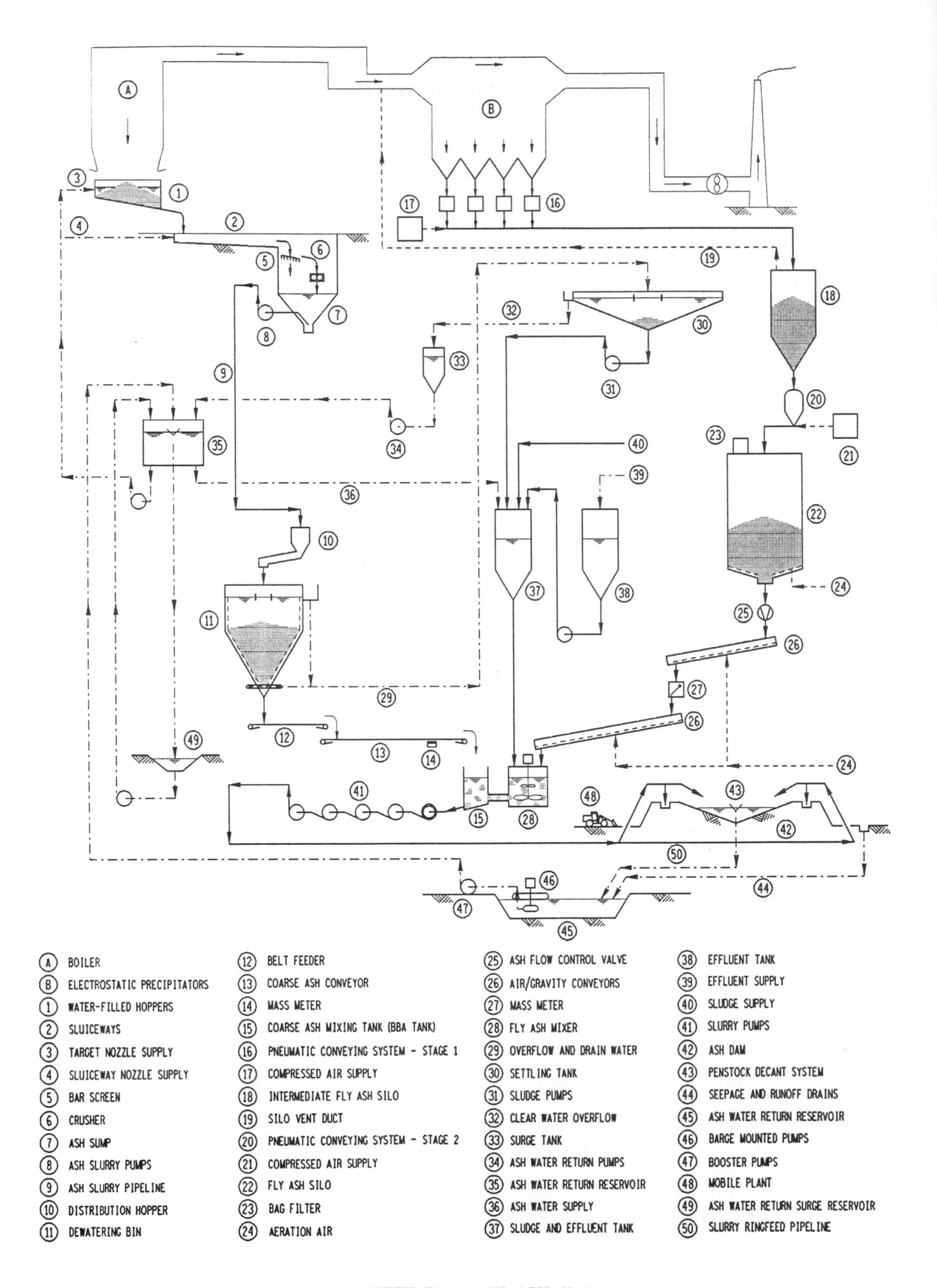
A BOILER
B ELECTROSTATIC PRECIPITATORS
1 WATER-FILLED HOPPERS
2 SLUICEWAYS
3 TARGET NOZZLE SUPPLY
4 SLUICEWAY NOZZLE SUPPLY
5 BAR SCREEN
6 CRUSHER
7 ASH SUMP
8 ASH SLURRY PUMPS
9 ASH SLURRY PIPELINE
10 DISTRIBUTION HOPPER
11 DEWATERING BIN
12 BELT FEEDER
13 COARSE ASH CONVEYOR
14 MASS METER
15 COARSE ASH MIXING TANK (BBA TANK)
16 PNEUMATIC CONVEYING SYSTEM - STAGE 1
17 COMPRESSED AIR SUPPLY
18 INTERMEDIATE FLY ASH SILO
19 SILO VENT DUCT
20 PNEUMATIC CONVEYING SYSTEM - STAGE 2
21 COMPRESSED AIR SUPPLY
22 FLY ASH SILO
23 BAG FILTER
24 AERATION AIR
25 ASH FLOW CONTROL VALVE
26 AIR/GRAVITY CONVEYORS
27 MASS METER
28 FLY ASH MIXER
29 OVERFLOW AND DRAIN WATER
30 SETTLING TANK
31 SLUDGE PUMPS
32 CLEAR WATER OVERFLOW
33 SURGE TANK
34 ASH WATER RETURN PUMPS
35 ASH WATER RETURN RESERVOIR
36 ASH WATER SUPPLY
37 SLUDGE AND EFFLUENT TANK
38 EFFLUENT TANK
39 EFFLUENT SUPPLY
40 SLUDGE SUPPLY
41 SLURRY PUMPS
42 ASH DAM
43 PENSTOCK DECANT SYSTEM
44 SEEPAGE AND RUNOFF DRAINS
45 ASH WATER RETURN RESERVOIR
46 BARGE MOUNTED PUMPS
47 BOOSTER PUMPS
48 MOBILE PLANT
49 ASH WATER RETURN SURGE RESERVOIR
50 SLURRY RINGFEED PIPELINE

Fig. 6.0: BASIC FLOWSHEET FOR NEW ASH HANDLING SYSTEM AT MATLA POWER STATION

C418/025

The transmission of vibration through wet particulate solids

H P VAN KUIK and A J MATCHETT, BSc, PhD, CEng, MIChemE
Department of Chemical Engineering, Teeside Polytechnic, Middlesbrough, Cleveland
J M COULTHARD
Department of Civil Engineering, Teeside Polytechnic, Middlesbrough, Cleveland
J PEACE
British Steel Technical, Teeside Laboratories, Middlesbrough, Cleveland

Abstract

A methology was developed to measure the transmission of vibration for beds of powders on a horizontally vibrating plane in a continuous and non destructive manner. This methology was used to monitor changes of vibrational transmission through damp caking materials in time. It was possible to deduct the mode of failure from the recordings. Friction coefficients were found to increase by at least a factor of three in a caking process with damp sand, rising from around 0.4 to more then 1.2. For damp coal a rise in friction coefficient from about 0.65 to more than 1 was found.

Introduction

Vibration is often used in materials handling systems to induce flow in hoppers and silos. It is also used to maintain and control the flow of powders in vibratory feeders. There are a large selection of vibrating devices for particulate systems available from manufacturers. Vibration is often proposed as a panacea to many materials handling problems. In spite of this, there is a lack of knowledge of the effects of vibration upon particulate systems, and many facilities are specified on an empirical basis. A review of the uses of vibration in materials handling systems has recently been made by the British Materials Handling Board[1].

The most comprehensive study of vibration in powders, to date, is that of Roberts[2]. He used a modified Jenike cell, in which vibration was applied transverse to the direction of shear. He demonstrated that if vibration is applied to a powder prior to shear, then the powder consolidates and the force required to induce shear is greater than the non-vibrated case. However, if vibration is applied during shear, then the force required to maintain shear is less than the non-vibrated case. Hence, vibration can be detrimental to powder flow in certain circumstances. Roberts attempted to apply his shear cell findings to a pilot plant hopper filled with sands. The results were confusing, and in some cases vibration was able to induce flow, but not in others. There was no explanation for these results[2].

Vibration systems are explicitly dynamic and vibration changes the magnitude and direction of particle-particle contact forces. In order to appreciate the nature and extent of these changes, it is necessary to collect data at a rapid rate compared to the time of one vibration cycle-for a 50 Hz system, a data collection rate of 1000 Hz will give 20 data points per vibration cycle. However, several workers have collected data at the relatively slow rates normally associated with Jenike cell operation[3][4]. Such results are some form of average of conditions over many vibration cycles. The effects of vibration upon internal and wall friction were investigated at these relatively slow rates of data collection. Thus vibration can reduce the force of friction by reducing normal contact forces over part of the vibration cycle. However, at other points in the cycle, the direction of acceleration is reversed and vibration increases normal and frictional forces. This has led to confusion about the effects of vibration upon friction. It was stated in [3] that vibration reduces the coefficient of friction. However, if the coefficient of friction were calculated allowing for the actual normal contact forces in the vibration cycle, it could be argued that vibration increased the coefficient of friction.

The limitations of analysis at these slow rates of data acquisition lead to oversimplified models of the effects of vibration. Matchett[5] demonstrated that the effects of vibration upon simple powders are very complicated, and that conditions within the material change during the vibration cycle. This must be taken into account in the collection of data and the analysis of results.

The work presented in this paper is an attempt to investigate vibrating particulate systems in a rational and rigourous manner. This involves the collection of data at two timescales. The SHORT timescale is that at which conditions change during a vibration cycle. It necessitates the collection of data at rates of hundreds or thousands of Hz. Above this timescale there are changes in packing and segregation effects, whose effects are only seen at intervals of seconds or minutes. This is the LONG timescale.

In this study,the effects of vibration upon wet particulates on a horizontal, approximately sinusoidally vibrating plane have been investigated. Motion of the plane and the surface of the powder have been monitored using triaxial accelerometers. A system of data collection has been developed in which bursts of data are collected at a rate of approx. 1000 samples per second for several seconds. These bursts of data collection are initiated at intervals of the order of minutes. Thus data is available at both the SHORT and the LONG timescale. Collection of data at 1000 samples per second over the complete duration of a run lasting up to an hour would necessitate the acquisition of an excessively large amount of data each run, even with modern, computer-based data collection and storage systems.

The systems studied are relatively simple, but have great industrial relevance. They simulate the interactions between materials and vibratory conveyors. When such equipment is used with damp solids, it has been found that materials will often stick to the conveyor surfaces as cake[1]. This cake has been shown to be densely packed, have a high moisture content and contain a greater proportion of fine particles than the parent material from which it was formed. This indicates that both solids and liquid segregation is occurring[6]. The cakes are retained on the bed of the conveyor in spite of the relatively high accelerations exerted by the conveyor. It will be shown that the accelerometer system is able to monitor changes in structure and mode of failure of the solid, on-line, as they are taking place, non-destructively and without stopping the experiment. Parameters such as the coefficient of friction can be estimated from the data, and the formation of cake can be explicitly detected.

Experimental Technique

A Triton vibratory feeder was used as a vibrating table. This feeder consisted of a mild steel feederpan mounted on four vertical compression springs. The feederpan was driven by two counter rotating motors with out-of-balance weights, mounted at opposite sides of the pan. The motors were mounted horizontally for these experiments, so that the principal vibration was in the horizontal direction.

A square box without top or bottom was placed on the vibrating table to contain the sample. A weighed sample was placed in the box and gently levelled off with a piece of metal sheet. The depth of the sample was measured by pushing a small metal rod in the sample, positioning a marker at surface level on the rod, taking the rod out again and measuring the distance up to the marker with a ruler.

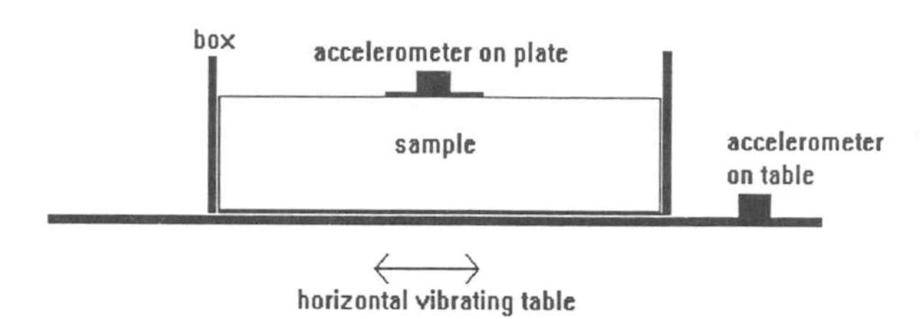

Figure 1: Experimental equipment

Two triaxial accelerometers were used to measure the vibrational accelerations. These accelerometers produced signals proportional to acceleration and provide acceleration in three dimensions as a function of time. One accelerometer was positioned on the vibrating table, close to the box with the sample. The other one was placed on a plate on top of the sample, approximately in the centre of the box. The bottom of the plate was lined with course sand to grip on the sample and inhibit the plate from sliding relative to the sand. See Figure 1.

The accelerometers were linked to an interface which sampled the data and stored them in temporary memories. The interface was controlled by a personal computer and the sampled data were stored in a file on the computers hard disk. Software was available to analyse and manipulate the sampled data. The interface was set up to sample at 500Hz for two seconds every ten minutes. The experiments were also visually checked every ten minutes by watching them with a strobe light. The light frequency was adjusted so that the table seemed to move very slowly. If no relative motion between the top of the sample and the table could be seen, it was concluded that the material had caked.

A separate device was available to measure the speed of the drive motors and to determine the acceleration in the horizontal direction, filtered within 1Hz of the drive speed. All the experiments were performed at a speed of 1100 r/min and a principal horizontal mean acceleration of 1.46g.

Both damp building sand and Pitson McClure coal were used as sample material. The material was first dried in a tray drier at 80°C for sand and 60°C for coal and then mixed with water in a ribbon blade mixer. The coal was sieved after drying so that all the particles were smaller then 1.2mm.

experiment	box	sample load	moisture content	time until caking	initial density	final density
1	1	1 kg	6.8	54 mins	1266 kg/m3	kg/m3
2	1	2	8.0	25	1266	
3	1	3	8.2	40	1229	
4	1	4	9.1	35	1266	
5	1	6	8.6	52	1211	1519
6	2	2	8.0	42	1300	
7	2	4	8.8	40	1217	
8	2	6	8.5	40	1226	1788

Table 1: Data of the experiments with building sand

experiment	box	sample load	moisture content	time until caking	initial density	final density
9	2	1 kg	9.4	no caking	596 kg/m3	kg/m3
10	2	1	13.5	40 mins	572	841
11	2	0.5	12.8	32	511	
12	2	2	12.7	23	540	817

Table 2: Data of the experiments with coal

box 1: 374*192*106mm internal
weight 1132g

box 2: 378*158*78mm internal
weight 335g

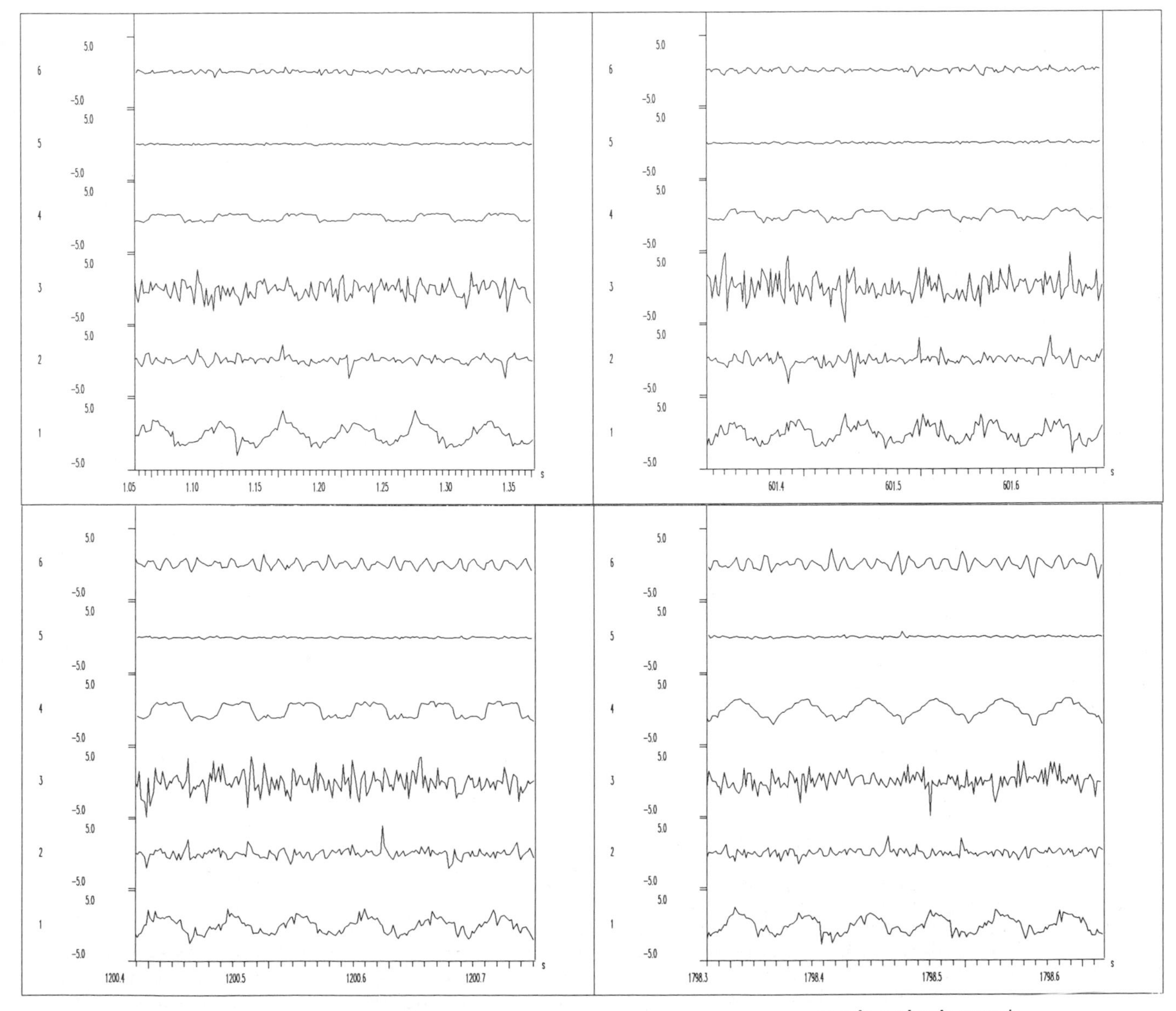

Figure 2: Accelerometer traces for experiment 2

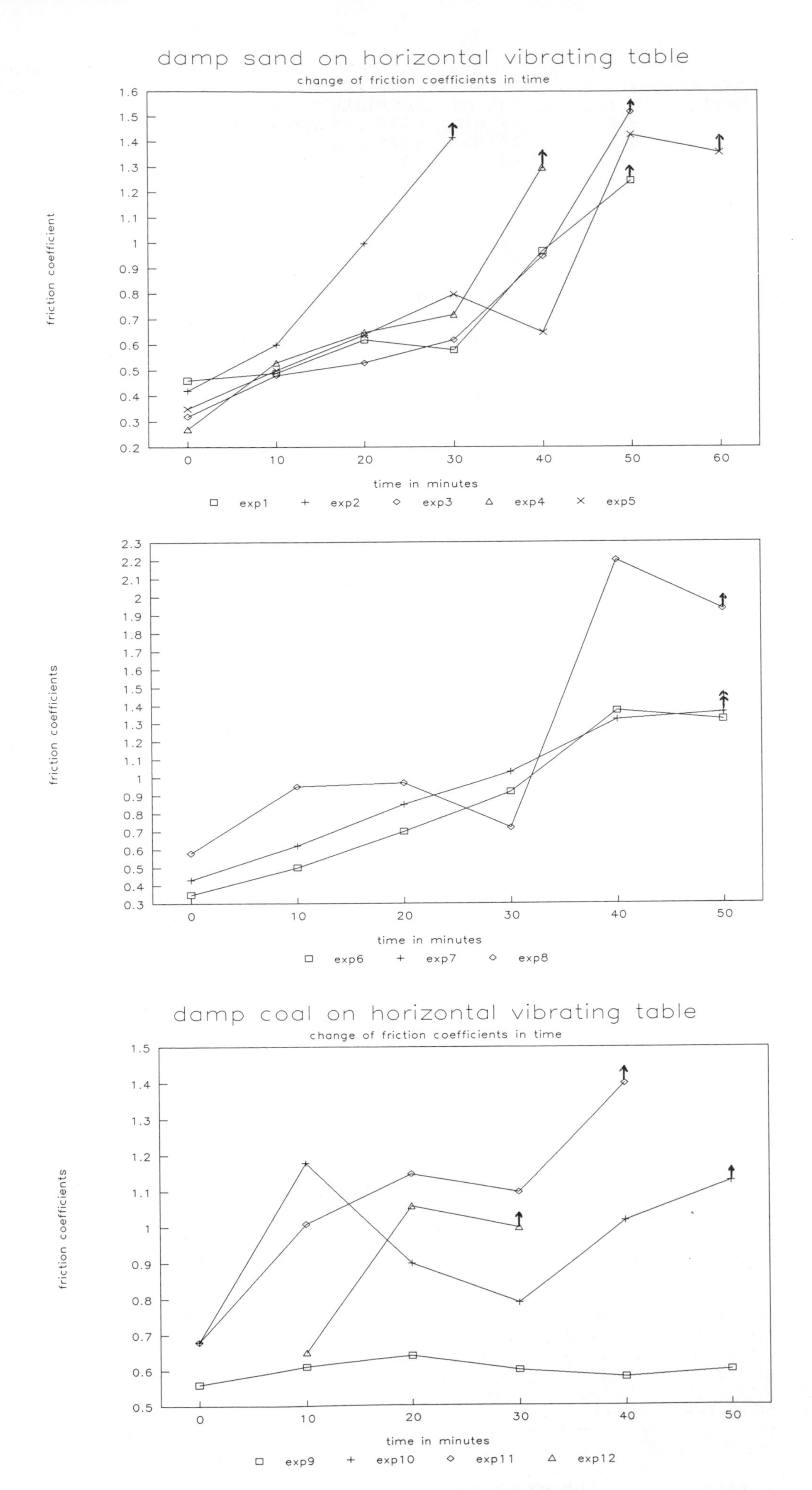

Figure 3: Change of friction coefficients in time

Several sample sizes were used in the range of 0.5 Kg to 6 Kg. A light and a heavy box were available and both were used.

Results and Discussion

The results are summarized in Table 1 and Table 2. The experiments were reproducible. Caking usually took place after 30 to 50 minutes. No significant difference was seen in the use of the different boxes. It is therefore likely that the box had no major influence on the behaviour of the sample.

An example of the accelerometer traces can be found in Figure 2. These graphs show the acceleration for the six channels in g against the time in seconds. Channel 1,2,3 correspond to x,y,z accelerations of the table, channel 4,5,6 correspond to x,y,z accelerations of the top of the sample. The x direction is the principal direction of vibration while z represents the vertical direction.

In the first seconds the whole sample slid over the table and the strobe light showed that failure was predominantly at the wall. It is not possible to deduce directly from the traces whether failure is internal or at the wall. The sample's acceleration in the x-direction was constant for almost half a period and then quickly reversed direction to become constant again for nearly another half period. This behaviour corresponds to material failing either at the wall or internally. In this situation the force exerted on the upper portion of the sample is the frictional force transmitted across the shear zone, which is constant. The acceleration of the upper accelerometer is this force divided by the appropriate mass, which will also be constant. The magnitude of these plateaus depends upon the frictional force, which is a function of the coefficient of friction (See Appendix).

There was a considerable damping of the high frequency components in the vertical direction.

The mode of motion and failure was very much the same at 600 and 1200 seconds but the level of constant acceleration was increased, indicating an increased friction coefficient. The vertical damping was decreasing in time, also indicating a change in the mode of failure and structure of the sample.

At 1798 seconds the material had caked. The sample moved in a different mode and the acceleration in x-direction was similar to the acceleration of the table except that the high frequency components in the x and y direction were damped. This indicates no major failure between the table and the top of the sample. After the experiment was stopped the sample was found to be a dense solid lump, rigidly stuck to the table surface. These readings corresponded to observations of cake formation using the strobe light.

The Appendix shows how to deduce friction coefficients from the accelerometer traces. Friction coefficients were calculated for all ten minutes intervals and the results are plotted in Figure 3. The coefficients were deduced from the average between the positive and the negative acceleration. In the case of caking the calculated friction coefficients are minimum values. The downward dips in the trends of experiments 1,5 and 8 were due to lack of grip between the top of the sample and the plate with the accelerometer. This slip was visible under the strobe light and was corrected after detection by repositioning the plate. This indicates that very careful experimentation is essential for vibrating systems.

Experiment 9 shows what happens when the material (coal) was not tending to cake, due to a lower moisture content. In this case the friction coefficient remained fairly constant throughout the test.

The friction coefficients were initially similar to conventional coefficients determined in a Jenike test. The friction coefficients increased at least by a factor of about three for damp sand during the experiments. For the caking coal samples an increase by at least a factor 1.5 to 2 was found.

The increase of the friction coefficients could have major implications to the design of material flow equipment. The results show that vibration can produce frictional effects which are quite different from conventional values. It is possible that these responses are indicating that the dynamic properties of powders may be quite different from the conventional 'pseudo static' behaviour at low rates of strain and low absolute strain. Therefore there is a need for further study of the effects of vibration on particulates.

Conclusions

The described technique was able to analyse the behaviour of damp materials on a vibrating plane on a continuous and non-destructive basis. It was possible to follow the mode of failure by the shape of the accelerometer traces. Values of friction coefficients could easily be deduced from the acceleration levels.

Caking occurred under these conditions with a large increase in both wall and internal friction, hence cakes can withstand high accelerations. The friction coefficients found in these cases were much higher then conventional values.

Future Work

The aim for the near future is to relate the changes in friction coefficient to changes in structure due to segregation and compaction.

References

1. BRITISH MATERIALS HANDLING BOARD, 'Vibration Guide-lines-The Effect of Vibration on Bulk Materials and Plant', BMHB,1990.
2. A.W.ROBERTS, 'Handbook of Powder Science and Technology', Chapter 6, Fayed and Otten(Eds), Reinhold,1986.
3. T.AKIYAMA, T.NAITO, T.KANO, Powder Technol., 45(1986), 215.
4. T.P.FISHER, D.S.COLEMAN, Powder Metall., 17(1974), 302.
5. A.J.MATCHETT, 'A Theoretical Model of Shear in a Vibrating, non-cohesive Particulate Solid', presented at Advances in Particulate Technology conference, Surrey University, April 1991.
6. A.J.MATCHETT, 'Vibration in Materials Handling Systems', Proc. Metall.Soc.Canada, 29th Annual Conf.Vol 18, 75, 1990.

Appendix Deduction of Friction Coefficients from Accelerometer Traces

The mass of powder = m
The mass of plate + accelerometer = M
Total mass above plane = (M+m)
The coefficient of friction = μ
Subscript w refers to wall friction
Subscript i refers to internal friction
Assume no vertical components (These are zero on average)
Let the mean acceleration in the slip region be a; use the absolute value if the acceleration is negative

Frictional Force = $(M+m)a = \mu(M+m)g$

Hence $\mu = a/g$

If failure at wall then $\mu_w=a/g$; $\mu_i>a/g$
If failure internal then $\mu_i=a/g$; $\mu_w>a/g$
If no failure then $\mu_w>a/g$; $\mu_i>a/g$

The values plotted in Figure 3 therefore represent either exact values of wall friction or exact values of internal friction, when there is failure. In case of failure at the wall the values represent minimum values for internal friction and for internal failure they give minimum values for wall friction. When cake has formed there is no failure and the values represent minima for both internal and wall friction, when the maximum acceleration sustained by the cake is taken as a basis for calculation.

C418/011

Techniques for assessing the flowability of sodium carbonate (soda ash)

D GELDART, PhD, CEng, FIChemE and R GAUTIER
Postgraduate School of Powder Technology, University of Bradford, West Yorkshire
N ROLFE, BSc, PhD
Research and Technology Department, ICI Chemicals and Polymers Limited, Northwich, Cheshire

SYNOPSIS Three tests have been studied using samples of light sodium carbonate (soda ash) having median sizes in the range 47–115μm. The deaeration or collapse test could be carried out quickly and reproducibly, and the standardised de-aeration time varied inversely with mean particle size. The Warren Spring Laboratories' Cohesion Tester also revealed differences between samples but tests took longer, required more skill, and the results were more variable. The Rotating Fluidized Bed was found to be unsuitable.

NOTATION

$\bar{d}_p$	mean particle size, eqn. 2
d_{pi}	average of two adjacent sieve apertures
d_{pm}	median particle size
$\bar{d}_{sv}$	mean surface volume size (eqn 1)
d_{svi}	surface volume size of ith size fraction
F_{45}	mass fraction of powder less than 45μm
H_D	intercept on vertical axis of collapse graph (height of dense phase of fluidized bed)
H_s	settled bed height after collapse process
U_c	rate of collapse of fluidized bed
U_{mb}, U_{mf}	minimum velocities of bubbling and fluidization
T_c	standardised collapse time
X_i	mass fraction of powder in ith size fraction

Greek

μ	viscosity of fluidizing gas
ρ_g	density of fluidizing gas
ρ_p	particle density (including pores)
σ	size spread

1 INTRODUCTION

ICI operate the Solvay ammonia-soda process for manufacturing sodium carbonate at three of their sites at Northwich, Cheshire, U.K. The overall reaction is:

$$NaCl + CaCO_3 \rightarrow Na_2CO_3 + CaCl_2.$$

The process is operated on a 1 million t/year scale, and produces two main products. The first is light sodium carbonate, so-called on account of its low bulk density of 0.56 g/ml, which is made by calcination of the sodium bicarbonate produced in the process. The second is heavy sodium carbonate, which has a bulk density of 1.09 g/ml, and is produced by recrystallising the light form to sodium carbonate monohydrate, followed by recalcination. Sodium carbonate is widely known as soda ash by both the producers and the customers.

The reasons for the difference in the bulk densities are two-fold. Firstly, heavy ash is much larger than light ash (median sizes of 350 μm and 100 μm respectively). Secondly, the shape of heavy ash leads to a different, closer, packing of the particles. These factors also result in heavy ash being much more free-flowing.

Whilst the inherent flow properties of heavy ash are good they can be seriously affected by poor handling either in the production process, or during delivery to the customer. ICI needed a rapid method of assessment of flow properties which could be used by their process operators to guarantee that the product supplied to their customers would always be satisfactory. Preliminary work showed that it might be possible to use the poured angle of repose for this purpose; at the end of a more detailed study (1) it was established that a simple piece of equipment, the ICI/Bradford vertical funnel, could be used to give reproducible values of poured angles of repose which could then be used for control purposes in the heavy ash process.

In the work reported in this paper, we turned our attention to devising suitable techniques for evaluating the flow properties of light ash.

2 LIGHT ASH HANDLING

Light ash (Na_2CO_3) is made at Northwich in ICI's Lostock Works and a proportion has to be transported 5 km to Wallerscote where it is made into heavy ash. The bulk handling system is shown diagrammatically in Fig. 1. On the whole, the light ash manufacturing process gives a product with median sizes in the range 80 to 115μm, and a very wide size distribution, but transporting the ash between sites can change both the mean size and the size distribution due, largely, to attrition. This in turn changes the flow properties and affects the mixing and tumbling behaviour of the powder in the inclined rotating recrystallisation drum used to make heavy ash. The behaviour deteriorates markedly when the median size falls below about 85-90μm.

3 BASIC POWDER PROPERTIES

A typical size distribution is shown in Fig.2, from which it can be seen that the spread σ, defined as

$$\frac{d_{p84\%} - d_{p16\%}}{2} \text{ is approximately}$$

60μm and the relative spread, σ/d_{pm} is 0.65, where d_{pm} is the median size. Although the median size d_{pm}, is useful for calculating the relative spread, and for reference purposes, it does not have any fundamental significance as do the surface/volume and equivalent volume mean sizes, $\bar{d}_{sv}$ and $\bar{d}_v$ respectively.

The most appropriate mean size of a powder which gives proper emphasis to the role played by the fine particles is:

$$\bar{d}_{sv} = {}^1/\sum \frac{X_i}{d_{svi}} \qquad (1)$$

For particles larger than 38μm, the most frequently used method of particle size analysis is sieving, and in that case equation 1 is rewritten as

$$\bar{d}_p = {}^1/\sum \frac{X_i}{d_{pi}} \qquad (2)$$

where d_{pi} is the arithmetic average of adjacent apertures and X_i is the mass fraction within each size range. (The relationship between $\bar{d}_p$ and $\bar{d}_{sv}$ for angular particles like soda ash has been discussed in detail elsewhere (2).)

However, a significant mass fraction (10-25%) of the powder is less than 38μm and unless further information on the sub-sieve fraction is available from some other source, equation (2) cannot be used with any confidence. A Malvern laser diffraction size analyser was used on the sub-75μm fraction so that results would overlap with the sieve analyses. Careful analysis and matching of size distributions revealed that the smallest particles in all the samples were 4μm, and that treating the fraction passing through the 38μm sieve into the pan as having a mean size of 21μm (= $^{(38+4)}/_2$) was satisfactory. Thus, sieving under standardised conditions with new sieves and a new Inclyno sieving machine, was used throughout the work reported here.

It is interesting to note that in the case of light soda ash, for all the 23 samples tested, there was a direct proportionality between the surface-volume size, $\bar{d}_p$, as determined by sieving, and the median size, d_{pm}. This can be seen from Fig. 3 in which

$$\bar{d}_p = 0.738\ d_{pm} \qquad (3)$$

The correlation coefficient is 0.97.

This proportionality implies that the shapes of the distribution curves (e.g. Fig. 2) remain approximately constant, merely shifting to the left or right. In that case the relative spreads should also be constant, and indeed they are, varying only between 0.61 and 0.72.

The particle density is about 1150 kg/m^3 but this is difficult to determine exactly because the particles are porous. The combination of mean particle size (as calculated using equn.2) and density places light soda ash in the middle of Geldart's Group A (3); however, because the size distribution is so large, segregation of the powder occurs rather readily when it is poured, or fluidized at low velocities, and the portion of the powder which becomes richer in fines tends to behave more like Group A/C, that is, it behaves more cohesively.

4 SELECTION AND DESCRIPTION OF TESTS

Tests based on the poured angle of repose, used successfully with the larger particles of heavy ash, proved to be unsuitable when applied to light ash because the powder blocked the funnel. Because earlier work (4) on cracking catalyst (a typical Group A powder) had established that the de-aeration or collapse test was very useful in predicting flowability, it was decided to apply it to light ash.

Tests carried out by ICI on light ash in 1976 using the Warren Spring Laboratories Cohesion Tester indicated that the device was of potential use for the present purposes, and this was also included in the programme. Finally, a rotating fluidized bed, used with some qualitative success in the early 1980's (5) at Bradford, was also a potentially useful candidate, since it simulates to some extent the powder flow in the rotating crystalliser in which the light ash sometimes causes problems.

4.1 Collapse Test

The theory behind the collapse tests has been discussed thoroughly in an earlier paper (6) and only the technique is described here.

Sufficient powder to give a settled bed depth of 500-600mm is put into a vertical column 100mm internal diameter, fitted with a porous distributor. About 2.3 kg of light ash is needed for each test. The bed is fluidized with air at 6-10 cm/s to thoroughly mix the powder and the gas supply suddenly cut off using a solenoid valve in the line between the flow meter and the column. The change in bed level is recorded on video and plotted as a function of time from the playback. The sequence of events is shown in Fig. 4. The basic information derived from a plot of bed height against time is the intercept, H_D, obtained by extrapolating the straight line back to t = 0; the collapse rate, U_c, which is the slope of the hindered sedimentation stage; and t_c, the critical time found by extrapolating the straight line forwards to the horizontal line which represents H_s, the height of the settled bed.

It should be noted that the straight line portion 'b' is obtained only for Group A and AC powders; Group C powders behave differently in that phase 'b' is much smaller, or non-existant and 'c', the consolidation stage is much longer.

The critical settling time, t_c, and H_D, the intercept, increase with initial bed height, in contrast with U_c which is independent of H_s. However, the ratios H_D/H_s and t_c/H_s (called the standardised collapse time, T_c) are independent of H_s; within Groups A and AC, T_c increases as mean particle size decreases. (6)

4.2 The WSL Cohesion Tester

This device measures the cohesive strength of samples of powders in various states of compaction. The shear stress required to cause a powder sample to fail is found using a radially-finned, annular, open-topped cell mounted on a vertical spindle. About 400g of light ash is needed for each test. A full description of the device and its mode of operation is described by Svarovsky (7). Each sample (11 in all) was consolidated in the cell for 3 minutes using a weight of 17lb (7.876 kg) before performing the test, and six experiments were done on each sample. The torque required for the sample to fail was calculated. Tests were repeated after an interval of 6 weeks to check reproducibility.

4.3 Rotating Fluidized Bed

This device is based on a design by Judd and Dixon (8), and developed further by Wong (9) who found distinct differences between powders in the various Groups. A porous gas distributor forms the curved wall of the cylinder and both the gas flow pulled through the distributor by a vacuum system and the speed of rotation of the drum can be varied. The intention in the present work was to measure the dynamic angle of repose in the expectation that this would vary sufficiently to act as a criterion for flowability.

5 RESULTS

5.1 Collapse Test

Preliminary tests showed that repeated experiments with the same sample of powder using the lab's compressed air supply produced progressively increasing values of the standardised collapse time, T_c. It was suspected that chemisorption of water vapour and/or carbon dioxide might be causing a change in the surface characteristics of the particles. It was also possible (though unlikely) that the size distribution of the powder was changing progressively during the tests. Three changes to the procedure were introduced. (a) a filter was fitted above the bed to ensure that no fines were lost. (b) the powder was dried in an oven at 160°C and (c) nitrogen was used instead of air. These steps solved the problem and good reproducibility was achieved thereafter. On switching back to air from nitrogen the same progressive change observed originally returned, showing that changes in PSD were not responsible.

Typical plots are shown in Fig. 5. The results are reproducible with little scatter and follow a straight line. In all, eight powders were tested, five of which were formed by blending soda ash taken from batches available in large quantities. This was done in order to cover a range of median sizes commonly encountered on the plant, but not available at the time the experimental programme was under way. The size distributions were also arranged to match those typically produced.

Results for the 8 batches of soda ash are shown in Figs. 6 - 11. Bed expansion ratio (H_D/H_s) decreases linearly as mean particle size increases (Fig. 6) whilst the collapse rate increases with increasing particle size (Fig. 7). It can be seen from Fig. 8 that the bulk density of the settled bed (calculated using H_s) decreases as particle size increases, though the data are scattered more than those in Figs. 6 and 7. The standardised collapse time (Fig.9) shows the best correlation with particle size and has slightly less scatter when plotted against median size, d_{pm}.

In two earlier papers (4, 6) it was found that T_c was proportional to $(U_{mb}/U_{mf})^{1.5}$ where U_{mb} and U_{mf} are the minimum bubbling and minimum fluidization velocities, respectively. Because of the very wide size distribution which caused segregation by size in the column at low velocities, it was not practicable to measure these two velocities. However, they can be calculated from equation 4, and in Fig. 10 T_c is plotted against U_{mb}/U_{mf} on log-log scales, where:

$$\frac{U_{mb}}{U_{mf}} = \frac{2300\ \rho_g^{0.126}\ \mu^{0.523} \exp\ (0.716\ F_{45})}{\bar{d}_p^{0.8}\ g^{0.934}\ (\rho_p-\rho_g)^{0.934}} \qquad (4)$$

The best fit line is given by

$$T_c = 3.06\ (U_{mb}/U_{mf})^{1.53}\ \ (s/m) \qquad (5)$$

The exponent is very close to the previously measured values of 1.5 but the constant is much lower. The term exp (0.716 F_{45}), which reflects the influence of the mass fraction of the powder below 45μm, varies between 1.08 and 1.2. If we neglect this then T_c should be proportional to $(\bar{d}_p)^{-1.22}$. A log-log plot of T_c versus $\bar{d}_p$ (Fig.11) gives the correlation

$$T_c = 1.36 \times 10^4.(\bar{d}_p)^{-1.59}\ (s/m) \quad (6)$$

where $\bar{d}_p$ is in microns. The scatter is small having a correlation coefficient of 0.97.

5.2 WSL Cohesion Tester

Results of measurement of the cohesive strength are shown on Fig. 12 with error bars drawn in. Although these show a good deal of scatter around the average value the reproducibility of the average values is good. There are two possible reasons for the scatter.

(1) The light ash has a wide size distribution so that in filling the cell a certain, unknown and variable amount of segregation may occur, giving differing amounts of cohesivity at the shearing plane.

(2) The experiment involves rotating a spindle manually at as constant a velocity as possible and this cannot always be achieved.

It was evident from these results and from other more extensive tests carried out in 1976 by ICI using a similar device, that the shear stress for failure levels off for samples with median sizes larger than about 90-95μm. We felt justified in drawing a horizontal line through the data relating to the four samples having the largest median sizes. The sample BB3 gave anomalous results probably because it was a blend in which ground-up fines were added. These had a different shape from fines produced 'naturally' on the plant. The results from BB3 have therefore been excluded from the least squares calculation.

The results are in accord with experience on the plant, in that problems start to be experienced when the median size drops below about 90μm.

5.3 Rotating Fluidized Bed

Unfortunately this device revealed no real differences between the finest and coarsest samples tested and we discontinued systematic work on it.

6 DISCUSSION

In view of our finding that flowability, as measured both by the standardized collapse time and the WSL cohesion meter, is directly related to mean and median particle size, why not measure PSD and use it as a control criterion? There are two arguments against this:

(i) Although $\bar{d}_p$ and d_{pm} are strongly correlated with flowability, they are not the only parameters of importance; particle shape and roughness also play a role, and these are not revealed in particle size measurements.
(ii) Measurement of PSD requires skill and well-maintained equipment - and this is as true of sieving as it is of the most up to date laser diffraction machine. Moreover, any method of PSA is only useful if sampling on the plant is representative; this is much more difficult to achieve if the sample size has to be 10g, as in the case of sieving, or 1-2g, as for laser diffraction. It is especially difficult, because of segregation, if the particle sizes range from 1000 to 4μm, as with light ash. Moreover, there is as yet no laser diffraction machine which can deal with such a wide size range.

Both the WSL test and the collapse test have advantages and disadvantages. The collapse test equipment was custom built in the University of Bradford workshops whereas the WSL machine is available from Ajax Equipment of Bolton and from Wykeham Farrance Engineering Ltd., Slough. At this stage of their development neither could be put in a process control room and left for shift operators to use on a routine basis since both operational skill and some data processing are required. However, the collapse test equipment is capable of being developed into a semi-automatic device for routine testing. A pressure probe/transducer system can replace the need for a video camera recorder, and the signals fed into a computer to give the standardised collapse times within a few minutes. The WSL device is also capable of further development and is certainly more suitable for finer, more cohesive powders which cannot be fluidized.

7 CONCLUSIONS

(1) The flowability of the powder in a process application depends on whether it is being handled in an aerated or a consolidated state, and the appropriate laboratory technique should be selected to characterise material.
(2) The properties of an aerated or partially aerated product, such as one being conveyed in a chemical plant, are best characterised by the deaeration test. For light sodium carbonate, the mean size of a sample was found to be inversely proportional to the standardised collapse time. The results were very reproducible. The deaeration device can therefore be used for a rapid assessment of mean particle size and its influence on flowability.

The technique is easy to use, and the device is both robust and cheap. It is capable of further development to provide a test suitable for use on process plant.

(3) When a powder is consolidated or compacted, its flow properties are better characterised by a cohesion tester. Handling products in a silo, or in a settled bed in a mixer are two typical examples of process applications. In the case of light sodium carbonate, the cohesive strength was found to be independent of median particle sizes greater than 90μm, but was found to increase rapidly below this size. This increase in strength was linear over the size range 47 to 90μm.

This test is more suited to laboratory than plant use; great care has to be taken when carrying out the measurements, and a minimum of six tests have to be made on a given sample in order to obtain a reliable average value.

REFERENCES

(1) GELDART, D. MALLET, M.F. and ROLFE, N. Assessing the flowability of powders using angle of repose. *Powder Handling and Processing*, 1990, 2, No.4, 341-345.

(2) GELDART, D. Estimation of basic particle properties for use in fluid-particle process calculations. *Powder Technology*, 1990, 60, 1-13.

(3) GELDART, D. Types of gas fluidization. *Powder Technology*, 1973, 7, 285-292.

(4) GELDART, D. and RADTKE, A. The effect of particle properties on the behaviour of equilibrium cracking catalysts in standpipe flow. *Powder Technology*, 1986, 47, 157-166.

(5) GELDART, D. HARNBY, N., and WONG, A.C.Y. Fluidization of cohesive powders. *Powder Technology*, 1984, 37, 25-37.

(6) GELDART, D. and WONG, A.C.Y. Fluidization of powders showing degrees of cohesiveness. Part II-Rates of de-aeration. *Chemical Engineering Science*, 1985, 40, 653-661.

(7) SVAROVSKY, L. Powder Testing Guide, 1987, BMHB/Elsevier Science Pub. Co. Inc.

(8) JUDD, M.R. and DIXON, P.D. *Trans. Inst. Chem. Engrs (London)*, 1979, 57, 67-70.

(9) WONG, A.C.Y. Fluidization of cohesive powders. 1983, Ph.D. Dissertation, University of Bradford.

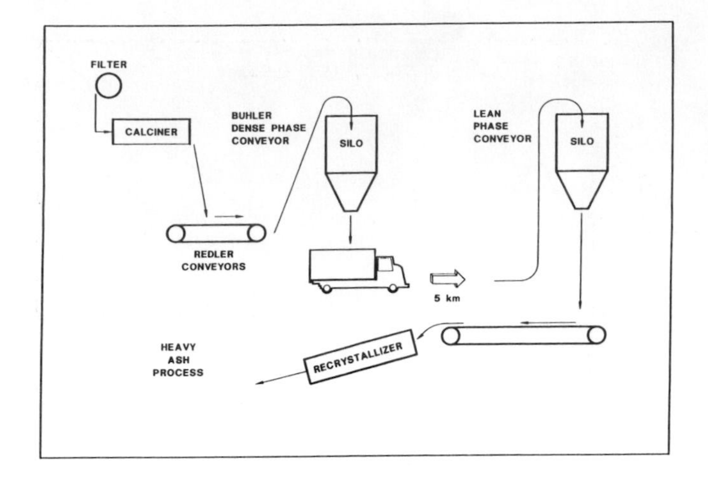

Fig 1 Flow diagram of solids handling system

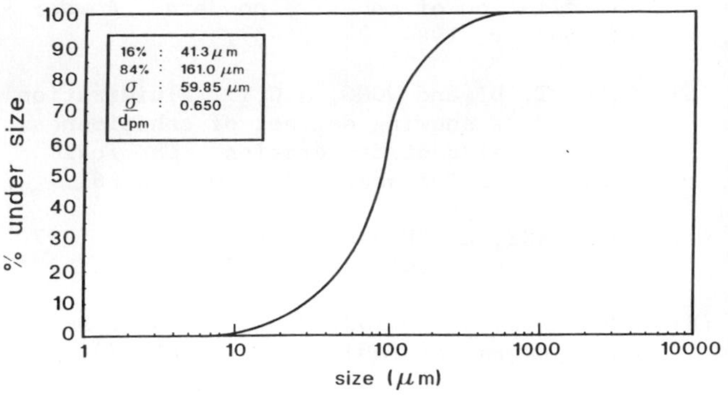

Fig 2 Typical size distribution of soda ash

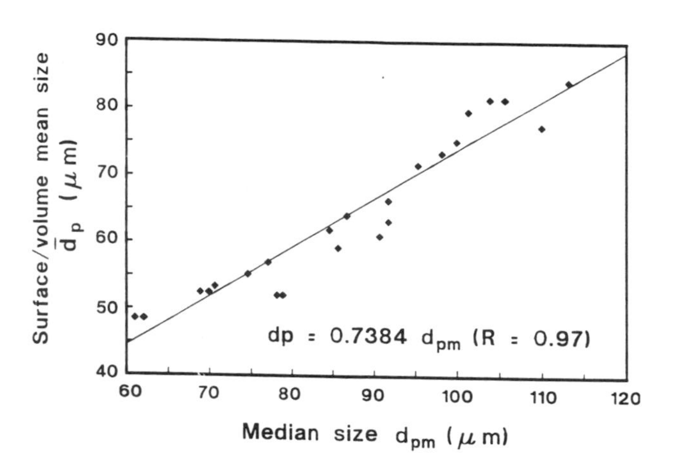

Fig 3 Comparison of median and surface/volume means sizes of light soda ash

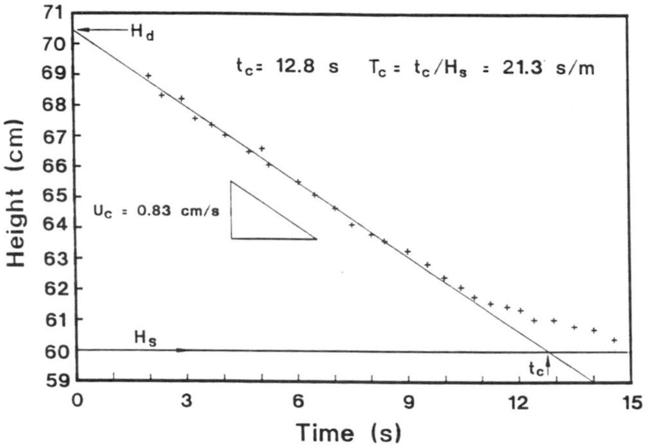

Fig 5 Collapse test for typical light ash showing (A) reproducibility (B) treatment of data

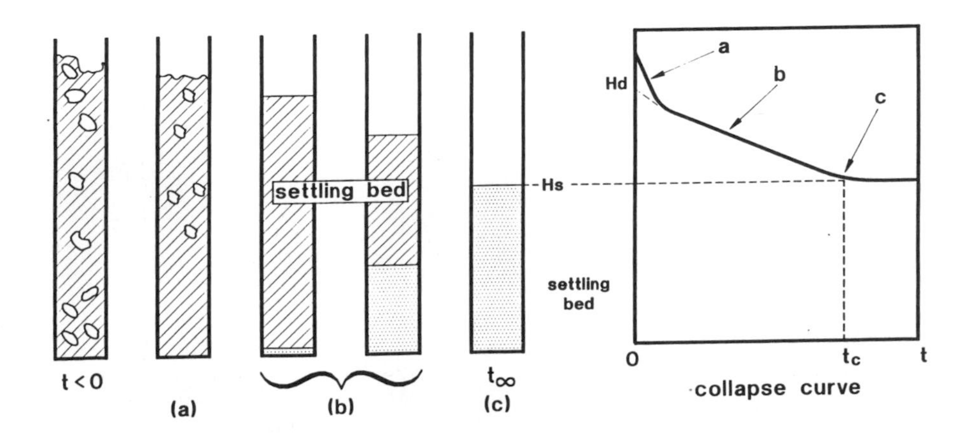

Fig 4 Collapse test

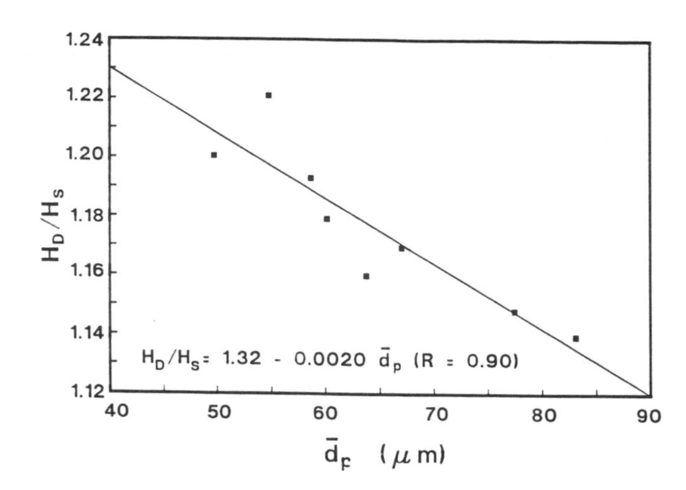

Fig 6 Bed expansion (H_D/H_S) as a function of mean particle size, $\bar{d}_p$

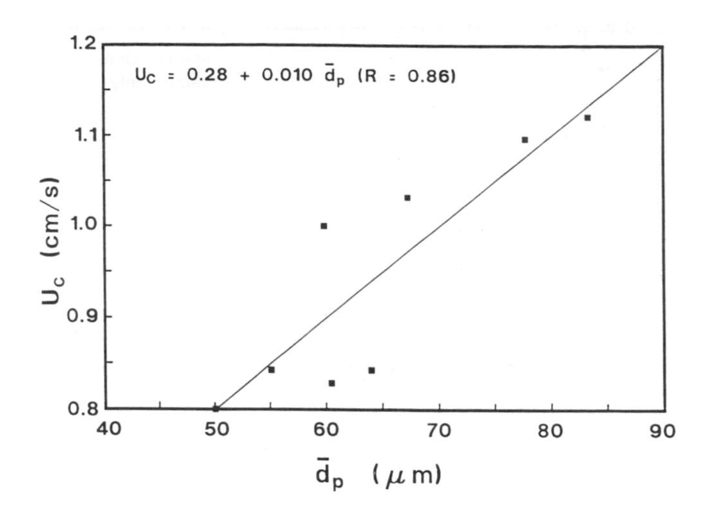

Fig 7 Collapse rate U_c as a function of mean particle size, $\bar{d}_p$

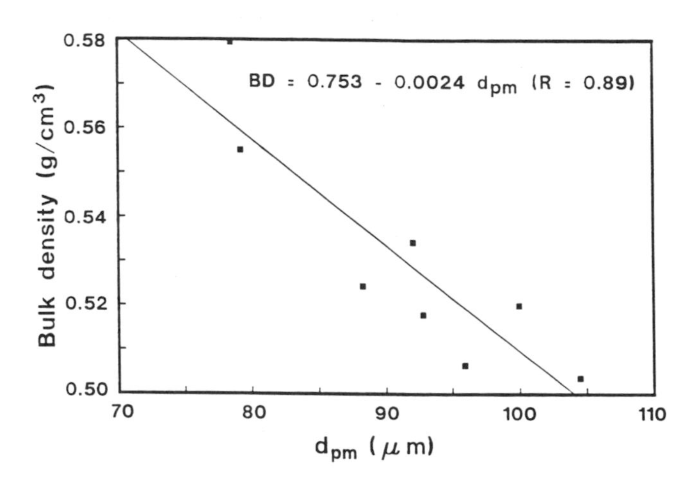

Fig 8 Settled bed density as a function of median size, d_{pm}

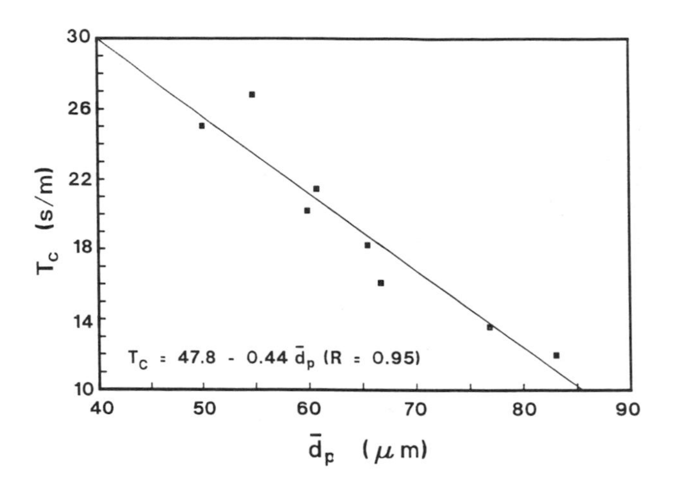

Fig 9 Standardised collapse time as a function of (A) mean particle size, $\bar{d}_p$ (B) median particle size, $\bar{d}_{pm}$

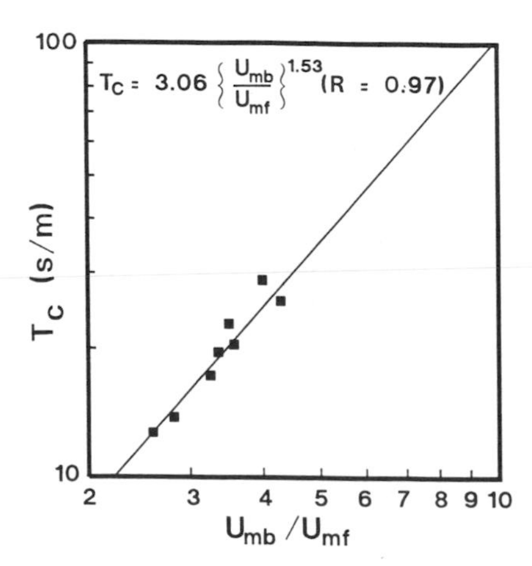

Fig 10 Standardised collapse time T_c as a function of U_{mb}/U_{mf}.

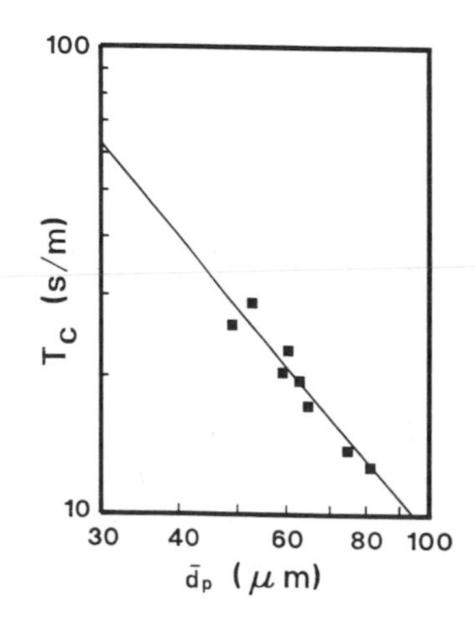

Fig 11 Standardised collapse time T_c as a function of mean particle size $\bar{d}_p$.

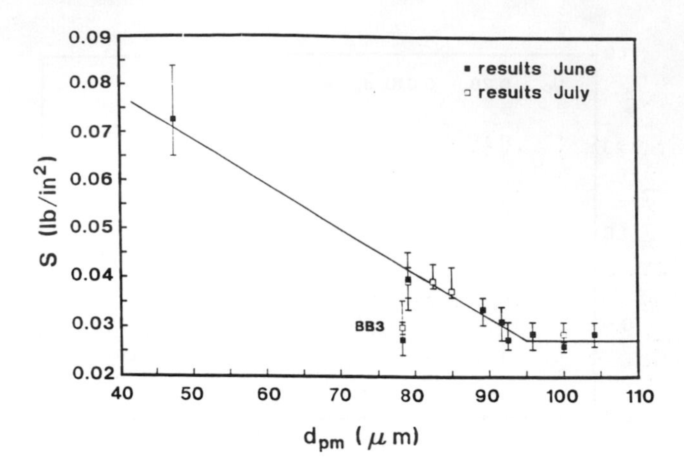

Fig 12 Shear stress for failure as a function of median size, d_{pm} (Warren Spring Laboratories Instrument).

C418/026

Friction and adhesion at bulk solids/wall interfaces

A R MASON, BEng and A J MATCHETT, BSc, PhD, MIChemE
Department of Chemical Engineering, Teeside Polytechnic, Middlesbrough, Cleveland
J M COULTHARD
Department of Civil Engineering, Teeside Polytechnic, Middlesbrough, Cleveland

Summary.

A novel shear cell, in which the plane of failure is not imposed, has been developed for testing damp bulk particulates at low normal stresses and high degrees of strain. Building sand has been tested against a stainless steel wall sample over a range of moisture contents. Wall friction angles within the range 23 to 33 degrees were observed. Over 85% of the strain occurred at the wall, rearrangement and internal failure accounting for the rest. The cell proved to be a versatile tool for material testing.

Introduction.

The measurement of the frictional properties of particulate materials is a complex and controversial field. The industry standard for materials handling is the Jenike cell [1]. This was developed from the soil mechanics direct shear box [2]. As a piece of equipment for detailed scientific research it leaves much to be desired. Limitations include:-

i) A limited range of strain.
ii) The cross section is circular and the area of shear changes as the test proceeds.
iii) The distribution of stresses is non-uniform, due to the loading mechanisms [3].
iv) The preconsolidation procedure is complex.
v) Strains are not measured, only stress.
vi) The equipment is said to have low reproducibility and to be operator dependent.
vii) The plane of shear is imposed by the design of the equipment.

The shortcomings of the Jenike cell have led several workers to develop improved methods for characterisation of frictional properties. Williams investigated the use of uniaxial compression [4] to measure the material yield stress directly. Arthur et al have developed a series of biaxial materials testers with the same aim [5]. Triaxial machines have also been investigated [6] [7]. These devices measure principal stresses directly, unlike the Jenike cell, and genuinely operate in 2 and 3 dimensional systems. They are also much more complex to use, as a result of this operational sophistication. Annular shear cells have been used, but the main aim of some work has been simply to calibrate them against Jenike cells, with resulting problems. The annular cell allows virtually limitless strain, but has other problems associated with it [8]. The Jenike cell and annular cell can also be used to measure wall friction, while the other devices are used primarily for internal friction measurement.

The research group at Teesside have been investigating the interactions between damp, bulk solids and retaining wall surfaces for several years, and have been active in their studies of 'caking'-the adhesion of particulate materials onto wall surfaces [9] [10], and have been seeking an appropriate shear cell for this purpose. An annular cell was originally used by Taylor [10], as it allowed large strains, after which the lid could be lifted to observe cakes adhering to the cell lid. Therefore, conditions under which caking occurred could be measured. However, the process of lifting the lid sometimes resulted in dislodgement of the cake, and in any case, upon lifting the lid the system was disturbed and the experiment ceased. A novel cell has been developed by Uesugi and Kishida [11] for use at small strains and high stresses in soil mechanics. This consisted of a direct shear system, but the particulate sample was held in a cell with thin laminated walls that could slide over one another. This enabled the sample to fail where it wanted, rather than imposing the plane of shear, and also, very importantly, enabled the strain distribution within the sample to be measured. A modified version of this cell was developed for use at relatively low stresses found in materials handling systems.

A long thin cell has been devised in which the effects of friction in the cell wall laminates is small compared with friction in the material under test. The laminated walls are mounted upon a base which is driven backwards and forwards by a motor, and the normal stress is imposed by a load placed on top of the sample.

The cell has been used to measure the friction between a stainless steel plate and damp sands, in order to demonstrate the use of this equipment.

The Novel Cell.

Figure 1 shows the main components of the shear cell, viewed in section. The walls of the cell are composed of a stack of plates which are free to slide over one another. Each plate is a 424mm by 124mm sheet of 2mm thick aluminium with a 400mm by 100mm area removed from the centre. 0.5mm thick PTFE, chemically etched on one side to facilitate adhesion, is glued to one side of each plate to reduce friction. The top plate is thicker than the others, being 10mm thick, to allow for compression of the sample. The particulate sample dimensions are 400mm by 100mm by approximately 25mm; as the test procedes the depth of the sample will reduce as the applied normal load is increased. This configuration for the cell enables the distribution of strain within the sample to be monitored throughout the experiment without interruption.

The stack of plates and the particulate sample rest on top of the wall sample which can be driven backwards and forwards by the drive shaft shown. As both the plates and the particulate sample are free to slide over the wall sample the area of contact remains constant throughout the experiment. A 304 stainless steel wall sample is used at present. Microswitches are fixed to the table that supports the cell, positioned so that they are contacted by the wall sample at the two extremes of its travel, approximately 24mm apart. The direction of motion of the wall sample is then reversed, allowing almost unlimited strain. The drive shaft is connected to a geared variable speed motor capable of delivering rates of strain of 0.0005mm/min to 2mm/min. A 0.5mm thick sheet of PTFE is glued to the top of the table supporting the cell, underneath the wall sample, to reduce friction.

A load cell (0 to 150lbs, tension and compression), connected to a chart recorder, is mounted between the drive shaft and the wall sample to measure the force required to propel the wall sample back and forth. From this measurement the shear stress acting upon the sample can be calculated.

Normal load is applied to the sample by means of an open-topped perspex box, measuring 506mm by 400mm by 100mm and being constructed of 6mm thick perspex, which is placed on top of the sample, inside the top aluminium plate, so that load is applied to the sample but not to the aluminium/PTFE plates. Iron shot is poured into this box when higher normal loads are required. The box is prevented from travelling backwards and forwards and from twisting as the wall sample moves by the four bearings shown in the diagram. A further two bearings are mounted, one in front of and one behind the box, to ensure that it remains vertical throughout the experiments. All of these bearings are held in place, against the perspex box, by a steel framework which fits around the cell: this is not shown on the diagram.

Testwork.

Samples of building sand were tested in the cell at moisture contents of 2,4,6,8,10,12,14,16,18% water by dry mass. The sand was prepared by drying to constant weight and then sieving to give a known maximum particle size of 1.18mm so that there were no particles present that were 'large' in comparison to the thickness of the aluminium/PTFE plates. Samples of approximately 1.5kg were placed into sealable plastic bags with the required amount of water. The samples were kneaded and left for at least 24 hours to equilibrate.

The load cell and digital recorder were switched on at least half an hour before the experiments commenced to minimise temperature effects. The wall sample, table and aluminium/PTFE plates were thoroughly cleaned to remove any sand that could increase friction within the cell and affect results.

The plates were stacked on the stainless steel wall sample and lined up so that the perspex box would fit inside the top plate. The sample was then poured into the cell; care was taken to break up any balled material.The sand was spread carefully in the cell to ensure that no empty spaces were left, especially in the corners. Great care was taken to prevent unnecessary preconsolidation of the sample at this stage. The perspex box was then lowered onto the sand.

The shear cell was started immediately after setting up had been completed. A rate of strain of 1.15mm/min was used. The shear force exerted by the wall sample was monitored continuously on a chart recorder via the load cell. At convenient moments, normally when the wall sample had reached the extent of its travel in either direction, the displacements of the aluminium/PTFE plates were noted. For each moisture content, tests were run at eight increasing normal stresses, in the range of about 1.1 to 4.7kPa. Pre-weighed bags of about 1.8kg of iron shot were poured into the perspex box to increase the load. The total amount of strain exerted upon any given sand sample was, approximately, 250mm.

Results.

Moisture content - normal stress - shear stress plots of the results for the forward and reverse directions of wall sample travel are shown in Figures 2 and 3 respectively. The normal stress - shear stress response of each sample yielded the wall friction angles and 'adhesion' values quoted in Table 1. The 'adhesion' values are the extrapolated vertical axis intercepts. Values are given for both the forward and reverse directions of wall sample travel. The bulk density of the sand was calculated at the start of the experiments, before any load was applied; these values are also shown, where available.

Table 1. Shear Test Results with Building Sand.

Moisture Content, %.	Initial Bulk Density, $g.cm^3$	Adhesion, kPa.		Wall Friction Angle, Degrees.	
		Fwd	Rev	Fwd	Rev
2	-	0.295	0.453	27.9	28.8
4	-	0.142	0.184	30.6	31.9
6	-	0.137	0.263	33.0	31.6
8	1.097	0.200	0.211	30.1	30.1
10	1.242	0.358	0.326	28.9	29.3
12	1.424	0.189	0.284	30.2	29.8
14	1.441	0.474	0.358	27.9	28.9
16	1.551	0.347	0.379	27.8	26.9
18	1.781	0.326	0.222	22.7	23.9

From the measured displacements of the aluminium/PTFE plates, the proportion of strain occurring at each available plane has been calculated. These values, averaged, for brevity, with respect to both wall sample direction of travel and normal stress, are shown in Table 2.

Table 2. Strain Distribution.

Moisture Content, %.	Failure Between Plates*:						
	Wall and 1.	1 and 2.	2 and 3.	3 and 4.	4 and 5.	5 and 6.	6 and Top.
2	-	-	-	-	-	-	-
4	0.860	0	0	0	0.116	0	0.025
6	0.794	0	0	0	0.113	0	0.093
8	0.852	0	0	0	0.088	0	0.061
10	0.801	0	0	0	0	0.132	0.068
12	0.903	0	0	0	0.067	0	0.030
14	0.792	0	0.114	0	0	0	0.095
16	0.854	0.027	0	0.070	0	0.015	0.033
18	0.954	0	0.018	0	0	0	0.027

* The plates are numbered from bottom to top.

Over the whole series of tests carried out, 85.1 percent of the strain occurred at the wall/plate 1 interface.

There is little fluctuation in the results illustrated in Figures 2 and 3, and the expected, linear normal stress - shear stress relationship is seen. There is no apparent trend in the relationship between shear stress and moisture content from this plot, although a slight decrease in shear stress is visible as saturation is reached. However, there is a much more clearly defined trend between wall friction angle and moisture content: there is a peak at around 4 to 8% moisture then a definite decrease in wall friction angle as moisture content increases, reaching a minimum value as saturation is reached, i.e. 18% moisture by dry mass for this sand. Throughout these limited strain experiments the forward and reverse coefficients of friction were within 6% of each other at each moisture content: see Figure 4.

The adhesion values, plotted in Figure 5, are scattered; no trend of adhesion with respect to moisture content is identifiable.

With reference to Table 2, it can be seen that the majority of strain was at the wall, as is to be expected when using a smooth stainless steel wall sample, although some internal failure was recorded. Variation in normal load through the depth of the sample due to the mass of the sample causes less normal stress to act upon the plane passing through the interface between the top, thicker plate and the top aluminium/PTFE laminate. Hence, frictional forces at this plane would be expected to be lower than in the rest of the sample, but this effect is small. Most of the internal failure was through this plane. The mobilisation of internal friction may have an effect on the distribution of strain throughout the depth of the sample, the more so due to the cyclic nature of the application of strain, and this experimental cell allows such effects to be visualised and measured.

Conclusions.

The shear cell has been successfully developed and used in a series of initial tests on sand. The capacity to quantitatively observe the strain distributions within a particulate sample, in the context of shear testing, makes this cell a versatile tool for the investigation of material behaviour.

Future Work.

Future work will include the optimisation of the cell design, especially regarding the minimisation of friction within the cell. The stress - strain characteristic of materials tested in the cell will be analysed in order to investigate the effect of the mobilisation of friction on strain distribution. Particle segregation and caking tendencies will be examined under conditions of extended strain.

References.

1. BMHB, "A Guide to the Design of Hoppers and Silos, BMHB, 1989
2. M.Bolton, "A Guide to Soil Mechanics", McMillan, 1979
3. F.J.C.Rademacher, G.Haaker, Powder Technology, 1986, 46, 33
4. J.C.Williams, A.H.Birks, D.Battacharya, Powder Technology, 1970/71, 4, 328
5. J.R.F.Arthur, R.K.S.Wong, Geotechnique, 1986, 36, 215
6. A.H.Gerritsen, R.Dekker, Powder Technology, 1983, 34, 203
7. D.Kolymbas, W.Wu, Powder Technology, 1990, 60, 99
8. H.Gebhard, Particle Technology, I.Chem.E.Symp.Ser.No.63, 1981, paper D3/T/1
9. P.D.Taylor, A.J.Matchett, J.Peace, Chem.Eng.Sci., 1987, 42, No.4.921
10. A.J.Matchett, P.D.Taylor, J.Peace, 'Adhesion in Materials Handling Systems', Problems of Industrial Dusts, Mess and Spillage, I.Mech.E, April, 1989
11. M.Uesugi, H.Kishida, Soils and Foundations, 1986, 26, 33

Figure 1: Novel Shear Cell.

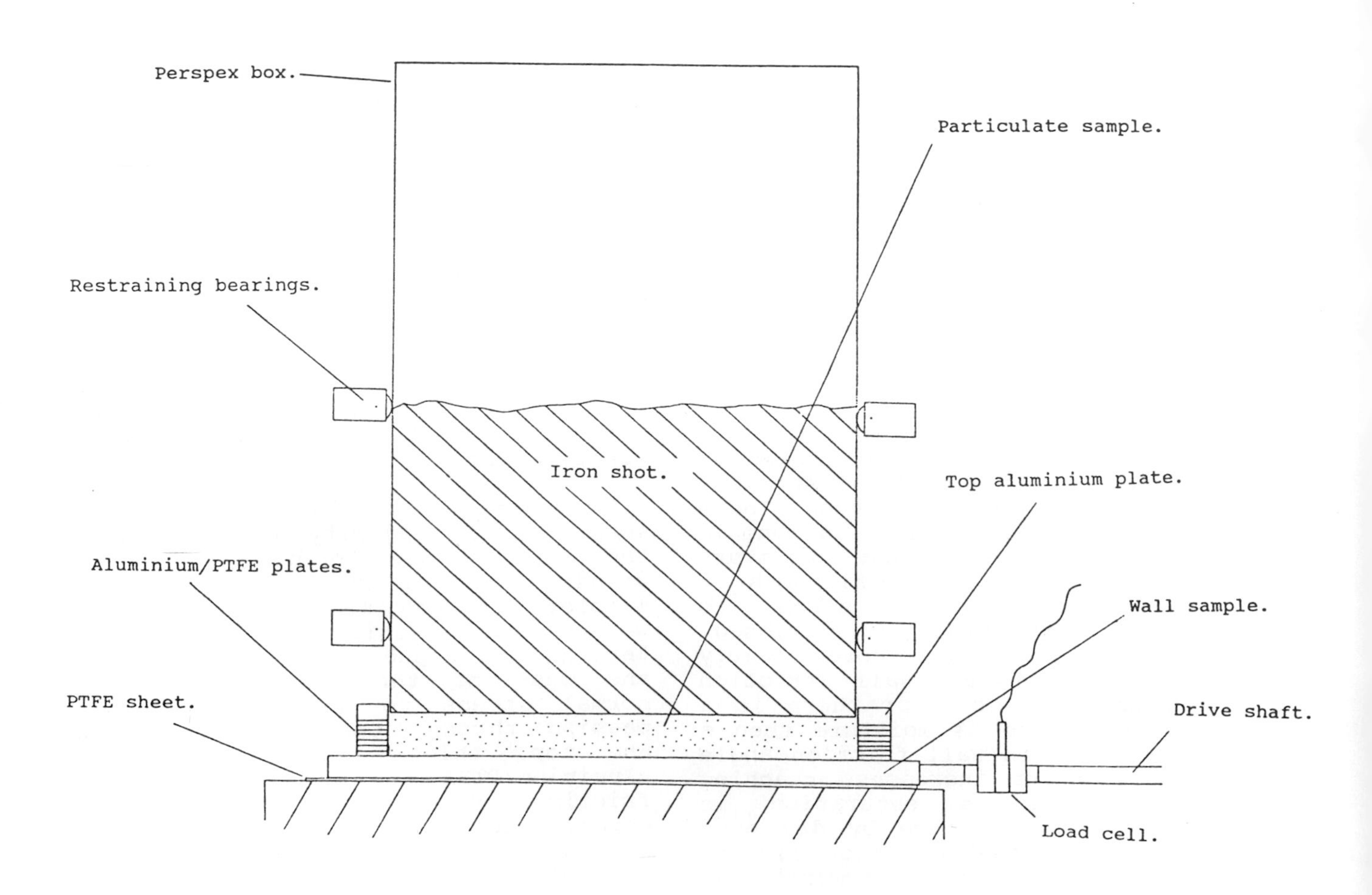

FIGURE 2. BUILING SAND TEST RESULTS.
FORWARD WALL SAMPLE DIRECTION.

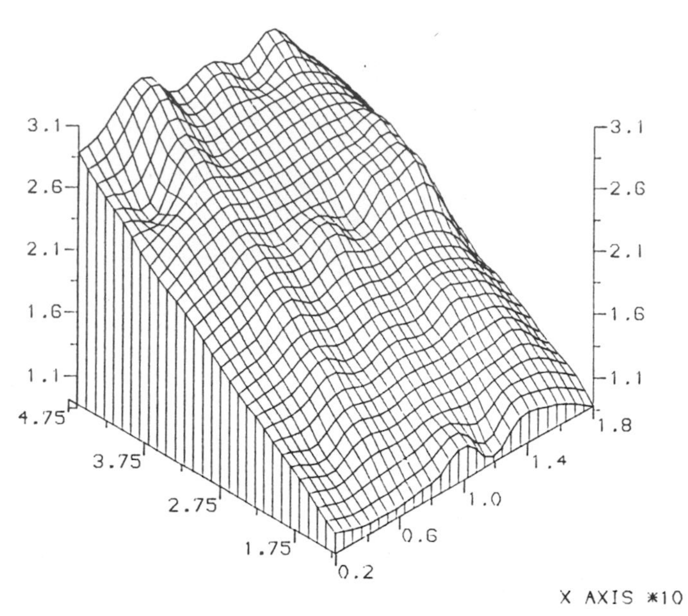

X AXIS : MOISTURE CONTENT, %. x 10
Y AXIS : NORMAL STRESS, KPA.
Z AXIS : SHEAR STRESS, KPA.

FIGURE 3. BUILDING SAND TEST RESULTS.
REVERSE WALL SAMPLE DIRECTION.

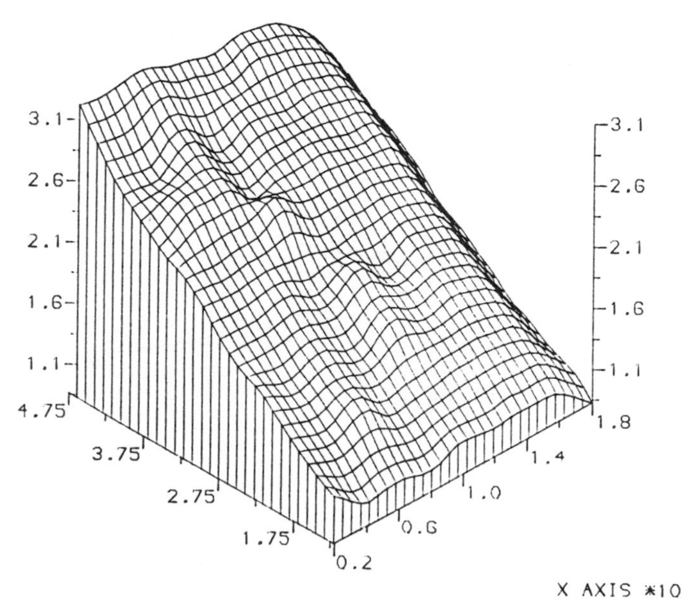

X AXIS : MOISTURE CONTENT, %. x 10
Y AXIS : NORMAL STRESS, KPA.
Z AXIS : SHEAR STRESS, KPA.

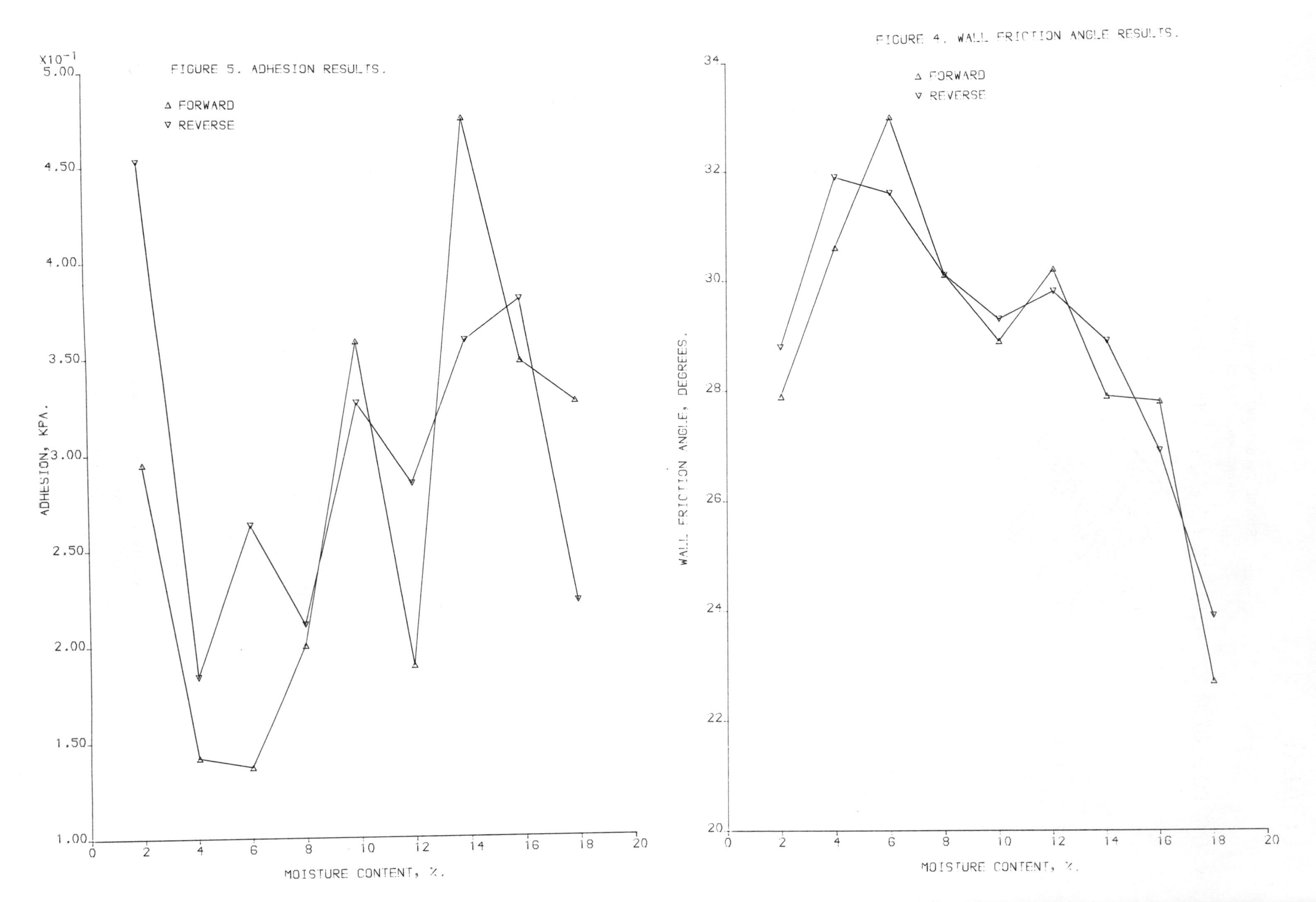

FIGURE 5. ADHESION RESULTS.

FIGURE 4. WALL FRICTION ANGLE RESULTS.

C418/028

Electrostatic problems in powder handling

D K DAVIES, BSc, PhD, CPhys, CEng, FInstP, FIEE and V EBADAT, BSc, PhD, AMIEE
Chilworth Technology Limited, Southampton, Hampshire

ABSTRACT Surface contact, separation and movement are essential features in powder processing. Furthermore, the majority of powders are not good conductors of electricity. Powder processing operations, therefore, contain all the necessary factors for electrostatic charge generation and retention. Processing problems such as powder agglomeration, adhesion to metal structures and variation in packing density, which can be attributed to electrostatic charging, have been investigated. The experiments are described and it is concluded that these problems can be controlled by correctly designed and located powered static eliminators.

1 INTRODUCTION

In all areas of industry, electrostatic charges hinder proper performance of processes. Because of the electrically insulating nature of many of the materials handled, manufacturers and processors are often faced with almost insoluble problems. In powder handling operations the product becomes electrostatically charged because of the frequent interparticle and container contact and separation. The main factors which determine the quantity and polarity of the charge are the chemical composition of the contacting surfaces, the contact area and the presence and nature of surface contaminants. Static charges are almost inevitably produced during operations such as sieving, pouring, spraying, grinding and pneumatic transfer, radically altering the process dynamics.

Whether or not an electrostatic problem arises depends on the equilibrium between the rates of charge acquisition and loss by conduction. Most powders are capable of retaining their charge when their volume resistivity exceeds 10^9 Ωm. One of the factors which influences the rate of charge loss is the ambient relative humidity. Water is adsorbed onto the surface of dielectric materials increasing the electrical conductivity. The extent to which water is adsorbed is, of course, dependent on the nature of the material and the relative humidity. Control of the humidity is a recognised method for reducing electrostatic problems.

Specific processes exhibiting typical electrostatic problems which have been investigated are sieving, tabletting and packing. These processes, and the conventional methods attempting to control the problems, are reviewed. The experiments are described in detail and it is demonstrated that the process problems can be controlled by properly designed and implemented powered ioniser systems.

2 POWDER PROCESS PROBLEMS

2.1 Sieving

In a typical sieving process, PVC powder wet with an electrically conductive liquid is dried in a hot air dryer and then transferred via a metal chute into a vibratory sieve. Particles smaller than a certain size pass through the sieve while the oversized powder is rejected and conveyed over the sieve into a collecting bin. Particle size analysis of the rejected material showed that the over-sized particles were caused by aggregation. This was correctly attributed to bipolar charging of the fine PVC particles which resulted in attractive and cohesive forces.

The manufacturer, being aware of the effect of humidity on the charge dissipation rate, introduced steam into the conveying chute upstream of the sieve and found this to be an effective way of reducing the charge induced aggregation. However, venting the steam required an open system and this produced a very unpleasant work environment.

Also, condensation of the steam on the sieve casing produced a wet powder cake which again reduced the sieving efficiency. Clearly a better method of controlling the powder charge was required.

2.2 Tabletting

Dry mixed powders are transferred from a small metal hopper through a powder feeder into the metal dyes of the tabletting machine. The powders are compressed in the dyes and then ejected in the form of solid tablets into a collector box. Charge is generated as the powder is transferred and compressed and as the tablets are ejected. Consequently, the charged tablets stick to the exit chute and the compression plugs causing blockage and crushing. Dust is produced leading to the eventual breakdown of the machine. At best the machine operates at about 30% of its production capability. Means for eliminating these problems are clearly desirable.

2.3 Packing Density Variation

Problems in collecting highly insulating powders in containers are familiar. In the extreme, highly charged powder can refuse to enter the container altogether, being repelled by the like charge of the earlier contents. More subtle problems have been experienced recently, however. Packs sold to the public must, of course, contain a precise quantity of the product but must also preferably, be packaged into bags without excessive empty headspace. The problem encountered is the self repulsion of charged powders which causes a reduction in packing density giving difficulty in filling to correct weight with subsequent settling leaving an empty pack space. There is a need for means to control the packaging problem.

3 THE EXPERIMENTS

The basis of these investigations was the measurements of powder charge-to-mass ratio and powder surface field coupled with examination of the effectiveness of powered electrostatic eliminators as a means of controlling powder charge. Detailed measurements have been made for each of problem circumstances described earlier using a Faraday pail technique. The method and the principle of powered ionisers are described briefly.

3.1 Faraday Pail Charge Measurement

The Faraday pail comprises a metal container enclosed by, but highly insulated from, a second earthed metal container. The capacitance of the inner container to ground is known and so the potential attained as the powder to be examined is poured into the inner container provides a direct measure of the charge. Weighing evidently provides the mass.

3.2 Ionisers

Application of an electric field in excess of about 3MV/m to air under normal conditions will result in electrical breakdown. Such high fields can be produced, locally, by applying voltages to sharp electrodes. The resulting discharges, known as "corona" discharges is also very local to the electrode as the field decreases rapidly away from the point to levels below those necessary to sustain the discharge. The corona discharge produces equal quantities of ions of both polarities. For D.C. applied voltages, the ions of like polarity to the applied potential are repelled away from the point but under A.C. stress bursts of both positive and negative ions are produced. In principle, therefore, ions of the necessary polarity to neutralise a powder charge can be produced by applying the appropriate potential and locating the electrode in a suitable position.

4. EXPERIMENTAL RESULTS AND DISCUSSION

4.1 Sieving

Table 1 shows the charge per unit mass and electric field measured on the powder in the conveying chute upstream of the sieve and on the sieve, respectively.

Table 1 Electrostatic Measurements

Location of Measurement	Charge per Unit Mass $C\ kg^{-1}$	Electric Field $V\ m^{-1}$
Conveying chute upstream of the sieve	-2.3×10^{-9} to $+2.3 \times 10^{-7}$	----
On the sieve mesh	----	0 to -3×10^{4}

The charge measurements show that the powder is partially charged with both polarities. It has been shown that bipolar charging of powders can result from particle size variations (1) and can be affected by the environmental conditions (2). In this case, fine particles charge positively and large particles negatively.

The observed positive charge was two orders of magnitude larger because the quantity of fines was much greater. The measured electric field can readily cause particle attraction (3).

AC eliminators were mounted in the conveying chute and also in the sieve housing above the sieve. The location of the ioniser is, however, crucial. If the spacing between the eliminator and the powder is too small zones of positive and negative charge can be produced in the flowing powder leading to enhanced attraction. Too great a distance means that the efficiency of the ioniser is reduced owing to recombination of the positive and negative ions before reaching the powder. Larigaldie et al (4) showed that the efficiency of such ionisers could be increased by decreasing the ion mobility by introducing water vapour into the ion stream, the free ions having been attached to the water droplets. In the present case electric field measurements as the powder entered the sieve showed a decrease to zero with the A.C. eliminators switched on together with a small flow of steam.

4.2 Tabletting

The charge-to-mass ratio on the tablets, determined by measuring the charge on a known number of tablets, at various locations in the tabletting machine is shown in Table 2.

Table 2 Charge Per Unit Mass Measurements

Measurement Location	Charge per Unit Mass C kg^{-1}
Feeder Hopper	$+0.4 \times 10^{-6}$
Entering exit chute	$+4.7 \times 10^{-6}$
Leaving exit chute	$+3.5 \times 10^{-6}$

In a separate experiment a small batch of the dry mixed tablet powder was placed on an earthed metal plate and charged by a corona discharge. The decay of charge was measured using a field meter positioned at a known distance from the powder surface and found to take about 22 minutes to decrease to 1/e of the initial value. This showed that the powder was highly insulating.

Active AC eliminators were found to be ineffective owing to the rapid recombination of the ions. Attempts to increase the efficiency by using blown air were unacceptable owing to the dispersion of settled dust. Balanced positive and negative dischargers which produce both positive and negative ions simultaneously from separate arrays of points driven by appropriate polarity power supplies was used.

Since the electrodes are physically separated the recombination rate is lower and the ion production rate, or current, (positive or negative) is shown as a function of distance from the ioniser in Figure 1.

Figure 1 Neutralisation Current as a Function of Distance between Eliminator and Charge Object

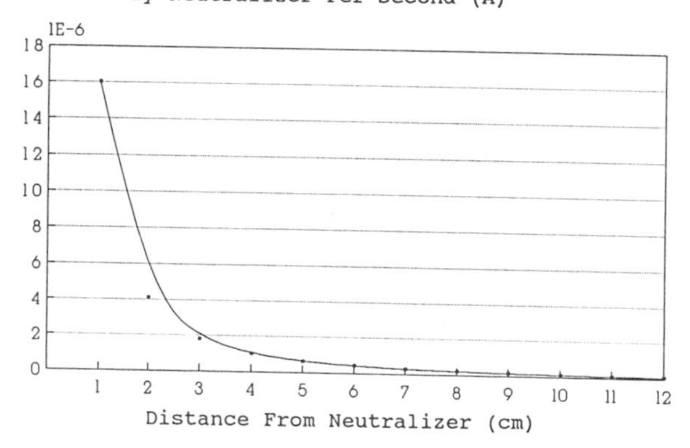

The tablets were produced at the rate of 42 per second. Given a tablet mass of 0.26×10^{-3} kg and a charge-to-mass ratio of 4.7×10^{-6} C/kg (from Table 2), the necessary neutralising current is 5×10^{-8} A. The tests on the balanced polarity ioniser showed that at a distance of 4 cm, for example, the eliminator could generate a neutralisation current of $\pm 1 \times 10^{-6}$ A (see Figure 1). This current is 20 times larger than what is needed to neutralise the electrostatic charge on 42 tablets per second.

4.3 Packing Density Measurements

In the real production environment the powder was packed directly from a powder mill. A small experimental mill was, therefore, set up directly above a large Faraday pail and the charge on the milled powder measured directly. The electrical and physical properties of the milled powder are given in Table 3.

Table 3 Properties of Test Powder

Volume Resistivity (Ωm)		**Charge Decay Time (sec)**		**Mass Median Diameter (μm)**
10% RH	**50% RH**	**10% RH**	**50% RH**	
4×10^{12}	4×10^{9}	1700	< 1	63 to 75

The data indicates that the powder is highly insulating at low humidities of about 10%

The packing density of the powder was measured for a number of humidities by collecting in a container which was weighed after careful levelling of the contents. The correlation between packing density and charge-to-mass ratio is shown plotted in Figure 2 despite a considerable scatter in the results. It is suggested that control of the packing density may be possible by using appropriate ionisers. This work is continuing.

Figure 2 Powder Packing Density as a Function of Charge Per Unit Mass

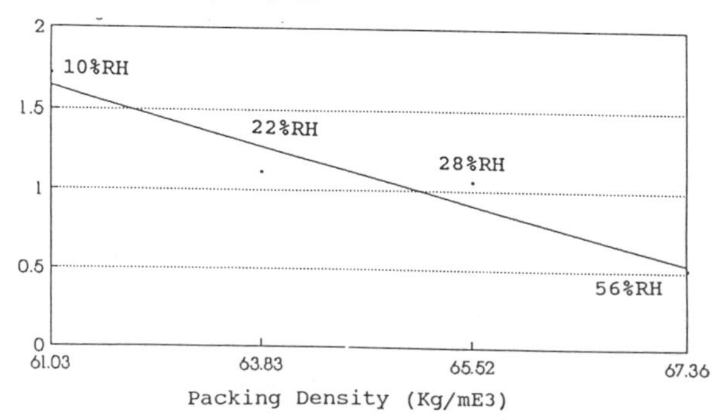

5 CONCLUSIONS

It may be concluded from these experiments that:

(1) There is a clear correlation between the several process problems and the electrostatic charges produced on the product; and,

(2) Control of powder charge and, hence, elimination of the process problem can be achieved by correct design and application of powered ionisers.

REFERENCES

(1) CARTWRIGHT, P., SINGH, S., BAILEY, A.G., ROSE, L.J., Electrostatic Charging Characteristics of Polyethylene Powder During Pneumatic Conveying. IEEE Transactions on Industry Applications, 1985, vol IA-21, No. 2. pp 545-546.

(2) EBADAT, V., SINGH, S., Importance of Humidity on the Control of Electrostatic Charging of HDPE Powder. Ppneumatech 4, Fourth International Conference on Pneumatic Conveying Technology, Glasgow, Scotland. 26-28 June 1990

(3) British Standard 5958: Part 1, Code of Practice for Control of Undesirable Static Electricity, 1980, p 27.

4. LARIGALDIE, S., GIBONI, N., A New Device for the Neutralization of Static Electricity on Pulvesulent Materials. Journal of Electrostatics, 10 (1981) 57-64.

C418/061

The relative economics of bulk transport by alternative modes

J M MANNING, BSc(Eng), FICE
RPT Economics Studies Group

SYNOPSIS A systematic comparison of the unit costs of bulk transport by road, rail, ship, conveyor, aerial ropeway and slurry pipeline. Curves are derived illustrating the effects of throughput volume and distance on unit cost and the probable cheapest modes in given circumstances.

1 NATURE OF COSTS

1.1 Introduction

The object of this paper is to compare the cost differences between a number of typical modes of transport on a systematic basis. Although comparisons of unit costs per tonne by mode are often made, these are only meaningful if the haul distance and total throughput are specified. Naturally a series of additional factors can be relevant in particular cases.

In this paper, an estimate of relative modal costs has been made for the transport of a hypothetical bulk commodity. Curves are derived which show the effect of both the volume of throughput and distance on unit costs. In principle, the aim is to assess total system costs though, in practice, it can be difficult to define precisely the boundaries of the system to be analysed. The main assumptions used in this paper are stated in Section 2.

The following modes are considered:

- road;
- rail;
- ship;
- conveyors;
- aerial ropeway;
- slurry pipeline.

Each of these modes of transport can prove the most economical in the right conditions. This fact is best explained in terms of the familiar distinction between fixed and variable costs. In principle, variable costs increase in line with the amount of transport activity undertaken, while fixed costs do not. The best general measure of transport activity is the number of tonne-kilometres transported.

In practice, the distinction between fixed and variable costs can never be precise, mainly because no cost is totally invariable with respect to the volume of transport. However, the important point is that these so-called fixed costs follow a stepped pattern and so are relatively invariable over given ranges of output.

The effect of the interaction between fixed and variable costs is shown in Figures 1(a) and 1(b). These take idealised cost per tonne relationships for three different scenarios and show the variations in derived cost per tonne-kilometre. The linear per tonne costs produce hyperbolic tonne-kilometre cost curves, in which fixed costs dominate at low distances whereas differences at greater distances are almost entirely controlled by the variable costs.

These cost relationships account for the difficulties in making inter-modal cost comparisons. At small distances, the cost differences are large but varying very rapidly so that small changes in haul distances can affect the cost ranking. At large distances the differences are stable but very small so that small changes in the assumptions (say, fuel prices) can easily change the results of the comparison.

1.2 Elements of Cost

Table 1 summarises the principal types of costs to be considered in a detailed comparison of system costs. Here the distinction between fixed and variable costs is largely one of time. For example, in the long run labour costs are variable though they become progressively more fixed the shorter the period that one is considering.

Terminal storage and handling arrangements can vary considerably - for example, open stacker/reclaimer stockpiles, silos, containers, palletised bags - and so will the costs. They are not part of the main haulage leg and therefore are not distance related but may typically constitute 5-10 percent of the system costs. If a terminal is part of a distribution chain, it is questionable whether the storage and handling should be included in the transport costs of the primary system being analysed, or taken to be a part of the secondary leg.

Table 1 PRINCIPAL COMPONENTS OF SYSTEM COSTS

Nature of Cost	DIRECT HAULAGE	OTHER		
		TERMINAL HANDLING	GOODS INVENTORY	INFRA-STRUCTURE
FIXED	Ownership Licensing Staff and Labour Insurance Attributable Overheads	Capital Costs Labour Insurance	Buffer Stocks	Provision Costs Staff and Labour
VARIABLE	Fuel/ Energy Maintenance	Energy Maintenance	Surge Stocks Goods in Transit	Maintenance

Any transport system involves a certain quantity of goods stockpiled and in transit. The value of these goods has to be financed since costs have been incurred in winning or processing which cannot be recovered until the goods are sold. A system which has larger stockpile requirements or longer transporting times therefore usually has higher financing costs, though this may sometimes be offset by a reduction in storage costs.

The biggest difference in 'other' costs between the modes occurs with the associated infrastructure. In the case of pipelines, conveyors and ropeways the transporting medium is itself the infrastructure and constitutes a high proportion of the costs. In the case of short sea shipments there are normally no infrastructure costs, assuming that costs of jetties and loading/discharging facilities are attributed to terminal handling.

Road and rail transport fall between these extremes. In both cases the infrastructure costs are significant but not always borne by the transporter to the same extent. In an independent mining operation, all the road or railway infrastructure would usually have to be constructed by the mining company, who would also bear the full costs of maintenance. It is more common in Europe for the rail transporter to have a private terminal and siding connected to the national rail network. Thus the transporter would pay the costs for his private siding but the infrastructure costs of the longer main line haul would be included in the general tariff charged to him by the railway company. It would be impossible to distinguish as part of the transport costs and may also carry an element of subsidy.

Similarly, in a developed environment, road users do not pay an identifiable element of main road infrastructure costs except through the medium of the licence fees. In a rigorous economic study such differences would be taken into account. In this paper comparisons are made on the basis of perceived financial costs to the users.

2. MODAL COSTS

2.1 General

Unit fixed costs are profoundly affected by throughput and utilisation. The required throughput (tonnes per annum, say) determines the capacity which must be provided and hence the level of fixed costs which are incurred. The utilisation of this capacity then determines over how many tonnes the fixed costs can be spread.

The actual throughput will change for a variety of commercial reasons and this will immediately affect utilisation and hence unit fixed costs per tonne. In the longer run, the utilisation rate can be seen as a measure of the efficiency with which the system is used. While throughput and utilisation rates are not implicit characteristics of transport systems, a modal system that is less flexible in its ability to adapt to changes in throughput and utilisation (i.e. one that has a fixed capacity, or a capacity that is only optimal over a narrow throughput range) will exhibit greater unit cost variations. This can affect the modal comparison.

Since the chosen system can seldom match the design throughput exactly, the actual unit cost will usually exceed the theoretical unit costs, as shown in Figure 2.

Provision, ownership or capital costs are usually calculated on an annual basis to derive unit fixed costs.

There are various ways of treating them, for example:

(a) Capital costs are divided by the estimated life, to derive an average annual cost. Interest charges on borrowed capital should be added during the amortization period, although this is often not done.

(b) Capital costs are valued as the interest foregone on the capital employed. As a variant, valuation may be on the basis of an arbitrary target return on capital. This assumes that capital is available and that there are no borrowing charges.

(c) The initial capital is repaid on an annuity, or mortgage basis, in which a fixed annual payment is calculated so as to repay the initial loan plus accrued interest on the diminishing balance.

(d) Actual provision costs are included as incurred, in the cost stream which is set against benefits. This method is only meaningful if the evaluation uses a discounted cash flow (DCF) approach.

Table 2 shows for methods (a), (b) and (c) the annual charge that would be derived by alternative methods for a capital cost of £100 with an interest rate of 10 percent and various asset lives.

Table 2 ANNUAL CHARGE FOR CAPITAL INVESTMENT OF £100 (Pounds)

Evaluation Method	Asset Life (years)		
	7	10	15
(a) Capital cost/ Asset life (excl. interest charges)	14.29	10.00	6.67
(b) Fixed interest payment (market rate)	10.00	10.00	10.00
(20% return on capital)	20.00	20.00	20.00
(c) Annuity	20.54	16.27	13.15

Though convenient for simple commercial appraisals, methods (a) and (b) have significant deficiencies. Method (d) is the most accurate, but (c) is useful to develop a feel for the impact of capital costs if precise financing conditions are not known.

Stockpiles are as often sized by experience and rule of thumb as by detailed inventory management or simulation studies. Ten percent of the annual throughput is a typical figure. Given that the value of the material may range from zero (domestic waste) to say £300 per tonne (high grade minerals or bulk chemicals), a general treatment is impracticable. The point for intermodal comparison is that the minimum size of any stockpile in the system must be at least the maximum consignment size or, practically, 1.5-2.0 times that size. Thus the stockpile financing charges for short sea transport with 3000 tonne consignments may be up to 100 times greater than a road transport system moving in 30 tonne lots. Stockpiling requirements for the continuous modes (conveyor,ropeway and pipeline) are quite small in respect of the transport alone but it is prudent to make larger allowances for system breakdowns.

2.2 Road Transport Costs

Road transport costs are covered in more detail here than the other modes both because more data are readily available and because it must be regarded as the benchmark mode. Not only does it carry more than half of all freight in the UK but it is the only mode that bears realistic comparison over the range from very short to moderately long hauls.

The numbers of trucks required for given throughputs can vary enormously with the operating assumptions made as to working days, shifts, speeds, utilisation efficiency etc.

Table 3 shows plausible minimum and maximum numbers of 25 tonne trucks for a 50km haul and three different throughputs.

Table 3 REQUIRED NUMBERS OF TRUCKS (50km haul; 25 tonne payload)

Throughput tonnes per annum	Number of Trucks		
	Min.	Max	Assumed for analysis
50 000	2	6	4
500 000	7	50	23
5 000 000	58	490	221

The numbers used in the analysis were derived from the mean of the following two sets of assumptions:

Working days per year	250	300
Shifts per day	1	2
Average trip speed, k.p.h.	70	30
Utilisation	75%	75%
Load/discharge time, hours	0.17	0.17

Assumed no backhaul traffic

Table 4 shows typical 'going rate' operating costs for 25 tonne payload trucks.

Table 4 TRUCK OPERATING COSTS (Pounds; 25 tonne payload)

Item	Fixed Costs per year	Percent
Licences	3100	10.4
Wage Costs	12000	40.4
Rent and Rates	1200	4.0
Insurance	1200	4.0
Capital costs	10480	35.2
Overheads	1700	
Sub Total Fixed Costs	29680	100.00
	Running Costs(a) per km	Percent
Fuel	0.172	56.3
Lubricants	0.010	3.6
Tyres	0.040	12.6
Maintenance	0.080	27.5
Sub Total Variable Costs	0.302	100.0

(a) For loaded vehicle

Table 5 shows approximate terminal handling costs assuming that such facilities are required at both ends of the transport leg.

Table 5 TERMINAL HANDLING COSTS (Pounds)

Throughput (tonnes per annum)	FIXED		VARIABLE (£ per tonne)
	Capital	Other	
50 000	50 000	40 000	0.05
500 000	100 000	60 000	
5 000 000	200 000	90 000	

Table 6 shows the estimated quantities of goods inventory, including one terminal stockpile and goods in transit. These are minimum requirements - much more may be stocked for reasons of production breakdown security or to even out production and distribution links but this is not necessarily a consequence of the transport system.

Table 6 GOODS INVENTORY (tonnes)

Throughput (tonnes per annum)	Length of haul (km)			
	20	50	100	200
50 000	300	300	320	350
500 000	1500	1600	1800	2200
5 000 000	9000	10000	12000	15000

An arbitrary value of £50 per tonne is assumed for the later analysis when the cost of the goods inventory is included in total system costs.

On the basis of the cost assumptions made in the previous section, Table 14 gives the total system costs for road transport for three different throughputs. Fig. 3 shows fixed and variable costs for road transport of 50 000 tpa and 5 000 000 tpa. The surprisingly small variation with throughput is because of the high flexibility of the road mode in adjusting to such changes.

Finally, Fig. 4 shows the characteristic tonne-kilometre cost curves for the road transport mode. The effect of fixed costs is shown by the increasing differences between the throughputs at hauls below about 50 km.

2.3 Rail Transport Costs

Estimation of rail transport costs starts, as for road, with an assessment of the fleet requirements, shown in Table 7. For consistent comparison, the same throughputs and distances have been used as for the other modes, although rail transport is unlikely to be considered for small throughputs and distances. If it were, lighter rails and trains would be used in practice, probably with some kind of shuttle system. However, to simplify the analysis, standard costs have been used throughout and are shown in Table 8. In line with the comments in Section 1, infrastructure costs have been included as though the operator were providing a complete railway line with all the civil engineering works in addition to trackwork. The actual cost used is a very arbitrary value. £1 500 000 per km would be a high cost for a single line standard gauge track laid in fairly flat terrain in undeveloped country. On the other hand, if the route traversed densely built up urban areas or mountainous terrain, the unit cost could easily be five times as great.

Table 7 ROLLING STOCK REQUIREMENTS FOR RAIL TRANSPORT

Throughput (tonnes per year)	Transport Distance (km)	Number of Trains per day	Number of Train Sets Required
50 000	50	0.25	1
	200		1
	500		1
500 000	50	1.70	1
	200		1
	500		2
5 000 000	50	5.57	2
	200		4
	500		7

Assumptions
(i) Wagon payload 50 tonnes
(ii) Maximum train payload per locomotive 1000 tonnes
(iii) Average running speed 50 kph
(iv) Load/discharge time 3 hours
(v) 300 working days per year
(vi) Maximum train consist 3 locos + 60 wagons

Rail system costs are shown in Table 15. It is immediately apparent that they are dominated by the costs of the infrastructure. Total tonne-kilometre operating costs for the higher throughputs and longer distances are significantly cheaper than all other modes, including sea transport. However, when the infrastructure costs are included on a reasonable economic basis (ten percent interest and 50 year life, in this case) the total system costs become relatively expensive. Since most national railways have long since written off the capital costs of the bulk of their networks, it is evident that a wide range of interpretations can be put on the real cost of rail transport. Also, when trackwork is shared by other services rather than being a dedicated line, the attributable costs will be less. What is presented here can be no more than a rough and ready assessment of representative values. Characteristic tonne-kilometre cost curves are shown in Fig. 5.

Table 8 ASSUMED RAIL TRANSPORT COSTS

CAPITAL COSTS	
Locomotive	£ 1 200 000 each
Wagon	£ 55 000 each
Infrastructure	£ 1 500 000 per km
OPERATING COSTS (per year)	
Fuel	£ 36 per journey hour per locomotive
Loco maintenance	6.0% of capital cost per year
Wagon maintenance	2.5% of capital cost per year
Track maintenance	0.15% of capital cost per year
Staff	2 per train set, plus allowance for terminal staff depending on throughput

2.4 **Ship Transport Costs**

Ship transport is characterised by high capital costs which increase significantly with throughput, although the unit costs continue to decline. The capital costs of berths and handling plant are also high, rising from about 25 percent to 50 percent of the ship costs at the larger throughputs. Determination of optimum provision of resources also becomes more critical at high throughputs, as indicated in Table 9.

Table 9SHIP TRANSPORT OPERATING SYSTEM REQUIREMENTS

Throughput (tonnes per year)	Transport Distance (km)	Assumed Ship Fleet		Estimated number of Berths Required (at each end)
		No.	Size (dwt)	
50 000	100	1	1000	1
	500	1	1000	1
	1000	1	1000	1
50 0000	100	1	1000	1
	500	1	4000	1
	1000	2	3000	1
5 000 000	100	4	5000	3
	500	5	10000	3
	1000	7	10000	4

Ship transport costs are usually considered in terms of capital and daily operating costs. The latter are relatively low and, in terms of the definitions in Section 1, themselves include a high proportion of fixed costs. Hence the distance variable costs are very low, making ship transport the preeminent mode for longer distances. Typical values are shown in Table 10.

Table 10 TYPICAL SHIP TRANSPORT COSTS (Pounds)

Ship Size (dwt)	Capital Cost (£)	Operating Cost		Capital Costs	
		Work Day (a)	Non Work Day (b)	Berth	Plant
1000	1 500 000	1600	1300	900 000	600 000
5000	5 700 000	3200	2300	2 400 000	1 080 000
10000	9 000 000	5000	3000	6 000 000	2 400 000

(a) At sea
(b) In port

The goods inventory requirements are higher than for other modes both because of the larger consignments and the slower speeds which mean that more goods are in transit. But at large throughputs and distances, such differences are outweighed by the magnitude of the goods flow and the minimum required stockpiles vary from two to five times those needed for road transport depending mainly on the ship sizes assumed.

Table 16 gives the total system costs for ship transport and Figure 6 shows the characteristic tonne-kilometre cost curves. Inland waterway transport costs have not been separately evaluated but could be expected to be broadly similar to short sea transport for the purposes of modal comparison.

An important point with this mode is that the loading/discharging port may not be the desired end point of the transport chain (as, to a lesser extent, with rail). Thus additional road distribution haulage may be necessary and should be included in the system costs. If the road haul distance(s) are significant, their costs may offset the ship transport saving. If wholly road transport is a feasible alternative (i.e. there is not actually a sea barrier uncrossed by a ferry) there would be a 'break-even' point between the prime origin and destination, where the costs of the two alternative systems would be equal.

2.5 **Conveyors and Aerial Ropeways**

Conveyors are extremely well suited to the transport of bulk materials, especially in their traditional role over relatively short distances - say, up to five kilometres. Nevertheless, there are examples of long distance conveyors of 50km or more, particularly where throughput is large and the terrain would be difficult for other modes.

Manpower requirements are low for this mode, as is the volume of material tied up in internal storage. However, energy costs are significant and annual maintenance costs can also be substantial. One disadvantage is that a breakdown will bring the whole system to a halt, which is not usually the case for rail or road.

Typical costs are not easy to establish since there are too few installations to provide a representative body of data. Table 11 gives the assumptions made for this paper, from which Table 17 shows the system costs and Figure 7 the characteristic unit cost curves for both conveyors and aerial ropeways.

Table 11 COST ASSUMPTIONS FOR CONVEYORS

Item	Cost	Remarks
Capital cost	£3000 per km per 1 tph of capacity	50 tph assumed as minimum practical capacity and 5km as minimum length to allow for non distance related works
Belt Replacement	£240 per km per 1 tph of capacity	Replacement every 8 years
General Maintenance	£0.004 per km per tpa	
Electrical Power	£0.004 per km per tpa	
Staff	Dependent on number of shifts required to achieve throughput with installed capacity	

Aerial ropeways are similar in their basic concept to conveyor systems. They tend to outperform conveyors at lower throughputs and are even better suited to difficult terrain.

Theoretically, the distance that can be covered is unlimited since it is usual to adopt successive flights of about 5-10km but it is believed that the longest currently in operation is 76km, from a manganese mine in Gabon. Typical costs are scarce, as for conveyors, and the similar cost assumptions are given in Table 12.

Table 12 COST ASSUMPTIONS FOR AERIAL ROPEWAYS

Item	Cost	Remarks
Capital cost	£2400 per km per 1 tph of capacity	10 tph assumed as minimum practical capacity and 10km minimum length, to allow for non distance related works
Rope Maintenance	£400 per km per 1 tph of capacity	Replacement after 10 years or 5 000 000 tonnes
General Maintenance	£0.005 per km per tpa	
Electrical power	£0.004 per km per tpa	
Staff	Generally 3 shifts and dependent on number of flights i.e. distance.	

2.6 Slurry Pipeline

A slurry pipeline transports solids as a finely ground powder in a moving stream of water. This can be cheap and effective, although it obviously cannot be used for all types of material. Furthermore, even after decanting (i.e. letting the solids settle out and pumping off the excess water) the mixture is likely to contain about ten percent water. This may require additional expenditure on heating/drying.

Pipelines have specific technical characteristics which in certain circumstances give them unique advantages over all other means of bulk transport. For example, they are even less affected by adverse climate and terrain than conveyors and ropeways. Pipelines can be buried under the ground (admittedly at the price of a substantial increase in capital and maintenance costs) so that their impact on the environment becomes minimal. Under certain circumstances such factors can become as important as cost differences in determining the optimal system.

Although the initial cost of the pipeline is relatively high and must include the costs of a water supply, operating costs are generally quite low. Overall unit costs become particularly economical at high throughputs and pipelines can be used over long distances. However, it is probably the mode that is most sensitive to throughput changes from the initial design conditions. Unit costs rise sharply with either decreases or increases from the design throughput, even if these are technically and operationally feasible. Cost assumptions are given in Table 13, system costs in Table 18 and characteristic cost curves in Figure 8.

Table 13 COST ASSUMPTIONS FOR PIPELINES

CAPITAL COSTS		
Required Throughput (tonnes per annum)	Assumed Pipe Diameter (mm)	Cost per km (pounds)
50 000	100	120 000 (It is
100 000	100	120 000 assumed
200 000	100	120 000 that 100mm
500 000	150	300 000 is the
1 000 000	200	480 000 minimum
2 000 000	250	670 000 practical
5 000 000	400	1 210 000 diameter)
OPERATING COSTS		
Item	Cost	
General Maintenance	£0.004	per km per tpa
Electrical Power	£0.005	per km per tpa
Staff	Constant for distance but dependent on number of shifts	

3. CONCLUSIONS

The total system costs of all the modes are compared in Figures 9 and 10 for various throughputs. In general, the variations are in line with commonly accepted experience of modal comparisons and the changes in optimal modes occur at the sort of distances that would be expected. It is again emphasized that most modal cost comparisons are pretty close and usually depend greatly on the particular circumstances. Final choices are often made for reasons which are not necessarily cost based, such as:

- compatibility/familiarity with existing installations,
- flexibility of operation,
- environmental considerations,
- customer requirements at receiving end,
- political pressures,
- handling requirements of the product.

Nevertheless, it is suggested that the data presented here enable obvious 'non-runners' to be discarded so that subsequent analysis and evaluation can be concentrated on the two or three most likely prospects.

Table 14 TOTAL SYSTEM COSTS FOR ROAD TRANSPORT (Pounds per tonne)

Throughput (tonnes per year)	50 000			500 000			5 000 000		
Transport Distance (km)	5	50	500	5	50	500	5	50	500
FIXED COSTS									
Direct Haulage	1.19	2.38	11.89	0.36	1.37	11.47	0.31	1.31	11.37
Terminal Handling	0.24	0.24	0.24	0.04	0.04	0.04	0.01	0.01	0.01
Goods Inventory	0.03	0.03	0.03	0.01	0.01	0.01	0.01	0.01	0.01
Sub-Total	1 46	2.65	12.16	0.41	1.42	11.52	0.33	1.33	11.39
VARIABLE COSTS									
Direct Haulage	0.12	1.16	11.62	0.12	1.16	11.62	0.12	1.16	11.62
Terminal Handling	0.05	0.05	0.05	0.05	0.05	0.05	0.05	0.05	0.05
Goods Inventory	neg	neg	0.02	neg	neg	0.02	neg	neg	0.02
Sub-Total	0.17	1.21	11.68	0.17	1.22	11.68	0.17	1.21	11.68
TOTAL COSTS	1.63	3.86	23.84	0.58	2.64	23.21	0.49	2.54	23.07
TOTAL COSTS (Pounds per tonne-km)	0.32	0.08	0.05	0.12	0.05	0.05	0.10	0.05	0.05

neg = Negligible

Table 15 TOTAL SYSTEM COSTS FOR RAIL TRANSPORT (Pounds per tonne)

Throughput (tonnes per year	50 000			500 000			5 000 000		
Transport Distance (km)	50	200	500	50	200	500	50	200	500
Capital Cost of Rolling Stock	9.06	9.06	9.06	0.98	0.98	1.63	0.42	0.81	1.43
Operating Costs	4.47	4.79	5.44	0.52	0.74	1.44	0.23	0.58	1.22
Terminal Handling	0.44	0.44	0.44	0.44	0.44	0.44	0.44	0.44	0.44
Goods Inventory	0.13	0.13	0.13	0.02	0.02	0.02	neg	0.01	0.01
Sub total Direct Haulage	14.10	14.42	15.07	1.96	2.18	3.53	1.09	1.84	3.10
Infrastructure, including trackwork	153.54	614.16	1535.39	15.35	61.42	153.54	1.54	6.14	15.35
TOTAL COSTS	167.64	628.58	1550.46	17.31	63.60	157.07	2.63	7.98	18.45
TOTAL COSTS (Pounds per tonne-km)	3.35	3.14	3.10	0.35	0.32	0.31	0.05	0.04	0.04
TOTAL COSTS EXCLUDING INFRASTRUCUTRE (Pounds per tonne-km)	0.28	0.07	0.03	0.04	0.01	0.007	0.02	0.009	0.006

neg = Negligible

Table 16 TOTAL SYSTEM COSTS FOR SHORT SEA TRANSPORT (Pounds per tonne)

Throughput (tonnes per year)	50 000			500 000			5 000 000		
Transport Distance (km)	20	50	500	20	50	500	20	50	500
Ship Transport Costs	14.43	14.45	14.85	1.49	1.52	3.50	0.68	0.68	1.79
Provision of Berths and Handling Plant	6.61	6.61	6.61	0.66	0.66	1.22	0.36	0.36	1.11
Terminal Handling	0.20	0.20	0.20	0.20	0.20	0.20	0.20	0.20	0.20
Goods Inventory	0.11	0.11	0.11	0.04	0.04	0.04	0.01	0.01	0.02
TOTAL COSTS	21.35	21.37	21.77	2.39	2.42	4.96	1.25	1.25	3.12
TOTAL COSTS (Pounds per tonne-km)	1.07	0.43	0.04	0.12	0.05	0.01	0.06	0.03	0.006

Table 17 TOTAL SYSTEM COSTS FOR CONVEYORS AND AERIAL ROPEWAYS
(Pounds per tonne)

Throughput (tonnes per year)	50 000			500 000			5 000 000		
Transport Distance (km)	5	10	50	5	10	50	5	10	50
CONVEYOR									
Capital	1.97	3.94	19.72	0.28	0.56	2.58	0.28	0.56	2.81
Operation	0.96	1.02	1.51	0.13	0.19	0.69	0.09	0.15	0.64
Goods Inventory	0.05	0.05	0.05	0.05	0.05	0.05	0.05	0.05	0.05
TOTAL COSTS	2.98	5.06	21.28	0.46	0.80	3.54	0.42	0.76	3.50
TOTAL COSTS (Pounds per tonne-km)	0.60	0.50	0.43	0.09	0.08	0.07	0.08	0.08	0.07
AERIAL ROPEWAY									
Capital	0.63	0.63	3.16	0.48	0.48	2.40	0.48	0.48	2.40
Operation	0.56	1.11	3.65	0.12	0.25	1.04	0.09	0.18	0.84
Goods Inventory	0.06	0.06	0.06	0.05	0.06	0.06	0.05	0.06	0.06
TOTAL COSTS	1.25	1.80	6.87	0.65	0.79	3.50	0.62	0.72	3.30
TOTAL COSTS (Pounds per tonne-km)	0.25	0.18	0.14	0.13	0.08	0.07	0.13	0.07	0.07

Table 18 TOTAL SYSTEM COSTS FOR PIPELINES
(Pounds per tonne)

Throughput (tonnes per year)	50 000			500 000			5 000 000		
Transport Distance (km)	5	50	500	5	50	500	5	50	500
Capital	1.58	15.78	157.77	0.39	3.94	39.44	0.16	1.59	15.91
Operation	1.49	1.89	5.94	0.53	0.93	4.98	0.09	0.50	4.55
Goods Inventory	0.06	0.06	0.08	0.06	0.06	0.08	0.05	0.05	0.07
TOTAL COSTS	3.13	17.73	163.79	0.98	4.93	44.50	0.30	2.14	20.53
TOTAL COSTS (Pounds per tonne-km)	0.62	0.35	0.33	0.20	0.10	0.09	0.06	0.04	0.04

REFERENCES

1. World Bank Technical Paper Number 38 - "Bulk Shipping and Terminal Logistics"
2. "Commercial Motor" Tables of Operating Costs
3. Department of Transport Statistics Bulletin (88) 38
4. Economics of Handling Materials in the Iron and Steel Industry, P.M. Worthington, Journal of the Iron and Steel Institute, Oct. 1962.
5. Long Distance Conveying, E.J. Fawcett, NCB
6. Long Distance Transport of Commodities in the Northwest, D. Vincent and C.G. Baker, Bulk Solids Handling, Feb. 1985.
7. Croners Operational Costings, Dec 1990
8. Various in-house studies

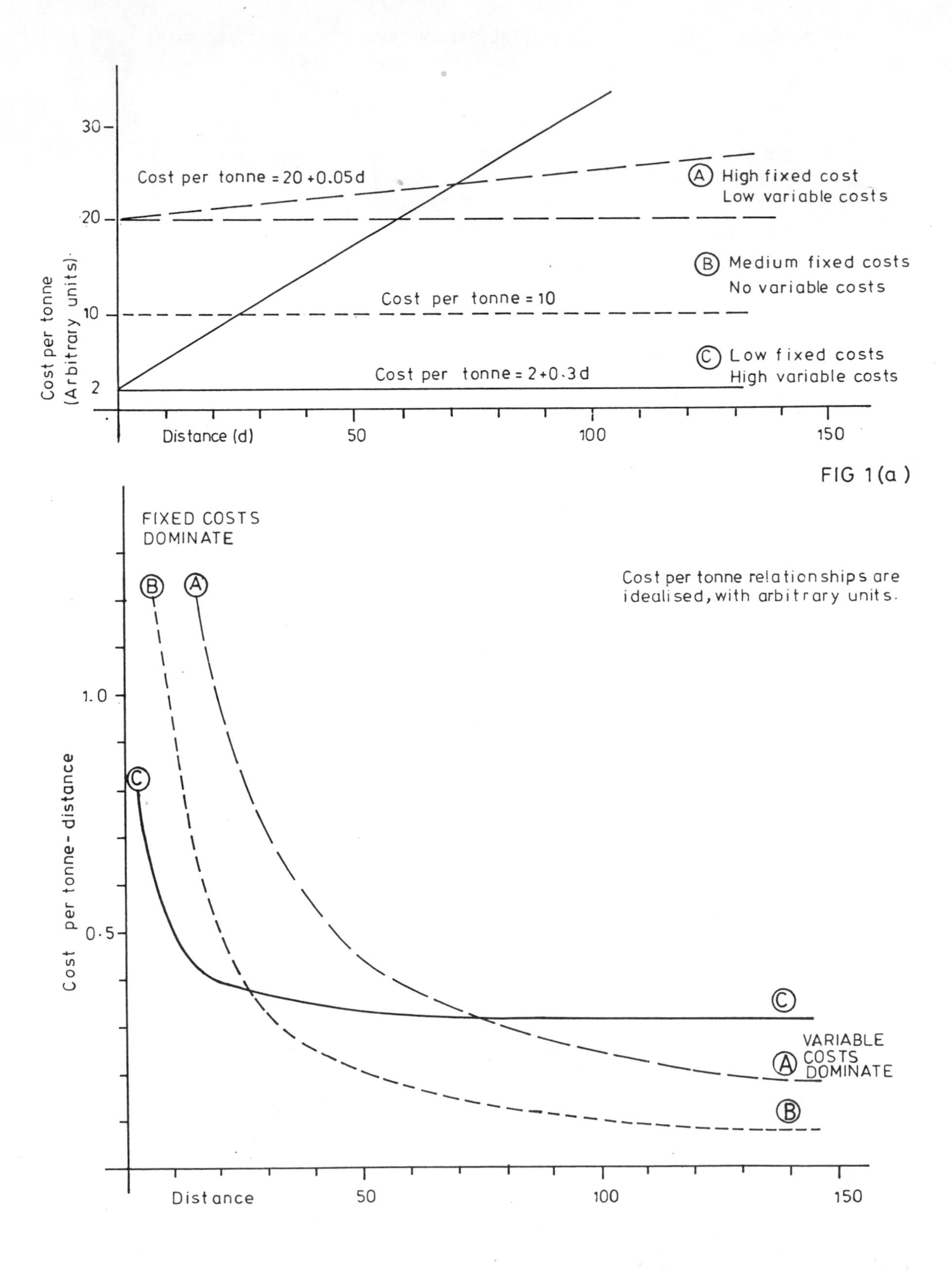

30
20
10
2
Cost per tonne (Arbitrary units)
Cost per tonne = 20 +0.05d
Cost per tonne = 10
Cost per tonne = 2+0·3d
A High fixed cost
Low variable costs
B Medium fixed costs
No variable costs
C Low fixed costs
High variable costs
Distance (d)
50
100
150
FIXED COSTS
DOMINATE
Cost per tonne relationships are idealised, with arbitrary units.
1.0
0·5
Cost per tonne- distance
VARIABLE
COSTS
DOMINATE
Distance
50
100
150

FIG 1(a)

FIG 1(b)

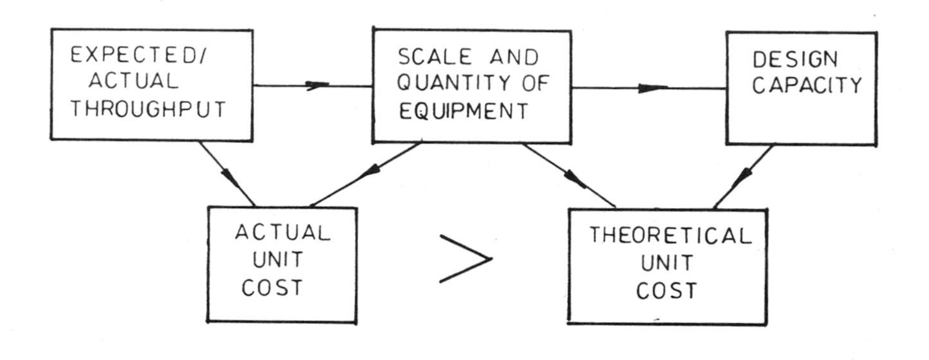

COST DERIVATION

FIG. 2

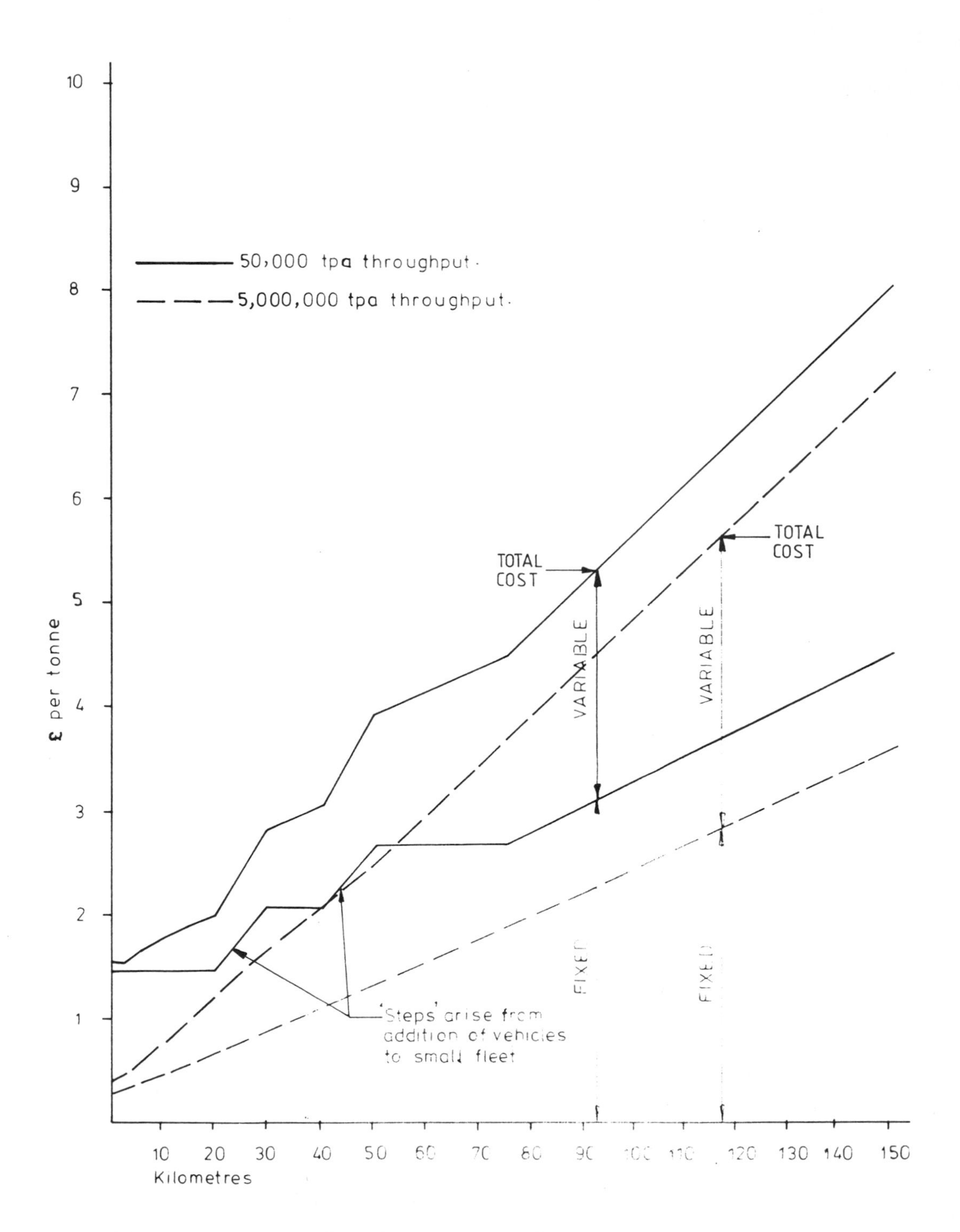

ROAD HAULAGE FIXED AND VARIABLE COSTS PER TONNE

FIG. 3

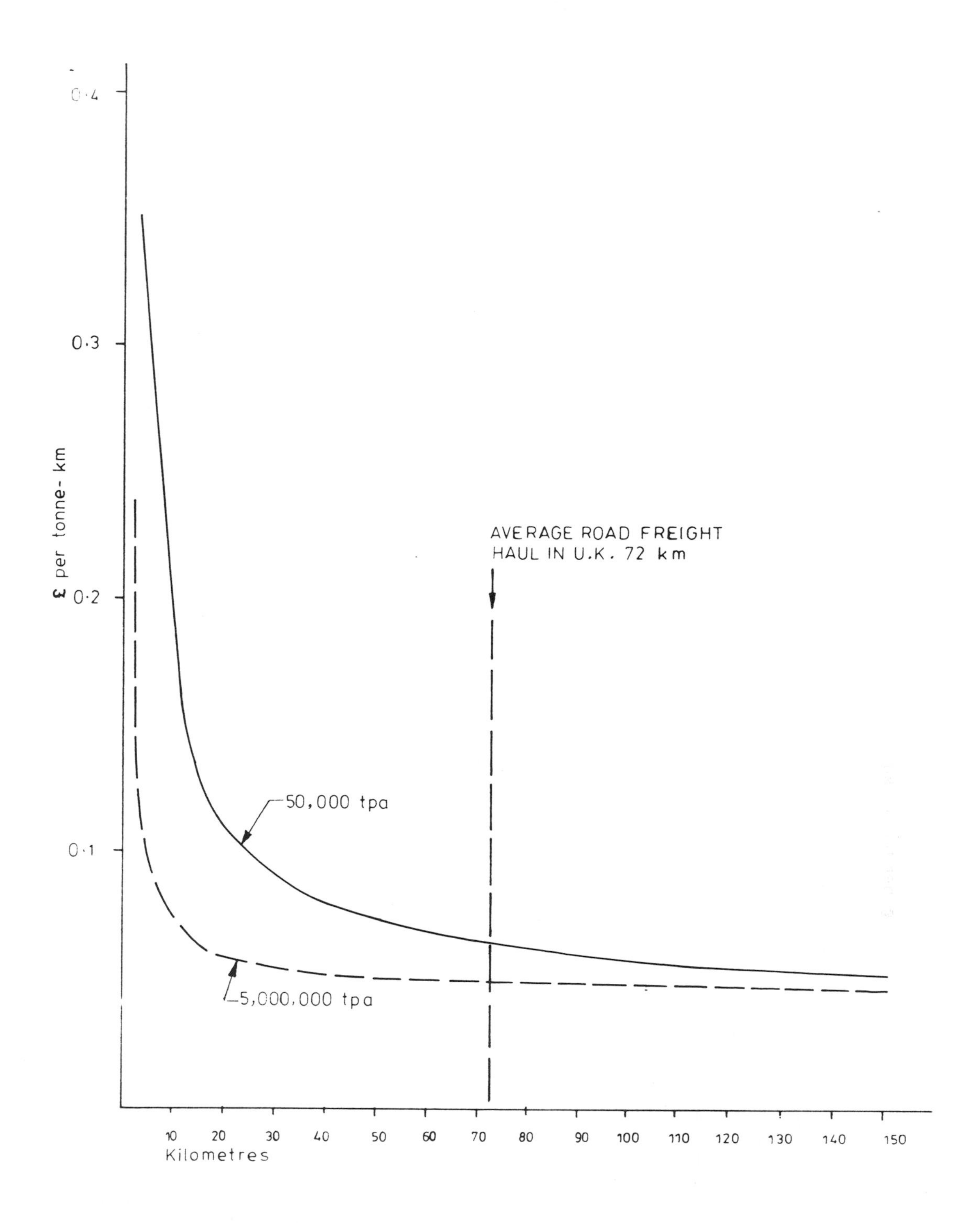

UNIT COSTS PER TONNE-KM FOR ROAD TRANSPORT

FIG. 4

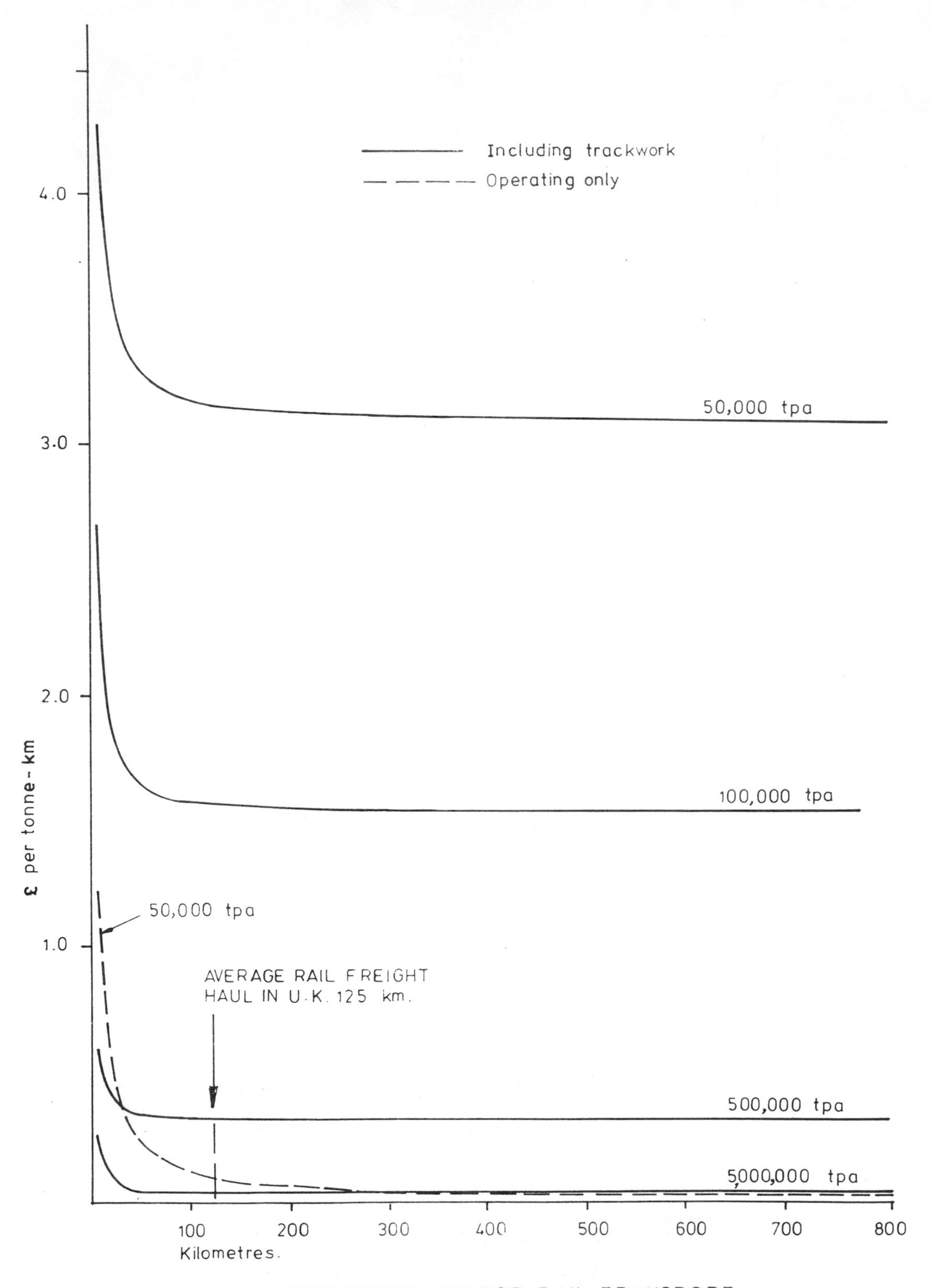

UNIT COSTS PER TONNE - KM. FOR RAIL TRANSPORT

FIG. 5

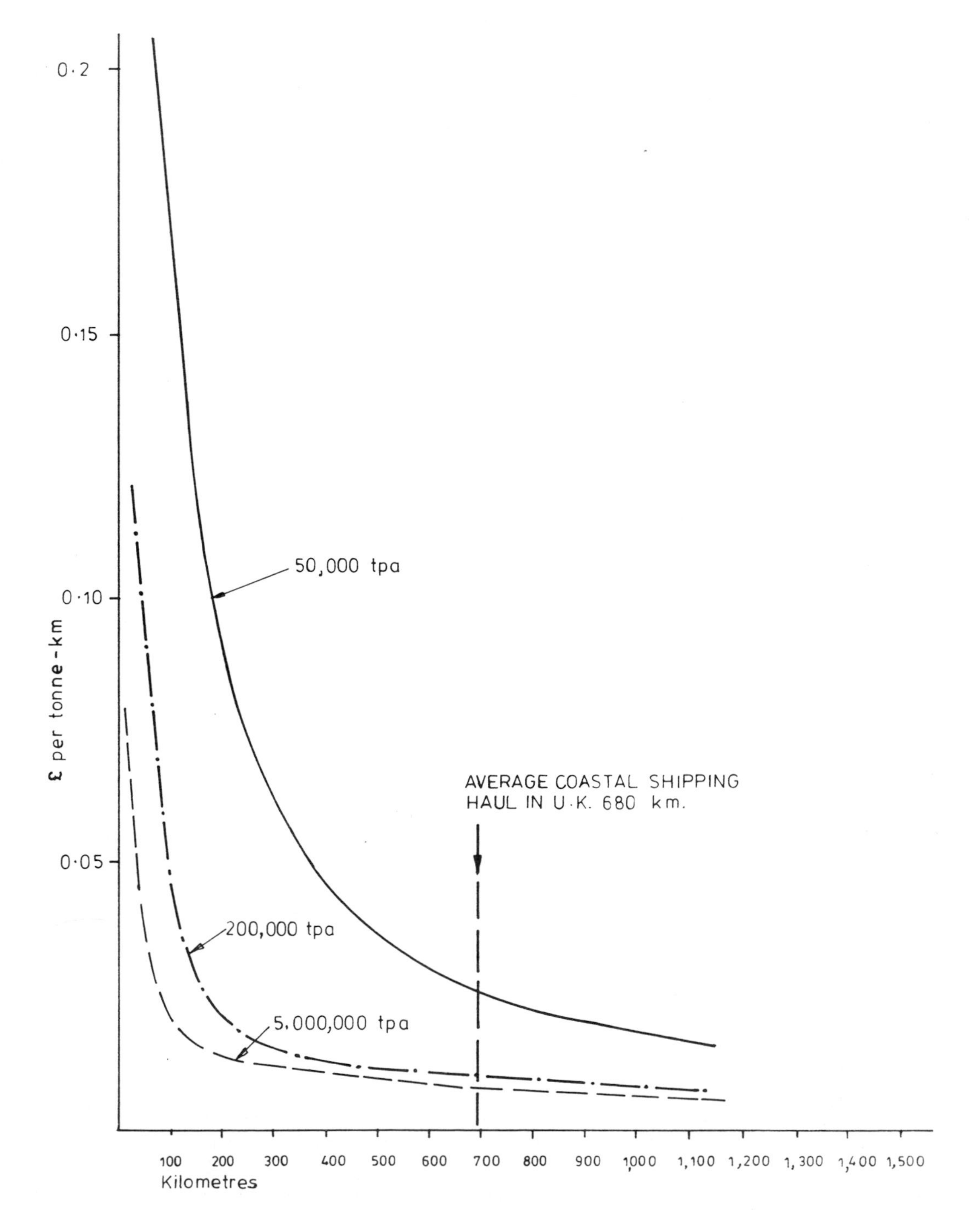

UNIT COSTS PER TONNE-KM FOR SHORT SEA SHIP TRANSPORT

FIG. 6

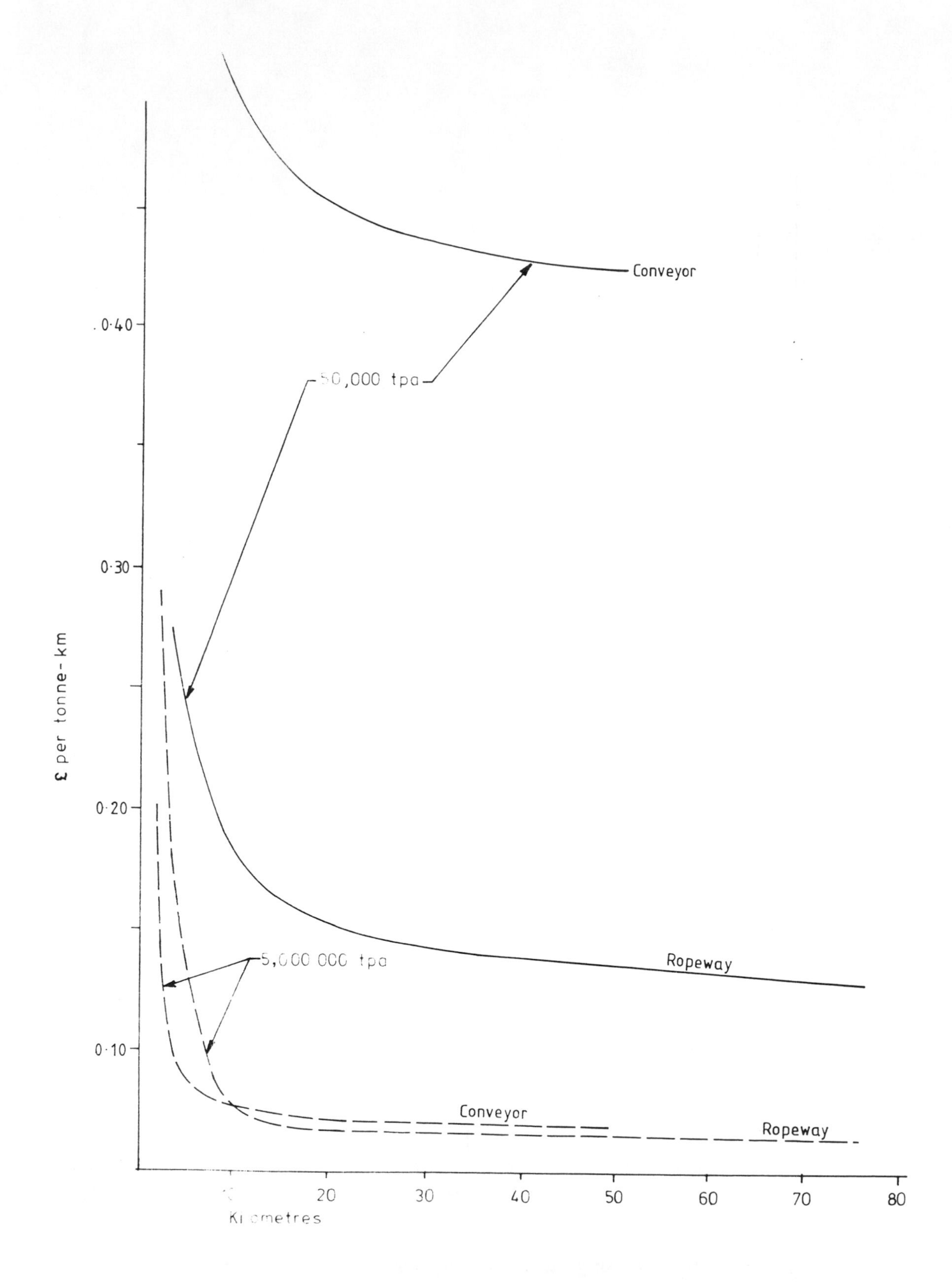

UNIT COSTS PER TONNE-KM FOR TRANSPORT
BY AERIAL ROPEWAY AND CONVEYOR

FIG. 7

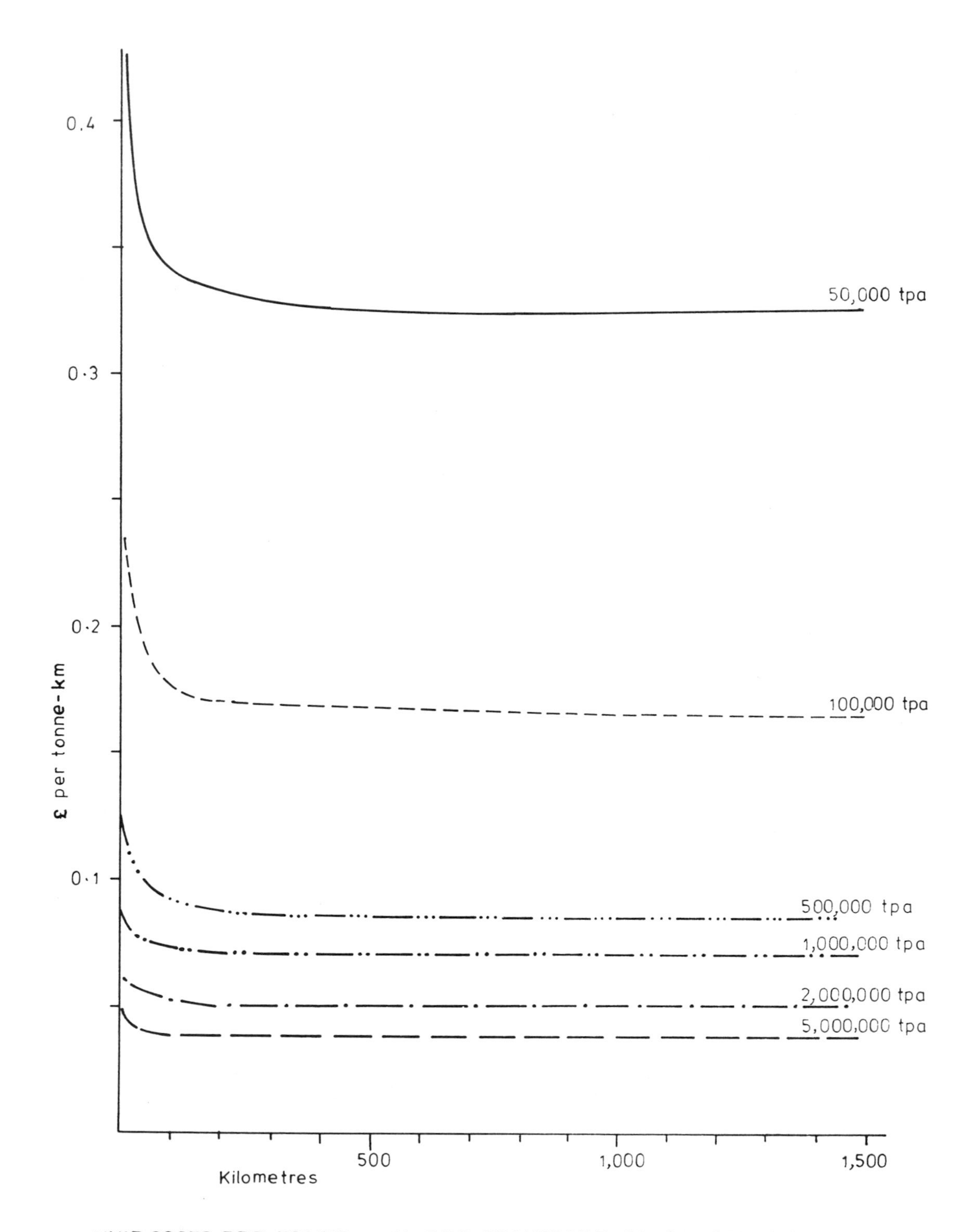

UNIT COSTS PER TONNE -KM FOR TRANSPORT BY SLURRY PIPELINE

FIG. 8

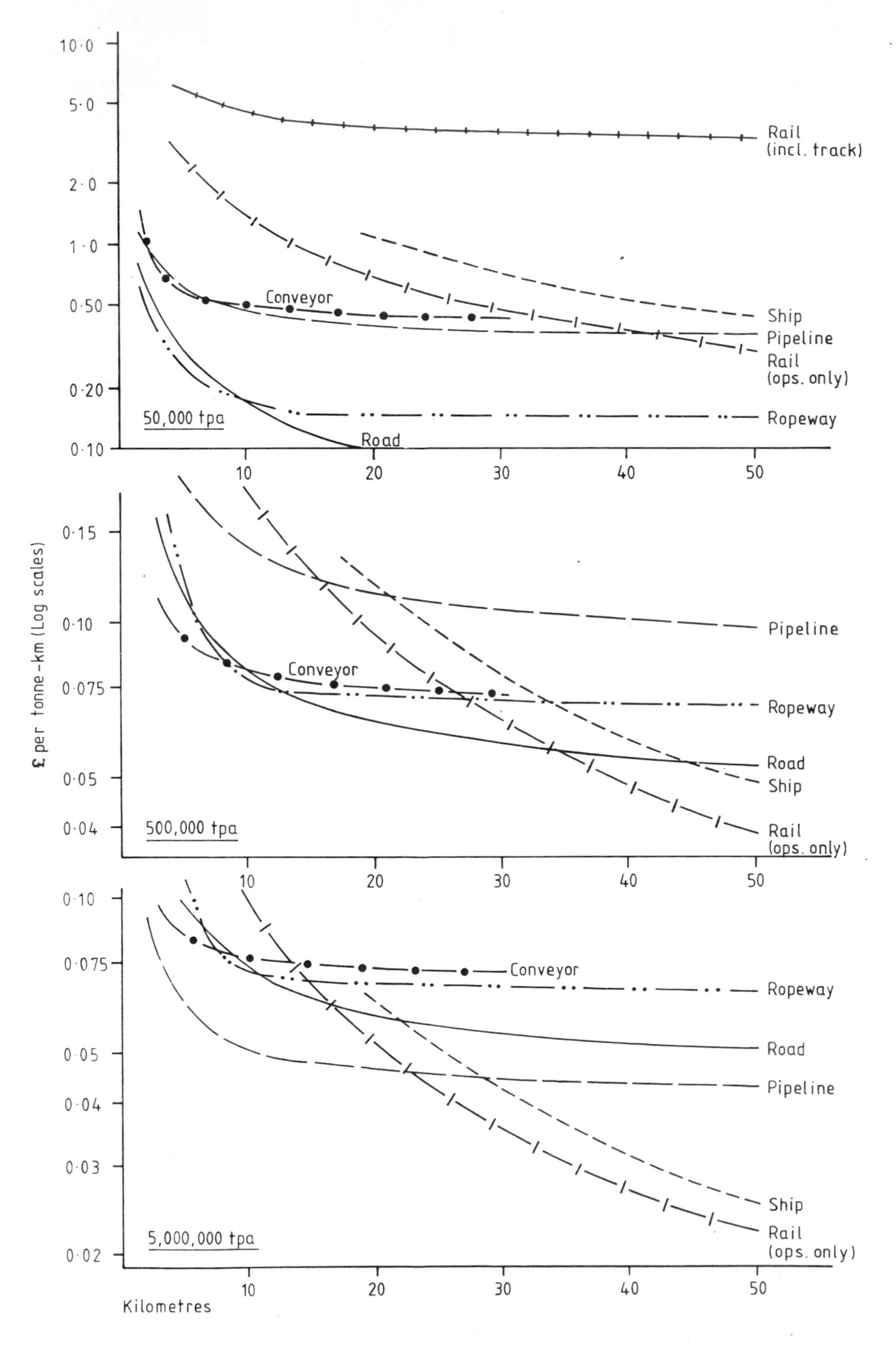

COMPARATIVE UNIT SYSTEM COSTS PER TONNE - KM
FOR DISTANCES UP TO 50 Km.

FIG. 9

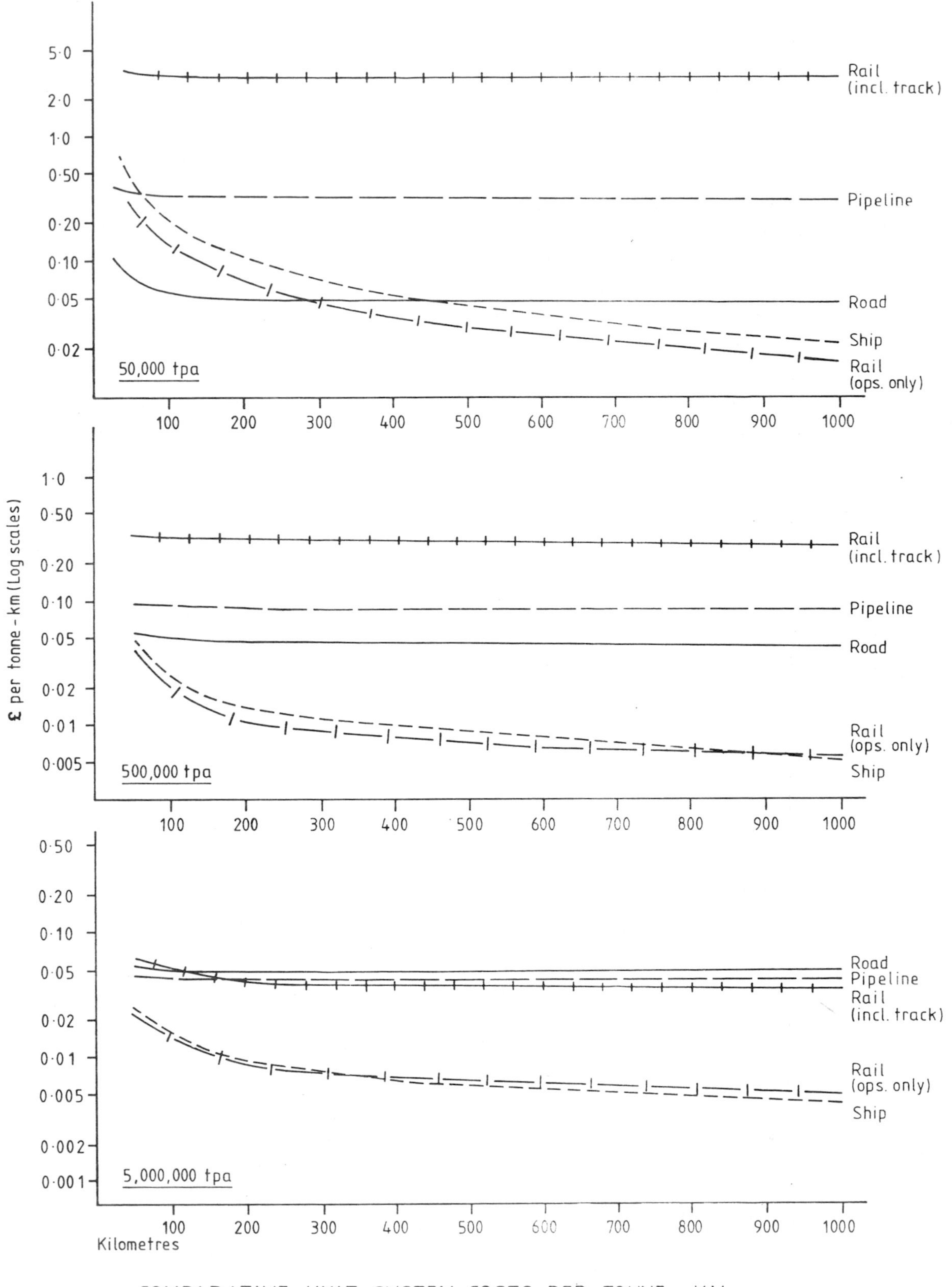

COMPARATIVE UNIT SYSTEM COSTS PER TONNE – KM
FOR DISTANCES FROM 50Km TO 1000Km.

FIG. 10

C418/021

Transportation of solids using vibratory troughs

G A GREEN, BSc, CEng, MIMechE and N HALL, BEng
Steetley Magnesia Products Limited, Hartlepool, Cleveland

SYNOPSIS The requirement to reduce production costs and meet changes in legislation relating to the working environment has made the identifying of replacements for standard belt conveyors essential. This paper describes some experiences in evaluating and implementing alternative equipment in the process industry.

1 INTRODUCTION

The transportation of solid materials in bulk, from between supply and discharge stations is a constantly recurring problem in industry, whether the solids are in the form of nuts/bolts or pharmaceutical powders. At Steetley Magnesia Products 180,000 tonnes of solid materials are moved annually, generally using conventional trough/belt conveyors with all their disadvantages of maintenance, spillage and environmental problems.

The materials transported around the Hartlepool Works vary from coal, petroleum coke, fine magnesia powders and magnesia pellets, to granular compounds of magnesite.

Today's business climate and the continually changing scene regarding environmental legislation, means production costs must be optimised by improving plant availability, reducing waste and eliminating those problems, such as dusts, that create unacceptable working conditions for employees.

The main process activity at Hartlepool is the manufacture of magnesia from seawater, which commenced in 1937.

The plant was the first of its kind in the world and remains the only operation in the U.K. Over the past 20 years the technology developed at Hartlepool has been utilised by many other manufacturers throughout the world. Their newer plants have highlighted the inefficiencies at the Hartlepool Works which are mainly due to age and the fact that original design made use of manual techniques to operate and control the process.

Many projects are underway in areas of construction materials, bearing technology, process control and, significantly, in evaluating other means of material transport rather than conventional rubber belt conveyors. Some of the methods being evaluated are pneumatic blowing, air slides, closed belts and vibrating troughs.

This paper has been prepared to inform on the uses, problems, advantages arising out of the experience gained from using vibrating troughs at Hartlepool Works.

2 THE CASE FOR VIBRATING TROUGHS

Any plant which operates continuously for 24 hours per day, 365 days per year, cannot suffer penalties because of engineering inadequancies. Breakdowns of a minor nature, e.g. bearing failure on conveyor drums, if on the main process routes, can be costly in terms of lost capacity, non-conforming product, increased production costs, apart from the actual cost of the repair. At Hartlepool the reliability of standard trough belt conveyors was poor with the degree of spillage and dust fall-out considerable and rubber belt wear above the norm. Failures were also more frequent than need be, thus demanding a major improvement in the performance of equipment used for transportation.

Taking the above statements in turn it is possible to offer comment as follows:

i) Reliability generally is improved if the number of working parts of an item are reduced and in the case of vibrating troughs vs. belt conveyors this is the case. Historically, in many industries, vibrating troughs have better records of reliability over standard conveyors and this was, therefore, seen as a positive advantage.

ii) Spillage and dust fall out is minimised, as there is no rotary motion and transfer chutes can be installed much more efficiently, thus in many instances, eliminating or considerably reducing this problem.

iii) Conveyor belt wear is totally eliminated as there is no need for a travelling "carrier" as the material itself is made to travel over the "carrier" by the induced vibration.

iv) Failures in the main, once design has been optimised, are confined to the motors used for creating the vibration. The optimising of the design, however, is most important as errors here can cause severe problems, some of which will be covered later in the paper.

3 BACKGROUND TO VIBRATING TROUGHS

The vibratory conveyors to be discussed in this paper have their origins in the simple manual screening systems of activities such as panning for gold, which then progressed to the automated screens developed for the cement and grain industries.

However, the use of vibration in bulk solid materials handling has only really developed this century, mainly due to the engineers' natural fear of this phenomenon. Most engineering practices are directed towards the elimination of vibration and the dreaded occurrence of resonance has often destroyed many machines and structures.

The evolution of vibratory conveying and the associated drive systems are discussed in more detail in G.D.Dumbaugh's review for the "Journal of Powder and Bulk Solids Technology" (Ref. 1).

Modern day vibratory conveyors are a logical development from the old "swing type" or shaker conveyors, which are usually mounted upon swinging arms and operate at a low frequency and large amplitude.

The theory of moving bulk solids by vibratory action is that particles of the material are conveyed in a series of "hops". The particles are carried forward and upward with minimal frictional losses, in the forward stroke, but they then fall near vertically when the direction of the conveying body's movement is reversed. The particle has, therefore, occupied a new position along the body and it has effectively moved forward. Each movement forward is known as a cycle and the distance moved is directly related to the drive stroke. The operating frequency is usually measured as the number of cycles per minute. Figure 1 shows this movement.

Figure 1 : Illustration of a Coil Spring Vibrating Trough Showing Bulk Solids Movement

The modern vibrating conveyors use an extension of this principle but, due to the higher operating frequencies, the particles of material are projected forward and upward at accelerations of 4-5g. When the conveying body is at the peak of its velocity, the body is decelerated, again, at 4-5g, but the particles of material in suspension deccelerate under normal gravity.

Theoretical claims relating to the specific frequencies, amplitudes and direction of amplitudes are calculated from fundamental equations of particle dynamics. Experimental comparisons give close results, but they tend to depart, with increasing amounts as the depth of material increases.

The basic elements of vibrating conveyors are very simple and comprise

a) The member containing and transporting the bulk solid, which is usually referred to as the trough, pan, deck or screen body.

b) A drive system to generate motion, and

c) Supports, in the form of "soft" mounting springs used to minimise the transmission of dynamic forces generated in the supporting structure. Most designs at "isolation springs" will expect to remove 80% of transferred vibration. Thes supports can be from below or above.

4 DRIVE SYSTEM

The capabilities of any vibrating conveyor are directly related to the type of drive system employed. The comparative review by Dumbaugh traces the history and compares several of the types of drives used over the years by a variety of companies.

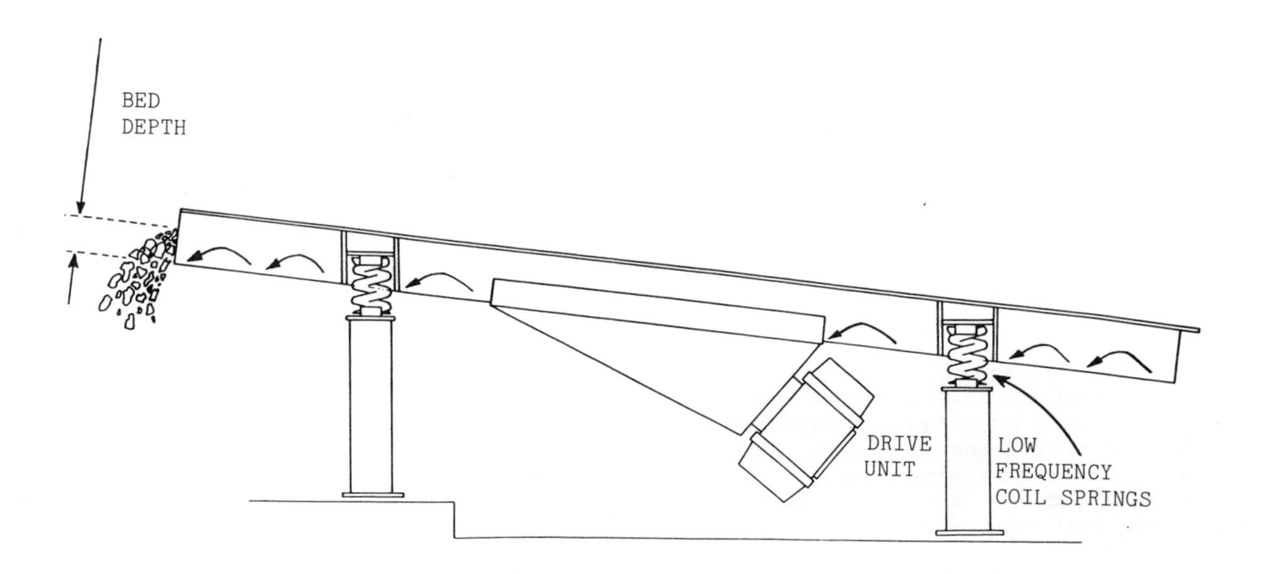

The type used by "Vibraflo" is not specifically covered but its basic principle is that of a vibrator motor with twin rotating masses. They rotate in opposite directions and are, in all instances, mounted in some form of resilient arm; giving naturally synchronous rotation. In this manner the forces acting along the plane at the body are combined and the forces acting perpendicular to the body are cancelled out. The result is a uni-directional force being applied to the vibratory conveyor body. This is illustrated in Figure 2.

Figure 2 : Action of Forces Within The Vibrator Motor

Forces 'A' being opposed are cancelled out

Forces 'B' combine to give uni-directional force

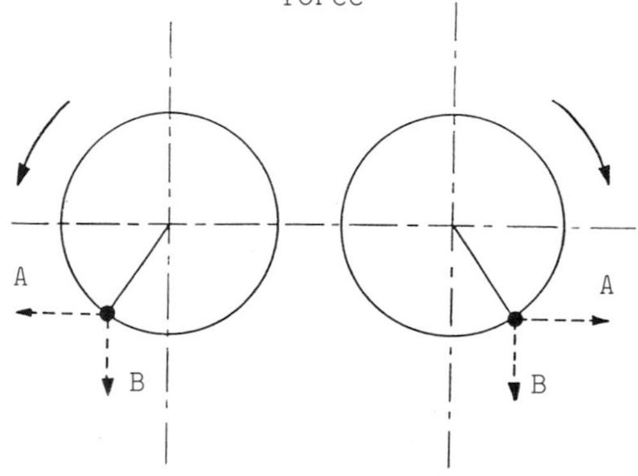

The original application of this principle was first claimed in 1957 by a German Company: Fiege & Joest, in the USA (Ref. 2).

Vibraflo was formed in 1968 and manufactured vibratory conveyors utilising the F&J technology.

5 TYPES OF VIBRAFLO CONVEYORS

a) Coil Spring Suspended Type

This type of conveyor is as shown in Figure 1 and involves a conveying body simply mounted on low frequency coil springs from the floor or supporting structure, with a uni-directional force applied at around 35°. Due to the soft mounting, the rotating masses can be rigidly fixed to the body, natural synchronisation being achieved through the mountings.

This type of conveyor was used for the first vibrating troughs to be utilised at S.M.P.L. but it does have limitations.

The length of the conveyor is limited by the "beam" strength of the body. Beyond certain lengths the inherent deflection of the trough causes damping, changing the direction of amplitude and the loss of conveying movement. In some extreme conditions it can even be reversed.

To achieve an acceptable beam strength the body has often to be oversized in relation to the required duty in order to achieve a reasonable length, causing secondary problems of support and cost.

b) Lath Mounted Base Type

This was the next development and the original system incorporated a trough being mounted on stiff spring laths which are located on a steel base frame. The stroke action can be applied by a drive unit attached at any reasonable point along its length, but the nearer to the centre the better.

Figure 3 : Basic Lath Mounted Base Conveyor

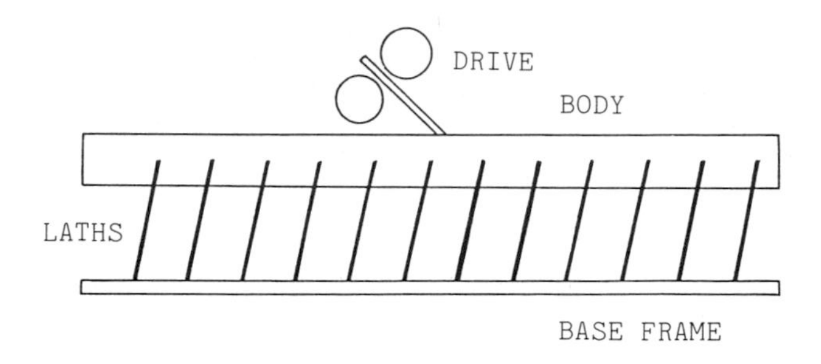

This type of conveying structure, as opposed to type (a), results in lengths of up to around 20m being achievable, but there are subsequent disadvantages.

i) The base frame has to be securely attached to a extremely rigid foundation, because the stiffness in the laths does not dampen the transmission of generated dynamic forces through the base.

ii) Due to this transmission of vibration and the need for a rigid structure for mounting, they could rarely be utilised at high levels. Resonance was a frequent problem, because the operation of the conveyor often increased the speed through, or ran near to, the natural frequency of the supporting structure, sometimes with dramatic effects.

iii) The application of the drive unit tended to cause the conveyor body to bow, pushing extra loading onto the support laths within the area of attachment. This would give variation in the body movement and subsequent conveying velocities. Hence in this form its industrial use is highly limited and further development ensued.

The result of progress into a base type vibratory trough led to the Compensated Base Type Vibratory conveyor, which has wide use at the Hartlepool site.

The main difference from that of the conventional base type conveyor is the substitution of the original fabricated base frame by a deep section, large mass base. This is then effectively supported and isolated from the transmission of vibration from the floor or supporting structure.

A further modification was to apply the drive in a horizontal place at either the front or rear of the unit. (See Figure 4)

Figure 4 : Compensated Base Conveyor

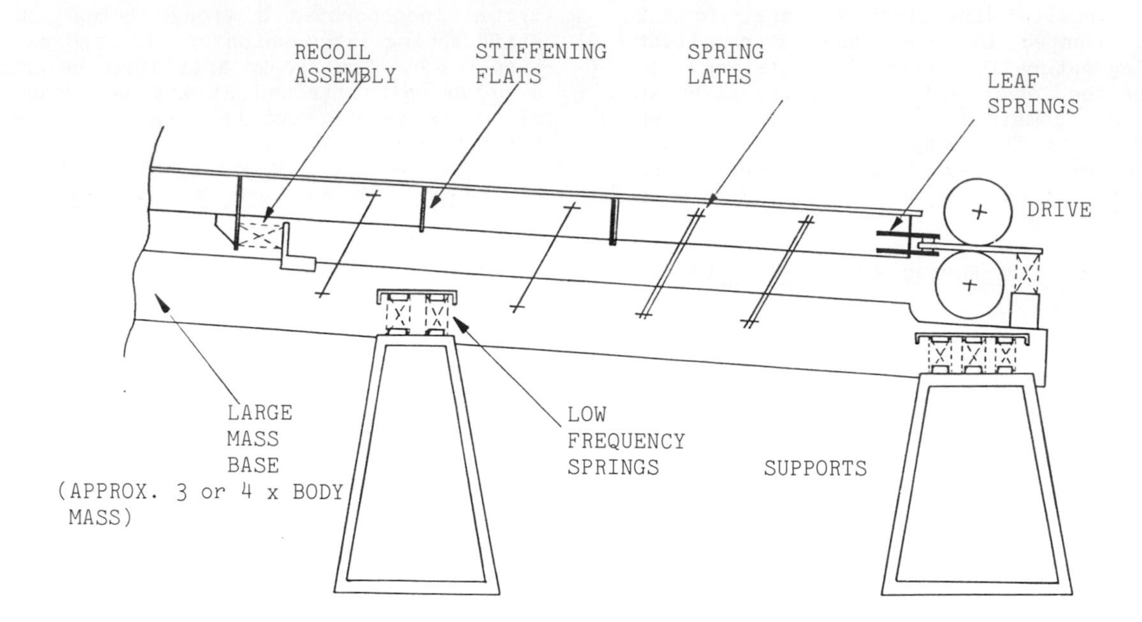

Lengths of up to 35m have since been achieved and the isolation of transmitted vibration is such that during tests the machine does not require fastening down to the floor, merely sitting on the base plates of the low frequency springs.

This type of unit can then be adapted over an increased range of applications, including spanning adjacent structure, making benefit of the beam strength of the base to minimise the number of supporting low frequency springs.

6 CHOICE OF CONVEYOR BODY

The body of the vibratory conveyor has two main functions:

a) To accept and correctly distribute the applied forces from the drive unit.

b) To handle at the desired rate, the material for transportation.

If point (a) is examined, it can be shown that the body is subjected to an alternating stress, which is at a maximum at the drive, diminishing to zero at the no drive end.

Figure 5 : Stress Distribution Along Conveyor

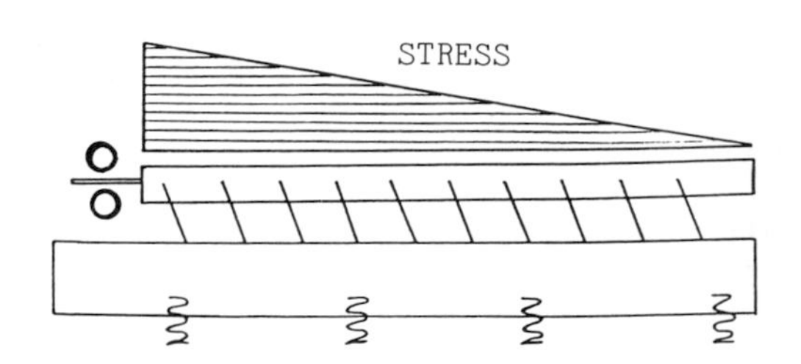

The body with the lath supports, under normal operation, is not prone to any appreciable bending moments. Thus, in the selection of the body section, the main requirements is for a resiliance to the alternating stress along its longitudinal axis.

The beam design best to accommodate this stress is obviously a tube, although rectangular or 'U' forms can be acceptable within certain units.

So far as point (b) is concerned, the nature of the material has to be ascertained, information being derived from experience and tests. Dry, free flowing materials present little risk of buildup and so an enclosed body is acceptable. This is desirable when the material is hazardous, as with many chemicals and dust type products, or when the process must be free from spillage or contamination.

The conveying rates also depend on the area of the body. If a large capacity of material is to be conveyed within a limited space, then a square or near square rectangle will offer the greatest cross sectional area.

This is because the circular tube can only be effectively filled to just over the half of its depth. Beyond that, the material is being projected over a progressively diminishing area and a subsequent loss in velocity occurs. This loss of area does not occur in a rectangular section, providing a free space of 25mm is allowed.

7 SPECIFIC APPLICATIONS AT STEETLEY MAGNESIA PRODUCTS LIMITED

7.1 First Application

Six horizontal rotary kilns are operated on the site, all of which can be fired with pulverised coal, petroleum coke or mixtures of both. The two fuels are delivered in bulk, stored, milled and classified as required. Any failure to supply fuels to the kilns precipitates two immediate penalties, viz: loss of production, or production at more cost due to the use of higher value fuels. The materials handling equipment through which all the fuels were handled, was prone to spillage, unreliability and thus a source of high maintenance expenditure. A decision to replace the equipment was taken with the following criteria used for the basis of selection:

a) Fewer items of equipment.
b) Enhanced reliability of units, i.e. few working parts.
c) Saving of space.
d) Improvement of working environment.
e) Elimination of belts.
f) The equipment would need to both transport and elevate.

Once it was accepted that material could be elevated within the limits of operation of vibrating troughs (12° incline) it was agreed the new system comprising of a coil suspended unit, bucket elevator and a lath mounted compensated base unit would be ordered and installed.

A schematic of the existing equipment and new equipment is shown below.

Figure 6 : Comparative Illustration, Showing Solid Fuel Handling Before and After the Use of Vibratory Conveyors

Following some tests with material mixes and prior to the commencement of major installation work the first feed conveyor, which was located in a pit 15 feet deep, was replaced with a trough to establish its suitability in a working situation.

This first phase was accomplished with few problems and a plan to remove the existing equipment and replace with the new system within a period of 72 hours was initiated.

Having overcome the normal difficulties which arise in jobs of this nature, the new system was generally installed to plan, and commissioned immediately. All initial signs were good and early feelings of euphoria grew, but then the fuels being fed to the system changed from being dry to damp. This caused numerous problems, all of which precipitated circumstances far worse than had previously existed.

a) Original layout, with reversible CV belts at positions 1, 2 and 3

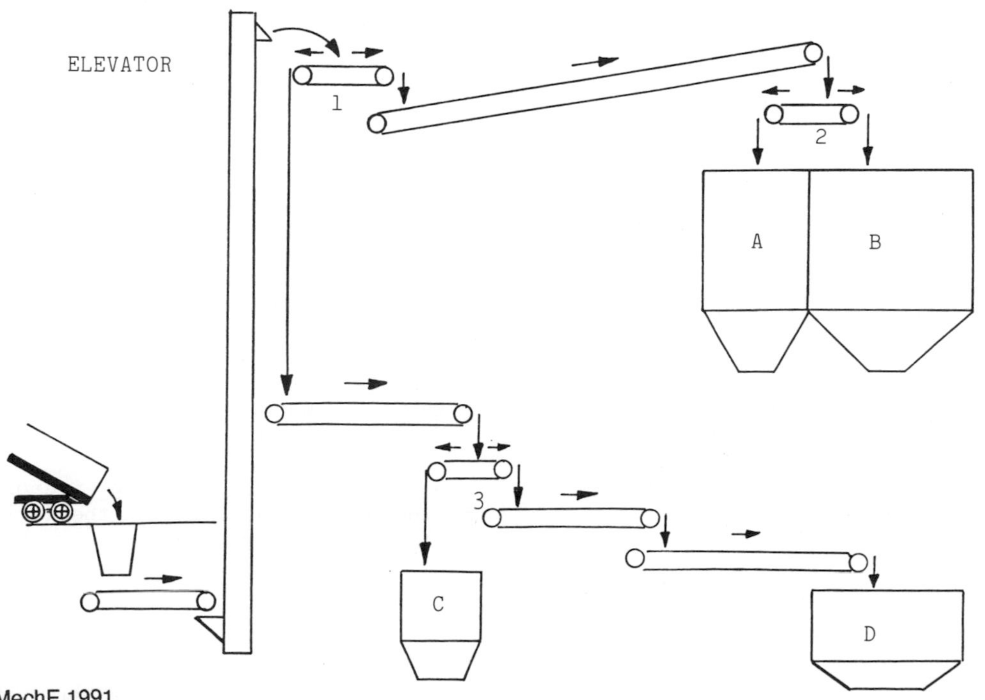

b) New layout, with "drop outs" at positions 1, 2 and 3

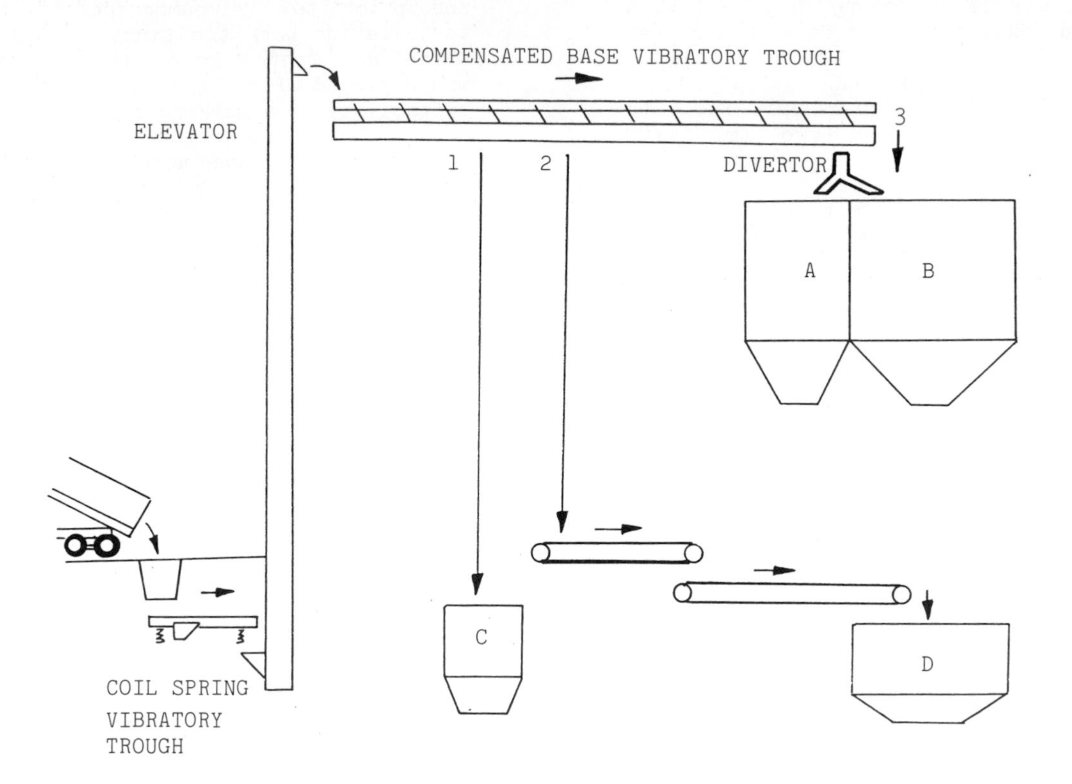

Firstly it became apparent that throughputs of the system became significantly less and this was found to be due to a layer of hard material building up on the bottom of the troughs. The vibrating action of the trough was causing a separation of fine and coarse material such that fine material ran underneath the coarse and due to its damp consistency was adhering to the surface and over a relatively short period of time hardening off. This had the effect of reducing the cross-sectional area of the trough and thus its transfer capacity was reduced. This caused a further degenerating response from the Operators who overfed the system to achieve higher throughputs. This resulted in severe spillage problems, due to dampening of the vibrational forces, which were far worse than before because material would spill from the open troughs and "rain" down through the supporting structure.

Attempts were made to eliminate this build up with the use of low friction coatings both in the form of sheet (Perplas) or applied coatings (polyurethane paints) none of which were very successful. In the final analysis Operators have had to be instructed to scar off the buildups as a routine duty, this only being necessary when the coal is damp.

The next problem to identify itself was the failure of the flap valves, initially by not sealing, which resulted in leakage (and ancilliary obstacles), and then complete breakdown of the valve due to vibration. Although this problem proved difficult to solve it was solely due to engineering design and soon remedied once all factors were taken into account.

A series of weaknesses were then identified as the equipment clocked up running hours, viz:

i) Severe cracking in the motor bedplate
ii) Spring failures
iii) Cracking in the body of the trough
iv) Cracking in the compensating base

A common cause of the cracking was due to the techniques of manufacture, albeit the supplier had not experienced similar faults with equipment delivered to numerous other customers. Methods of manufacture had created areas of high stress brought about by flame cutting, welding and sharp changes in contour. It must, however, be said that all of the above had been aggravated by the peculiar working circumstances when the overfilling of the troughs had occurred.

A further contributing factor to the widespread cracking was the repositioning of a permanent magnet.

The design of the vibrating trough incorporated a section fabricated from stainless steel to facilitate the installation of the magnet for the purpose of extracting tramp metals from the material being conveyed. When the repositioning of the magnet was carried out it was placed above a section of the trough which was fabricated from normal carbon steel which, over a period of time, itself became a magnet. In fact, the field of magnetism was apparent over a larger area and included the spring laths and the compensated base. The magnetism had the effect of combating the vibratory forces and thus loads and stresses experienced by the unit were completely different to those it had been designed to withstand.

A number of modifications have been undertaken to prevent a repetition of these failures which, to date, have all proved satisfactory. In summary this has included:

i) the redesign of the motor bedplate
ii) the complete redesign and manufacture of discharge flaps
iii) the extended use of stainless steel sections to remove problems of induced magnetism

The operation of vibrating troughs in this application fell far short of our expected goals, although enthusiasm remained undaunted and further uses for the equipment was sought in other areas of the process.

7.2 Finished Products

The initial use of vibrating troughs to transport coal and their eventual success resulted in the replacement of conventional rubber conveyor belts in other areas. The next logical progression was to that of finished products.

At Steetley the Kilnhouse is fairly confined for space, with the presence of six rotary kilns, six coolers and the associated feed, finished product, fuel, exhaust gas ducting systems and other ancillary equipment.

The finished product systems invariably involve a large use of conveying equipment and they are, unfortunately situated for the greater part in pits and restricted access areas. For this reason spillage is costly both in terms of cleaning and in direct monetary loss. (The quality assurance of our products insists upon spillage being classified as waste).

Any failures in the FP conveying also has consequential effects on the operation of the rest of the plant.

The first application of a coil spring vibrating trough in the FP area was to replace a rubber belt taking chrome clinker product from our No. 1 Kiln cooler. The products is hot (200°C+) and abrasive, and burning belts often caused the belt to fail, meaning the cooler and kiln had to be shutdown causing production losses.

The product is of a dry, circular, pelletised form, lending itself to vibratory transport and the steel construction of the trough meant there were no problems with heat.

The trough has run virtually trouble free for two years with the only real problems being motor failure. This was caused by flooding due to cooling water and it was in no way related to the motor's performance.

The trough runs at an angle of around 7° to take it from the cooler discharge chute to an inclined rubber skirted belt, by which time the product has cooled sufficiently to prevent burning. There have never been any wear problems, with the effective "bouncing" of the pellets causing little abrasion.

A vibratory trough was then installed from a second cooler, which transports Spinel, our most expensive product, and thus a most critical area to ensure a trouble free maintenance environment. Again, the product is abrasive, hot and free flowing. The pellets are, however, larger and heavier and the drop height of cooler discharge to the trough has caused wear upon impact. This has, however, been resolved by replaceable liner plates and a modification to the chute work should eradicate the problem.

A third trough has now been installed on our No. 5 Kiln finished product transport system, yielding equivalent benefits.

7.3 Feed Systems

The next logical progression was to utilise vibrating troughs on our feed systems. The feed material to No. 1 Kiln is of a pellet nature, formed from compacting prefired magnesia powder.

The original system was to feed into the kiln via. a rubber conveyor belt. The pellets are discharged into a hostile environment and they tend to have a quantity of fines with them, causing environmental problems.

Failure of the rotary valve on the feed system results in hot gases coming up the spout causing inevitable burning of belts. These belts were beginning to fail on a weekly basis and so an improvement was imperative. The whole system was high on cost, time and maintenance, not only in replacing belts but with the associated equipment of bearings, drive drums etc.

A vibratory trough of a lath mounted compensated base type was installed with high expectations, but the design had a severe weakness. The attempt to abolish the environmental effects of dust had led to the choice of a totally enclosed rectangular conveyor body. This immediately presented a major problem. The inherent phenomenon of a vibrating action meant that the fines of the pellets separated out (stratification) and their hygroscopic tendancy resulted in the 8" body trough being totally blocked off within 8 hours. The Operators could not get at the material to clear the blockage so an open rectangular trough was soon installed. However, the operation of the trough became labour intensive with continual cleaning being required. Against all advice and technical understanding a loose rubber lining was installed. The rubber lining vibrated at a different frequency to the body and no powder buildup occurred. This is a great example of the spirit of adventure in engineering, and our "experiment" prompted Vibraflo into an investigation which discovered a design of a rubber bottomed trough from several years earlier, on a "sticky" product. This has since been utilised on our chrome feed systems.

Figure 7 : Illustration of Rubber Lined Trough

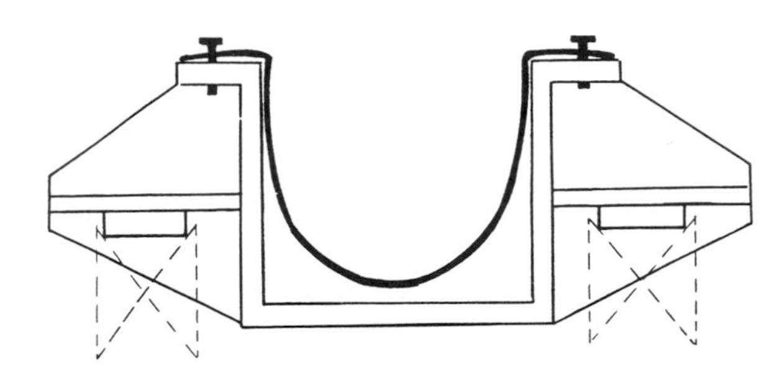

7.4 Spinel System

The culmination of our experiences gained with vibratory troughs at Steetley has recently lead to a large capital project to install a complete new finished product system for our Spinel products, utilising Vibraflo conveyors and bucket elevators.

The choice of conveyors was based on:

a) The material was free flowing, dry and of low capacity.

b) The units were fairly long (around 19m) and therefore required a body capable of accepting the maximum stresses of the size of drive unit applied.

c) The installation was at a high level in a relatively light gantry structure, where the transmission of vibration could not be tolerated.

From these factors the machines selected were 350 mm Ø tubular compensated base type machines. The 350 mm size being selected upon strength parameters and being oversized for the conveying duty required. The high level application of the conveyors, above the kilns, was accommodated in gantries not much heavier than would have been required for conventional belt conveyors.

A modification to the conventional design was to apply one rear mounted drive followed by one front mounted drive along two machines to work in tandem without a loss of height at the transition point. Figure 8 shows a schematic arrangement of the Spinel layout.

8 CONCLUSION

The real benefits obtained relating to fewer moving parts, improved reliability, reduced spillage, lower dust emissions, resulting in decreased maintenance costs, less wear and higher availability has made the project most rewarding. It is the intention of the Company to pursue further uses of such equipment in other process areas.

9 REFERENCES

(1) **DUMGAUGH, G.D.** A comparative review of vibratory drives for bulk solid handling system.

Journal of Powder & Bulk Solids Technology 8(1984)2, 1-17.

(2) **JOEST.** Improvements in or relating to Vibration Drives.

Complete specification in U.S.A. May 1957. Ref. AWB/dav/RPH/KX/C17/5.

10 ACKNOWLEDGEMENT

Mr.B.Scarth, Vibraflo, Pudsey, Yorkshire.

Steetley Magnesia Products Limited

Figure 8 : Schematic Illustration of the Spinel Finished Product System

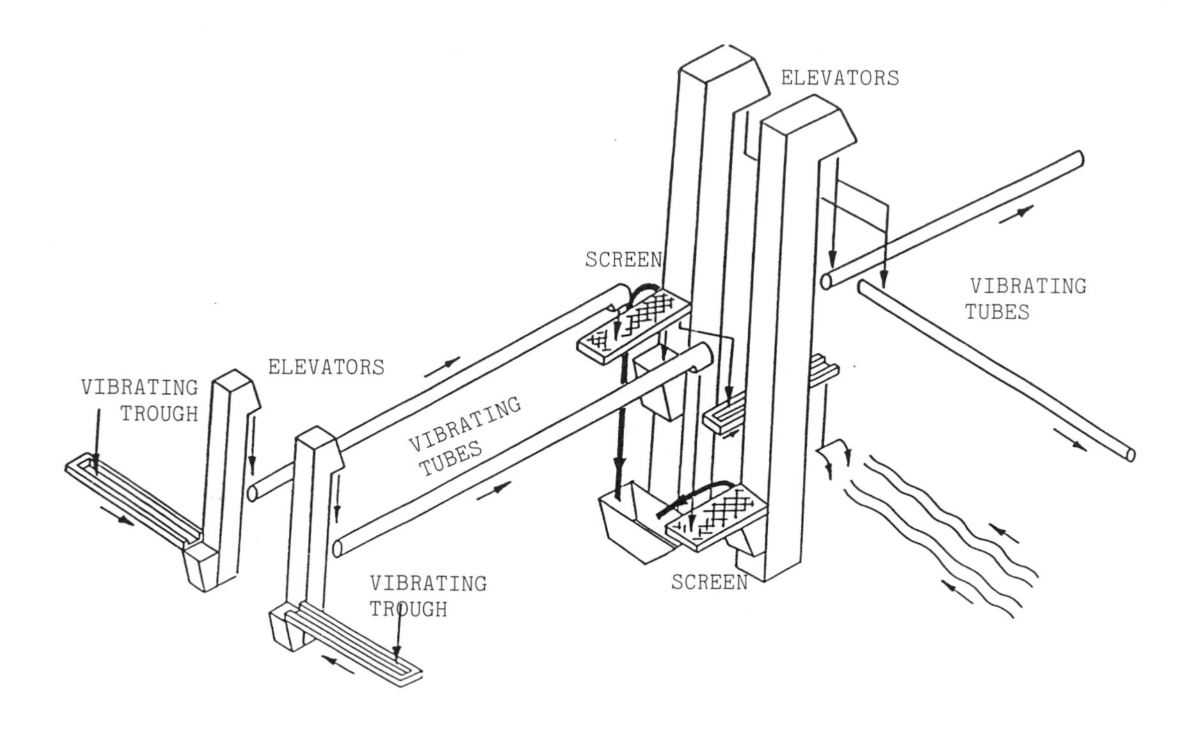

C418/058

The study of the relationship between belt speed, pulling force and the motor power of the belt conveyor

G-B LI, MEng
Mechanical and Electric Department, Lianyungang Chemical and Mining College, Jiangsu, People's Republic of China

SYNOPSIS Belt speed, pulling force and power of motor are three important parameters for a succesful design of belt conveyor. This paper presents analysis of some factors which influence belt speed, pulling force and power of motor of conveyor. According to the regression theory, their mathematical expression has been carried through. From this expression, reasonable belt speed has been put forward at which the pulling force and power of motor are minimum. The results calculated by this approach agree with the practice.

NOTATION

W_1, W_2 resistances of troughed and return belt
F part resistance
q, q_0 material and belt weight per unit length
q_1, q_2 troughing and return idler weight per unit length
L conveyor length
β angle of conveyor
ω motion friction coefficient
Q mass-flow rate
V belt speed
B belt width
V_0 reasonable belt speed
G_1,G_2 troughing and return idler weight
l_1,l_2 troughing and return idler spaceing
K material cross-sectional coefficient
γ unit weight of material
C belt angle coefficient
a,b coefficients
W total resistance
P pulling force
N power of motor

1 INTRODUCTION

With the increasment of vast quantitics of bulk solids transported over long-distance today, more and more attentions will be paid to the high speed conveyor. In type-chosen design of belt conveyor, choice of belt speed is often encounted at the begining and should be dealt with seriously.It is known that if belt speed is higher,pulling force of pulley is smaller but power of motor is greater.On the contray, pulling force is greater and power of motor is smaller with belt speed getting lower. But so far, the analyses of their exact relations have not yet been undertaken well. In order to obtain reasonable belt speed at which the pulling force and power of motor are minimum under given conditions, it is necessary to analyse their relations first of all.

2 DISCUSSION ON PULLING FORCE AND POWER OF MOTOR WITH REGRESSION THEORY

2.1 Various resistances of belt conveyor

As we know, belt motion resistance includes the following ones:

(1) Resistance of troughed belt is

$$W_1 = (q + q_0 + q_1)L\omega \cos\beta \pm (q + q_0)L\sin\beta$$

(2) Resistance of return belt is

$$W_2 = (q_0 + q_2)L\omega \cos\beta \mp q_0 L\sin\beta$$

(3) Part resistance

Part resistance F includes resistances of clearing device, feed chute, feeder, pulley, etc.
So, total resistance is

$$W = W_1 + W_2 + F$$

$$= (q + 2q_0 + q_1 + q_2)L\omega \cos\beta \pm qL\sin\beta + F \quad (1)$$

In type-chosen design, L, β, Q, ω are generally given parameters, so the key to analysis is to find the relation among q, q_0, q_1, q_2, B and V.Some of the basic expression are given in the following:

2.2 Expression of material weight and belt speed

It is listed as follows:

$$q = \frac{Q}{3.6V} \quad (a)$$

2.3 Expressions of idlers weight

It may be calculated as follows:

$$q_1 = \frac{G_1}{l_1} \quad (b)$$

$$q_2 = \frac{G_2}{l_2} \quad (c)$$

From Tab.1 [2], the relation between idlers weight and belt width is illustrated in Fig.1. Then, with regression theory, it is easy to find their property is a quadric curve, that is

$$G_1 = m_1 B^2 \quad (d)$$

$$G_2 = m_2 B^2 \quad (e)$$

and $$m_1 = \frac{\Sigma B_i G_i}{\Sigma B_i^3} = \frac{492.6}{23.912} = 20.6$$

$$m_2 = \frac{\Sigma B_i G_i}{\Sigma B_i^3} = \frac{412.2}{23.912} = 17.5$$

So substituting $m_1 = 20.6$, $m_2 = 17.5$ into the expression (d), (e) yields:

$$G_1 = 20.6B^2 \quad (f)$$

$$G_2 = 17.5B^2 \quad (g)$$

On the other hand, there exists abasic equation

$$B = \sqrt{\frac{Q}{K\gamma\ C} \cdot \frac{1}{V}} \quad (h)$$

Substituting (f), (g), (e) into (b), (c) leads to

$$q_1 = \frac{20.6Q}{K\gamma\ Cl_1} \cdot \frac{1}{V} = \frac{C_1}{V} \quad (i)$$

$$q_2 = \frac{17.5Q}{K\gamma\ Cl_2} \cdot \frac{1}{V} = \frac{C_2}{V} \quad (j)$$

Where $C_1 = 20.6Q/(K\gamma\ Cl_1)$

$C_2 = 17.5Q/(K\gamma\ Cl_2)$.

Tab.1 The relation between G_1, G_2 and B

Roller type	Belt width (cm)						
	80	100	120	140	160	180	200
	Roller weight (kg)						
T-R	14	22	25	47	50	72	77
R-R	12	17	20	39	42	61	65

Where T-R & R-R mean troughing & return rollers

2.4 Expression of belt weight

A lot of factors, such as L, β, Q will influence the choice of belt, namely, the choice of q_0. For given L and β, the relation between q and q_0 is approximately linear based on regression theory, as shown in Fig.2. It is satisfied by the equation

$$q_0 = a + bq \quad (k)$$

Analysis shows coefficients a, b can be approximatedly expressed as:

$$a = (17.7 - 0.0011L)\cdot(1 + 0.002\beta) \quad (l)$$

$$b = (5.7L - 3100)\cdot(1 + 0.708\beta)\ 10^{-5} \quad (m)$$

Substituting (a), (b), (i), (j), (k) into (1) generates:

$$W = \left[LCos\beta \left(\frac{Q}{3.6} + \frac{bQ}{1.8} + C_1 + C_2 \right) \pm \frac{QLSin\beta}{3.6\omega} \right] \cdot \frac{\omega}{V} + 2aL\omega\ Cos\beta + F \quad (2)$$

Because pulling force is used to overcome total resistance, equation (2) should be equal to pulling force. With longer belt conveyor, resistance along the length of the conveyor plays an important roll compared with part resistance so that it can be neglected.

Thus, simplified equation of (2) in this case is

$$P = W = L\omega \left\{ \left[Cos\beta \left(\frac{Q}{3.6} + \frac{bQ}{1.8} + C_1 + C_2 \right) \pm \frac{QSin\beta}{3.6\omega} \right] \cdot \frac{1}{V} + 2aCos\beta \right\} \quad (3)$$

Equation (3) showns that pulling force changes greatly in certain range of belt speed. It indicates that using high speed conveyor to reduce belt forcr is favourable to modern conveyor.

Also, power of motor can be expressed as

$$N = \frac{PV}{102} = \frac{L\omega}{102} \left\{ \left[Cos\beta \left(\frac{Q}{3.6} + \frac{bQ}{1.8} + C_1 + C_2 \right) \pm \frac{QSin\beta}{3.6\omega} \right] + 2aCos\beta\ V \right\} \quad (4)$$

Equation (4) make clear that power of motor increases linearly as belt speed increases, but the variation value of power will not be as large as that of pulling force except that $2aL\omega\ Cos\beta$ is greater. In term of above discussion, we found that belt speed can not be infinitely enhanced to minimise pulling force, it is restricted by power of motor and operating costs, etc. It follows that there exists a reasonable belt speed at which pulling force and power are minimum.

However, with short distance belt conveyor, the change of pulling force and power will be more complex for part resistance will take larger part.

3 REASONABLE BELT SPEED AND CASE STUDIES

A brief review of some case studies is presented to illustrate the solution of reasonable belt speed and its contrast with actual belt speed (neglecting part resistance).

3.1 Study No.1 is a 1101 m long conveyor installed in 1968. The conveyor is basically horizontal, transporting 3500 t/h with a belt speed V = 5.3 m/s. Other parameters are as follows: $\gamma = 1\ t/m^3$, K = 458, C = 1, $l_1 = 1.5$ m, $l_2 = 3$ m, $\omega = 0.03$. From equation (i),(j),(l) and (m), we calculate a = 16.5, b = 0.032, $C_1 = 105$, $C_2 = 44.6$. Substituting above data into (3), (4) yields

$$P = 1089\ \left(1 + \frac{35.9}{V} \right) \quad (5)$$

$$N = 10.7\ (\ 35.9 + V\ ) \quad (6)$$

For further explanation and calculation, Fig.3 is plotted which shows variation of pulling force and power. In Fig.3, we can find the reasonable belt speed V_0 = 5--6 m/s at which pulling force will reduce slightly and power of motor increase less in that speed range. Obviously, V_0 is in close agreement with actual belt speed (5.3 m/s). It may be furthermore concluded that the longer conveyor is and the more mass-flow rate is, the greater the reasonable belt speed will be under the same condition. Therefore, in the future, the use of high speed conveyor over long distance is feasible technically and economically.

3.2 Study No.2 is a 3750 m long belt conveyor operated in Cassinga Iron Mine. The angle of conveyor is 1.6° , it transportes 1500 t/h with a belt speed 3.67 m/s, γ = 2 t/m^3, other data are supposed the same as those of study No.1. Similarly, we can calculate a = 13.6 , b = 0.15, C_1 = 22.5, C_2 = 9.6 and

$$P = 3058.8 \left(1 + \frac{35.4}{V} \right) \qquad (7)$$

$$N = 30 (35.4 + V) \qquad (8)$$

Their properties can be shown in Fig.4, and it illustrates that the reasonable belt speed V_0 = 4--5 m/s is slightly greater than actual belt speed (3.7 m/s).

3.3 Study No.3 is a 1580 m long belt conveyor installed at a coal port in 1984. The conveyor is basically horizontal, transporting 6000 t/h with a belt speed 4.5 m/s. Other data are the same as those of study No.1. By calculating, we obtain a = 15.96, b = 0.059, C_1 = 180 , C_2 = 76.4 and

$$P = 1516.8 \left(1 + \frac{66.2}{V} \right) \qquad (9)$$

$$N = 14.9 (66.2 + V) \qquad (10)$$

The characteristic of two formulas are plotted in Fig.5 and we can see its reasonable belt speed V_0 = 6--7 m/s. It is greater than actual speed (4.5 m/s). The reason is that actual belt speed is chosen not only from the point of minimum pulling force and power, but also from material type, use condition, belt running stability, etc. So, actual speed is often less than reasonable speed. As shown in Fig.5 , with reasonable belt speed, the pulling force decreases by 7970 kg and power of motor increases by 37.3 kw, compared with actual belt speed. On the contray, if reasonable speed reduces to V_0 = 4--5 m/s which accords with actual speed, mass-flow rate should be reduced to about 4000 t/h.

4 CONCLUDING REMARKES

This paper analyses the relation between pulling force, power of motor and belt speed of conveyor with the guidance of regression theory.The reasonable belt speed calculated by this approach agrees with actual speed. This provides strong basis for type-chosen design of belt conveyor. This approach and conclusions are more suitable for belt conveyor of longer distance and higher mass-flow rate. The practice has proved it.

Finally, it is necessary to state that reasonable speed is discussed here from the viewpoint of power and pulling force, not considering material type, use condition, belt running stability, belt wear, etc. So, in practice, the belt speed we design should not be higher than the reasonable speed.

5 REFERENCES

[1] Chinese Mining University , Mining Conveying Machinery, Coal Industry Publishing House. 123--128, 1980

[2] Technical Centre of Hoisting and Conveying Design of Chemical Industry Ministry, Conveying Machinery Handbook, Volume 1., Chemical Industry Publishing House. 444--447, 1981

[3] Yang Fuxin, Structure, Principle and Calculation of Belt Conveyor, Volume 1,Coal Industry Publishing House. 181--193, 1983

[4] A.Harrison and A.W.Roberts, Techical Requirements for Operating Conveyor Belts at High Speed, bulk Solids Handling, Volume 4 ,100--103, 1984

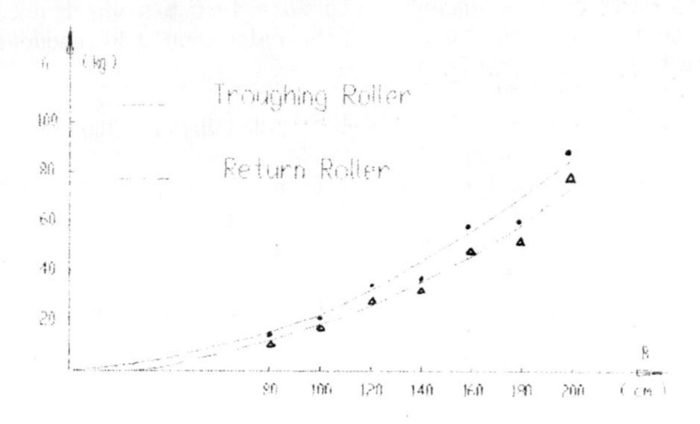

Fig.1 G as a function of B

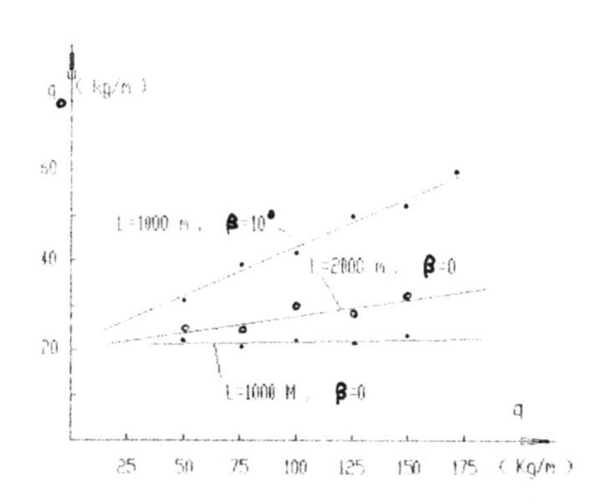

Fig.2 q_o as a function of q

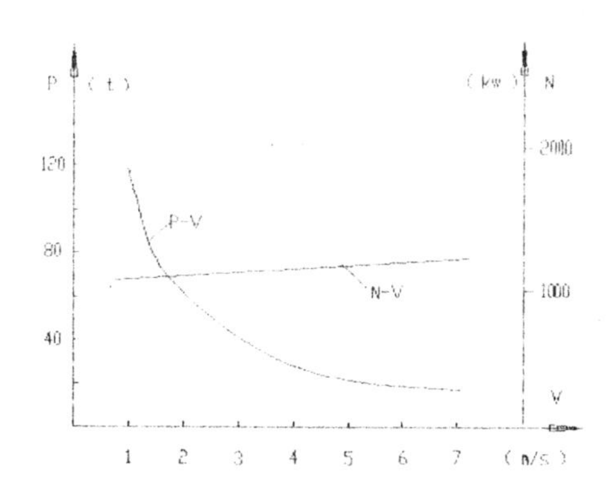

Fig.4 Pulling force and power VS. speed

--study No.2

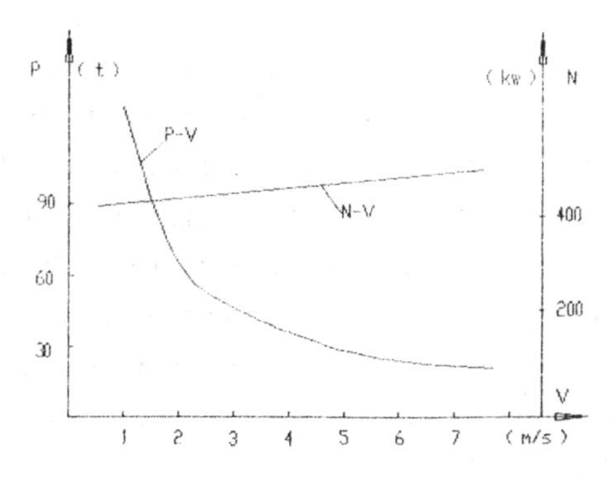

Fig.3 Pulling force and power VS. speed

--study No.1

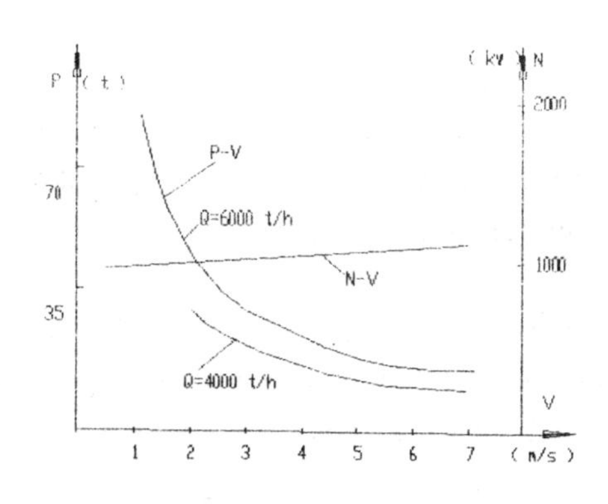

Fig.5 Pulling force and power VS. speed

--study No.3

C418/055

Fine powder flow phenomena in bins, hoppers and processing vessels

T A ROYAL, MS, MASME and J W CARSON, PhD, MASME
Jenike and Johanson Incorporated, North Billerica, Massachusetts,
United States of America

SYNOPSIS Fine powders often exhibit significant two phase (solid/gas) interactions when they are handled in bins, hoppers and processing vessels. As a result numerous and more complex flow problems can occur which are not observed with larger particle bulk solids.

The powder's flow pattern has a major effect on the development of certain flow problems, especially flooding. This can usually be overcome by using mass flow designs. Unfortunately, when mass flow is achieved the powder's discharge rate through a hopper outlet may be severely restricted.

These phenomena are explored along with related topics such as:

- Steady flow phenomena in bins and hoppers
- Air permeation systems
- Unsteady flow phenomena (e.g. settlement)
- Fluidized handling of powders

1 TYPICAL FINE POWDER FLOW PROBLEMS

Many bulk solids exhibit one or more of the following flow problems:

- *no flow* due to arching or ratholing
- particle segregation
- limited live or useable capacity (usually the result of a ratholing problem)
- degradation (spoilage, caking, oxidation) which is usually the result of a first-in-last-out flow pattern
- structural failure of bins due to loads being applied to them which they were not designed to withstand.

Materials for which the mean particle size is less than, say, about 100 μm (termed a *fine powder* in this paper) can also exhibit the following additional flow problems:

- flooding or uncontrolled flow which is often the result of a collapsing rathole in a funnel flow bin
- limited discharge rate through a bin's outlet
- unsteady flow phenomena such as settlement as the powder is charged into a bin.

2 HOW FLOW PATTERNS AFFECT FLOW PROBLEMS

2.1 Funnel flow bins

When a fine powder is stored in and discharged from a bin having a *funnel flow* pattern, ratholing and flooding problems are almost inevitable. The reason for this is that when a powder becomes deaerated in the stagnant region of a funnel flow bin, it usually develops sufficient cohesive strength that its *critical rathole diameter* (the flow channel diameter at which a rathole becomes unstable) is larger than the size of a flow channel which might reasonably develop.(1) Thus a rathole develops whose diameter is approximately equal to the diameter of the flow channel. In the case of a funnel flow bin with a circular outlet, this is roughly equal to the outlet diameter. A funnel flow bin having a square or rectangular outlet will tend to develop a circular flow channel (and resulting rathole) whose diameter is approximately equal to the length of the diagonal of the outlet.

When a rathole develops in a funnel flow bin, e.g. as the bin level drops when the in-feed rate is slower than the discharge rate, a condition is set up which can lead to flooding. The scenario is as follows: Since the sides of a rathole are often relatively unstable, the rathole occasionally collapses due to ambient vibrations in the plant, restart of filling of the bin, or someone hitting on the sides of the bin. When this happens the fine powder drops through a column of air (or whatever gas is in the bin) causing the particles to become fluidized. Most bin feeders (e.g. screws, belts, vibrating pans, rotary valves) are generally designed to handle a solid, not a fluid. Thus when this fluidized material reaches the outlet, it often flows uncontrollably resulting in what is termed *flooding*.

A similar phenomenon can occur in a funnel flow bin even if a rathole does not develop. The reason for this is that the narrow flow channel provides only a short residence time for material entering the bin. Since this material is often in a fluidized state just due to falling into the bin or coming from a pneumatic conveying line, it has only limited time to deaerate which can also result in a flooding condition at the outlet.

2.2 Mass flow bins

Flooding and ratholing problems can be overcome by using bins having a *mass flow* pattern. While this is always a significant improvement in the handling of fine powders, it is not without problems of its own. Arching over the outlet may still occur, as with a funnel flow bin, although it is less likely because of the steeper and less frictional hopper walls. The other problem which can occur is one of limited discharge rate through the bin outlet. It may seem strange that the same fine powder which can flood through an outlet uncontrollably in a funnel flow bin can, at the same time, flow through a mass flow bin at a rate which is lower than desired for process requirements. The reasons for these phenomena will be described below.

2.3 Segregation

Fine powders can also segregate, usually by one of two mechanisms: (2)

- *fluidization* (air entrainment) which often results in vertical striations within a bin. The larger and/or more dense particles become concentrated below the finer and/or lighter particles.
- *particle entrainment* in an air stream which can result in unusual segregation patterns depending on where particles drop out of suspension.

As with larger particle materials, fine powders can exhibit significant levels of segregation as they discharge from a funnel flow bin. This is the result of the segregation pattern which develops as the bin is filled combined with the first-in-last-out flow pattern of a funnel flow bin. Unlike coarser particle material, however, many fine powder segregation problems cannot be often overcome by simply converting to a mass flow pattern. This is particularly true of fine powders which segregate by the fluidization mechanism. In fact, since materials discharge from a mass flow bin in a first-in-first-out flow sequence, the effect of a vertical striation profile can be much worse than experienced in a funnel flow bin. Fortunately there are ways that this can be minimized, e.g. by charging the material into the bin in such a way that the larger and/or more dense particles are not driven down into the bed of material.

3 STEADY FLOW PHENOMENA IN MASS FLOW CONTAINERS

The maximum flow rate of a fine powder through a bin outlet can be several orders of magnitude lower than that of a coarser particle material. The primary reason for this limitation is the air pressure gradient that forms naturally at a bin outlet as air flows up into the hopper section. This pressure gradient acts upward, counter to gravity, thus reducing the rate of material discharge. The magnitude of the pressure gradients are very low with coarser particle materials because they are more permeable. In other words, air can easily flow through their voids without resulting in much of a pressure drop.

Fine powders exhibit significant two-phase flow effects due to the movement, however slight, of interstitial gas as the powders compress or expand during flow. The following is a description of the three flow-rate dependent modes that can occur in a mass flow bin, depending on the solids flow rate.

1. The first mode of flow is characterized by the steady gravity flow of partially deaerated material controlled by a feeder. The limiting steady state condition occurs when compaction in the cylinder forces too much gas out through the material top surface. This causes a slight vacuum to form as the material expands while flowing through the converging portion of the bin. The result is a gas counterflow through the bin outlet, which forces the solids contact pressure to drop to zero and limits the *steady* solids flow rate.

2. The second mode occurs at flow rates somewhat greater than the limiting rate. This mode of flow is characterized by an erratic, partially fluidized powder discharging from the bin which can be controlled by some feeder arrangements. At these flow rates, a steady rate from the bin can best be achieved by the use of an air permeation system at an intermediate point in the bin to replace the lost gas.

3. The third mode of flow occurs when the flow rate is too high to allow much, if any, gas to escape from the material voids. In this extreme, the material may be completely fluidized and flood through the outlet unless the feeder can control fluidized solids.

Testers are available to measure the *permeability* (3) and *compressibility* (4) of powders and other bulk solids. From such tests one can calculate critical, steady state flow rates through various outlet sizes in mass flow bins. With this information, an engineer can determine the need for changing the outlet size and/or installing an air permeation system to increase the flow rate. Furthermore, one can determine the optimum number and location of air permeation levels, and estimate the air flow requirements.

3.1 Example

The types of conditions which can occur are best illustrated by means of an example. Using a proprietary, two-phase flow computer program (5) we have analyzed the bin shown in Fig. 1. For this analysis we have assumed that this mass flow bin is filled with a dry sludge powder having the following flow properties:

- Wall friction angle:
  35° along carbon steel walls in cylindrical section
  20° along 304 2B surface finish stainless steel walls in conical hopper section
- Minimum outlet diameter to prevent arching in a mass flow bin: <75mm
- Effective angle of internal friction: $\delta = 50°$
- Bulk density/consolidating pressure relationship:

$$\gamma = \gamma_0 \left(\frac{\sigma}{\sigma_0}\right)^{\beta}$$

where: $\sigma_0 = 0.62$ kPa
$\gamma_0 = 0.577$ tonne/m^3
$\beta = 0.05$

- Permeability/bulk density relationship: $K = K_0 \left(\frac{\gamma}{\gamma_0}\right)^{-a}$

where: $K_0 = 0.00335$ m/sec
$a = 6.4$

- Minimum bulk density: 0.545 tonne/m^3
- Specific gravity of a particle: 1.29

When one analyzes these values of wall friction angle, effective angle of internal friction, and hopper angle, the Jenike theory (1) indicates that a mass flow pattern will develop.

3.2 Effect of different levels of powder

For steady flow into and out of the bin at a rate of 7 tonne per hour, Fig. 2 shows the resulting interstitial air pressure distribution which develops while maintaining the cylinder full, half full or quarter full of material. This is for boundary conditions of atmospheric pressure at both the top of the bin as well as at the outlet. Note that just above the outlet, an air pressure gradient due to upward flow of air exists which tends to reduce the average vertical solids contact stress in this region. The limit is when this solids contact stress goes to zero which indicates the formation of a free surface, i.e. an arch.

The calculated air pressure gradients and solids contact stresses at the outlet for the different levels in the bin are as follows:

Material level in cylinder	Air pressure gradient divided by bulk density	Average vertical solids contact stress
Z,m	$\frac{dp/dz}{\gamma}$, dimensionless	σ_v, kPa
1.52	-1.13	0.106
3.05	-1.22	0.067
6.10	-1.31	0.032

Note that the air pressure gradients are negative which indicates that they act upward, counter to flow of the powder. Their magnitudes are higher for higher levels of material which leads to lower solids contact stresses at the outlet. This is because the material is compacted more in the cylinder at high levels, the longer retention time allows more gas to escape through the top, and so more air must enter through the bottom as the powder flowing towards the outlet expands.

In fact calculations reveal that if the cylinder height were increased to 9.4m the average vertical solids contact stress would approach zero. For this level of material or higher we expect that flow would be unstable at the desired rate of 7 tonne per hour.

3.3 Effect of outlet size

A similar result is shown in Fig. 3 in which the same 7 tonne per hour discharge rate is used with a full cylinder but now three different outlet diameters are shown: 457, 304 and 203 mm. Note that as the outlet becomes smaller, the magnitude of the air pressure gradient near the outlet increases; therefore the solids contact stress decreases. In fact, a limiting rate will occur for the smallest opening size.

3.4 Effect of powder flow rate

The results of a number of runs are shown in Fig. 4. Here we have assumed a full cylinder of material and a 305 mm diameter outlet. This figure vividly demonstrates the effect that flow rate has on both interstitial air pressure and solids contact stress. Note that below 7.5 tonne per hour, the bottom boundary condition of atmospheric gas pressure can be met. Above this rate we have an unstable flow condition since the solids contact stress goes to zero. The only way that this boundary condition can be satisfied is to have a vacuum condition existing at the outlet.

One way to correct this problem would be to increase the size of the outlet as shown in Fig. 3. However there is a limit to how large the outlet can be and still be reasonable in terms of the size of a feeder which might be used to control the discharge rate or the size of interfacing equipment.

Another way to correct this would be to use a standpipe below the outlet but above the feeder. However, there is a limit to the amount of vacuum that can be maintained in the standpipe.

The most effective way this flow rate limitation could be overcome is through the introduction of a small amount of air using an air permeation system (7,8) as shown in Fig. 5. The location as well as the quantity of air must be closely examined; otherwise too much or too little air may be added which will cause either a flooding condition or will not be sufficient to achieve the desired flow rate. As shown in Fig. 5 addition of 2.55 or 3.4 m^3/hr at the transition results in stable flow.

The other area of interest is when the flow rate is very high. In this case material does not have time to deaerate in the bin and remains fluidized. Handling fluidized materials is discussed below.

4 UNSTEADY FLOW PHENOMENA

4.1 Firing of an air blaster

In addition to the steady flow phenomena described above, fine powders can exhibit various types of unsteady flow phenomena as well. For example, when an air blaster is fired or a vibrator turned on, material flow is likely to be unsteady for some brief period of time as previously nonflowing material starts to flow. A similar phenomenon occurs with coarser bulk solids as well. The effects of this energy input, however, are often quite different with a fine powder, especially when an air blaster is fired into a bin. With both a fine powder and a coarse bulk solid a solids stress wave propagates through the material. If the material is very compressible and/or frictional, the effect of the stress wave is dissipated quickly.

Another effect is an air pressure gradient imposed on the bulk solid because of the introduction of excess air. For a fine powder this pressure gradient can be several orders of magnitude higher than for a coarse bulk solid. The magnitude of the air pressure gradient imposed is approximately proportional to the square of the distance from the blaster. As a result, an air blaster is often more effective in breaking arches that occur near its nozzle, but less effective in breaking arches that occur some distance away. Detailed computer calculations based on the flow properties of the material are necessary for determination of number, size and location of air blasters in a particular bin. (5)

4.2 Settlement

Another interesting unsteady phenomenon with fine powders is that of *settlement.* (9) Nearly all bulk solids settle as they are filled into a container because of the vertical compression of the material lower in the container as new material is added on top of it. With coarse bulk solids, the amount of settlement is usually quite small, and as soon as filling is stopped, there is little if any additional settlement. A fine powder behaves quite differently both in the amount of settlement as well as its duration. The reason for this has to do with the inability of air and other gases to flow freely through the voids of fine powders. Therefore, when a fine powder is compressed (such as when it is being filled into a container), not only do the void spaces shrink, but the gas pressure within the voids increases. Thus, during filling, an upward gas pressure gradient continually exists which tends to support the fine powder and limit its compaction. At the end of the filling cycle, this upward gas pressure gradient is still present and gradually dissipates over time as the gas escapes through the top surface. Thus, the settlement of a fine powder takes place much more slowly than that of a coarse bulk solid and, in general, the amount of settlement is much greater. In fact, it is not unusual to find a fine powder experience 300 mm or more of settlement which occurs over several days or longer after the filling cycle has ended.

As an example of this phenomenon consider the bin shown in Figure 1 for storage of dry sludge powder having the same flow properties as previously given. The question is how fast the bin can be filled and how soon it can be used after filling without encountering flooding due to the material being somewhat fluidized in the bin. Figure 6 gives the results of calculations for a fill rate of 91 and 455 tonne/hr. As can be seen the air pressure in the voids of the solids increases during filling and then dissipates over time as the air escapes through the top surface and settlement occurs. The pressure after filling is sufficient in both cases to cause problems with feeders not designed to handle material in a fluidized state. However in both cases this pressure is completely dissipated after about 40 minutes.

In the case of the 455 tonne/hr fill rate the maximum air pressure at the bottom of the cylinder is 11.9 kPa and the top surface settles down 245 mm as the air escapes. At 91 tonne/hr. the maximum gas pressure is only 3.2 kPa and the material settles only 45 mm.

The phenomenon of settlement becomes more pronounced with taller bins handling finer, less permeable and more compressible materials. It can also be important in loss-in-weight hoppers, where extremely high fill rates are required to minimize the amount of time the hopper is in a volumetric mode.

5 FLUIDIZED HANDLING OF POWDERS

Sometimes it is more practical to handle fine powders in a fluidized state rather than to allow them to deaerate and try to handle them by gravity alone. Through the use of air pads, air slides and/or air nozzles, some or all of the contents of a bin can be fluidized. Such material cannot form a stable arch or rathole. In addition fluidized material will discharge through a bin outlet several orders of magnitude faster than a deaerated fine powder.

Some of the considerations involving fluidized handling are as follows:

1. The bulk density of the fine powder upon exiting the bin will be low and nonuniform. Thus if the material is being used to fill another container (e.g. a truck or rail car), it may not be possible to fit the required mass of material into that container. If, on the other hand, the material is going into a process where close control of flow rate is important, the nonuniformity of bulk density may create major control problems.

2. The cost of energy required to fluidize the bin may not be insignificant. This is particularly true if dry air must be used because the material in the bin is hygroscopic.

3. Particle segregation may be improved or made worse through the introduction of air. While there are some blenders on the market which use air as the motive force to blend materials, simply putting an air pad or an air slide into a bin is more likely to segregate a fine powder rather than blend it, especially if the material segregates by the air entrainment (fluidization) mechanism. Fines end up on the top and more dense or larger particles end up on the bottom.

4. It is not always clear how much of the bin volume needs to be fluidized. It is generally best to fluidize the entire contents of the bin, but this may be neither practical nor necessary if the bin is relatively large. If only a portion of a bin is fluidized, one must be concerned about the potential for a rathole forming in the deaerated material outside of the fluidized volume. In addition void space must be provided for the material to expand.

5. Another concern is how long to leave on the fluidizing air. This question is important not only during discharge, but more importantly, when there is no discharge taking place. Fine powders which are cohesive when deaerated become very difficult to refluidize after they deaerate. If this is the case, intermittent fluidization during periods of no discharge may be necessary.

6 SOLUTIONS TO FINE POWDER FLOW PROBLEMS

Fine powder flow problems can usually be minimized if not eliminated by using one or more of the following techniques:

1. Use bins designed for mass flow rather than funnel flow. This will automatically eliminate problems of ratholing, limited live capacity, and the first-in last-out flow pattern which often causes degradation of the powder being handled. In addition, mass flow designs minimize particle segregation problems provided that the segregation pattern within the bin is of a side-to-side as opposed to a top-to-bottom type.

2. Make sure that the outlet from the bin is large enough to prevent arching.

3. Ensure that the design of the feeder is compatible with that of the bin so that the full area of the outlet is *live*.

4. Whenever a pneumatic conveying line terminates at a bin, arrange the interface such that the material enters the bin tangential to the cylinder wall. While this may result in some side-to-side segregation of particles, it will minimize any top-to-bottom striations which are difficult to overcome.

5. Based on the permeability and compressibility of the powder, calculate the critical, steady-state flow rate through the bin outlet. If possible operate the feeder such that the discharge rate from the bin is always less than this critical rate.

6. If a higher discharge rate is needed, enlarge the size of the outlet and feeder, lower the level of material in the bin, and/or use an air permeation system.

7. Place a standpipe between the outlet of a bin and the feeder so as to create a vacuum condition at the outlet. This will result in somewhat higher rates of material discharge.

8. Beware of the settlement time of powders when placed in a bin. If sufficient time is not allowed for settlement, problems of flooding through the outlet can occur because of entrapped gas.

9. Use fluidized handling of powders if the material characteristics and operational requirements allow it.

REFERENCES

(1) JENIKE, A. W. Storage and flow of solids. Univ. of Utah Eng'r. Exp. Station Bulletin 123, 1964

(2) CARSON, J. W., T. A. ROYAL and D. J. GOODWILL Understanding and eliminating particle segregation problems. *Bulk Solids Handling*, 1986, 6, 139-144.

(3) ANON. Testing series #4: Estimating flow rates of fine powders. *Flow of Solids® Newsletter*, 1989, IX, No. 2.

(4) ANON. Testing series #3: Compressibility measurements serve many purposes. *Flow of Solids® Newsletter*, 1988, VIII, No. 2.

(5) JOHANSON, J. R. Two-phase flow effects in solids processing and handling. *Chem. Eng.*, January 1, 1979, 77-86.

(6) CARSON, J. W. and D. S. DICK Seals improve coal flow into a pressurized environment. *Power Eng'r.* March 1988, 24-27.

(7) BRUFF, W. and A. W. JENIKE A silo for ground anthracite. *Powder Tech.*, 1967, 1, 252-256.

(8) CICCARIELLO, A. and J. POWERS Pyramidal bins ease flow of Kaolin clay at Filtrol. *Chem. Processing*, 1986, 49, No. 6, 30-31.

(9) JOHANSON, J. R. and A. W. JENIKE Settlement of powders in vertical channels caused by gas escape. *J. Applied Mech.*, 1972, 39, 863-868.

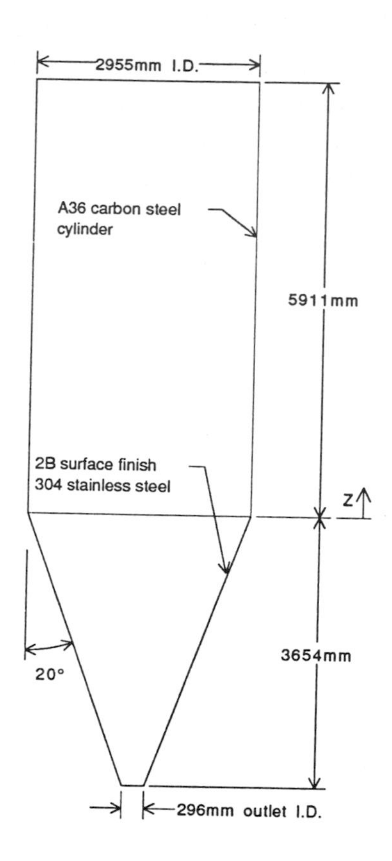

Fig. 1 Dry sludge powder bin

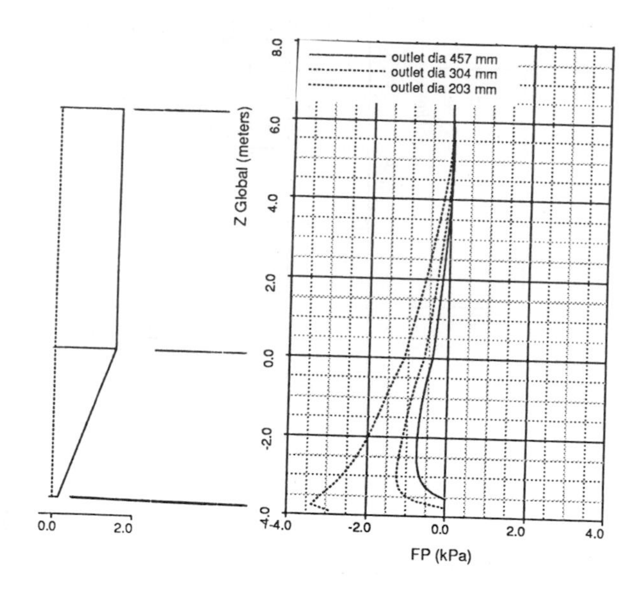

Fig. 3 Interstitial air pressure vs. outlet size for 7 tonne/hr

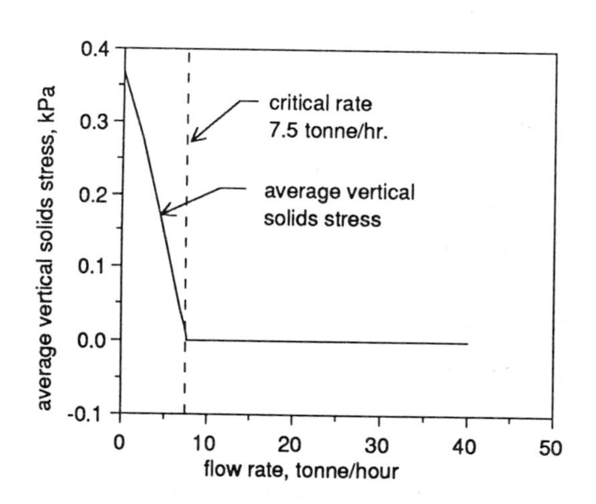

Fig. 4A Average vertical solids stress at outlet vs. flow rate

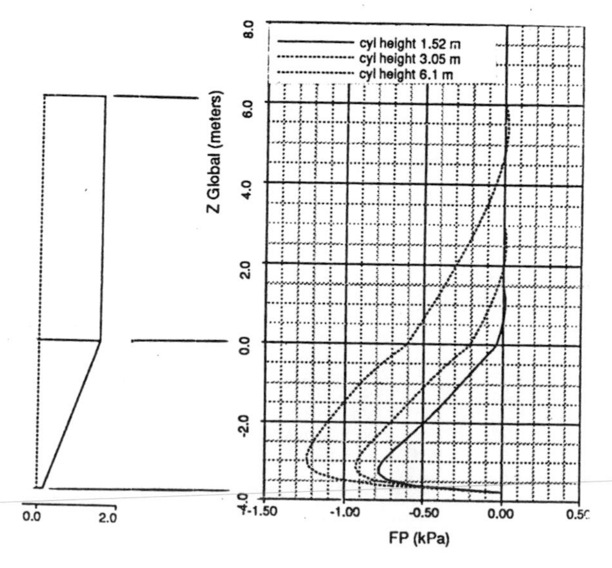

Fig. 2 Interstitial air pressure vs. material level for 7 tonne/hr

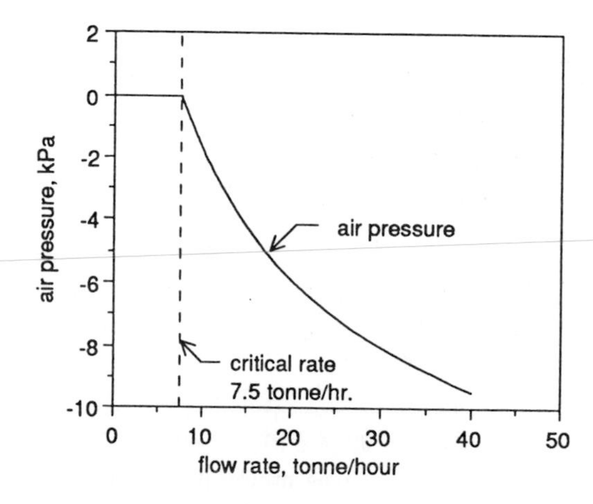

Fig. 4B Air pressure at outlet vs. flow rate

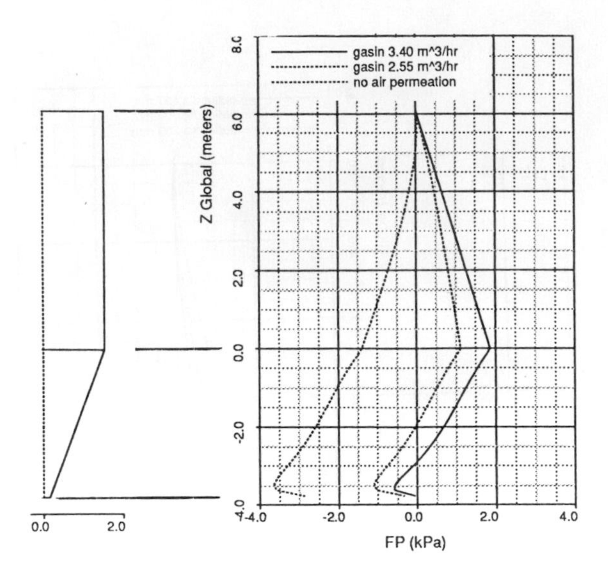

Fig. 5 Effect of air permeation system for 12 tonne/hr discharge

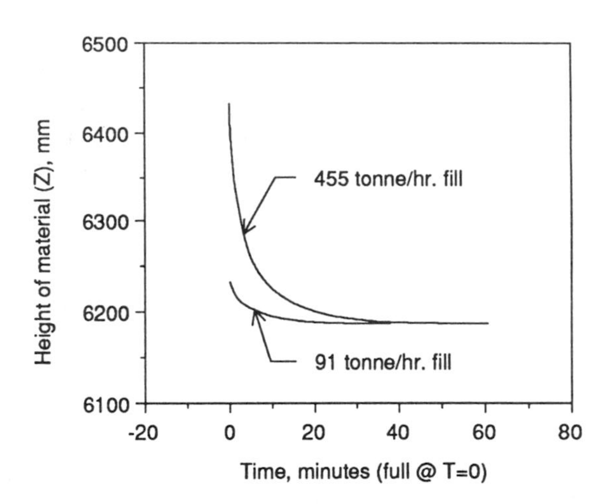

Fig. 6A Settlement of material vs time

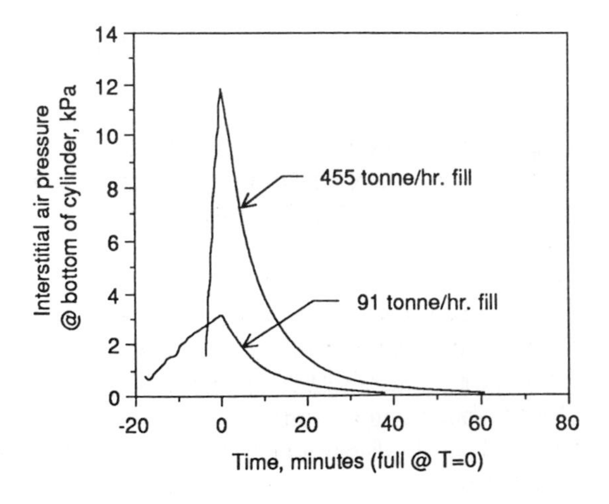

Fig. 6B Air pressure vs time

C418/038

Design of large underground ore bins

O J SCOTT, BE, ME, MIEAust
Department of Mechanical Engineering, University of Newcastle,
New South Wales, Australia
P R CHOULES, BApplTechMech, IMAust
M and A Mining and Industrial Consultants Pty Limited, Brisbane,
Queensland, Australia

SYNOPSIS The various requirements for the design of large underground storage bins for the handling of ore over a wide range of moisture contents are presented in the context of a design for the continuing upgrade of a large copper mine in Irian Jaya.

1. INTRODUCTION.

As part of an ongoing upgrade program at a copper mine and mill site in Irian Jaya a major rearrangement of the ore handling was proposed. Included in this were very large underground ore bins, associated ore passes and conveyor systems. The authors are part of a consultancy team that evaluated various arrangements and prepared designs for the underground bins and associated discharge systems. The concept involved up to nine bins with initially proposed cross sections ranging to 17m x 7m excavated as enormous caverns to 70m high each with a potential total capacity in excess of 15 000T and live capacity to 8500T. These storage bins with vertical infeed ore passes to 450m height and associated crushing and conveyor systems would help overcome dependency on an overhead cableway system that transported all ore from the mine at 3700 metres elevation to the mill site stockpile at 2700 metres. Additionally the upgrade would allow for increased mine output and development of ore reserves. The proposed overall annual throughput will be in the order of 18 MT.

While there are numerous instances of underground ore bins in use throughout the world, the scale of the proposed bins and their remote location in an area of very high rainfall inspired a degree of caution in the design. With an average daily rainfall of 19mm the possibility of major water ingress into the ore stream or directly into the bin via aquafiers was a possibility to be addressed. Prediction of the behaviour of the ore under these circumstances and of possible extreme load cases for the bin and associated equipment if filled with saturated ore had to be prepared.

Paramount at all times was the need to prepare a design which would provide reliable flow of ore under all circumstances while having the necessary structural integrity. Experience at the site suggested that the ore was at times difficult to handle. Blockages in the reclaim area of the mill feed stockpiles were not uncommon while underground stope draw points and dump locations had a long history of obstructions requiring in some cases continuing manual intervention to maintain flow.

2. STORAGE BIN DESIGN PHILOSOPHY

All too often in the past the design and operation of storage and handling systems for bulk granular materials has been treated empirically. There have been many instances of costly production interruptions when bulk solids have formed cohesive arches across the outlets of storage bins, stable ratholes have formed in stockpiles or build up has occurred in transfer and underscreen chutes. In most cases these flow interruptions are directly attributable to incorrect design with little regard to the flow properties of the materials being handled.

Over recent years significant advances have been made in the development of theories and associated design procedures for bulk solids within storage facilities. These procedures enable bins, feeders and transfer chutes to be engineered for greater reliability and predictable behaviour.

A logical process for the design of storage and handling equipment for granular materials has a number of basic steps:

- Define the production requirements
- Define the local constraints, including in this case the geotechnical limitations of where the bins are to be located.
- Obtain representative samples and perform laboratory tests to determine the strength and flow properties of the bulk solid to be stored over a range of likely ambient conditions
- Determine the geometric limits for a bin to give the desired capacity and reliable discharge. Select a general arrangement, including feeder, that conform to these requirements within the local constraints.

- Estimate the loads on the bin structure, feeders and other associated equipment.
- Design and detail the handling plant including the structure.

Prior to the involvement of the authors, mine site personnel had defined the production requirements, evaluated the local constraints and proposed a number of potential arrangements for the bins. While the general theories of the gravity flow of bulk solids are well defined [1 - 6], it is pertinent to discuss the salient aspects with respect to these underground ore bins where local constraints can be far more restrictive than for surface installations.

3. BIN FLOW PATTERNS

Two basic modes of flow have been recognised in a storage bin; mass-flow and funnel-flow. Mass flow is the more predictable mode of flow and occurs when the hopper walls are steep and sufficiently smooth to ensure flow adjacent to the walls. Mass flow has advantages in minimising segregation, and the prevention of hang-ups, surging and channelling. However when handling an abrasive ore in vast quantities through an underground bin, mass flow could be considered undesirable due to wear and spalling of bin walls by the sliding ore. Maintenance of this damage would be difficult and in most cases impractical. Spalled sections of ore could be large and cause outlet and downstream blockages.

Funnel-flow occurs in flat bottom bins or when the hopper walls are rough and too shallow. The flow can be erratic and segregation can take place. In funnel flow bins there is the likelihood of substantial quantities of bulk material remaining within the bin after discharge has ceased. Stable ratholes can develop in non-flowing regions of the granular material if the stagnant ore consolidates sufficiently to remain static after the flow channel has emptied. However in some instances funnel flow can be advantageous in minimising wear in coarse ore bins.

4. FLOW PROPERTIES

In order to determine the critical rathole diameters, cohesive arching dimensions and critical angles for hopper inclinations it is necessary to measure the flow properties of the ore from representative samples that were likely to display the worst potential handling characteristics. At the mine site the moisture content of the ore is uncontrolled and there are various pockets of clay and other minerals making predictions of the likely composition of handled ore difficult. To date a total of twelve test samples of the ore have been taken from the four ore bodies being worked.

While the ore to be stored and handled is primary crushed with a top size of 250mm in the flow of a bulk material, the standard convention for testing the cohesive strength is to examine the -4 +0 mm sub-sized sample. This is because during the flow of a bulk material with a wide range of particle sizes, the large particles move bodily while the material shears across the fines component. The coarse particles are a passive agent which do not develop cohesive strength without the fines to bind them.

To cater for variations in ore composition and condition, test samples were prepared to appropriate moistures up to saturation. Jenike style shear tests were conducted to determine their instantaneous and time related behaviour. Figure 1 shows the band of bulk strengths for various samples from the four different ore bodies over a range of moistures and undisturbed storage times up to 72 hours. Corresponding data on bulk density, effective angle of internal friction and wall friction angles were also obtained.

An assumption of an upper limit to strength equates to a minimum cohesive arching dimension of 1.6 m for a symmetric outlet and 0.85 for a long slotted outlet for any expected ore, moisture, and undisturbed time combination. Thus if the outlet is rectangular in form and has a length greater than three times the width then the critical arching dimension is about half that of a symmetric outlet. This important advantage in a long slotted outlets is dependent on the whole of the outlet having material drawn through it.

These dimensions do not take into account the possibility of a mechanical arch forming across an outlet. This is generally taken to be six times the maximum lump size to prevent interlocking [7]. This ratio is dependent on the particle shape, e.g. it can be ten times for some angular products but only four or five times for smooth particles not far removed from spherical. It is also dependent on the shape of the outlet with the ratio closer to five in long slotted outlets while in a symmetric outlet it should be at least seven to eight times.

In funnel flow designs the critical rathole diameter can be calculated from a knowledge of the unconfined yield strength at higher values of consolidation pressure as shown in Fig 2. The critical rathole diameter is dependent on the geometry (height, diameter and wall friction) of the bin as well as the characteristics of the ore. For the underground bins proposed the upper bound value ranged between a low of 3.5 m and a high of 9.2 m with the higher value being for the most cohesive ore after 4 hours of undisturbed storage at a high level of moisture. With the majority of the test results giving a maximum critical rathole dimensions of around 5.5m, it was decided to recommend a maximum diagonal for the outlet of 6.5m. This was considered satisfactory for ore up to 12.5% moisture if the bin had no discharge for periods of up to 24 hours.

Thus one form of a preferred arrangement would have a mass flow hopper installed in the floor of a flat bottom funnel flow bin. The hopper would have a minimum width of 0.85m increasing in the direction of feed, with a diagonal dimension of 6.5m. Wall Friction tests conducted on Bisalloy wear resistant plate, widely used as an abrasion resistant liner at the mine and mill site, showed that if the hopper chutework was lined with this material the sides of the chute needed to be inclined at 67.5° and the ends at 80° to ensure mass flow and its subsequent advantages in the outlet zone. Figure 3 shows a schematic of this arrangement including the use of gusset plates in the valleys of the chute to assist in reducing the build up of dead ore in the corners of the chute.

5. DESIRABLE FEATURES

While a number of initial options for the underground ore bins showed inverted pyramidal hoppers feeding to symmetric or near symmetric outlets the difficulty of mining such hoppers, their unlikely chance of operating in mass flow and the advantages of handling an abrasive ore in funnel flow mode suggested that a flat bottom bin operating in funnel flow had significant advantages. Though this would minimise wear, it was essential for stability in the downstream mill process that the inherent segregation characteristics of funnel flow were not aggravated. For this reason centralised charging of the bin was preferred as any offset in loading would induce significant segregation depending on the differential between infeed rate and discharge rate for the bin.

Centralised discharge was highly desirable so that a stable compacted column of fines did not develop below the fill point. This phenomena had been observed in coarse ore storage bins with multiple outlets where the fill point is located above a dead area, substantially reducing the live capacity. While multiple feeders could be arranged to ensure symmetrical draw, a single centrally located feeder was considered the most desirable arrangement. If maintenance, physical and operational constraints had enforced the use of multiple feeders, then these would need to be arranged so that the flow channels forming in the bulk ore above the feeders intersected at or near the top of the floor slab. This necessitates a proximity for multiple feeders that can be difficult to achieve within structural constraints.

A single central outlet was the preferred arrangement. For this outlet to operate satisfactorily it was necessary that the flow channel established in the bin would collapse and not form a stable rathole. This is achieved by ensuring that the maximum dimension of the outlet of the funnel flow is at least equal to the critical rathole diameter determined from the laboratory tests on the ore samples. The minimum dimension of the outlet needed to be sized to ensure that neither cohesive nor mechanical arching occurred.

6. FEEDER SELECTION

A range of feeders were initially considered for control of ore discharge. After review apron feeders were preferred, though the problems of high capital costs, difficulty in underground installation and site access were recognised. Local experience with apron feeders at the mine and mill suggested that they had acceptable maintenance features as well as the robustness necessary for the proposed location. An apron feeder suited the correct proportions for the preferred slotted outlet and could be designed to withstand the high vertical loads associated with filling the storage bin. One drawback with the use of apron feeders is the spillage of fines experienced during operation. This could be overcome by installing the feeder above the receival conveyor which then operates as a spillage collector.

Vibratory feeders were not able to take advantage of the reduced width of slotted outlets and were difficult to arrange to ensure the collapse of ratholes developing in the ore. Vibratory feeders would need to be at least 2.2m wide to ensure that cohesive arches did not form and would need to be arranged in multiples in such a way that the combined flow channels had dimensions greater than 6.5m. It was also considered possible that they would suffer from the application of high vertical loads exerted by the ore during filling to the high surcharges proposed for the storage bins. If the feeder supports were fully compressed by these loads the feeders could be difficult to start. Previous underground on-site experience with vibratory feeders was associated with minimal storage capacities under surcharges no more than 15m. In some underground locations significant problems had been experienced in their use largely due to mis-matches in feeder selection with performance requirements and inappropriate chute design.

Reciprocating plate feeders were also considered but were not recommended. As the plate feeder strokes only partially across the outlet, ore is drawn from the rear, restricting the development of a large flow channel. Their high wear, potential for spillage, high cyclic discharge rate and inability to completely empty the flow channel and thereby provoke the formation of a stable rathole were significant constraints. They had in their favour good resistance to the impact loads experienced in high surcharge storage bins and a lower initial capital cost.

Thus an apron feeder installed under a plane flow hopper centrally located in the slab floor of a flat bottom funnel flow bin conformed to the requirements of both the tests results and the preferred options for an ore bin. In general it is preferable that the slot increase in width in the direction of feed at a divergent angle of about 2.5° per side, and with the requirement for 3000T/hr discharge a 1500mm wide feeder was selected. The general arrangement of hopper and chutework is shown in Figure 4.

It is of interest to note that the feeder components and hydraulic drive were sized to cater for a wide range of chain tensions expected with the various operating conditions. The predicted tensions ranged from 238kN at a normal operation of 0.23m/s up to a high of 812kN at feeder startup if the bin was filled from empty with the feeder stopped. Reduced Factors of Safety ($F_{of}S = 4.8$) were accepted for the maximum tension which should have a very low frequency of occurrence.

Subsequent to the construction of the initial five bins complying with the basic dimensions shown in Figure 4, an additional four bins are under construction using apron feeders of 1800mm width to allow for an increase in discharge rate to 4500T/hr

7. BIN PROPORTIONS

It was considered that the most desirable shape to the bin would be of circular cross section rather than the rectangular cross sections initially proposed. A circular funnel flow bin is most efficient in terms of potential live storage to excavated volume and has the least segregation problems of any cross sectional shape if loaded symmetrically.

The height and diameter of a storage bin to provide the required volume of ore are important parameters. On strictly material flow characteristics there appears to be little restriction on the height or diameter of a storage bin, though a minimum diameter may exist below which the ore may arch across the walls of the bin. Bins of smaller diameter have an advantage in that the vertical pressure generated by the granular ore on the bin floor and the feeder pan during the initial filling would be lower. However for a constant storage capacity, the hydrostatic pressure of a possible saturated ore, would increase as the bin diameter reduced. This would significantly offset any reduction in the granular loads for the structural design.

For normal operating conditions, i.e where the ore is not saturated, a bin of smaller diameter would reduce the vertical loads on the feeder and thus reduce the power requirements of the feeder on start-up when the bin has been filled without any prior discharge occurring. The loads on the feeder, once an arched discharge stress state is established in the hopper chutework, should not be unduly influenced by the surcharge and thus would be independent of the bin height. Once the bin has had some discharge, laboratory and field tests indicate that the reduced load is generally retained while ever a cushion of material is left above the feeder tray, even if the feeder is stopped or the bin refilled.

The major impact of height/diameter of the bin will be in the potential live storage capacity. With a constant excavated volume the smaller the bin diameter, the greater will be the proportional live capacity of the bin given constant opening dimensions. If it is assumed that the opening in the floor of the bin has dimensions of 2.25 metres by 6.1 metres as shown in Figure 4 then a 10 metre diameter bin, 45 metres high would have a potential minimum live capacity of 5500 Tonnes. For a bin of 8m diameter and a height of 70m a minimum live capacity of about 6000T can be achieved.

8. SATURATED ORE

A major part of the initial planning and investigation centered on the prediction of design loads on the bin floor structure, feeder and associated steelwork. With the high average daily rainfall, the unstable nature of some of the mine terrain and the possibility of the ore passes and bins intersecting major aquafiers it was possible that the ore could become saturated. While there are various procedures available to predict the design loads for a granular material within a bin or silo, little has been published on the loads when the ore is saturated. One of the basic assumptions underpinning the theory of bulk solids flow and prediction of wall loads is that there is a variation between the lateral and vertical pressures within a bulk granular material which can be defined relative to the effective angle of internal friction. This variation is dependent on the ore behaving as a plastic rather than a visco-elastic continuum and that the test data to determine the cohesive strength and the effective angle of internal friction is independent of the rate of shear and dependent on the mean stress acting within the solid [3]. Once an ore becomes saturated it is possible that these basic assumptions no longer hold and the granular flow theories may no longer apply.

When granular flow theories were used to predict the vertical pressure on the bin floor and feeder then for a 70m high bin of 8m diameter a design pressure of 285kPa was calculated. If however the ore was considered to be a liquid with a maximum density of 2300kg/m^3 then a hydrostatic pressure of 1580kPa results. Unlike a liquid, a saturated ore will most likely transmit shear forces under static conditions and thus a purely hydrostatic pressure based on the density of the mixture of ore and water was considered unlikely. The walls of the bin will provide some shear support for the mass of saturated ore. For this case it was assumed that a more realistic upper bound pressure to be used in the design was one based on the granular material pressure component using a buoyant bulk density plus a hydrostatic component for the interstitial water. This gave a design loading for the floor and associated structure of 850kPa for a 8m dia bin of 70m height.

Having established these pressures it was decided to design the bin floor steelwork using code allowable stresses and deflections for the 850kPa pressure while the stresses could be allowed to reach $0.93F_y$ for the hydrostatic (1580kPa) case. All bin floor steelwork was supported from the drift floor rather than relying on wall shear connections.

To ensure that the maximum hydrostatic pressure associated with a 70m surcharge was not exceeded it was recommended that an access drift be driven into the upper level of the bin. This would allow discharge down the drift in the event that super saturated ore continued to be loaded into the in-feed ore pass. The arrangement and location of this drift to ensure that it did not become choked with ore was important. Novel methods of limiting the hydrostatic pressures by using self clearing drifts at intermediate levels were considered as a means of increasing the capacity of the bins by increasing their height but were not proceeded with.

A concrete infill floor was provided to seal the base of the bin, complement the depth requirement of the hopper and provide some damping of impact from falling ore on initial filling.

The potential problems in handling saturated ore within a bin of 3500 m^3 capacity does not end with increased floor loadings. There is the possibility of a "mud-rush" into the drive below the bin if super-saturated ore is presented to the feeder. However after careful consideration and despite the fact that a system could be designed to contain such a rush, the client opted not to proceed on the premise that accurate prediction of the onset of such an event was not possible. Instead the highest priority was given to preventing the ingress of water into the ore during its transfer to, or storage within the bins. Interseam water needed to be contained wherever possible and the use of water at transfer points to facilitate handling of the ore needed be limited or prohibited. The test results highlighted that there was potential for substantial improvements in the performance of all handling equipment if ore at lower moisture content was handled, though the practical difficulties in achieving this at the mine site had to be recognised.

The mine is presently considering a proposal to instal structural load monitoring on the bin floor steelwork. Such sensors could alert personnel to pressure increases thereby protecting the structure from excessive loads and warning operators of the potential for a mud rush.

9. CONCLUSION

The end result of the investigation was the proposal that underground bins of a form typically shown in Figure 5 should be constructed with an outlet as shown in Figure 4 utilising an apron feeder discharge. The use of an inclined drive into the upper section of the bin at the ore pass entry point was recommended to ensure that the hydrostatic loads did not exceed those predicted. The installation of access drifts into the bin at floor level would enable the bin to be emptied for maintenance of the feeder and associated inlet chute and liners. The drifts would also allow access for blasting of any stable compacted ore if this was necessary.

Recommendations were also made to operate the apron feeder at a low discharge rate during the initial filling of the bin or any subsequent refilling from empty to reduce the feeder startup loads. Once an arched discharge related stress state is established in the hopper chutework the high vertical loads associated with the initial fill are not expected to re-occur. It was also recommended that when ore with a high moisture was present in the bin that every effort be made to ensure that it did not remain undisturbed for long periods of time.

Since the incoming ore pass was only 2.1m diameter it was considered undesirable to fill the bin into the ore pass as blockages may occur if the ore pass was run full. Experience has since confirmed that blockages will occur requiring considerable effort and risk in dislodgment. Recommendations were made to use ultrasonic level detection equipment backed up by mercury tilt switches to prevent overfilling. These have been of limited success and a microwave system [10] of level detection is currently under investigation.

The first of these bins was excavated and the bottom sections site fabricated and installed in mid 1988. Following successful commisioning and subsequent operation, additional bins were considered. Currently five underground ore bins with the general arrangement detailed have been constructed with two more currently being built and a further two planned.

10. REFERENCES

1. JENIKE, A.W., "Storage and Flow of Solids". Bul. 108, Utah Engng. Expt. Station, Univ. of Utah, 1961

2. JENIKE, A.W., "Storage and Flow of Solids". Bul. 116, Utah Engng. Expt. Station, Univ. of Utah, 1962

3. JENIKE, A.W., "Storage and Flow of Solids". Bul. 123, Utah Engng. Expt. Station, Univ. of Utah, November 1964.

4. ARNOLD, P.C., McLEAN, A.G. and ROBERTS, A.W., "Bulk Solids: Storage, Flow and Handling". The University of Newcastle Research Associates (TUNRA) Australia, 2nd Edition, 1982.

5. ROBERTS, A.W., "Modern Concepts in the Design and Engineering of Bulk Solids Handling Systems". The University of Newcastle Research Associates (TUNRA)

6. "Standard Shear Testing Techniques for Particulate Solids using the Jenike Shear Cell", ISBN 0 85295 232 5. Published by The Institution of Chemical Engineers, European Federation of Chemical Engineers, 1989.

7. "Silos - Draft Design Code for Silos, Bins, Bunkers and Hoppers" Published by BSI in conjunction with the British Materials Handling Board. 1987.

8. HAMBLEY, D.F., PARISEAU, W.C. and SINGH, M.H. "Guidelines for Open-Pit Ore Pass Design. Vol 1." Bureau of Mines Open File Report 179 (1)–84, 1984

9. HAMBLEY, D.F., and SINGH, M.H. "Guidelines for Open-Pit Ore Pass Design. Vol II - Design Manual." Bureau of Mines Open File Report 179 (2)–84, 1984

10. SALTER, J.D., DOWNING, B.J., RIX, G.M. and MARAIS, M.G. "Development of rock pass level monitors for Finsch diamond mine, South Africa." *unknown source*, 1990

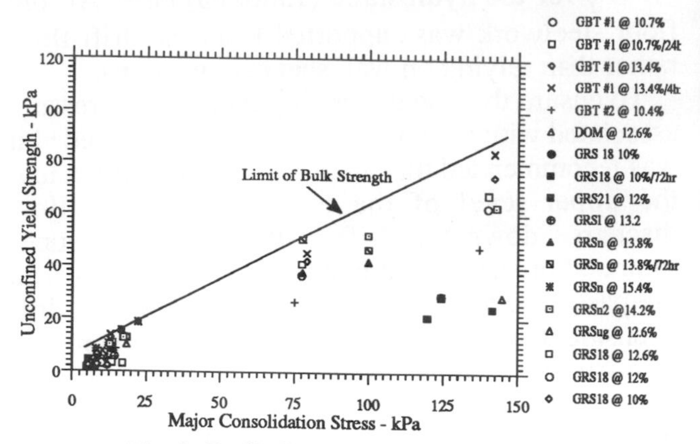

Fig 2. Bulk Strength of the Ore at High Consolidation Pressure

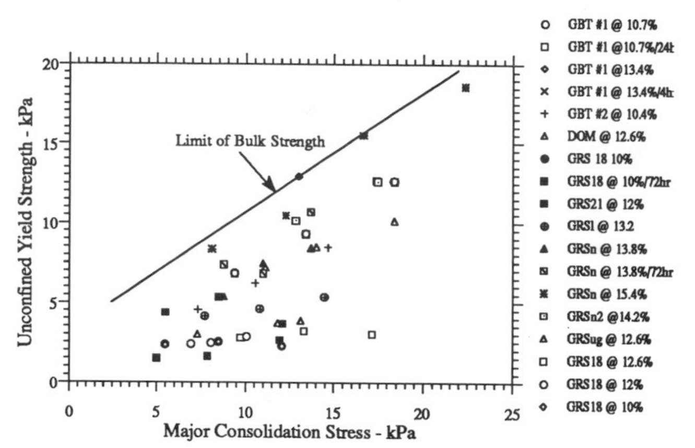

Fig 1. Bulk Strength of the Ore at Low Consolidation Pressure

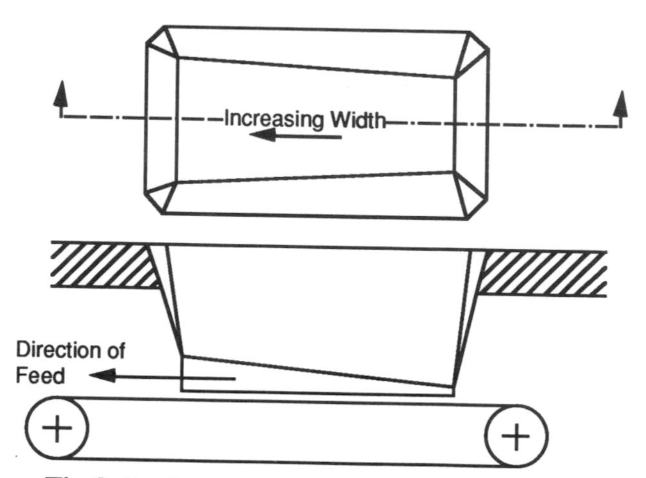

Fig 3. Preferred Outlet Arrangement in Flat Bottom Bin

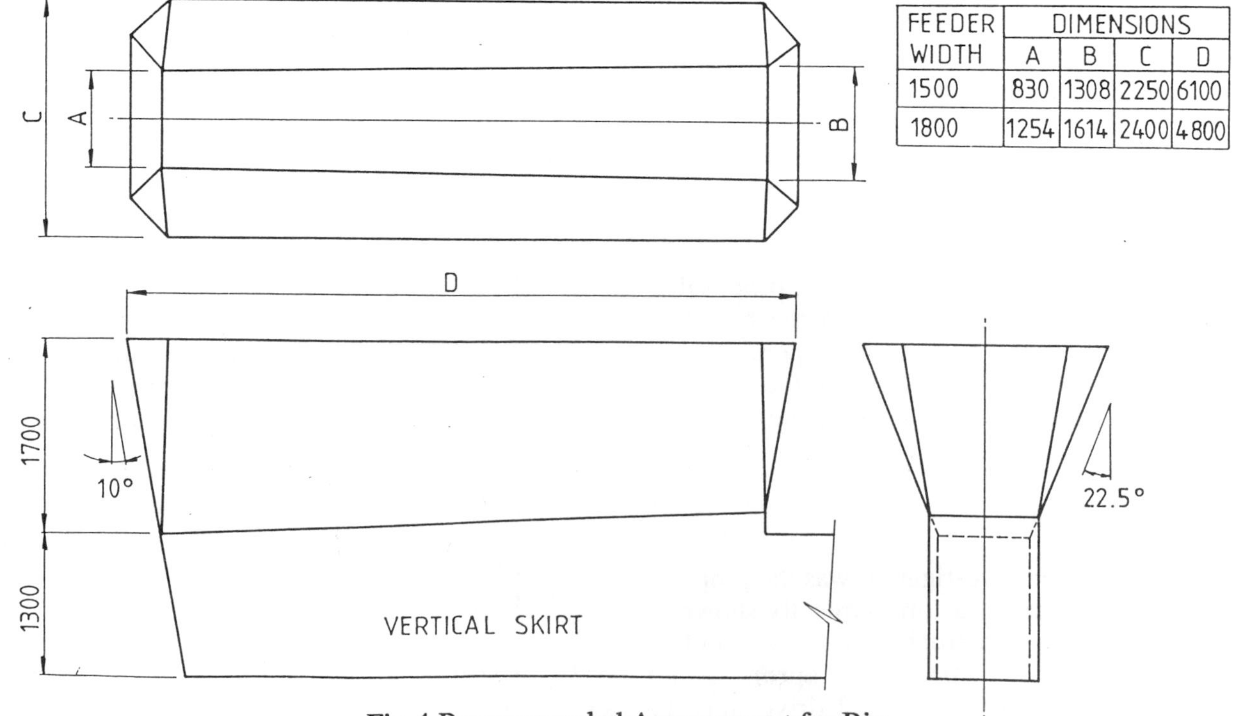

FEEDER WIDTH	DIMENSIONS			
	A	B	C	D
1500	830	1308	2250	6100
1800	1254	1614	2400	4800

Fig 4 Recommended Arrangement for Bin Chute

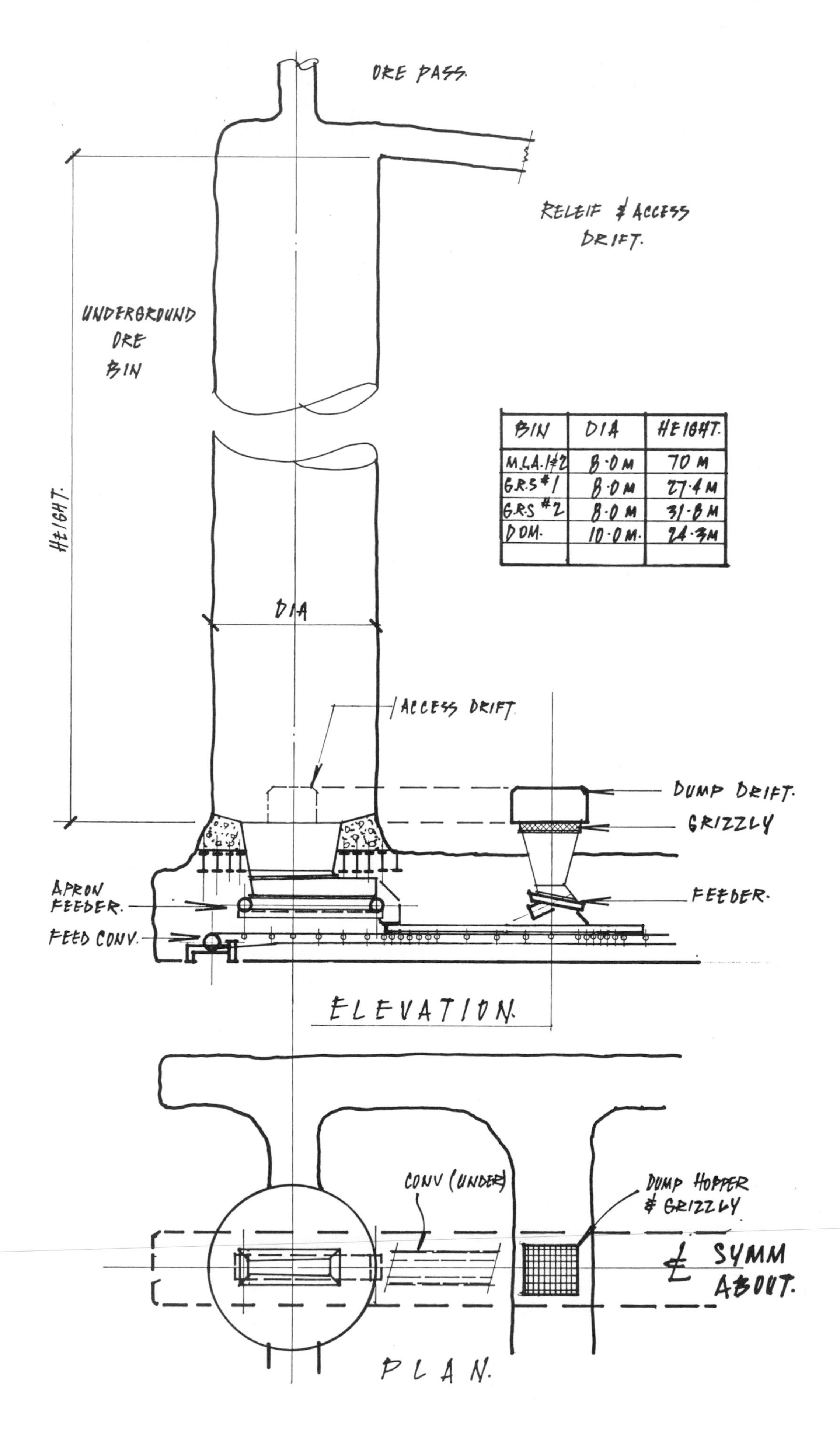

BIN	DIA	HEIGHT.
M.L.A.1#2	8·0 M	70 M
G.R.S #1	8·0 M	27·4 M
G.R.S #2	8·0 M	31·8 M
DOM.	10·0 M.	24·3 M

Figure 5. General Arrangement of Bin System

C418/046

Determining screw feeder geometry for specified hopper draw-down performance

A W ROBERTS, BE, PhD, ASTC, CEng, MIMechE, FTS
Institute for Bulk Materials Handling Research, University of Newcastle, New South Wales, Australia

SUMMARY

The design of screw feeders requires detailed consideration of the geometry at the interface of the screw and the feeder. This paper focuses on the determination of the required screw geometry to achieve specified draw-down performance objectives. A theory is presented which takes into account the screw geometry and variable volumetric efficiency along the screw in order to predict the required parameters of pitch, screw diameter, and core or shaft diameter to meet the stated objectives. Specific examples are highlighted and the optimisation of screw geometry to achieve a required objective such as uniform draw-down is presented.

1. INTRODUCTION

Screw feeders are used extensively in the bulk materials handling and powder processing industries to control the feed from gravity flow, mass-flow hoppers. Because of their positive displacement mode of operation, they have the particular advantage of providing good volumetric feed control; in the case of the feeding of fine powders, they also permit 'automatic' sealing of the hopper against flooding and uncontrollable discharge. This is achieved through the extended section of the screw that projects beyond the hopper at the discharge end. This is illustrated in Figure 1.

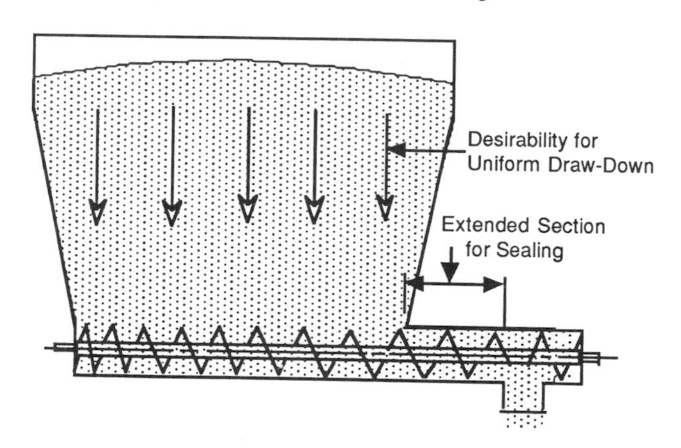

Figure 1. Screw Feeder

While screw feeders are sometimes fitted to conical hoppers, they are most commonly used in conjunction with plane-flow hoppers. To this end, it is most important that they be correctly designed and matched to the hopper. The screw feeder geometry must be carefully selected in order to give uniform draw-down in the hopper, which requires the capacity of the screw to increase in the direction of feed. If this objective is not met then the draw-down will not be uniform and this can have detrimental results, particularly with respect to segregation and degradation of the product. For example, it is well known that constant pitch, constant diameter screws are most unsatisfactory since the screw space is fully charged at the rear end of the hopper. Since no more material can then flow into the screw space, the draw-down is via a rathole at the rear end of the hopper, with bulk of the material in the hopper being 'dead'. The correct design of screw feeder will involve the selection of combinations of variable screw diameter, shaft diameter and pitch to give the necessary volume increase along the screw.

A detailed study of entrainment patterns of hopper and screw feeder combinations has been studied by Bates (Ref[1]). More recently a theory has been proposed by Roberts (Ref.[2]) which permits the interrelation between geometrical variables of the screw feeder and hopper draw-down patterns to be examined. This study is part of a joint research project into screw feeder performance being conducted by the POSTEC Research group in Norway and the University of Newcastle in Australia. The general principles involved in abovementioned theory for designing screw feeders for uniform draw-down in hoppers is presented herein.

2. VOLUMETRIC CAPACITY OF SCREW FEEDER

2.1. Screw Geometry

The relevant geometry of screw feeders is shown in Figure 2.

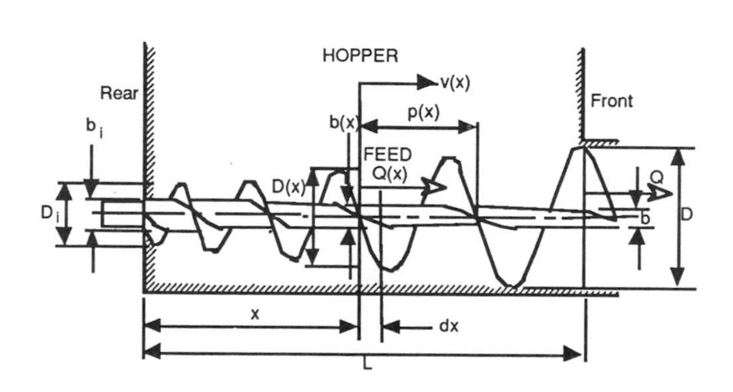

Figure 2. Screw Feeder Geometry.

(a) Screw Geometrical Variables

At any location x, the relevant screw geometrical variables are,

$p(x)$ = pitch
$D(x)$ = screw diameter
$b(x)$ = core or shaft diameter

When $x = 0$, the initial values apply. That is,

p_i = initial or starting pitch
D_i = initial screw diameter
b_i = initial shaft diameter

When x = L, the geometry at the outlet is defined. That is,

D = screw diameter at the outlet
p = pitch at outlet
b = shaft diameter at outlet

(b) Characteristic Dimension

The screw diameter and pitch at the outlet, together with the rotational speed, control the throughput or feed rate of the feeder. This implies that the volumetric capacity specified at the outlet is equal to or greater than the volumetric capacity at any location x. It is usual to select the screw diameter D as the geometrical characteristic dimension for the screw feeder and express the other geometrical variables as ratios of D.

2.2 Volumetric Flow Rate

At location x, the volumetric feed rate is given by

$$Q(x) = A(x)\, v(x)\, \eta_v(x) \qquad (1)$$

where, for location x,

$A(x)$ = Cross-sectional area of feeder
$v(x)$ = Feed velocity
$\eta_v(x)$ = Volumetric efficiency

The axial conveying velocity of the screw is related to the pitch p(x) by

$$v(x) = \frac{\omega}{2\pi}\, p(x) \qquad (2)$$

Where

$$\omega = \frac{d\theta}{dt} = \text{angular velocity of screw}$$

θ = angular displacement

Hence (1) becomes

$$Q(x) = \frac{\omega}{2\pi}\, A(x)\, p(x)\, \eta_v(x) \qquad (3)$$

Cross-Sectional area A(x) is given by

$$A(x) = \frac{\pi}{4}\,[\, D(x)^2 - b(x)^2\,] \qquad (4)$$

2.3 Feeder Capacity

At the feeder outlet, when x = L, the capacity of the feeder is given by

$$Q = \frac{\omega}{2\pi}\, A\, p\, \eta_v \qquad (5)$$

where

$$Q = \frac{\pi}{4}\,[\, D^2 - b^2] \qquad (6)$$

A = Cross-sectional area of feeder at outlet
η_v = Volumetric efficiency at outlet

2.4 Volumetric Efficiency

The volumetric efficiency at location x is given by

$$\eta_V(x) = \frac{Q(x)}{Q_T(x)} \qquad (7)$$

where

$Q_T(x)$ = Maximum theoretical throughput at location x with the feeder running completely full and the bulk solid moving at the feeder speed without slip and/or rotational motion.

The actual throughput Q(x) will be less than $Q_T(x)$ due to the following reasons:

(i) The axial velocity of bulk solid will be less than ideal or optimum velocity owing to rotary motion imparted by the screw.

(ii) Some slip may occur in the clearance space between the screw and the casing.

(iii) The 'fullness' of screw will decrease as rotational speed increases. That is the screw has the capacity to feed bulk solid faster than the bulk solid can flow by gravity in the hopper to fill the screw space.

(iv) The thickness of the screw blade will reduce the 'fullness'. However this effect is negligible.

Research has shown, (Refs. [3-5]), that the volumetric efficiency is composed of two components:

$$\eta_v = f(\eta_{VR} \cdot \eta_{VF}) \qquad (8)$$

where

η_{VR} = Rotational Effect - this decreases with increase in screw speed making for a more efficient screw feeder.

η_{VF} = 'Fullness' Effect - this decreases with increase in screw speed making for a less efficient screw feeder.

At lower speeds consistent with those of screw feeders, the η_{VF} effect is small. At higher speeds, as in screw conveyors, η_{VF} decreases at a faster rate than the decrease in η_{VR}. Hence the volumetric efficiency decreases with increase in speed.

It is to be noted that η_{VR} depends on the screw helix angles:

(i) At each location x, η_{VR} increases from the outside to the core since the helix angle increases in this direction.

(ii) Along the screw, the screw geometry changes - hence η_{VR} changes.

3. DISTRIBUTION OF THROUGHPUT ALONG SCREW

Consider the plane-flow hopper and screw feeder combination shown in Figure 3. The long slotted opening of the hopper requires particular attention to be given to the design of the feeder. The choice of correct feeder geometry in relation to the hopper and feeder interface is necessary in order to achieve the desired draw-down pattern in the hopper. This requires a knowledge of the characteristics governing the distribution of feed rate along the screw.

3.1 Variation of Throughput along Screw

Since the screw capacity increases in the direction of feed, the throughput or feed rate increases by dQ(x) over distance dx. Hence the gradient of the feed rate or screw capacity at location x is

$$Q(x)' = \frac{dQ(x)}{dx}$$

Substituting for Q(x) from equation (3), the gradient of Q(x) with respect to x becomes

$$Q(x)' = \frac{\omega}{2\pi}\,\frac{d}{dx}\left\{\, A(x)\, p(x)\, \eta_v(x)\,\right\} \qquad (9)$$

or

$$Q(x)' = \frac{\omega}{2\pi}\left\{p(x)\,\eta_v(x)\frac{dA}{dx} + A(x)\frac{dp}{dx} + A(x)\,p(x)\frac{d\eta}{dx}\right\} \quad (10)$$

The total throughput at location x is

$$Q(x) = \int_0^x Q(x)'dx \quad (11)$$

The units are $Q(x) = m^3/s$ and $Q(x)' = m^2/s$

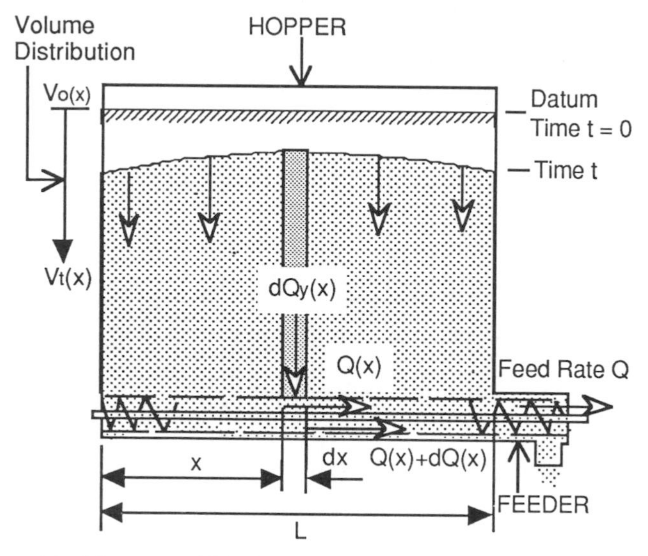

(a) Hopper and Feeder Interaction during Draw-Down

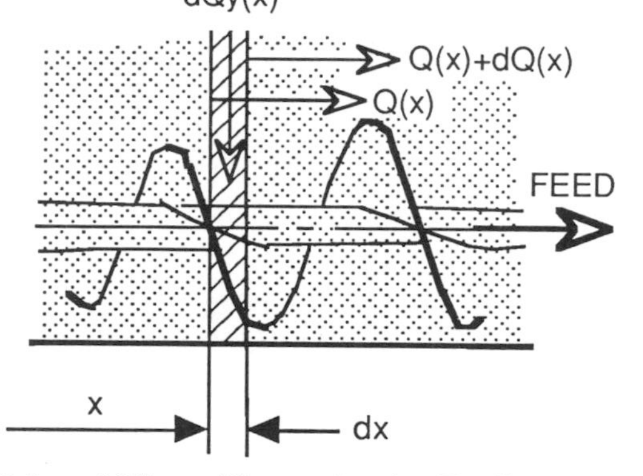

(b) Enlarged View of Screw showing Feeding Mechanism

Figure 3. Operation of Hopper and Screw Feeder System

3.2 Draw-Down Pattern in Hopper

Referring to Figure 3, at location x, the continuity of feed from the hopper to the screw is such that

$$dQ_y(x) = dQ(x) = Q(x)'dx \quad (12)$$

or

$$Q_y(x)' = Q(x)' \quad (13)$$

where $dQ_y(x)$ = incremental volume feed rate in hopper due to gravity flow

$Q_y(x)'$ = volume feed rate from hopper per unit length (m^2/s)

At location x, the total volumetric feed V(x) per unit length ($m^3/m = m^2$) from the hopper after time t has elapsed is

$$V(x) = \int_0^t Q_y(x)'dt = \int_0^t Q(x)'dt \quad (14)$$

If the screw feeder operates at constant speed, then $Q(x)'$ is constant and equation (14) becomes

$$V(x) \text{ at time } t = Q(x)'\ t \quad (15)$$

The draw-down pattern may be plotted for different or increasing values of time t as follows:

$V_o(x)$ = Initial volume per unit length at time t = 0. (m^2).

Volume distribution in hopper after time t has elapsed during constant feed

$$V_t(x) = V_0(x) - Q(x)'t \quad (16)$$

Expressing "t" in terms of screw revolutions. The time T for one revolution is

$$T = \frac{2\pi}{\omega} \quad (17)$$

Then (16) can be written as

$$V_{nT}(x) = V_o(x) - \frac{2\pi n}{\omega} Q(x)' \quad (18)$$

$$n = 0, 1, 2, 3.....$$

where n = number of screw revolutions.

3.3 Uniform Draw-Down

The application of equation (18) will enable the draw-down pattern in the hopper to be determined for a given screw feeder geometry. Ideally, the feeder geometry may be selected to achieve a desired draw-down pattern in the hopper. In most cases, the aim is to obtain uniform draw-down, for which the feeder should be designed so that

$$Q(x)' = \frac{dQ(x)}{dx} = \text{Constant} \quad (19)$$

4. NON-DIMENSIONAL REPRESENTATION

It is convenient to present the foregoing performance parameters in non-dimensional form. The hopper feeder representation is as shown in Figure 4.

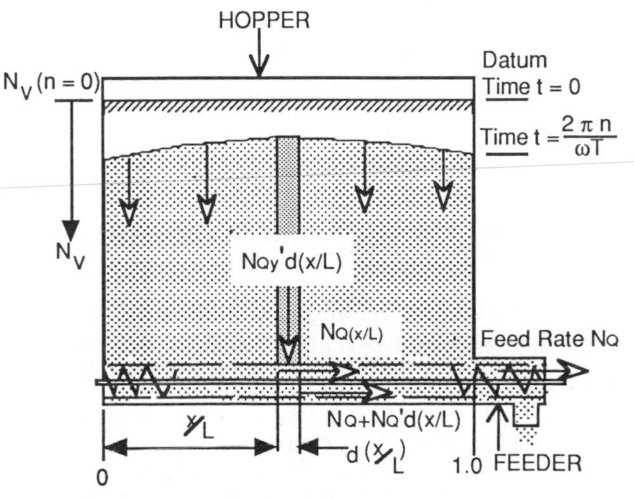

Figure 4. Non-Dimensional Representation of Hopper and Feeder

The screw diameter D at the outlet and feeder length L are selected as the characteristic dimensions; all other screw geometrical variables are expressed as functions of D and the coordinate distance x is expressed as a function of L.

4.1 Non-Dimensional Coordinate N_x

$$N_x = \frac{x}{L} \tag{20}$$

4.2 Non-Dimensional Throughput N_Q

$$N_Q = \frac{Q(x)}{\omega D^3} \tag{21}$$

where ω = Angular Velocity of Screw rad/sec

$= \frac{\pi s}{30}$ where s = rev/min

Substituting for Q(x) from (3) to

$$N_Q = \frac{1}{2\pi D^3} \left[A(x)\, p(x)\, \eta_V(x) \right] \tag{22}$$

4.3 Non-Dimensional Gradient of Throughput N_Q'

$$N_Q' = \frac{Q(x)'}{\omega D^2} \tag{23}$$

Substituting from equation (10)

$$N_Q = \frac{1}{2\pi D^2} \left\{ p(x)\,\eta_v(x)\frac{dA}{dx} + A(x)\,\eta_v(x)\frac{dp}{dx} + A(x)\,p(x)\frac{d\eta}{dx} \right\} \tag{24}$$

4.4 Draw-Down Pattern N_V

From equation (18),

$$N_V = \frac{V_{nT}(x)}{D^2} \tag{25}$$

N_V is the non-dimensional volume per unit length which is computed at discrete time values to indicate the draw-down patterns.

4.5 Non-Dimensional Speed N_s

For corresponding speeds – from Ref. [3]

$$N_s = \frac{g}{D\omega^2} \tag{26}$$

5. VARIATIONS IN SCREW GEOMETRY - GENERAL CASE

The application of the foregoing will now be examined in relation to specific screw geometrical configurations. For practical purposes, linear variations in screw geometries are considered.

5.1 Geometrical Relationships

It is assumed that all screw variables are expressed as linear functions, with the screw diameter D at the outlet being chosen as the characteristic dimension.

(a) Screw Diameter

$$D(x) = D\left[C_{11} + (1 - C_{11})\frac{x}{L} \right] \tag{27}$$

where

$$C_{11} = \frac{D_i}{D} \tag{28}$$

(b) Shaft or Core Diameter

$$b(x) = D\left[C_{21} + (C_{21} - C_{22})\frac{x}{L} \right] \tag{29}$$

where

$$C_{21} = \frac{b_i}{D} \quad \text{and} \quad C_{22} = \frac{b}{D} \tag{30}$$

Note that $C_{21} \leq C_{11}$

(c) Pitch

$$p(x) = D\left[C_{31} + (C_{32} - C_{31})\frac{x}{L} \right] \tag{31}$$

where

$$C_{31} = \frac{p_i}{D} \quad \text{and} \quad C_{32} = \frac{p}{D} \tag{32}$$

5.2 Area of Cross-Section

In this case, equation (4) becomes

$$A(x) = \frac{\pi D^2}{4} \left[K_1 + K_2\left(\frac{x}{L}\right) + K_3\left(\frac{x}{L}\right)^2 \right] \tag{33}$$

where

$$\left. \begin{aligned} K_1 &= C_{11}^2 - C_{21}^2 \\ K_2 &= 2\left\{ C_{11}(1 - C_{11}) + C_{21}(C_{21} - C_{22}) \right\} \\ K_3 &= (1 - C_{11})^2 - (C_{21} - C_{22})^2 \end{aligned} \right] \tag{34}$$

5.3 Derivatives of Geometrical Relationships

$$\frac{dA}{dx} = \frac{\pi D^2}{4L} \left[K_2 + K_3\left(\frac{x}{L}\right) \right] \tag{35}$$

$$\frac{dp}{dx} = (C_{32} - C_{31})\frac{D}{L} \tag{36}$$

$$\frac{dD}{dx} = (1 - C_{11})\frac{D}{L} \tag{37}$$

5.4 Volumetric Efficiency

As discussed in Section 2.4, the volumetric efficiency is influenced by several factors. At low speeds of operation, the rotational motion of the bulk solid has a major influence. Measurements of volumetric efficiency in screw conveyors handling grain have indicated volumetric efficiencies in the order of 80% at low speeds for screws of pitch equal to the screw diameter. Rotational motion in the feeder is a function of the effective screw helix angle which, in turn, depends on the screw pitch to diameter ratio. The properties of the bulk material will also influence the volumetric efficiency. Current research is directed at determining, more precisely, the volumetric efficiencies of hopper and screw feeder combinations. For the purpose of the present analysis, the following simple expression for the volumetric efficiency is proposed:

$$\eta_V = 1 - C_E \frac{p(x)}{D(x)} \qquad (38)$$

where the constant C_E is assumed to be $C_E \leq 0.2$.

The derivative of η_V with respect to x is

$$\frac{d\eta_V}{dx} = \frac{C_E}{[D(x)]^2} \left\{ p(x) \frac{dD}{dx} - D(x) \frac{dp}{dx} \right\} \qquad (39)$$

The relevant expressions for D(x) and p(x) may be substituted into equations (38) and (39).

5.5 Non-Dimensional Parameters

The various expressions developed above may be substituted into equations (22), (24) and (25) for N_Q, N_Q' and N_V respectively. In this way, the influence of variations in screw geometry may be conveniently examined.

6. TWO SCREW FEEDER EXAMPLES

To illustrate the theory presented, the two screw geometries illustrated in Figure 5 are examined.

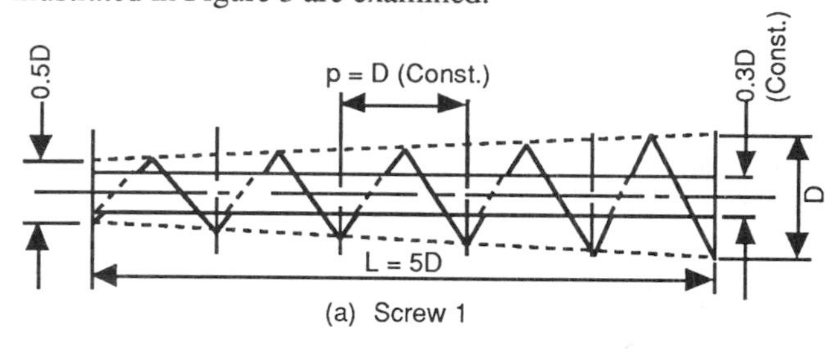

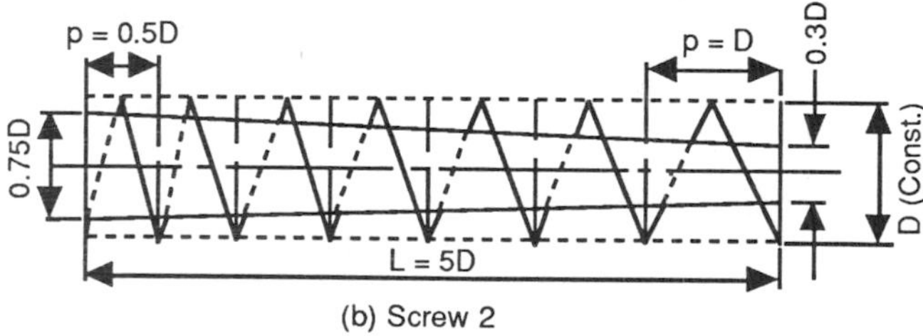

Figure 5. Screw Feeder Examples

(a) Case 1- Screw Geometry of Figure 5(a)

This screw has variable diameter and constant pitch and shaft diameter. The relevant constants are

$C_{11} = 0.5$ $C_{21} = 0.3$ $C_{22} = 0.3$ $C_{31} = 1.0$ $C_{32} = 1.0$
$C_E = 0.2$

The performance is presented in Figures 6 and 7. Figure 6 shows the non-dimensional parameters N_Q and N_Q' as functions of x/L while Figure 7 shows the the volumetric feed per unit length N_V at non-dimensional time increments n = 0, 10, 20 and 50. The N_V graphs indicate the draw-down patterns. As can be observed, the feed is biased towards the front of the hopper. This is also indicated by the N_Q' graph which shows an increase with x/L.

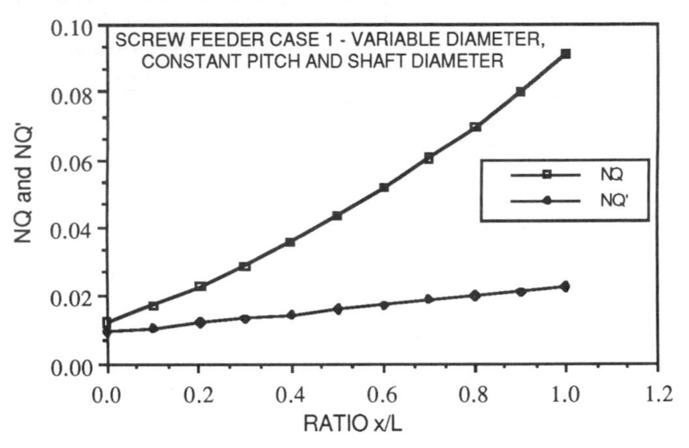

Figure 6. Screw Feeder Case 1. Non-Dimensional Parameters

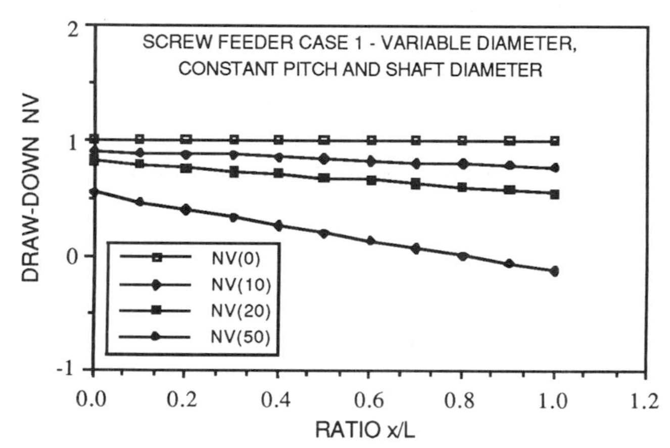

Figure 7. Screw Feeder Case 1. Draw-Down Patterns

(b) Case 2- Screw Geometry of Figure 5(b)

This screw has constant diameter and variable pitch and shaft diameter. The relevant constants are

$C_{11} = 1.0$ $C_{21} = 0.75$ $C_{22} = 0.3$ $C_{31} = 0.5$ $C_{32} = 1.0$
$C_E = 0.2$

The performance is presented in Figures 8 and 9. Figure 8 shows the non-dimensional parameters N_Q and N_Q' as functions of x/L while Figure 9 shows the the volumetric feed per unit length N_V at non-dimensional time increments n = 0, 10, 20 and 50. In this case N_Q' is almost constant and the draw-down is virtually uniform. The superiority of screw 2 is clearly demonstrated.

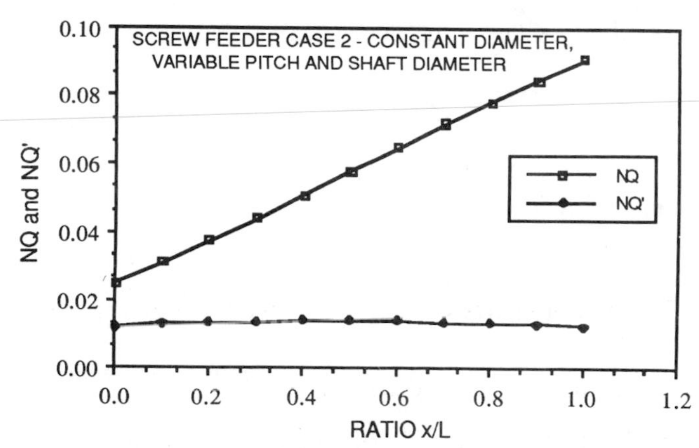

Figure 8. Screw Feeder Case 2. Non-Dimensional Parameters

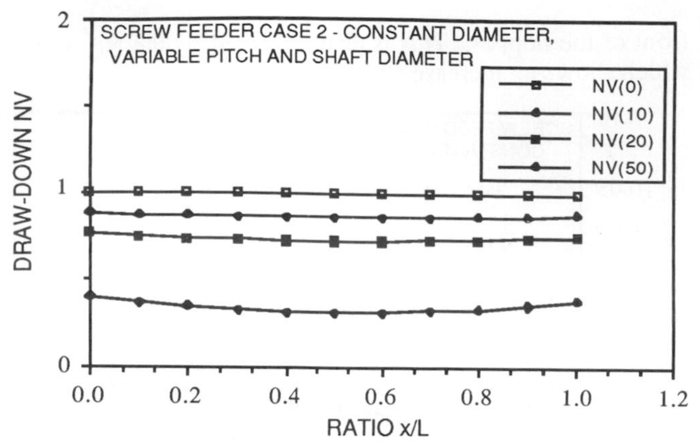

Figure 9. Screw Feeder Case 2. Draw-Down Patterns

7. 'OPTIMUM' SCREW GEOMETRY

It is theoretically possible to determine an optimum screw geometry to achieve a specified performance criterion such as uniform draw-down so that $N_Q' = 0$. By way of illustration, consider the example of determining the outside screw diameter profile for constant draw-down given that the screw has a constant pitch p(x) = D and constant shaft diameter b(x) = 0.3D. The initial diameter is $D_i = 0.5D$. As a first order approximation, it is assumed that the volumetric efficiency $\eta_V(x)$ is also constant.

With $N_Q' = 0$, it may be shown that the outside diameter is given by

$$D(x) = D \sqrt{(1 - C_{11}^2)\frac{x}{L} + C_{11}^2} \qquad (40)$$

where $C_{11} = D_i/D = 0.3$

The screw diameter profile ratio D(x)/D is shown in Figure 10, while the actual screw is shown in Figure 11. As can be seen, the outside diameter profile does not depart a great deal from the straight, tapered screw profile. Whilst this example is somewhat 'academic', it does illustrate that optimum geometries may be chosen which may not be all that difficult to manufacture.

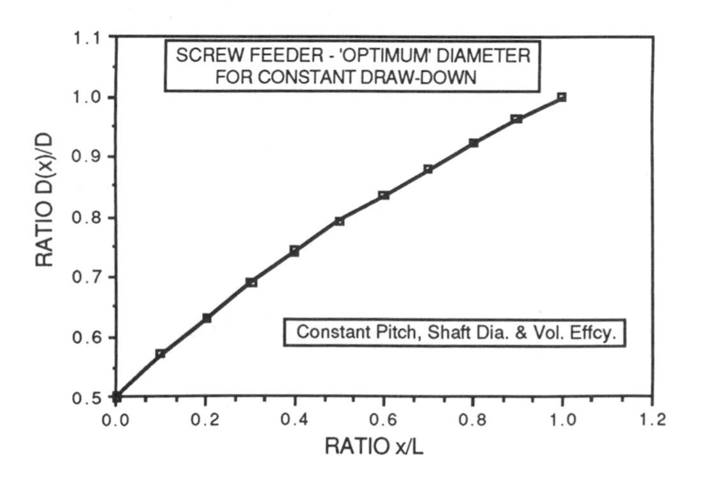

Figure 10. Ratio of D(x)/D for Uniform Draw-Down

8. CONCLUSIONS

This paper has presented a theory and methodology which may be applied to the design of screw feeders, either as a diagnostic tool to analyse existing or given screw geometrical configurations, or as a means of determining the required screw geometrical relationships to achieve a specified performance in terms of volumetric capacity. In general the attainment of uniform draw-down in the hopper is a desired goal if segregation of feed is to be avoided. While the theory presented takes into account the variations in screw diameter, pitch and shaft or core diameter, a simplified assumpion has been made with respect to the likely variation in volumetric efficiency along the feeder. This is generally satisfactory to indicate the application of the theory to screw feeder design; however, it is necessary for further information be obtained from experimental studies to refine the equations for predicting volumetric efficiency. Nonetheless, it is believed that the theory presented in this paper provides a satisfactory foundation for designing screw feeders.

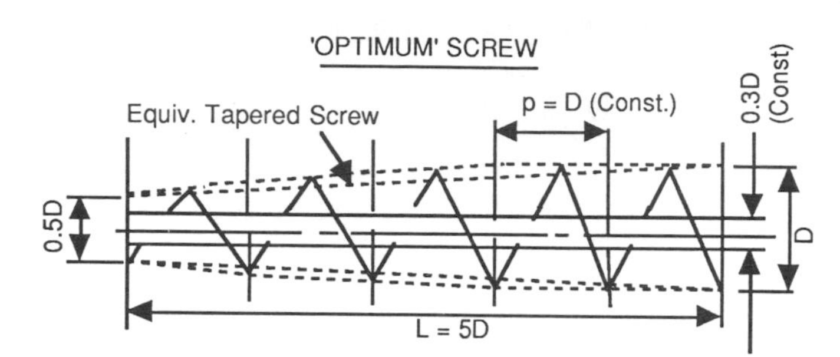

Figure 11. Actual Geometry for Optimum Screw giving Uniform Draw-Down

9. ACKNOWLEDGEMENTS

The concepts presented in this paper were initiated when the author was a Norwegian Government NTNF Senior Visiting Scientist at POSTEC Research, Porsgrunn Norway during April, May 1990. The support of NTNF is gratefully acknowledged. Particular thanks are due to Professor Sunil de Silva, Managing Director, and to Dr. K.S. Manjunath, Scientist, of POSTEC for the opportunity of visiting POSTEC and being able to contribute to the research being undertaken in screw feeder performance.

10. REFERENCES

1. Bates, L. "Entrainment Patterns of Screw Hopper Dischargers". Trans. A.S.M.E., Jnl. of Engng for Industry, Vol. 91, Series B, No.2, May 1969. (pp.295-302).

2. Roberts, A.W. "Volumetric Capacity of Screw Feeders". Interim internal Report to POSTEC Research, Porsgrunn, Norway, May 1990.

3. Roberts, A.W. and Willis, A.H. "Performance of Grain Augers", Proceedings of the Institution of Mechanical Engineers, Vol. 176 (8), 1962. (pp.165-194)

4. Roberts, A.W., "An Investigation of Grain Vortex Motion with Relation to the Performance within Vertical Grain Augers". Proceedings of the Institute of Mechanical Engineers, Vol. 178 (1), Nr. 12 (1963-1964).

5. Roberts, A.W., "Screw or Auger Conveyors Revisited", Paper for presentation at the Powder and Bulk Solids Conference, Chicago, USA. 1991

C418/044

Pneumatic conveying technology improving Australian bulk handling operations

O C KENNEDY, BE, P W WYPYCH, PhD, CPEng, MIEAust and P C ARNOLD, BE, PhD, FIEAust
Department of Mechanical Engineering, University of Wollongong, New South Wales, Australia

Synopsis This paper discusses the application of pneumatic conveying technology to materials handling, in particular focussing on recent developments in this field and highlighting the particular benefits use of this technology can bring to industry.

Nomenclature

D	Internal diameter of pipeline, m
Fr	Froude number, $Fr = V_f (g D)^{-.5}$
L	Total length of pipeline, m
L_v	Total length of vertical lift, m
m_f	Mass flow rate of air, kg s^{-1}
m_s	Mass flow rate of solids, kg s^{-1}
m^*	Mass flow rate ratio, $m^* = m_s m_f^{-1}$, kg kg^{-1}
p_f	Gauge air pressure, Pa g
P_{atm}	Atmospheric air pressure, Pa abs
P_f	Absolute air pressure, $P_f = P_{atm} + p_f$, Pa abs
Δp_f	Air-only pressure drop, Pa
Δp_t	Total pipeline air pressure drop, Pa
Re	Reynolds number
t	Operating air temperature, °C
T	Absolute air temperature, K
V_f	Superficial air velocity, $V_f = m_f (\rho_f A)^{-1}$, m s^{-1}
v_s	Conveying velocity of particles, m s^{-1}
v_∞	Free settling velocity of particle, m s^{-1}
ε	Pipe roughness, m
λ_f	Air-only friction factor
λ_s	Particle friction factor
λ_s^*	Pipe friction coefficient
μ_f	Absolute air viscosity, Pa s
ρ_f	Air density, kg m^{-3}
ρ_f^*	Air density ratio, $\rho_f^* = \rho_{fi}/\rho_{fe}$
ρ_s	Solids particle density, kg m^{-3}

Subscripts

e	Final (or exit) value at end of pipeline (or section of pipe)
i	Initial value at start of pipeline (or section of pipe)
m	Mean value for pipeline (or section of pipe)
min	Minimum value
j	Value relating to pipe no. j (j = 1 designates pipe at end, j = n designates pipe at start)
n	Maximum number of different pipe diameters

1. Introduction

The specific advantages of pneumatic conveying as a bulk materials handling process are attracting an expanding application in industry. These advantages include:

- The ease and versatility with which pneumatic conveying pipelines can be routed compared with alternate systems (e.g. conveyor belting, bucket elevators), particularly when considering the upgrading of existing installations.

- The enclosed nature of this conveying system minimises the opportunity for contamination of either the environment and/or the material being transported with obvious advantages for the handling of irritant or even toxic substances, as well as finding many applications in the food industry. This characteristic also offers increased security in comparison to other systems (e.g. open conveyors etc).

- Its suitability to automated operation, and hence low manning requirement.

Associated with, and indeed linked to, this expanding use of pneumatic conveying throughout industry has been a steady research effort focusing on the underlying principles of this technology and potential enhancements to the technology. As a result, numerous advances have been made over recent years.

This paper is intended to discuss a number of these developments, indicate the potential offered by their application, and in some cases, refer to experience with these systems in industrial situations.

2. Long-distance and/or high-throughput applications

Distinctive features of these types of applications include:

- A high pressure drop across the pipeline
- A limited range of reliable conveying conditions to meet the operational requirements of the system (i.e. with respect to short-distance and/or dense-phase systems)
- Difficulty in obtaining test rig data on a comparable scale to the proposed installation

Consequently, attempts have, and continue, to be made to generate reliable scale-up procedures that allow test rig data to be used to predict the performance of the full-sized system with reasonable accuracy and confidence. For example, refer to the scale-up equations proposed by Mills et al [1]. Inherent in this type of design procedure is the appropriate selection of critical elements of the conveying system, e.g. pipe diameter, air requirement and the total system pressure to be anticipated. Obviously an incorrect selection of these parameters would have a severe impact on the efficient and reliable operation of the final system.

Furthermore, to make the pneumatic conveying solution attractive for long distance and/or high throughput applications the pipeline in most cases should be stepped, with the diameter increasing towards the outlet. The primary benefits of this strategy are:

- Control the air and solids velocities along in the pipeline. Single-diameter pipeline systems that require high upstream pressures will generate very high exit velocities. This, in turn, is associated with excessive pipeline wear and product degradation. Stepping up the pipe diameter allows this velocity increase to be reversed at selected points (while maintaining an acceptable minimum velocity).
- Optimise the air usage required for the application.
- Optimise the capital cost of the system.

Therefore procedures that combine the reliable scale-up of test rig data with the optimal selection of a stepped diameter pipeline layout are particularly significant.

2.1 Long-distance applications

A procedure that combines these two attributes for long distance applications has been developed at the University of Wollongong, it relies on two major components:

i) Extensive test rig data, that establishes the behaviour of the material over a range of pipeline lengths and diameters.
II) Use of this data to determine, by correlation, the best values for characterising parameters in a pneumatic conveying model.

Having established the model parameters specific to this material it is possible to then use the model to predict the performance of the material in a specified conveying system. This procedure is suited to analysis by computer, and given a desired set of criteria (for example specific values of total pipe length, m_s, a total pipeline air pressure drop, a range of pipe diameters to use and the number of steps to consider) it is possible by a series of iterative computations to determine an optimum pipeline design (i.e. the specific pipe diameters to use and the appropriate length of each section).

The model utilises a friction factor approach (Darcy-Weisbach equation) and may be characterised by the following equations:

$$\Delta p_t = \lambda \frac{\rho_{fm} V_{fm}^2 \Delta L}{2 D} \quad \text{..... (1)}$$

The friction factor (λ) is split into a solids (λ_s) and air (λ_f) component:

$$\lambda = \left(\lambda_f + m^* \lambda_s\right) \quad \text{..... (2)}$$

The air-only friction factor (λ_f) may be estimated from [2]:

$$\lambda_f = \frac{1.325}{\left[\ln\left(\frac{\varepsilon}{3.7\,D} + \frac{5.74}{Re^{0.9}}\right)\right]^2} \quad \text{..... (3)}$$

An expression for the solids friction factor is determined by applying a correlation analysis to the test rig data. Various forms for this expression have been used [3], in each case basically consisting of a power function relationship of significant dimensionless parameters. For the examples presented here the relationship shown in equation (4) has been used.

$$\lambda_s = \mathbf{A}\,(m^*)^{\mathbf{b}}\,(Fr_m)^{\mathbf{c}}\,(\rho_f^*)^{\mathbf{d}} \quad \text{..... (4)}$$

where ρ_f^* is the ratio of the air densities (ρ_{fi}/ρ_{fe}), and the values of **A**,**b**,**c** and **d** are determined from a correlation analysis of the test rig data, and are specific to the material.

For this correlation analysis to provide meaningful results (in terms of scale-up accuracy), it is best that the test data comprise a wide range of phase densities (m^*), air pressures, pipe lengths and diameters.

In applying this approach to a stepped pipeline design, these equations are used to determine the air pressure drop for a given m_s and m_f, working back from known conditions at the outlet (usually atmospheric pressure) and determining the pressure drop for each pipe section in turn with the total pressure drop of the system being the summation of these individual pressure drops for each pipe section, i.e.

$$\Delta p_j = \frac{\lambda_j \rho_{fm} V_{fm}^2 \Delta L_j}{2 D_j} \quad \text{..... (5)}$$

$$\Delta p_t = \sum_{j=1}^{n} \left(\Delta p_j\right) \quad \text{..... (6)}$$

where the subscript "j" in equations (5) and (6) refers to the j'th pipe section in a stepped pipe line system containing "n" steps.

Applying this procedure for a range of solids and air mass flow rates results in the determination of the conveying characteristics of the material through a specified pipeline configuration (i.e. given the pipe diameters and lengths for each step). To optimise the pipeline configuration, two further aspects need to be considered:

- The design point needs to be selected (i.e. the system will be optimised for a specific conveying rate, or m_s value).
- A criterion for minimum reliable conveying criteria (say associated with the minimum air velocity in each pipe section - at the upstream end) needs to be selected.

The design point should be clear from the requirements of the application (the point here being to note that a system optimised for one particular design point will not generally be operating under optimised conditions if say, it is desired to run the system at a higher mass flow rate).

Minimum reliable conveying conditions have been the subject of considerable research, frequently concentrating on the selection of a suitable minimum superficial air velocity (V_{fmin}) to be maintained in the pipeline (a number of these relationships are presented in [3]). A related approach that has been found to be worthwhile for materials such as pulverised coal [4] is to base the minimum conveying conditions on the Froude Number, selecting an appropriate minimum value for this parameter after analysis of testwork on the material. This simple expression may be incorporated readily into the analysis procedure and, when computerised, becomes part of an iteration procedure that allows the pipeline configuration to be optimised.

As an example of this procedure the following results for pulverised coal are presented. Testwork was conducted on a 948 m pipeline, consisting of four pipe diameter sections as indicated in Table 1.

Table 1: Test Rig Pipeline Arrangement

Pipe Diameter (mm)	60	69	81	105
Pipe Length (m)	147	390	261	150

Applying test data from this rig during the conveying of a pulverised coal (further details in [4]) to the correlation analysis noted previously resulted in values for the constants in equation (4) of:

$$\lambda_s = 2.713\,(m^*)^{-0.596}\,(Fr_m)^{-1.576}\,(\rho^*)^{-0.06}$$

Using this result to optimise the design of a 2000 m pipeline to convey the material at a rate of 2 kg/s (using pipes with internal diameters of 178, 154 and 127 mm) generates the conveying details listed in Table 2.

Table 2: Optimised Pipe Configuration & Conveying Conditions

Pipe I. D. (mm)	Pipe Length (m)	Δp_j (kPa)	V_{fi} (ms^{-1})	V_{fe} (ms^{-1})
178	833.3	88.9	7.9	14.9
154	508.3	82.9	7.4	10.6
127	658.4	168.9	6.7	10.8
Design Point:	m_f = 0.445 kg/s m_s = 2.0 kg/s Δp_t = 340.7 kPa			

The conveying conditions listed as the design point in Table 2 represent one point on the conveying characteristics diagram for this material, the collection of these points (again readily generated by the model described earlier) allows the conveying behaviour of this material for the proposed pipeline to be plotted, as shown in Figure 1. Note, the region to the left of the "Fr . = 6" line would not be considered to provide reliable conveying.

In comparison a single diameter pipe configuration set up to transport 2 kg/s at approximately the same pressure drop would have the operating conditions outlined in Table 3 (as predicted by the model):

Table 3: Conveying Conditions for Single Diameter System

Pipe I. D. (mm)	Pipe Length (m)	Δp_j (kPa)	V_{fi} (ms^{-1})	V_{fe} (ms^{-1})
154	2000	373.2	7.4	34.6
Design Point:	m_f = 0.774 kg/s m_s = 2.0 kg/s Δp_t = 373.2 kPa			

This design would be expected to operate with a 75% higher air consumption, and generate solids velocities up to 120% higher than would occur in the stepped pipeline design. Not surprisingly the power requirement (as estimated by the approach suggested in [5]) is also markedly higher for this configuration, up by approximately 80%. These factors (in addition to reduced pipe costs) show the kind of benefits that may be achieved by the appropriate use of optimised stepped pipelines.

2.2 High-Throughput Applications

Another area where the consideration of stepped pipelines is often worthwhile is in high conveying rate applications. An example illustrating this aspect drawn from testwork recently conducted at the University of Wollongong involved the transport of crushed R.O.M. coal (sub 35 and sub 50 mm).

The purpose of the testwork was to investigate the possibility of conveying the material at a rate of 300 t h^{-1} over a distance of 150 m. Testing the performance of the material at conveying rates close to this level was not possible, however substantial testwork in 105 mm and 155 mm I.D. pipelines over distances ranging from 52 m to 164 m was able to be carried out (further details on the test rigs, results and materials may be found in [6]). The test results enabled the suitability of a model based on equations presented recently by Chambers and Marcus [2] to be evaluated.

This model considers various pressure loss components (due to acceleration, bends, vertical lift, solids and air alone) in determining the total pressure loss, and has been found useful in predicting the lean-phase conveying of relatively coarse products (such as the R.O.M. coal being investigated in this case). A significant adjustment made to the equations has been the use of mean pipeline conditions (e.g. ρ_{fm}, V_{fm}) rather than exit conditions (e.g. ρ_{fe}, V_{fe}) when calculating the pressure drop, as the exit conditions (with the implied higher air velocities) have been found generally to predict higher pressure losses than actually observed. Details of the equations used presently are given in [6] (noted as the Weber A4 model).

The model produced values that fell within a range of ± 11.5% of the experimental data, and hence provided a reasonable level of confidence for proceeding to use the equations to estimate the performance at 300 t h^{-1}.

Minimum conveying conditions for the proposed system were again required, and were part of the consideration in the original testwork. This established that a minimum Froude number of 16 was a reasonable conveying limit for this material and mode of conveying. The model was used to estimate the conveying performance of a single pipe diameter system, and a stepped pipe system (two pipe diameters), with the results presented in Table 4.

Table 4: Crushed R.O.M. Coal - Estimated Conveying Performance for 300 t h^{-1} System

Single Diameter System		
Pipe Dia. (mm)	337	
Pipe Length (m)	150	
m_f (kg/s)	5.084	
Δp_t (kPa)	75	
V_{fi} (m/s)	29	
V_{fe} (m/s)	50	
Stepped Pipe System		
	Up Stream Pipe	D/Stream Pipe
Pipe Dia. (mm)	337	387
Pipe Length (m)	78	72
Δp_n (kPa)	50	19
V_{fi} (m/s)	29	31
V_{fe} (m/s)	41	37
m_f (kg/s)	4.912	
Δp_t (kPa)	69	

The results shown in Table 4 indicate some gains in terms of air consumption and pressure drop, however the 20% drop in maximum superficial air velocity with the stepped pipeline system would reduce significantly the pipe wear rate of the system.

3. Plug-Phase Conveying

The transport of material in a plug phase, i.e. where the material is conveyed in one plug of approximately fixed length (equivalent in capacity to the charge in the pressure vessel) per cycle, has been found to provide a useful materials handling solution to many situations in Australia. The particular attributes of this mode of conveying include:

- Relatively low average solids velocities can be achieved (e.g. 3 to 5 ms^{-1}).
- Consequently, wear on system components and the degree of product degradation can be minimised.
- A mix of coarse and fine particles can to some extent be tolerated, though where there is a wide particle size range combined with a high proportion of fines, difficulties may be experienced in setting up reliable operating conditions - a problem which may be exaggerated if the mix of the material being fed into the system is variable. Booster air may need to be introduced into the pipeline to maintain the motion of the plug in these cases.
- Where batch loading of a number of materials (in specific proportions) into the blow tank has resulted in a layering of these materials in the pressure vessel, checks on the material delivered to the silo have indicated that a reasonable amount of blending takes place during the conveying operation.

Some examples of successful applications of this technology are listed below:

The transport of a diamond ore concentrate, described as a heavy, coarse, highly abrasive and friable material [7]. This product had a wide particle size range, and during the commissioning of the plant it was found necessary to inject additional air at a number of points along the pipeline to keep the plug moving overcoming blockages at these points. The abrasive nature of the material presented problems in maintaining effective seals on the blow tank fill valves - a problem that was overcome by the design of an inflatable seal type valve that has proven very useful for handling abrasive and gritty products. Subsequently, the plant has operated successfully for a number of years, though the maintenance programme needs to take account of the steady wear rate associated with handling this type of material for example, the discharge section of the pressure vessels was found to be gradually eroding, eventually calling for a redesign to incorporate replaceable lined sections in this region.

The transport and in-line blending of a range of foundry sands. This application was concerned with the transport of zircon, various sands, resin-coated sands and zircon and mixtures of these components. Important aspects included minimising the degradation of the products, minimising the segregation that may occur during the handling process, and also sufficient blending of the various ingredients. Testwork indicated that reliable conveying conditions could be established for each of the materials and material mixtures. Where mixtures of material were loaded separately into the blow tank and the conveyed material checked for segregation it was in fact found the conveying process had actually done a reasonable job of blending the products during the conveying operation.

4. Low-Velocity Conveying

Low velocity transportation (e.g. solids velocities between 1 and 2 ms^{-1}) gradually is attracting interest as a preferred alternative handling option, particularly in the food industries where the traditional advantage of pneumatic conveying in providing a contamination free transport mechanism (previously normally utilising a high-velocity dilute-phase system) can be combined with a much gentler conveying operation that significantly reduces the losses due to product degradation.

This mode of conveying may be applied usually to materials that retain some permeability when packed in the pipeline (i.e. do not form relatively impermeable plugs, and consequently should not contain a mixture of coarse and fine products). The product shape should also be generally rounded and not too readily disposed to the formation of interlocked conglomerates. Also, lower density materials also are usually more suited to this technology. Examples of products successfully tested, and with systems subsequently being installed to handle them, with this mode of conveying include: refined sugar, soybean, wheat, green barley, malted barley,spray-dried milk powder, cracked rice, plastic pellets. Particularly friable products, such as granulated coffee, have also been conveyed successfully with low rates of degradation. Table 5 lists typical operating conditions obtained for a number of products, using this mode of transport. Note the solids velocities generally fell in the range 0.5 to 1.5 ms^{-1}.

Table 5: Typical Operating Conditions, Low Velocity Applications

Material	Pipe Details L'th x Dia. (m)	m_f (kgs^{-1})	m_s (kgs^{-1})	Δp_t (kPa)	V_{fm} (ms^{-1})
Milk Powder	52 x .105	.0166	3.9	135	1.0
	"	.0215	5.0	140	1.2
Coffee:					
Powdered	52 x .105	.019	4.0	75	1.3
Granulated	52 x .105	.009	1.5	90	.6
Refined Sugar	52 x .105	.0110	3.4	245	.5
	"	.0220	8.0	310	.8
	"	.0310	12.0	345	1.1
	96 x .105	.0178	4.5	460	.5
	"	.0210	6.0	515	.6
	116 x .105	.0165	3.8	485	.5
	"	.0230	5.0	555	.6
Broad Bran	94 x .105	.125	.85	45	9.9
Semolina	94 x .105	.068	.85	75	4.8
Puffed Rye	52 x .105	.038	.24	9	3.5
Rye	52 x .105	.060	1.7	100	3.9
Wheat	52 x .105	.058	1.9	127	3.4

Other significant advantages of this technology include:

- The ability to stop and restart the conveying operation at any stage during the cycle, as the permeability of the material allows air to pass through the pipe full of material, maintaining sufficient pressure differential at the front end of the plug to cause the leading section to break away - recommencing the cycle.
- Product mixtures can be transported with very little in-line segregation (particularly in contrast to lean-phase systems) as this conveying process involves minimal particle interaction (especially around bends).

Until recently, the typical feeding mechanism for this type of system was a bottom discharge blow tank (incorporating a fluidising zone at the outlet to "condition" the product on discharge). Also, a tandem system or a piggy-back installation was used for continuous conveying. New developments in rotary valve design, allowing them to be used in high pressure applications (≈ 300 kPa_g), have increased the attractiveness of this mode of conveying - particularly where it is being considered as an alternative to an existing dilute-phase system. Installation requirements with regard to headroom are substantially reduced, installation time and disruption to existing equipment and operations may be minimised, as may be the actual installation time required during the conversion of a pre-existing system. The initial capital outlay for the system is also lower than an equivalent blow tank system, though the maintenance costs for the rotary valve should be taken into account. Use of a rotary valve fed system also has the capacity control and adjust the solids mass flow rate directly.

5. Screw-Feeding Blow Tanks

With the increasing popularity of dense-phase conveying systems some attention has been given to the potential of directly linking the transport operation into an on-line process, such as feeding pulverised fuel into a boiler. This has the obvious advantage of eliminating an additional operation (and its associated equipment) from the plant, and is made possible by the very small proportion of air involved in conveying material in the dense-phase mode over short distances (e.g. 50 < m* 200).

An essential element in ensuring that this type of application will be successful is the achievement of accurate and readily responsive solids mass flow rate control. A new development in blow tank technology (N.E.I. John Thompson (Aust.)) that has demonstrated potential in this field is the screw-feeding blow tank. Figure 2 shows a general arrangement of a typical screw-feeding blow tank. Significant aspects of this design include:

- A screw feeder to control the discharge rate from the blow tank
- A variable speed drive (attached to the screw-feeder) to allow precise and automated control of the solids mass flow rate to the process
- A "balance" pipe that equalises the pressure across the screw-feeder minimising its wear and negating the tendency the product may otherwise have to be pneumatically conveyed through the screw flights.

Industrial applications successfully utilising this technology include:

i) 1.6 m^3 vessel (with a 0.5 m^3 piggy-back blow tank attached to allow continuous conveying) transporting pulverised coal over 100 m to a Direct Ignition System. Four way flow splitting is included, and the flow can be selected in the range of 1 to 6 t h^{-1}.

ii) 0.2 m^3 vessel transporting sub 12 mm coal washery fines 230 m directly into a boiler, at a solids mass flow rate of 2.5 t h^{-1}.

iii) A continuous transport system using tandem 2 m^3 blow tanks to inject sub 20 mm crushed coal into molten lead.The conveying rate varies with process requirements within the range of 0.2 to 4 t h^{-1}, over a distance of approximately 30 m.

6. Conclusions

The advances in conveying technology and the developments in analysis procedures and modelling techniques presented in this paper demonstrate the expanding potential pneumatic conveying has to offer innovative, viable and competitive materials handling solutions to industry.

Significant aspects noted in this paper include:

i) The use of stepped pipelines in long-distance and high-throughput applications, and the importance of using careful testwork and accurate correlation analysis in optimising the pipeline design.

ii) The potential of plug-phase conveying systems to transport materials at moderate velocities with low rates of degradation, and in the case of mixtures to provide some in-line blending during the process.

iii) The use of low-velocity conveying, particular in the case where product degradation is of critical importance.

iv) The advent of technology such as the screw-feeding blow tank injector offers the possibility to incorporate the conveying system directly into the on-line process.

7. References

[1] Mills, D., Mason, J.S. and Stacey, R.B., Solidex 82, Harrogate, U.K., 1982, pp C1 - C75.

[2] Chambers, A.J. and Marcus, R.D., Pneumatic conveying calculations, Second International Conference on Bulk Materials Storage Handling and Transportation, Wollongong, Australia, 1986, pp 49 - 52.

[3] Design Workshop on Pneumatic Conveying, Short Course Notes, ITC Bulk Materials Handling, University of Wollongong, Wollongong, Australia, Nov. 1990, Ch. 10.

[4] Wypych, P.W., Kennedy, O.C. and Arnold, P.C., Pneumatic conveying of pulverised & crushed R.O.M. coal, Pneumatech 4, Glasgow, Scotland, 26-28 June, 1990.

[5] Pneumatic Conveying of Bulk Materials, Short Course Notes, The Wolfson Centre for Bulk Solids Handling Technology, Thames Polytechnic, London, U.K., Oct./Nov. 1989.

[6] Wypych, P.W., Kennedy, O.C. and Arnold, P.C., The future potential of pneumatically conveying coal through pipelines, Bulk Solids Handling, Vol. 10, No. 4, 1990, pp. 421-427.

[7] Wypych, P.W. and Arnold, P.C.,Recent engineering developments in the application of pneumatic pipeline transport of bulk solids to Australian industry, National Engng. Conf., Perth, W.A., Australia, 10-14 April, 1989.

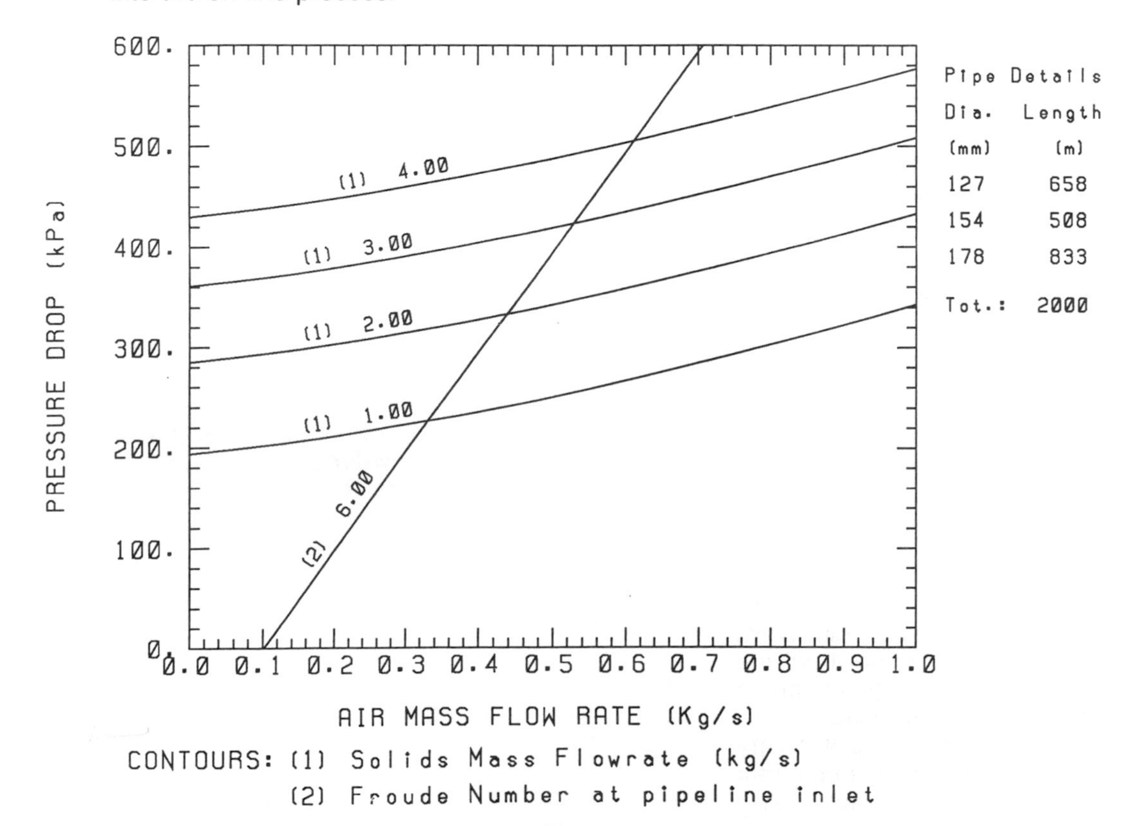

Fig 1 Conveying characteristics for a pulverised coal transported a distance of 2000 m, using the stepped-pipeline configuration set out in Table 2 (as determined by the model of section 2.1)

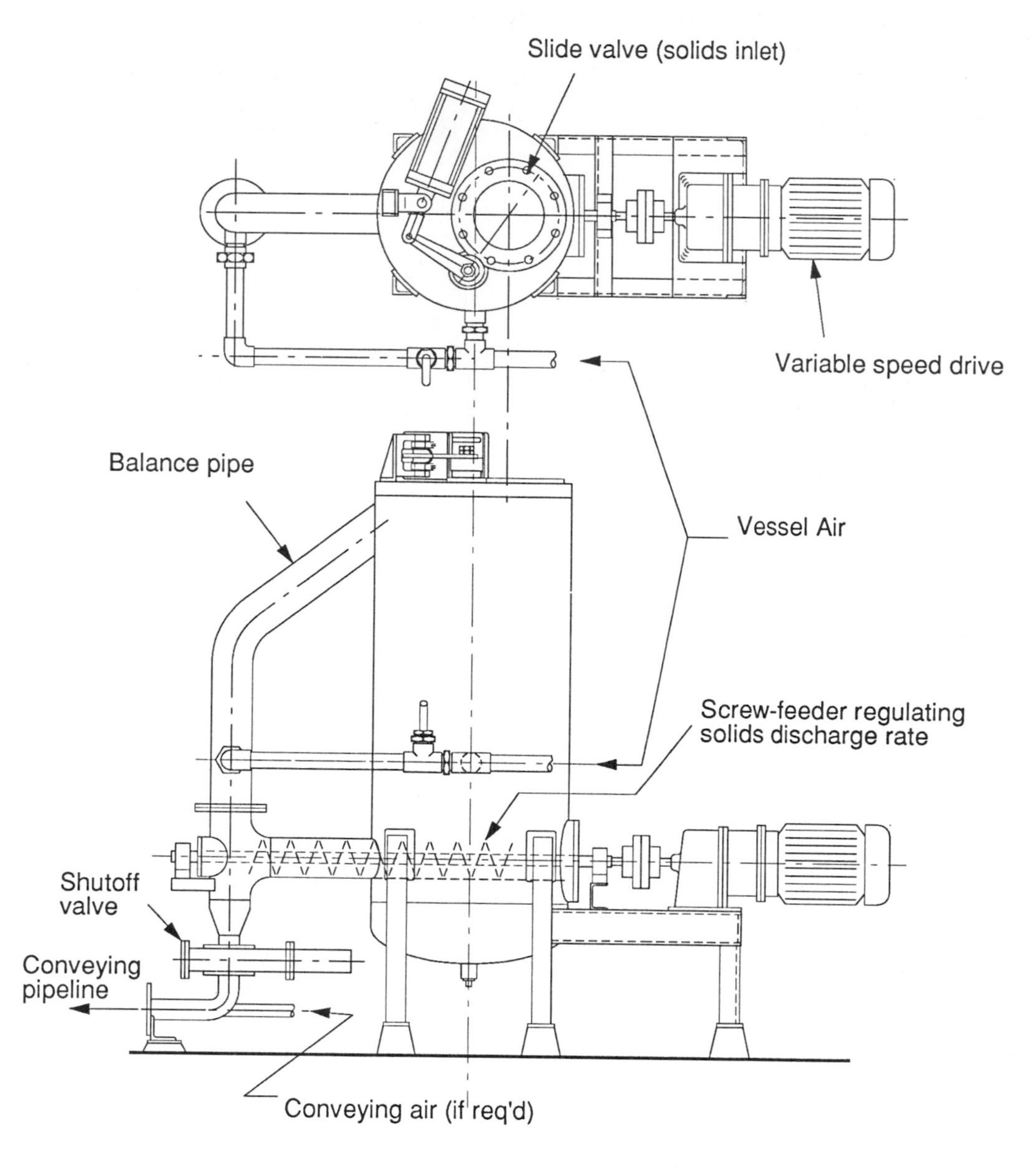

Fig 2 General arrangement of a screw-feeding blow tank

C418/053

Latest techniques for the design of reliable and energy efficient pneumatic conveyors

M S A BRADLEY, BSc, PhD and A R REED, BSc, PhD
The Wolfson Centre for Bulk Solids Handling Technology, Thames Polytechnic, London

SYNOPSIS This paper outlines some of the advances in the design of pneumatic conveying systems that have been made in the United Kingdom in recent times. Given the demands on industry to design systems having ever increasing throughputs, conveying distances and general complexity, it is clear that accuracy of design is an important requirement. Also,since the physical characteristics of a product and their interaction with the conveying medium has a significant effect on the behaviour of a material in a pneumatic conveyor, it is clear that the design technique should take account of this important effect. It is proposed that the design procedures outlined in this paper go a long way to satisfying these requirements.

NOTATION

ΔP	=	pressure drop caused by bend
ρ_s	=	suspension density
c	=	superficial air velocity
K	=	coefficient.

1. INTRODUCTION

The suitability of pneumatic conveying for modern industrial processes and the economic benefits of handling powdered and granular solids in bulk have led to a worldwide growth in this form of materials handling. As a consequence, the demands of such handling operations are continually increasing in terms of throughputs, conveying distances and general complexity. Clearly this provides challenges for the designers of such systems. Thus, the purpose of this paper is to focus on the various techniques that have been developed in the UK in recent times to assist engineers to design pneumatic conveying systems that will operate in a reliable and energy efficient way.

2. SYSTEM DESIGN AND THE ROLE OF PRESSURE DROP

The work of Duckham (1) in developing pneumatic systems to unload wheat grain from ships and barges to flour mills in the London docks nearly a century ago was a major contributing factor in the development of pneumatic conveying technology as we recognise it today. Since those early days developments have been such that there are now a wide range of options available to potential users with respect to basic types of systems, system components and mechanisms of conveying materials in pipelines. In this respect, the selection of appropriate components which, when arranged into a system, will interact in such a way that the system operates in a reliable and energy efficient manner, depends to a large extent on the designer making an accurate prediction of the drop in pressure of the air/gas in the conveying pipeline that will occur in the resultant system. This dependence is becoming even more important with the general trend towards systems that are increasing in terms of throughputs, conveying distances and general complexity. Therefore, it is clear that making accurate predictions of this pressue drop is fundamental to the business of designing systems to operate in a reliable and energy efficient way. However, before it is possible to introduce the latest thinking in this area, it is important to appreciate the advantages and disadvantages of the other approaches for predicting this pressure drop that are in use at the present time. By this way, the benefits of the latest techniques in this area and their contribution to the process of design will become apparent.

3. PREDICTION OF PRESSURE DROP METHODS CURRENTLY IN USE

There are two basic methods of predicting the pressure drop to be expected in pneumatic conveying pipelines in common use. These may be classified as,

i) The global testing and scaling approach, and

ii) the piecewise analytical approach,

the salient features of which are described in the following sections.

3.1 The Global Testing and Scaling Approach

One method that is used for predicting pressure drop involves testing a sample of the product which is to be conveyed in the final system in a pilot scale rig over a wide range of operating conditions, and measuring the product and air flow rates and resulting pressure drops. The obtained data are then scaled by experimentally determined factors to predict the pressure drop in the projected system. As such it is an approach that the authors' organisation took an active role in developing during the early 1980's, details of which are described in a recently published book (2).

The approach has the advantage that real test data on the product to be conveyed in the projected system are used for the design work. This is important because it provides useful information on the conveyability of the product, as well as determining valuable data on the minimum conveying velocity etc. However, it is generally recognised that a problem arises with the approach when the final pipeline has a different number and/or distribution of bends from the test line. Originally, it was hoped that an equivalent length approach for pipe bends could be used in conjunction with the scaling procedure for pipeline length. Unfortunately, attempts to determine the necessary values of equivalent length have shown that they are very dependent on conveying conditions (air velocity and solids concentration), as well as being affected by product type and bend geometry. For example, a programme undertaken at Thames Polytechnic (3) found values varying between 8 and 20m for one product and pipeline in a lean phase, suspension flow system. Another programme (2) revealed between 2 and 20m, again for a single product and pipe bore, with a strong correlation with air velocity at inlet. The result of this, coupled with the fact that falling pressure along a conveying line leads to increasing velocity, means that the true equivalent length of a bend will be dependent upon its position in the conveying line as well as other factors. The implication of this is that this approach makes accurate prediction of pipeline pressure drop difficult, particularly when there is a significant difference in the number or position of bends between the pilot rig and plant system.

3.2 The Piecewise Analytical Approach

The alternative way of dealing with the pipeline as a whole is to treat each of its features separately, starting from known flow conditions at one end of the pipe and estimating the pressure loss and change in flow conditions caused by each bend and straight length in turn, progressing along the pipe and thus finishing up with a value for the total pressure drop. Such a piecewise approach may by employed were suitable models describing relevant pipeline features exist. By working in this was, the effect of bends in the line can be analysed using the true conveying conditions prevailing at the point where they are located.

Recent work on analysing and modelling the complex processes in conveying spherical, mono-sized particles by air through straight horizontal pipelines in a suspension mode of flow is encouraging, (4). However, the considerable difficulties in accounting for the characteristics of the particles constituting materials of industrial interest, eg. size distribution, shape, surface texture as well as the density differences present when conveying blends of products through different pipeline features and in different modes of flow, means that for the foreseeable future it will be difficult to predict pressure loss reliably by such means. A brief survey of the literature will show that a vast amount of work has been undertaken in this direction, with little agreement even for lean phase, suspension flow conditions, let alone for the much more complicated cases of non-suspension flow.

4. AN IMPROVED METHOD OF PREDICTING PRESSURE DROP

From the foregoing it is apparent that both approaches have some advantages; the Testing and Scaling approach because it uses real data for the conveyed product, thus giving a high certainty level about the effects of product type, and the Piecewise Analytical approach because the effects of pipeline features (especially bends) can be examined in detail. From this it can be seen that if a method is developed whereby pressure drop predictions could be made using test data from the actual product, but using a piecewise approach rather that a global one, then it would share the advantages of both the methods mentioned above without suffering from the drawbacks. Since the details of such an approach have been recently published elsewhere (5), only the salient features of the method will be discussed here.

In essence the method is a three step approach incorporating:

1. Testing the product to be conveyed in a rig designed to obtain data on the effects of individual pipeline features (straight lengths and bends etc).

2. Entry of the data into a specially developed storage system designed to be quick and easy to use.

3. Recall of the data from the storage system and synthesizing the performance characteristics of the proposed pipeline.

The requirements for the method are therefore a suitable design of test rig, useable systems for storage and recall of the data, and a means of using the data for the synthesis of pipeline conveying characteristics, from which appropriate design parameters can be selected.

By employing this basic approach it is argued that the confidence level associated with determining such design parameters can be improved considerably.

5. SALIENT FEATURES OF THE IMPROVED METHOD

From the foregoing it is clear that the basis of the improved method is one of measuring the pressure drops caused by the features comprising a pipeline, ie. bends, horizontal and vertical runs etc, when conveying the material in question over a wide range of operating conditions.

5.1 Pressure Drop Caused by Bends and Straights

To measure the steady pressure gradient along a straight pipe presents no great difficulty, requiring simply pressure tappings at appropriate intervals along the pipe. Measurement of the pressure drop caused by a bend is a little more difficult. It has been demonstrated for both single- and multi-phase flow that most of the pressure drop occurs not within the bend itself, but in the straight pipe downstream where the disturbed flow is sorting itself out. However, by obtaining pressure profiles along the straight pipes adjacent to a bend it is possible to establish a value for a step change in pressure equivalent to the loss caused by the bend, as illustrated in Fig. 1. By this way it is possible to account for the effect of bends. However, since it has been shown that bend geometry can have a significant effect of the pressure loss (6), it is important that tests are undertaken on a bend of the same basic geometry to those to be used in the final system. Further details of the type of experimental set-up required to obtain the necessary data are given in (5).

The pressure traces obtained from the test work are essentially the same as that shown in Fig.1, with the shape and length of the curved section varying with the concentration of solids in the pipeline. A typical such profile is shown in Fig. 2. From such data it is possible to account for the effects of bends and straight sections of piping.

5.2 Data Storage Systems

5.2.1 Bends

Research has shown that the following is a suitable approach for accounting for the effects of bends,

$$\Delta P = \tfrac{1}{2} K \rho_s c^2$$

where,

ΔP = pressure drop caused by bend, in bar,

ρ_s = suspension density, ie. kg of product flowing per m^3 of conveying air (using true volume flow rate of air at pressure in the pipe, not 'free-air' conditions),

c = superficial air velocity, calculated from true volume flow rate of air and pipe cross-sectional area,

K = coefficient.

This is similar to the approach used for bend losses in single-phase flow, where the loss is taken to be proportional to the dynamic pressure of 'velocity head' of the flow, with the value of the coefficient depending only on bend geometry and to some extent, pipe bore.

For the case of gas-solid flows it has proved impossible to make the coefficient independent of the variables in the equation. However, it has been found that it can be represented on a single graph for each bend and product type, against either air velocity or suspension density. Examples of this are shown in Fig. 3. By this way it was found that neither air density nor pipe bore had any significant effect on the values for K. A detailed analysis of the effect of bend geometry on the pressure drop caused by bends is given in (6).

5.2.1 Straights

The approach for dealing with the pressure drop caused by straight sections of pipe is to sum the pressure gradients attributable to the air only in the pipe (calculated from appropriate single phase flow methods) and the additional gradient caused by the addition of the solid particles in the air. To facilitate this, models are developed from the test data to account for this additional pressure gradient.

5.3 Synthesis of Pipeline Conveying Performance Characteristics

Having obtained the necessary data, it can be used to synthesize the performance characteristics (ie. the relationship between air and product flow rates, and the conveying line pressure drop) for pipeline systems of virtually any configuration, from which it is then possible to select suitable design parameters.

5.3.1 Procedure

To employ this approach as a design tool, the first requirement is a datum for pressure. For a positive pressure system this will be at the end of the conveying line where the product is discharged usually into a receiver at around atmospheric pressure or occasionally into a process operating at a known pressure.

The mass flow rate of product will be a primary design parameter, so will be known. Values for mass flow rate of air and pipeline bore are chosen, in the first instance by an educated guess. From this information it is possible to calculate the superficial air velocity and suspension density at the end of the pipe, and use these values in the equation for straight pipe pressure drop to estimate the pressure loss in the final straight section. This gives the pressure at inlet to this straight section, from which new values for velocity and suspension density are calculated. These values are then used to estimate the pressure drop caused by the bend at this point from which the pressure and thus velocity and suspension density at inlet to the bend can be calculated. The procedure is simply repeated for each straight and bend in turn working back to the start of the pipeline, to obtain the total pressure drop along the pipe and the velocity at inlet.

Should this process result in an air velocity that is too low to satisfactorily convey the product, then the whole process should be repeated using increased air flow rates until an appropriate value is achieved.

In the case of a vacuum system, the datum for pressure is at the beginning of the pipeline and so, in this case, the procedure starts at this point and progresses forwards along the pipe towards the receiver.

Clearly, this simple procedure is amenable to the use of a computer, utilising the basic flow chart shown in Fig. 4. Using such a programme the whole procedure can be repeated quickly and easily to explore the effect of different values of pipe bore and air mass flow rate. By this way it is possible to build up a clear picture of possible systems to meet the required duty, from which a choice can be made on the grounds of economy of installation and operation.

6. STEPPED BORE PIPELINES

The problems often quoted relating to pneumatic conveying systems such as high energy consumption, high maintenance through erosive wear and undesirable particle degradation all emanate from the high velocities that can result from the expansion of the conveying air within single, uniform bore pipelines. From this it follows that there are advantages to be gained by controlling these velocities to within certain limits, which may be achieved by increasing the bore of the conveying line in one or more carefully controlled steps.

Whilst there is nothing new about employing stepped bore conveying pipelines, there is little doubt that the benefits to be accrued from the approach have not been generally appreciated by industry. One reason for this is that the design procedures for such pipelines have been more uncertain than those of single bore systems, and, as a result, the reliability of such systems cannot always be guaranteed. In this respect it has been the positioning of the steps that has been the problem. Obviously, the step(s) should not be too soon otherwise the air velocity may drop to a value below that required to satisfactorily convey the product. Under such circumstances the pipeline will block. Conversely, the step(s) should not be too far down the pipeline since high velocities will result and the benefits of the approach will not be fully realised.

6.1 Positioning of Steps

There are two approaches that can be used to assist in positioning of the steps in pipeline bore. The approach that has been in widespread use on mainland Europe for many years, and more recently in Australia (7), is based on ensuring that at the change in bore a number of the Froude form does not fall below a certain value. The argument behind this approach is that it accounts for the variation of minimum conveying velocity with changing pipeline bore. The other criterion that is sometimes used to locate the position of the step in bore, is that at this position the superficial air velocity at the increased bore should not fall below a value identified as necessary for successful conveying. Recent work undertaken by The Wolfson Centre at Thames Polytechnic suggests that the latter approach can be used with confidence to locate the position of the step. The implications of this is that using the Froude type number approach for this purpose will lead to a workable system, but one that does not realise the full benefits in terms of minimising velocities and operating pressure.

Irrespective of which criterion is employed to position the step(s) in bore, a key aspect of designing such pipelines is predicting the pressure profile so that actual air flow rates and velocities are used in the appropriate procedures. With this in mind it is clear that the improved method of predicting such pressure profiles discussed in the previous sections enhances the confidence levels associated with designing such systems. For further information on this aspect of pneumatic conveying the reader is referred to (8).

7. CONCLUDING REMARKS

It has been the intention of this paper to outline some of the advances in the design of pneumatic conveying systems that have been made in the United Kingdom in recent times. Whilst attempts to model gas-solid flows are proceeding, it is recognised that the difficulties of modelling the complex phenomena present when conveying particles of industrial interest means that the development of this approach into a reliable design tool is extremely unlikely in the foreseeable future. However, given the demands on industry to design systems having ever increasing throughputs, conveying distances and general complexity, it is clear that accuracy of design is an important requirement. Also, since the physical characteristics of a product and their interaction with the conveying medium has a significant effect on the behaviour of a material in a pneumatic conveyor, it is clear that the design technique should take account of this important effect. It is proposed that the design procedures outlined in this paper provide an approach for meeting these requirements.

References

(1) ANON 'Duckham's pneumatic grain elevator and conveyor', The Engineer, 20 July. 1894

(2) MILLS, D. 'Pneumatic conveying design guide', Butterworths, 1990.

(3) WESTAWAY, S. F, 'An investigation into the performance of venturi eductors for the transport of solid particles in pipelines', MPhil. Thesis, Thames Polytechnic, London, 1987.

(4) MASON, D. J, MARKATOS, N. C and REED, A. R, 'Numerical simulation of the flow of gas-solids suspensions in acceleration regions of pipelines', Proc. 2nd. Int. Phoenics Users Conf., London, Nov. 1987.

(5) BRADLEY, M. S. A and REED, A. R, 'An improved method of predicting pressure drop along pneumatic conveying pipelines', Powder Handling and Processing, Vol. 2, No. 3. (Sept. 1990).

(6) BRADLEY M. S. A, 'Pressure losses caused by bends in pneumatic conveying pipelines : effect of bend geometry and fittings', Powder Handling and Processing, Vol 2, No 4, (Nov. 1990).

(7) WYPYCH, P. W and REED, A. R, 'The advantages of stepping pipeline bore in pneumatic conveying systems', Proc. 4th Int. Conf. on Pneumatic Conveying Technology, Glasgow, June 1990.

(8) BRADLEY M. S. A, 'Stepped bore pipelines - a new lease of life for dilute phase pneumatic conveyors', Proc. IMechE Conf. BULK 2000 (Bulk Materials Handling - Towards the Year 2000), London, Oct. 1991.

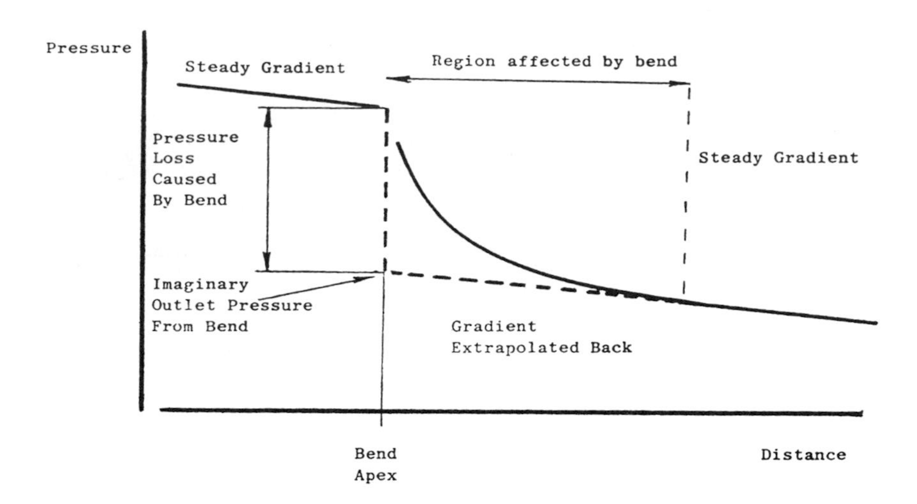

Fig 1: Schematic of the pressure distribution adjacent to a bend, showing the region in which the pressure drop caused by the bend is developed.

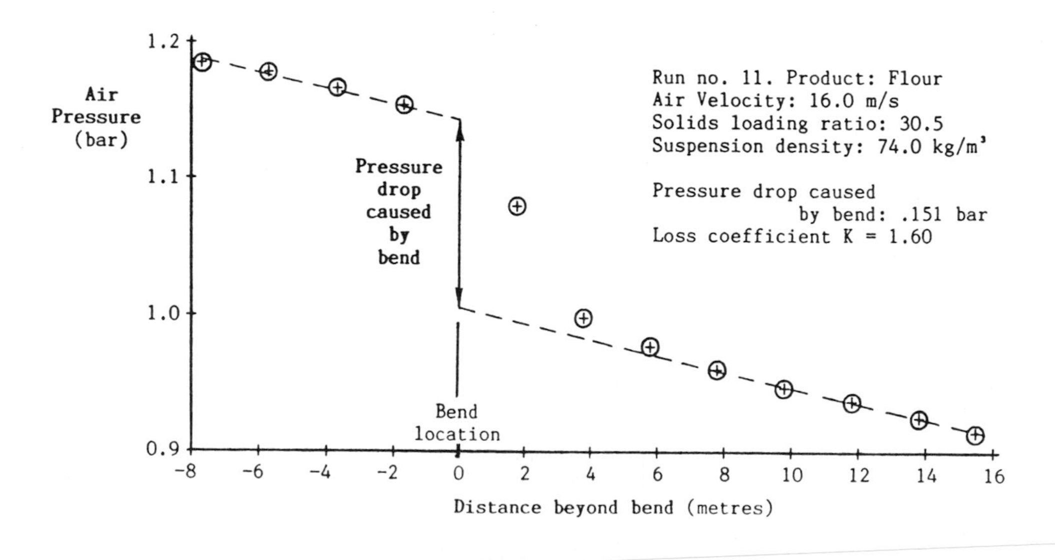

Fig 2: Example of pressure distribution measured adjacent to bend. N.B. distance from bend measured from intersection of centre-lines of adjacent straight pipes.

Short Radius Bought-out Bend with Sockets, 2in.NB, Wheat Flour. Ranges of Suspension Density at Bend Outlet shown.

Short Radius Bought-out Bend with Unions, 2in.NB, Polyethylene Pellets. Ranges of Suspension Density at Bend Outlet shown.

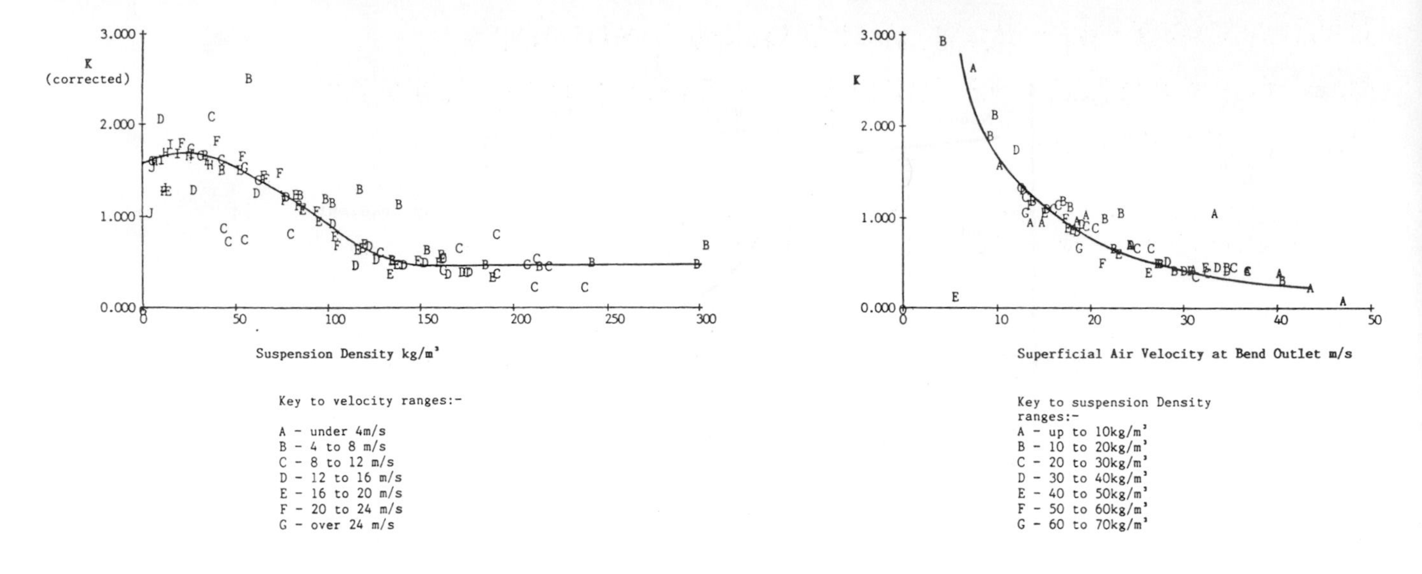

Fig 3: Graphs of bend loss coefficent K vs. suspension density and superficial air velocity for two products, from experimental results.

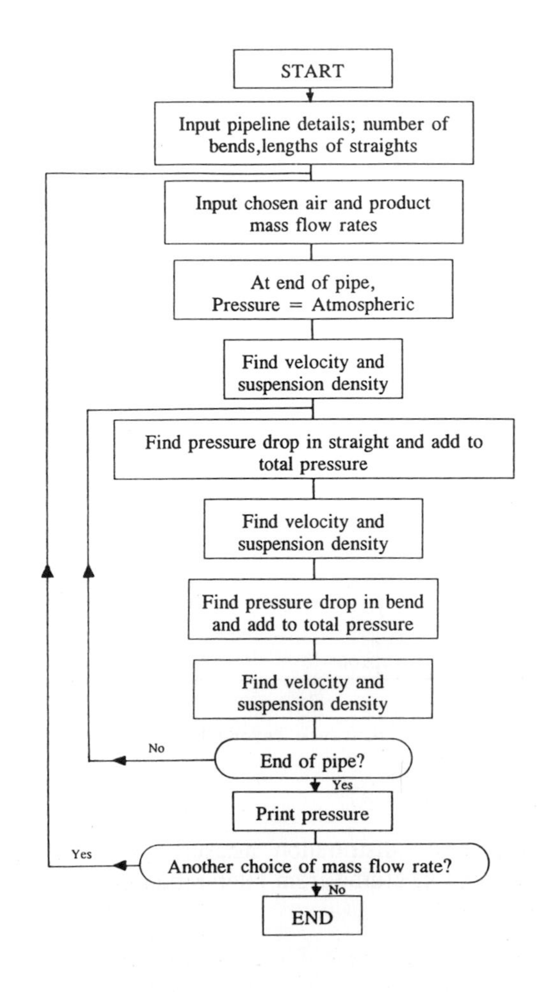

Fig 4: Computer program for assessing pipeline pressure drop using method described

C418/019

Evaluation of a flow control system for granular materials using the electromechanical valve for solids

C M MARTIN, BASc and M GHADIRI, MSc, PhD, CEng, MIChemE
Department of Chemical and Process Engineering, University of Surrey, Guildford, Surrey

SYNOPSIS The operation and performance of a novel device, the Electromechanical Valve for Solids (EVS), for control of flow of granular materials are described. The EVS applies a d.c. electric field to the granular material in a dense-phase state. This induces electrical interparticle forces which retard and can ultimately halt the flow, hence providing an actuating mechanism. The EVS has a very fast response time which makes it very useful for fast flow control. The response characteristics of the EVS, the performances of two weighing devices used in this work to determine the flowrate, and the results of flow control using a simple PID control action are reported. It is shown that the performance of the EVS is limited by the slow response of the weighing devices. To obtain accurate measures of flowrate or weight gain, sampling frequencies slower than 1 Hz are necessary, while the EVS has so far been operated at frequencies up to 20 Hz.

1 INTRODUCTION

The control and measurement of the flow of particulate solids is an important stage in many manufacturing processes. However, while fast, reliable and accurate valves and flowmeters have long been available for fluids, similar devices are not yet available for granular materials.

Flow control of particulate solids is accomplished mainly by mechanical means such as a screw conveyor or a rotary star feeder, in which the material is translated forward at a rate depending on the angular speed of the screw or star. Although commonly used in industry, these devices have serious drawbacks which stem from the inherent features of their design and operation. They are susceptible to wear when handling hard and abrasive materials and to seizure for soft materials, and cause attrition of particles when handling delicate and friable materials. Furthermore, these devices are slow in response and have short term fluctuating characteristics and long term drift, which make tight control and fast processing difficult with granular materials.

Recently some interesting non-mechanical concepts for flow control have emerged, such as the L, J and V valves based on the aeration principle (1,2) and the Magnetic Valve for Solids (MVS) (3). These devices have narrow ranges of application because of severe limitations on their conditions of effective operation. For example, the valves based on the aeration principle, while responding relatively quickly to an increase in the aerating fluid flow, respond very slowly to decreasing fluid flow. For the Magnetic Valve, it is necessary to add a small quantity of magnetic particles, and this is often unacceptable for the process. Furthermore, in the processing of bulk solids it is unlikely that one technique will ever be sufficient to handle all materials, and to cover a wide range of flowrates.

At the University of Surrey, we have pursued the use of d.c. electric fields for controlling the dense-phase flow of granular materials. This has led to the development of the Electromechanical Valve for Solids (EVS) for on/off operation and flow control of granular materials (4). The EVS has no moving parts and is mounted in situ with no external mechanical connections, so it does not suffer from the problems associated with mechanical devices. It can be operated with a wide range of granular materials without the need for dosing with special materials, as in the case of the MVS, and it may be used under adverse conditions such as elevated pressures and temperatures. Most importantly, the EVS provides a wide turndown ratio with a very fast response time.

In a complete flow control system, the control valve must be accompanied by a device for measuring the flowrate. Both devices must respond quickly, and with good accuracy and reliability to produce an effective control system. At present, the flowrate of granular materials is monitored mostly using load measuring devices such as weighing scales, loss in weight feeders, and continuous weigh belt feeders. Often, continuous flowrate measuring and closed loop control are avoided altogether by manually weighing out materials in batches which are then transferred to process units. Therefore, a real need exists for the development of a complete control system to provide fast, reliable and accurate measurement and control of flow of particulate solids.

The fast response of the EVS has inspired us to develop such a system, but the lack of a suitably fast flowrate measuring device has hindered this work. Here, a brief description of the design and operation of the EVS is given, followed by presentation of the results obtained so far from tests on the EVS control system. The performance of the EVS for a simple PID control action is described, and the performances of two types of weighing device are compared for measuring continuous flowrate.

2 DESCRIPTION OF THE EVS

The EVS consists essentially of a set of two electrodes which is installed within the duct in the path of the flowing material. The electrodes are connected externally to a high voltage supply (EHT) unit for establishing an electric field. The design of the electrodes forms an important feature for satisfactory operation of the EVS and only certain configurations are suitable (4). Typical electrode configurations are shown in Fig. 1(a) for cylindrical and rectangular ducts. The upstream electrode may be in the form of a few wires stretched across the duct or may simply consist of pins protruding into the duct. The downstream electrode may conveniently be in the form of a wire mesh or closely spaced parallel wires fixed across the duct. For the EVS to operate, it is essential that the flow in the space between the electrodes remains in the dense-phase form so that continuous interparticle contacts are maintained. Therefore, the upstream electrode must not significantly obstruct the flow. The downstream electrode must consist of sufficiently small openings, since the halting of the flow on the application of the electric field is brought about by arching of the particles over the individual openings of the downstream electrode.

The primary electrical mechanism responsible for the valve action is the formation of electroclamping forces (5). This results from the constriction of electric current near the contact points between particles. The resulting enhancement of the electric field in the gap immediately surrounding the contact point produces attractive forces between particles. Provided that the particles are not too conductive so as to draw excessively high currents above the power limitations of the EHT, nor too resistive so as to draw such low currents that there is insufficient effect, the forces generated may be very large and comparable in magnitude with external forces, such as those due to gravity.

The feasibility of the EVS for flow control of granular materials has been demonstrated in two cylindrical columns of 64 and 115 mm i.d., and in a narrow rectangular slot of 200 mm x 12 mm. A large number of materials have been tested so far including sand, coal, salt, FCC catalyst powder, seeds, spray dried powders, and various other chemicals and foodstuff. As the action of the EVS is to impede the flow of free flowing materials, for very cohesive materials discharge aids such as aeration or vibration may be needed. Experimental work on fine and cohesive powders has shown that once the material is made to flow, the EVS performs just as for non-cohesive materials, as described below.

Fig. 1(b) shows a schematic diagram of the test rig incorporating the electrode assembly. A slide valve or flap door is needed downstream of the electrodes, so that it can be closed during the initial filling of the column to obtain the material in a dense-phase form. Thereafter, the flap door is left open as the bed of material can be supported by the electric field. The EVS may be operated using either a continuous field, in which case a steady flowrate is obtained which is a function of the field strength, or alternatively a pulsed electric field may be used. The performance of these two modes of operation is discussed below for NaCl salt with a sieve size range of 90-600 μm, angle of repose of 32°, and poured bulk density of 1.25×10^{3} kg m^{-3}.

2.1 Continuous DC field

If an electric field is first set up to a suitably high level before opening the flap door, the particles between the electrodes will be 'frozen', thus supporting the material above the electrodes. If the voltage is gradually reduced, a point is reached when the material begins to flow. The voltage at this point describes the condition for flow initiation and is referred to as the static holding voltage, V_{sh}. For voltages lower than V_{sh}, a steady-state flowrate is attained which is a function of the applied field. Extensive testing has been carried out to determine the sensitivity of the mass flux to the geometry of the upstream electrode, the separation between the electrodes and the size of the column (6). When no electric field is applied, the mass flux does not appear to be affected by any of these factors to a discernible degree. With an electric field present, changes in any of the above factors produce some changes in the mass flux through the column. However, there is an overall trend for the flowrate response to the electric field strength. This is shown in Fig. 2(a) for a typical EVS performance curve, where NaCl salt has been used in the 64 mm column with a 20 mm gap between the electrodes and 5 mm square mesh openings in the downstream electrode. The flowrate curves have a flat, almost horizontal, section at very low voltages, and then become increasingly more steep at higher voltages. As the flowrate decreases to below about 50 per cent of the full flow, the response to voltage becomes unstable and a range of voltages exist where the flow may or may not be stopped. The data in Fig. 2(a) show that here the flow was halted at 10 kV, but during continued testing, it was found that the flow could be stopped at voltages ranging from 8 to 12 kV. Consequently, for good flow control at very low flowrates, where the response is unstable, operation with a continuous field is unsuitable and is best handled by operating in the pulsating field mode, as is discussed below.

2.2 Pulsating DC field

In this mode, the pulse amplitude, frequency and width can all be varied in order to obtain a desired flowrate. Fig. 2(b) shows the flowrate response to frequency for NaCl salt in the rectangular column with a 20 mm gap between the electrodes and 5 mm wide slots in the downstream electrode. A pulse amplitude of 12 kV was used to ensure complete halting of the flow between pulses. The flowrate response to pulse width has been found to follow a similar shape (6). Therefore, the flowrate responses to varying frequency and pulse width are opposite to that of the continuous field. For high flowrates, the curves are very steep, providing poor control, whereas for small flowrates, the curves become nearly linear with a small slope. Thus the two modes of operation complement each other to provide an overall wide turndown ratio for practical operation.

In the pulsating field mode, the maximum frequency is limited mainly by discharging the large capacitor of the EHT through the material, since the field is pulsed by inhibiting it at the low voltage input to the EHT. The maximum frequency at present is about 20 Hz. Much higher frequencies will be possible by switching the electric field at the high voltage output of the EHT, i.e., shorting the EHT's capacitor. Such a high voltage switch is currently being tested.

3 CLOSED LOOP FLOW CONTROL WITH THE EVS

The potential of the EVS for setpoint flow control in a closed loop system has so far been investigated only for the continuous field mode of operation. The test rig shown in Fig. 1(b) was used for dispensing NaCl salt from the 64 mm column. The control operation was carried out by a programme developed in-house using a Sinclair QL computer which was interfaced with the EHT for controlling the voltage, and with two weighing devices for monitoring the flowrate.

In our work, two different weighing devices with similarly high performance specifications were available for measuring the flowrate response of the EVS. One was a tension load cell model TRFP-0100K-111, manufactured by Industrial Transducers Limited, with 100 kg capacity and a continuous analog output which was digitised by a fast 16 bit A/D converter at the computer interface. The load cell is physically designed to weigh objects which are suspended from it. The second weighing device was a weighing platform model F150S, manufactured by Sartorius Limited, with 150 kg capacity and a maximum digitised signal output frequency of 10 Hz.

The overall system response consists of the combined responses of the EHT, the weighing device, charging and discharging of the material, and transportation time of the material from the electrodes to the collecting bin where it is weighed. The charge relaxation time of the particles is typically in the order of milliseconds, and the transportation time of the material can be easily varied. For the experimental setup used here, where the height between the valve and the bottom of the collection bin is about 1.0 m, the transportation time has a minimum value of about 0.4 s, as estimated by the free fall of the particles. Therefore, it remains to determine the response times of the EVS and the weighing device. To isolate the response characteristics of both the load cell and weighing platform from the response of the EVS, some simple measurements were first made without involving perturbations to the EVS.

The weighing devices were then used to measure the flowrate response of the EVS to voltage step changes and to a PID control action for setpoint control. The results of these tests are given below.

3.1 Comparison of performances of weighing devices

Two types of tests were carried out on the weighing devices to assess their effectiveness for continuous flowrate monitoring. In the first test, each device was loaded with a single 1 kg weight and the response with time was measured. In the second test, the variability or 'noise' in the flowrate measurements of each device was determined for varying sampling intervals.

The results of the first test are shown in Fig. 3. For the weighing platform, the test was performed by dropping a 1 kg weight from a few centimetres height onto the platform. The maximum sampling rate of 10 Hz was used for recording the response of the weighing platform on the Sinclair computer. For the load cell, the weight was suspended about 30 cm below it with a stiff rope and then supported by hand before being let to fall through a few centimetres to its full extension. The continuous analog output of the load cell was recorded using a digital storage oscilloscope. The results in Fig. 3 show relatively long stabilisation times, compared to the maximum sampling frequencies available, and large overshoot in the readings. For the weighing platform, the test was repeated five times and in four of the runs, the readings showed overshoot of the 'true' weight before reaching the final correct value, with the worst case showing almost 20 per cent overshoot. The times required for the response to stabilise to the final correct weight were 0.2 to 0.4 s. The response of the load cell in Fig. 3 shows oscillating behaviour with a maximum overshoot of about 140 per cent. The stabilisation time of the load cell was about 0.2 s. Quantitatively, the responses shown in Fig. 3 are certainly specific to the rather rough method of applying the weight. However, qualitatively, the overshoot effect is likely to exist to some extent whenever there is a step change in the applied weight, as when a change is made in the flowrate of granular material.

Fig. 4 shows the results of the noise tests. At each sampling interval, twenty flowrate readings were obtained from which the average and the relative standard deviation were determined. The flowrate was calculated from the change in accumulated weight over the sampling interval. Fig. 4 shows that for both weighing devices, it is necessary to use a sampling interval of about 1.0 s to reduce the relative standard deviation to 10 per cent. Therefore, although both devices are capable of much faster output frequencies, for reliable measurements in this testing rig, it is not advisable to use frequencies greater than about 1 Hz. It is obvious that this response is the combined response of the weighing device and the transportation time.

In conclusion, both weighing devices are

capable of fast sampling rates. However due to their relatively long stablisation times, these sampling rates can only be taken advantage of at the expense of accuracy and precision. Therefore, these devices are suitable for measuring steady flowrates with slow fluctuations where slow sampling rates are tolerable, but they are essentially unsuitable for monitoring processes where fast transients are involved.

3.2 Response of EVS system to voltage step changes

For the reasons outlined above, the measurement of the response characteristics of the EVS was essentially limited by the response of the weighing device. The results given above showed a large degree of variability in flowrate measurements made at frequencies greater than 1 Hz. This is too slow for assessing the capability of the EVS for fast flow control. An alternative sampling procedure was also used to improve the precision of the readings. This involved measuring and recording the accumulated mass, rather than the flowrate, as shown in Fig. 5(a) and 5(b) for the load cell and the weighing platform, respectively. In this way, the signal processing involved in calculating the flowrate was eliminated, and a frequency of 5 Hz was used to record the response of the weighing devices to a step change in the voltage applied to the EVS. In Fig. 5, the flowrate through the EVS is represented by the slope of the line and therefore, a change in the flowrate in response to a change in the voltage is shown by a change in the slope. Before and after a voltage step change occurs, the points on both curves in Fig. 5 follow closely straight lines, indicating little fluctuation from steady-state. In Fig. 5(a), the voltage on the EVS was decreased from 7 to 3 kV at the time of 4.0 s. The straight lines, which represent the flowrates before and after the step change, were calculated from two separate best linear fits to the data points from 0.0 to 3.8 s and from 5.2 to 6.4 s. In Fig. 5(b), the voltage was increased from 0 to 6.5 kV at the time of 1.0 s, and the straight lines were calculated from two separate best linear fits to the data points from 0.0 to 0.8 s and from 2.0 to 3.2 s. From Fig. 5, the total time taken from the initiation of the voltage step change to the point when a new steady-state flowrate was attained was approximately 0.6 to 0.8 s for both cases. This result is in agreement with previous measurements, since the response of the system is dominated by the response time of the weighing device, about 0.2 s, and the transportation time of the material, about 0.4 s. Therefore, no additional precision was gained from this method of analysis, as compared to the previous method where actual flowrates were calculated. Finally, in Fig. 5(b), it is worth noting the overshoot followed by an apparent 'loss in weight' shown by the response of the weighing platform to the voltage increase.

Similar tests were carried out in an attempt to decouple the transportation time and the weighing device response time using a high speed ultraviolet chart recorder to continuously monitor the analog output of the load cell. In these tests, two voltage steps of different magnitudes were applied to the EVS, with each step being applied both as a voltage increase and a voltage decrease. As in Fig. 5, the response times of the system to these steps were determined from the changes in slope of the weight versus time chart traces. The results are shown in Table 1, where transportation time is defined as the time elapsed between the initiation of the voltage step and the first indication of a change in flowrate, and the weighing time is defined as the time elapsed from the start of the change in flowrate to the point when the new steady-state flowrate is attained.

In Table 1, the transportation times range from 0.3 to 0.5 s, agreeing with the estimated time of 0.4 s for the salt to fall from the valve to the bottom of the bin. For the two voltage decreases of different magnitude, the weighing times can be considered to be essentially equal within the error of determining the actual points where slope changes start and end from the chart traces, and within the natural variation expected in normal operation. However, Table 1 shows that the weighing times for the two voltage increases differ significantly from one another, with the response time being larger for the step of larger magnitude. Also the weighing times for the voltage increases are much greater than for the voltage decreases. A possible cause for this behaviour is thought to be the overshoot in the voltage. By monitoring the potential of the upstream electrode using a digital storage oscilloscope, it has been observed that when the voltage is increased, it initially overshoots the required value. Therefore, the flowrate is essentially affected by two changes in voltage, rather than one, and this additional perturbation to the flow may affect the time required to reach steady-state.

3.3 PID control with EVS

The potential of the EVS for flow control has been examined so far using a simple PID algorithm operated from the Sinclair computer. The load cell was used as the flowrate measuring device and initially, a frequency of 1 Hz was used for the flowrate sampling rate. However, it became immediately evident that while 1 Hz was sufficiently slow to produce stable measurements for steady-state flow, an unacceptable amount of instability was produced in the case of transient flows produced by controlling to a setpoint. It was therefore necessary to base the PID control on a moving average flowrate calculated from five flowrate measurements at one second intervals. Using a moving average, to represent the 'instantaneous' flowrate, slowed down the system response significantly. For example, for a voltage step change from 7 to 3 kV, the transportation time remained about 0.5 s, but the effective weighing time was increased to 5 s. However, using the moving average method at least allowed the response limitations of the load cell to be overcome in such a way that the control abilities of the EVS could be observed.

The response of the EVS to three changes in flowrate setpoint using a constant gain of

0.0678 kg s^{-1} kV^{-1} is shown in Fig. 6. While the response is slow, the final setpoint error is negligible. For effective control, a PID method with constant gain is really only suited to linear processes. As shown by the EVS performance curve in Fig. 2 for the same NaCl salt and electrode configuration used in the control experiments, this process is not linear over the full flowrate range. However, by taking a limited flowrate range over this curve for control with constant gain, the nonlinearity is small and acceptable control is obtained. The gain used for control in Fig. 6 was determined by trial and error tuning for the flowrate range of 0.15 to 0.45 kg s^{-1}. In Fig. 2(a), the setpoint flowrate of 0.2 kg s^{-1} lies on the steep part of this range, where the flowrate is very sensitive to any change in voltage. Therefore the response of the EVS in Fig. 6 for the setpoint of 0.2 kg s^{-1} shows significant oscillation. The setpoints of 0.3 and 0.4 kg s^{-1} move progressively onto the flatter part of the curve in Fig. 2(a), so that the oscillating response of the EVS to these setpoints becomes damped in Fig. 6. Therefore, an obvious improvement to this control method would be the introduction of a self-tuning gain which is updated according to the flowrate response in the range of the setpoint.

4 CONCLUSIONS

The performances of an ITL load cell and a Sartorius F150S weighing platform were compared for measuring flowrates of granular materials. The devices are acceptable for measuring steady-state flowrates where sampling rates less than about 1 Hz are tolerable. However, both devices are unsuitable for measuring flowrates with fast transients due to their relatively long stabilisation times.

Using a simple PID control programme in a computerised closed loop system, the EVS produced good flowrate control with respect to final setpoint error, but the fast response of the EVS could not be exploited due to the limitations of the weighing devices for measuring the flowrate and due to the long transportation time.

Overall, the EVS has considerable potential for use in a variety of bulk solids processes. In on/off operation, the EVS can be used for dosing and fast packaging. For controlling a continuously flowing material in a metering or feeding process, the EVS can achieve a wide turndown ratio by using a continuously applied voltage for large flowrates and a pulsating voltage for small flowrates. Compared to conventional mechanical valves, the EVS offers numerous advantages, including reduced wear, fast control, smaller space requirements and potential savings in operating and capital costs.

ACKNOWLEDGEMENTS

The authors would like to thank Messrs. T.K. Lo and C. Mann for their assistance in this work.

REFERENCES

(1) Knowlton, T.M., in Gas Fluidization Technology, ed. D.G. Geldart, 1986, Ch.12, p.341, John Wiley & Sons.

(2) Leung, L.S., Chong, Y.O. and Lottes, J. Operation of V valves for gas-solid flow. Powder Technol., 1987, 49, 271-276.

(3) Jaraiz-M., E., Levenspiel, O. and Fitzgerald, T.J. The uses of magnetic fields in the processing of solids. Chem. Eng. Sci., 1983, 38, 107-114.

(4) Ghadiri, M. and Clift, R. European Patent Application No. 87308304.2, 1987.

(5) Martin, C.M., Ghadiri, M. and Tüzün, U. Effect of the electrical clamping forces on the mechanics of particulate solids. Powder Technol., 1991, 65, 37-49.

(6) Ghadiri, M., Martin, C.M. and Morgan, J.E.P. An electromechanical valve for solids. Submitted to Powder Technol., 1991.

Table 1 Response of EVS system to voltage step changes

Voltage step kV	Voltage decrease		Voltage increase	
	Transportation time, s	Weighing time, s	Transportation time, s	Weighing time, s
3 ⇔ 7	0.3	0.4	0.5	0.7
1 ⇔ 7	0.3	0.3	0.3	1.2

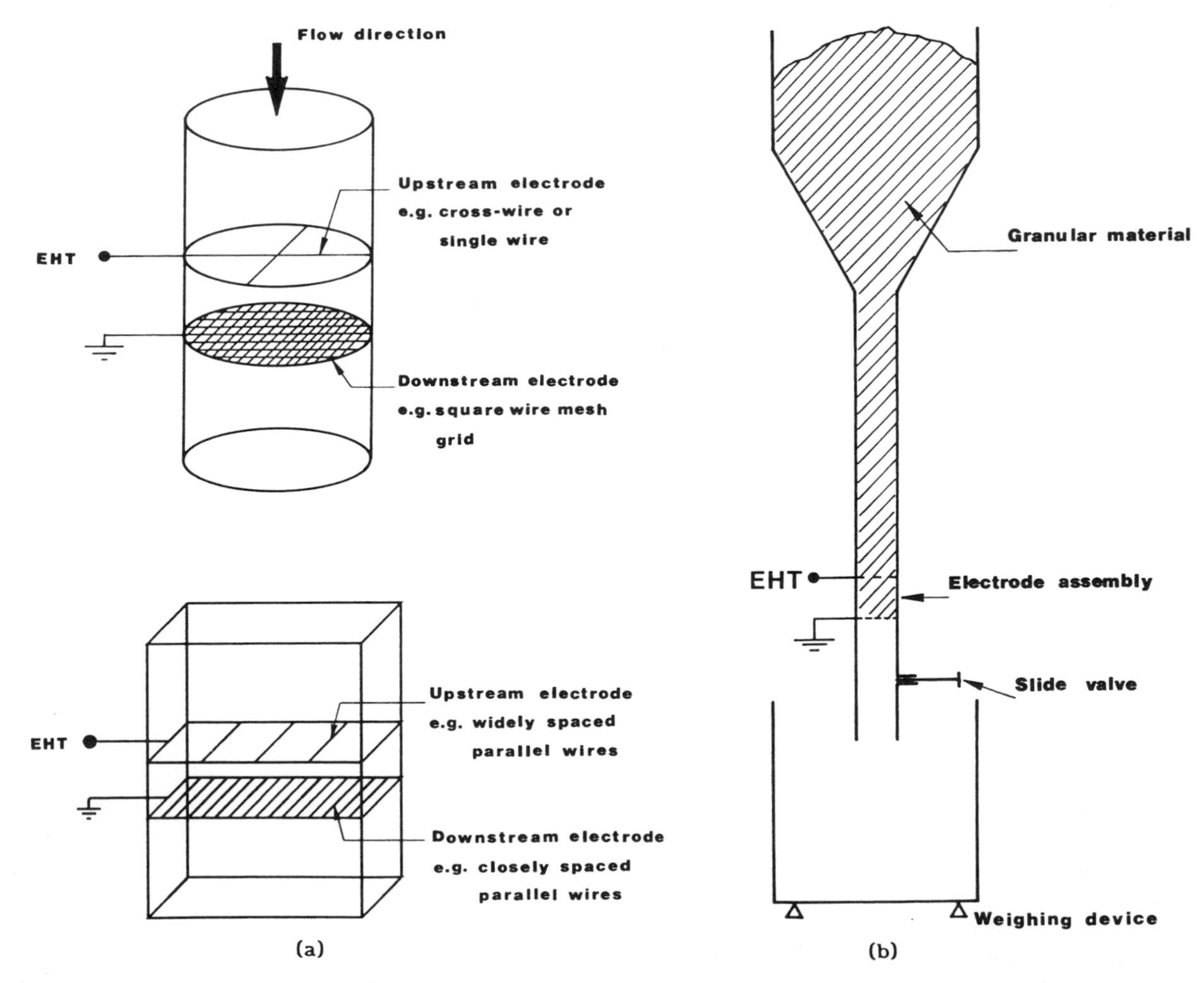

Fig. 1 (a) Examples of electrode configurations for cylindrical and rectangular columns. (b) Experimental rig for testing EVS for flow control.

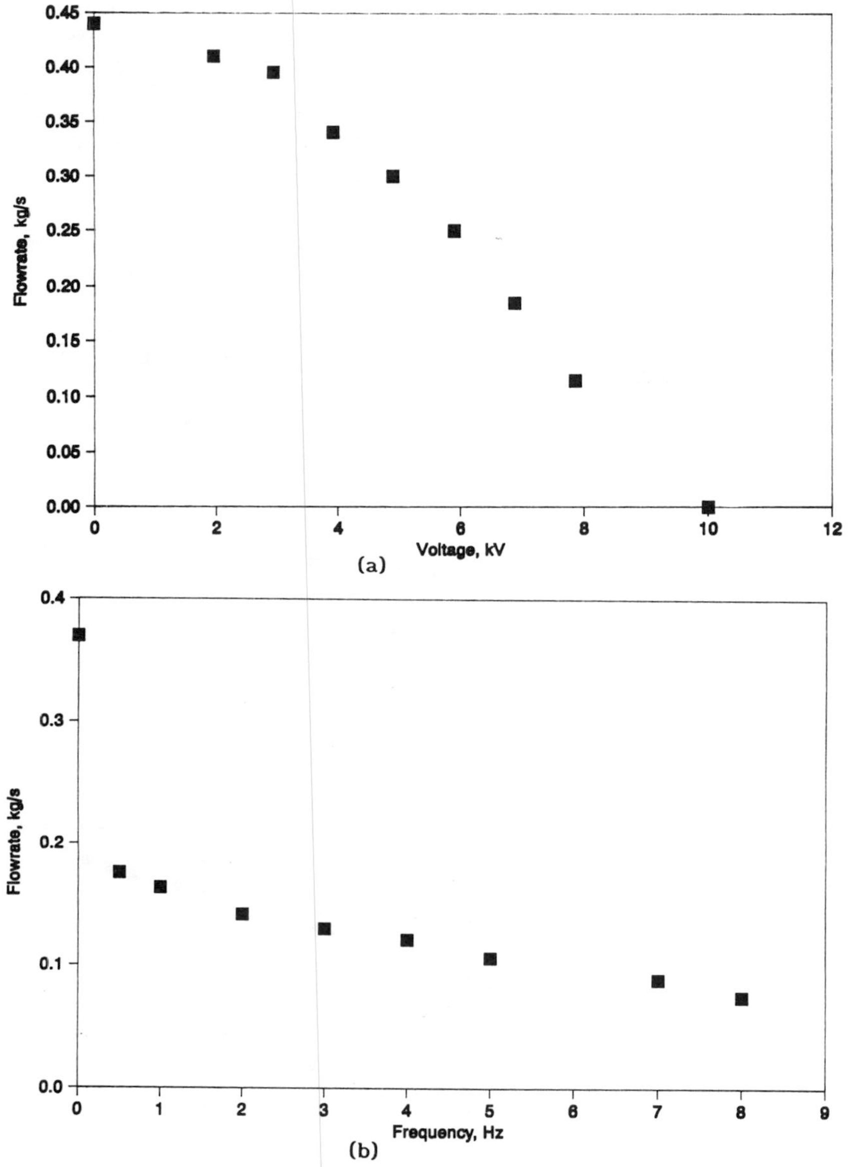

Fig. 2 (a) Effect of continuously applied voltage on flowrate through EVS for NaCl; 64 mm diameter column, 20 mm inter-electrode gap, 5 mm square openings in downstream electrode. (b) Effect of frequency of pulsed electric field on flowrate through EVS for NaCl, 50 per cent pulse width, 12 kV amplitude; rectangular column 200 mm x 12 mm, 20 mm inter-electrode gap, 5 mm wide slots in downstream electrode.

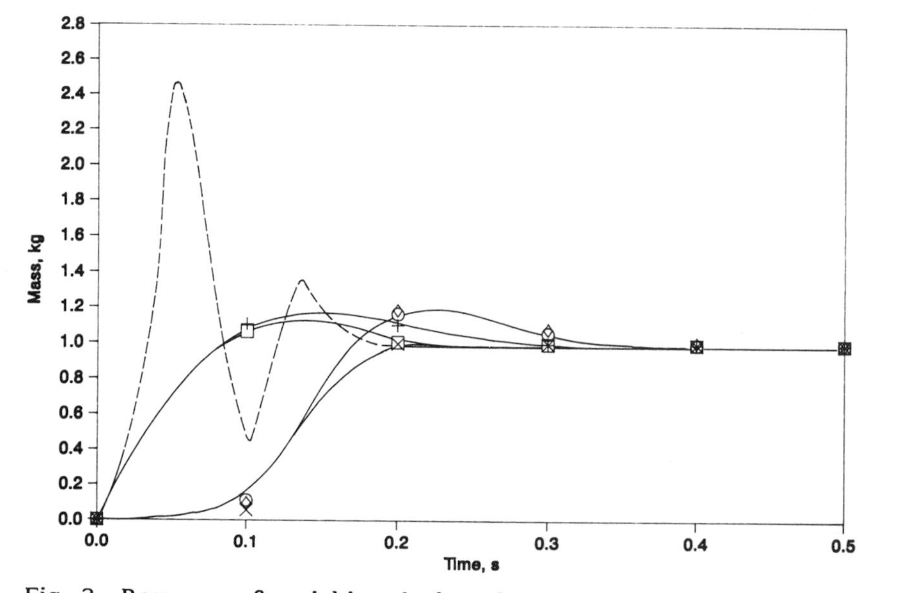

Fig. 3 Response of weighing devices for loading with 1 kg weight: weighing platform (five runs), ◇, +, ✕, ○, □; load cell: – – – –.

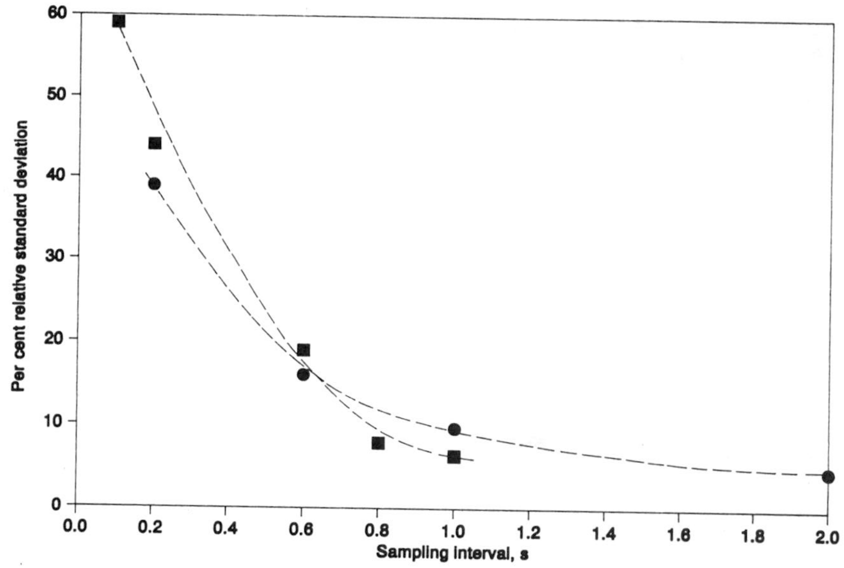

Fig. 4 Effect of sampling interval on per cent relative standard deviation for flowrates measured using weighing devices interfaced with Sinclair QL computer: load cell, – –●– –; weighing platform, – –■– –.

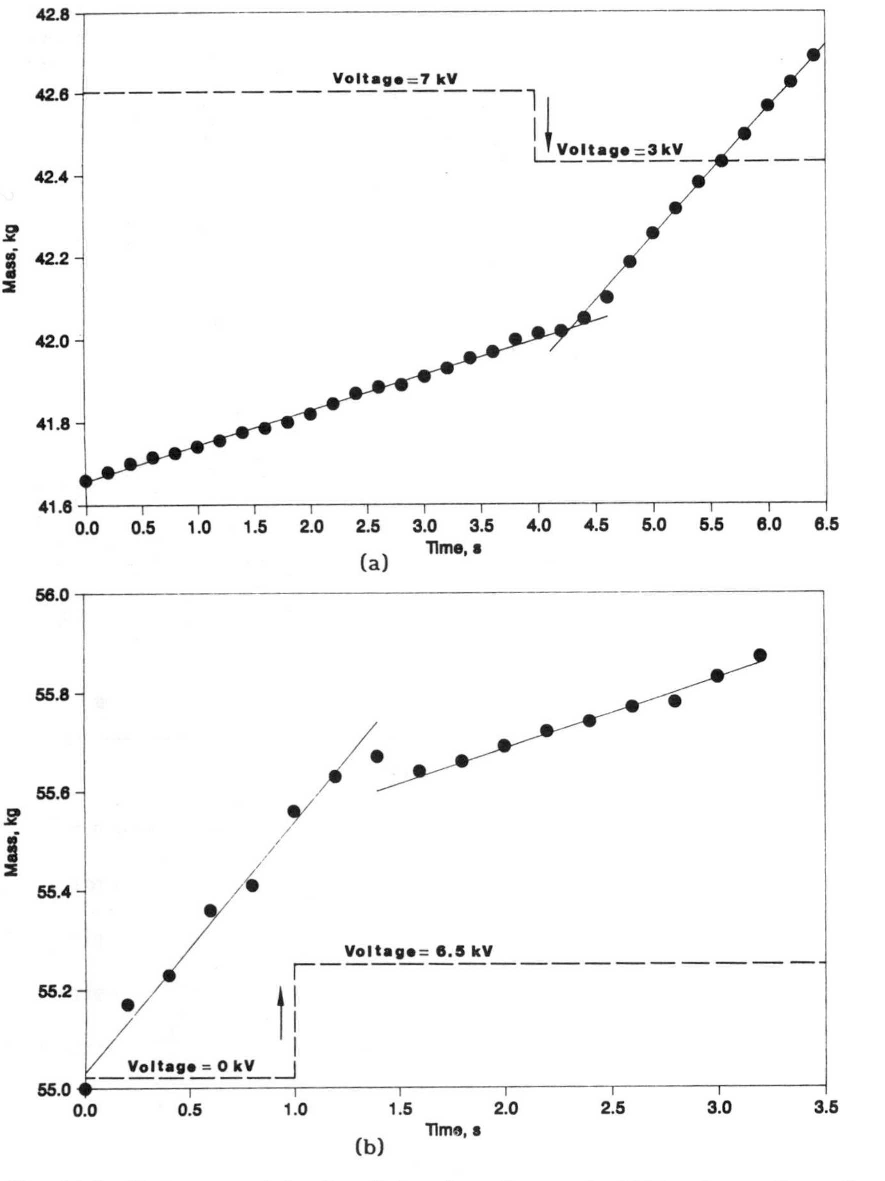

Fig. 5(a) Response of load cell to step change in EVS voltage from 7 to 3 kV made at 4.0 s. Symbols show experimental data. Solid lines show best linear fits to data over ranges of 0.0 to 3.8 s and 5.2 to 6.4 s. (b) Response of weighing platform to step change in EVS voltage from 0.0 to 6.5 kV made at 1.0 s. Symbols show experimental data. Solid lines show best linear fits to data over ranges of 0.0 to 0.8 s and 2.0 to 3.2 s.

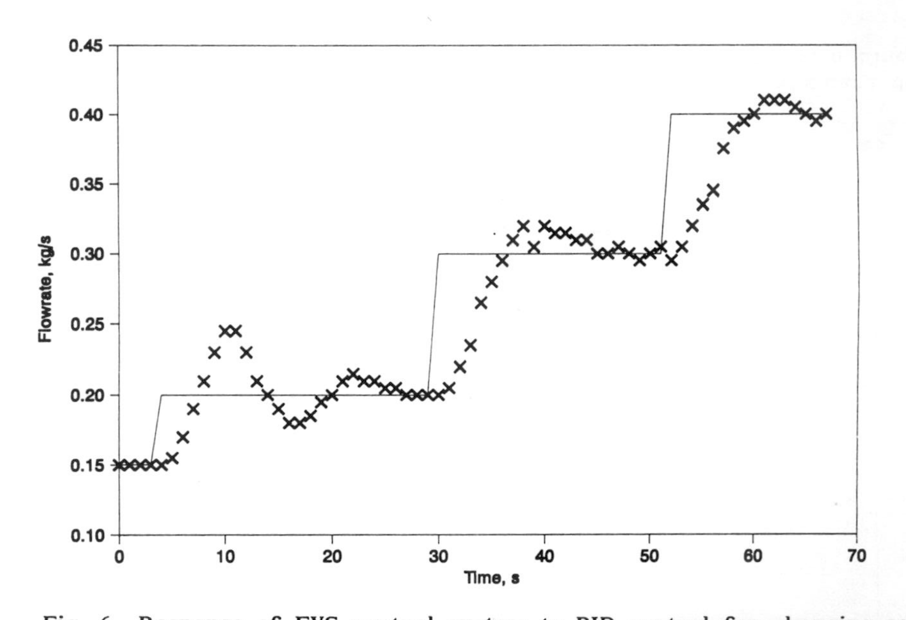

Fig. 6 Response of EVS control system to PID control for changing setpoint: setpoint, ——; moving average flowrate measured using load cell, ×.

C418/024

Bulk weighing – value for money

J P KELLY, BSc, CEng, FIEE
Inflo Bulk Control Systems Limited, Orpington, Kent

SYNOPSIS The design and application of continuous weighing systems is crucial to many aspects of the operation of large-scale bulk terminals and process plants. The importance of such weighing equipment is examined with respect to trading, process control, legislation and terminal development.

1/ INTRODUCTION

The international trade in bulk commodities is now at a greatly increased level compared with previous decades, and has been influenced by economic factors to concentrate the transportation and handling processes into extremely large-scale facilities. Vessels and terminals, both loading and discharging, are now at levels which would have been unforeseeable only a few years ago, and there are now several sites which can handle bulk materials in the range of 50 mtpa and at flow rates well in excess of 10000 t/h.

Similarly, the plants which both produce and consume these bulk materials have experienced the necessity to operate at optimum efficiency, which usually involves a degree of automation, and to reflect contemporary demand for environmentally-friendly operations.

The contribution that continuous weighing can make to these endeavours is quite significant, and needs to be carefully considered at as early a stage as possible in the planning of such facilities.

2/ TRADE

Consider the volume of the trades in the main bulk commodities, using estimated figures for 1990:

Iron Ore	355 mtpa
Coal [all types]	330 mtpa
Grains	185 mtpa
Phosphate Rock	40 mtpa
Bauxite/Alumina	50 mtpa
Minor Bulks	135 mtpa

(Figures ex Internat. Bulk Journal)

The commodities are produced in areas where either they occur naturally or where the correct climatic/agricultural conditions exist, and are predominantly destined for the more advanced industrialised economies, with the exception of grain, which goes to a wider range of destinations. Clearly this gives a strong international flavour to the trade pattern, although this in turn poses problems in the legislative sense where no effective International standards or controls exist to supervise the trade.

The values of these transactions are quite enormous as can be seen by considering just the first two of the above list:

	Ore	Coal
Value/tonne	$55	$40
Trade volume	$20 bn	$13 bn

Truly, then, the need for "value for money" is clear, and careful consideration of the basis for determining value should pay handsome dividends.

Bulk materials are traded on the basis of:

- Quality
- Quantity

Quality is determined by analysis of samples of the materials being handled, and this can be done in the traditional manner using laboratories either on-site or at remote service centres, or more recently by using on-line analysis techniques. The latter have only become available in the last decade and are based usually on nucleonic interrogation techniques.

They are more expensive than conventional sampling and analysis techniques, if the costs of the service laboratory are ignored, but offer the outstanding merit of yielding real-time data on quality as opposed to the conventional method where data is only available historically, usually after a number of days. In the increasingly competitive world of today, the value of real-time data can offer outstanding advantages to the purchaser and user of bulk materials, which will eventually reflect onto the producer.

Quantity is determined by a variety of means, but most often either by surveying or weighing, or both. It must seem strange to those not involved in such trade, that if it is the weight that is of interest, why should any technique other than weighing be used? The reason for this anomoly is largely historical, for when the original trades began, bulk weighing was either unavailable or of a performance that was considered unacceptable, or inappropriate for the materials involved.

The grain trade was probably the first to use bulk weighing techniques, at which time, only static weighing was possible as belt weighers were yet to be developed. Consequently, even today, the grain trade largely continues to use static weighing techniques, even though belt weighers are available to the highest, certified standards. Indeed, the process of handling and storing grains at the majority of bulk terminals is strongly influenced by the necessity to elevate the material in order for it to be weighed by static dump scales. This is a classical example of "skewed" logic whereby the most sensible and feasible technique is ignored and benefits forfeited, all in the cause of convention. (see figure 1)

Other, more recent trades have proven that it is quite feasible to trade in materials to far higher values and volumes using continuous weighing techniques which are as accurate and carry legal approval as static weighing. The lack of competition in the grain trade due to the influence of the few, very large and private grain companies is cited as a factor in this peculiar part of the market. (1)

In contrast, the co-operative grain handlers in the Southern hemisphere have shown that it is eminently possible to trade on a more modern basis, and the whole shape of their terminals reflects the advantages to be gained by current, as opposed to 19th century, thinking on these matters.

In the non-grain trades, however, a further anomoly exists in that the dominance of draft survey as the means for weight determination is despite the fact that modern, legally-certified weighing techniques exist and are proven in the leading terminals. Draft survey is unrecognised by any of the Legal Metrology Authorities throughout the world, and yet is tolerated in International trade to a significant degree. Despite the claimed performance of the surveyors who earn a handsome living from their services, any critical examination of the technique reveals the obscurity and lack of traceability of the method and procedures such as to render as absurd, the claims made for accuracy. In those cases where true comparisons are possible between weighing and surveying, experience has shown that the draft survey technique is seriously flawed, and often biassed to an alarming degree. (2),(3) & (4)

If specification is to mean anything in this context, it is vital to have a trail of traceability as regards both quality and quantity, and the only way in which weight can be traced, is to conduct legally-controlled weighments. The quantification in monetary terms of this traceability, is such that, for example, the significance of an error of only 1.0% in a shipment of 100 000 tonnes of coal, is approximately $30 000. If that error can be reduced or controlled to the level of 0.1%, then a "saving" of $27 000 is on offer for that one consignment. Multiply that for an annual throughput of say 10 mtpa, and the total saving is of the order of $ 2 700 000.

If trade is to be fair, and seen to be fair, then the function of specification and determination is vital, and MUST be traceable to known and proveable standards. Legally-approved bulk weighing has a vital role to play in that endeavour, and should ideally, be built in to the facility as an intrinsic component from the very outset, and not added as an afterthought. The enhanced requirements of the current legislation for belt weighing equipment serve to consolidate that point as there is a clear definition of the complete system into which the belt weighing equipment is to be built [see below]

3/ PROCESS CONTROL

In the production and utilisation of bulk raw materials, there is invariably a degree of processing taking place, aimed at achieving a target specification for the material concerned. This might be as simple as sizing the material or as complex as washing and blending a coal mixture to achieve target parameters for ash content, sulphur content and calorific value.

A variety of measuring devices might be used to gain knowledge of the process condition so that control action can be taken such as to achieve the targets. Without that knowledge based on measurement, the control will not be achieved, other than by coincidence.

Once again, therefore, it is seen that the role of continuous weighing equipment is crucial to the correct performance of the overall process, and its performance must be critically examined and specified, in relation to the overall control strategy. In many processes, the principal parameters are referenced to weight, eg calorific value and ash content, and the correct determination of weight therefore has a direct influence on the principals.

The drive for Quality Assurance on the part of the leading processors demands that a traceability trail exists for the process, and again that has a direct impact on the definition and proof of performance of the weighing equipment. If the overall Quality specification for the processed material is set to a tolerance of say, 1.0%, then each of say, three variables within the process, must be controlled to within +/- 0.57%, requiring the instrument involved in measuring that variable to be repeatably accurate to at least that level. That is a degree of precision that is high by any standards, but in the environment usually experienced in the bulk solids industries, it is doubly so, as the conditions are mostly unfavourable - noise, vibration, climatic changes etc..

The same economic factors which have brought about fewer, larger plants, have also required lower operating costs, which in turn means fewer human operatives and more automation. Fortunately, the progress made in the microelectronics industry has brought huge advantanges to hand in the form of PLC and computer facilities, but unfortunately, there has been no similar easing or relief available to the weighing instrument supplier. Certainly, the availability of microprocessors has improved the electronic processing of transducer signals, but in contrast, the main area of difficulty, namely the conveyor itself, has almost certainly become less benign.

The designer and builder of conveyors has a much greater knowledge of structures, and the computing power to make use of that, to limits not approached previously. The effect of this is to make conveyor structures more "bendy" than hitherto, and whilst still perfectly safe for conveying purposes, they are much more difficult on which to weigh continuously and repeatably. The supplier of weighing equipment has to be much more careful to achieve conditions in which structural stability is achieved over the whole range of loadings, and is often required to stiffen structures which are otherwise perfectly satisfactory. (see figure 2)

Likewise, belt speeds are increasing to levels which can create "noise" problems in a signal sense, which can interfere with the parameter being measured unless careful control is taken. This may mean, in addition to stiffening of the structure mentioned above, extra attention being paid to the ramping and alignment of the conveyor over the entire weighlength. At the highest speeds currently in use, in the range 6 - 9 m/s, this will require the weighlength to be extended by perhaps 10 metres either side of the actual weightable, and only the highest quality idler sets to be used. Any eccentricity and out-of-balance of the idlers can have unwanted effects at lower speeds. At higher speeds, these factors become even more significant. The aim must be for the belt to glide over the entire weighing area in the smoothest possible manner, in order to avoid spurious forces affecting the scale itself, and to avoid the material on the belt becoming airborne as a result of striking a proud, or mis-aligned idler at high speed.

The role of computers and PLC's in process plant automation has been quite dramatic in the last decade, and huge operating gains have been made as a result. Control regimes which previously were the preserve of the skilled operator, have routinely become vested in software, and today, even expert systems have made their debut in bulk materials processing, particularly for kiln control in the cement industry. These control packages are based on rules which are defined by the operators, and can allow multi-variable processes to be controlled in a much more relaxed manner than previously, and by perhaps as few as a single operator, where previously there would have been a task-sharing team on duty.

However, any control system is only as good as the knowledge it has of the actual conditions in the process, and so we remain totally dependant on the accuracy and repeatability of the instruments providing the inputs to the control system.

The application of continuous weighing to such plants takes on an added dimension, as consideration has to be given to the response and the time delays inherent in the process, which will influence the location and number of weighing instruments being used. The higher the target performance for the process as a whole, the more important will be the quality of the information obtained from the instruments in the plant. In addition to the conventional weighing parameters of accuracy and repeatability, the significance of resolution is little understood by bulk handling specialists, but can dramatically affect quality of the signals from the plant.

In the case of belt weighers, which are integrating instruments, the accuracy of the instrument is dependant on a number of factors, one of which will be a function of the length of the belt itself. An integrating instrument requires time to arrive at a result which is acceptably accurate. Instantaneously, its output can vary enormously and is of only secondary value for control purposes. Any attempt to control the process by reacting to the instantaneous output of a belt weigher would be doomed to failure, as it would swing wildly in a totally erroneous fashion, trying vainly to follow the weigher. This is an unstable condition and can lead to complete loss of control.

The minimum time for a belt weigher to arrive at any sensible result is one complete revolution of the belt, and thus if the weigher is unthinkingly located in a long belt, there will certainly be severe frustration experienced by the control system, which could be avoided by the correct selection and location of the weighing units. If the weigher itself is to be a part of the controlling action, then it is usual to incorporate it in a short-centred conveyor called a weigh feeder. Such a device includes a variable speed drive such that the flow rate can be controlled by reference to the measurement from the scale and a setpoint or target value.

If the flow rate is to be controlled over a very wide range, more than 10:1, then in addition, it will be necessary to control the preceeding feed device in conjunction with the weigh feeder speed, such that the combined effect of the two can maintain the degree of control required. The preceeding device might be a vibro-feeder, or a plate or apron-feeder, on the outlet of a storage bunker. Extreme care has to be taken in specifying and selecting these elements in the system if the overall target performance is to be met.

By their nature, automated plants are capital-intensive - they are designed to replace labour with capital so as to improve the economics of the operation as a whole. However, the very absence of the human-being with his powers to observe and exercise rational judgement, renders the plant totally dependant on the validity of the information being gathered from sensors and processed by the automatic control equipment. Where the pattern of the process variables can be identified, then rule-based systems have an important part to play, but in many areas of bulk materials, the behaviour is random and therefore unpredictable, and must be constantly measured if control is to be maintained.

The influence of weighing equipment performance can thus be seen to be a vital input to this process, and it follows that the performance that is required should be correctly specified and proven.

4/ PERFORMANCE AND THE LAW

It is a sad fact that with the exception of the legal requirements for trading by weight, there are no industry, national or international standards for the performance of continuous weighing equipment. It is not uncommon, therefore, for the users and suppliers of such equipment to invoke the legal standards for trading by weight for defining performance as a "default menu" even though the equipment might not be intended for trading use. It would be useful therefore, to consider the use and extent of these standards or regulations, as they have such a direct bearing on performance.

If performance is to be achieved and quality assured, it is essential to have traceability of the means adopted, and that requires definition and reference to available standards. If this is done, then there is also the correct basis for contract between user and supplier, and it is possible for the former to exercise control over the latter. Much of the disenchantment so often expressed by users as regards continuous weighing equipment can be directly laid at their own door by their failure to correctly contract for what they require. Their failure to make reference to available standards allows the incompetent or the unscrupulous to promise that which cannot be delivered, and without recourse by the user - a state of affairs which is rarely tolerated in most other areas of engineering.

An examination of the regulations that apply to weighing equipment used for trade , must first consider what that use means. In common with most countries, the UK has its own national regulations enshrined in the Weights & Measures Act of 1985 and in that Act there is a definition of so-called "use for trade". This is as follows:

"where the transaction is by reference to quantity or is a transaction for the purposes of which there is made or implied a statement of the quantity of the goods to which the transaction relates"

"a transaction is the transferring or rendering of money or money's worth in consideration of money or money's worth, or the making of a payment in respect of any toll or duty"

Thus we see that where money or its kind changes hands in return for the supply of goods or services, and that supply is based on the WEIGHMENT of a material, the determination of that weight must be by legally-approved means. Nowadays, that requires the use of a weighing machine which has been pattern-approved and certified as fit for use for trade, and which is controlled in service by the local Trading Standards Officer [TSO].

NB: It is permissible to trade in a bulk material where the quantity is arrived at by a means other than weighment, and such means are uncontrolled by any legal body - so-called "private treaty" arrangements. This is the territory served by the surveyors and arbitrators

Pattern Approval

This is conducted by the central Weights & Measures Laboratory of the country concerned, and in the UK is the NWML at Teddington. Their responsibility is to examine and test the instrument intended to be used for trade, in order to prove compliance with the appropriate regulations. This involves the most detailed testing of the mechanical and electronic components in the instrument, and with the imminence of the Single Market in 1993, will also embrace the requirements of more general Directives on Machinery and Electro-magnetic Compatibility which are applicable to all products to be sold in future.

The fundamental aim of the examination is to prove the "Metrological Security" of the instrument under test, i.e. that it is fit for the purpose and cannot be used in a fraudulent manner. The regulations state quite clearly what the requirements are for accuracy, repeatability etc., but are necessarily less precise in the area of fraudulent use. Suffice it to say that the Inpectors job is to prove the immunity of the instrument to any interference, whether intentional or not, that might cause an incorrect measurement to occur.

This will cover the obvious areas of mechanical and electrical stability and accuracy, temperature dependance, etc., but will also dwell on the less obvious areas of software security and immunity to external influences by whatever means.

All the tests at the pattern approval stage are conducted under carefully controlled conditions, such as are to be expected in a National Laboratory, but they only form a part of the overall testing procedure. The final phase of the testing takes place at the first practical application of a weigher which has been pattern-approved, and is known as the "in-situ testing". This is the first time that the weigher is examined together with a real conveyor, belting, structure etc., as the physical size and nature of a conveyor belt scale and supporting structure largely precludes the possibility of laboratory examination of a "real" unit. As such, the laboratory testing is a simulated programme, and is done to levels which are probably immeasureable at the "in-situ" phase.

The site testing concentrates on performance parameters such as:

- zero stability
- sensitivity to small load changes
- material testing for linearity, accuracy and repeatability

These tests require the site to be equipped with material testing facilites which are equal to the task, and include most importantly, a check or reference scale which is at least 5 times more accurate than the belt weigher under test. Naturally the procedures for transferring the weighed test loads to and from the test instruments must also be carefully controlled, and the whole procedure can well take several days and cost a large amount of money in labour and facilities.

Assuming that all these tests are successful, then the belt weigher is approved fully, stamped for that particular site and can be freely marketed for similar applications.

However, it is not as simple as it sounds, for the very good reason that the regulations now in force, and planned for the future, extend to far more than the instrument itself. They prescribe that the complete installation, including the weigher and the conveyor, be specified in certain important respects. For example, the location of the scale is considered in relation to feed points, tangent points etc., and the structural stability is controlled to the extent that it should be rigid and not subject to undue deflection under loading conditions. Restrictions are imposed as regards the type and location of the tachometer forming part of the scale. The selection of scale intervals for the totalisers and the limits on the duration of certain tests forces the selection of shorter rather than longer conveyors, which is all to the good.

Suffice it to say that the regulations cover most of the practical points which will yield benefits if adhered to, as regards the application engineering of the installation, and as such, provide a valuable benchmark for any application, whether or not it is intended for trade use. An increasing number of users are therefore requiring compliance with these regulations as an indication of the ability of the supplier to deliver performance to a level which can be adequately defined for contractual purposes.

The UK regulations will be supplemented in 1993 by the EEC Directive on Automatic Weighing Equipment and supporting Regulations, which are currently in the final phase of preparation. By then, it will be possible to refer to the EC regulations for belt weighers, meaning that there will be one standard for compliance across the whole Community. This will also be co-ordinated with the International agreement on regulations which is co-ordinated by the Organisation for Legal Metrology [OIML], and which will then provide a realistic basis for specifying performance across the world.

When that happy day is reached, we might see the demise of the so-called "Private Treaty" that is so widely practised in international trade today, whereby consenting adults are free to trade on the basis of weight,which can be determined by whatever method those involved choose. It is in this area that the technique of draft survey predominates, as discussed previously. It is to be hoped that with the issue of common and high standards across the world, the traders will be persuaded that it is in everyones interest to see trade conducted in a fair and correct manner, backed by the Law of whatever country is involved.

5/ TERMINAL DEVELOPMENT

How can the development of terminals be affected or influenced by the use of continuous weighing equipment?

By using legally-approved weighing equipment at such facilities, it will be necessary to follow the provisions of the regulations and ensure that the instruments are correctly located, in suitably designed and constructed conveyors, and that the necessary material test facilities are built into the plant from the outset.

This last point can be most significant both technically and economically, as it is not often realised that the testing facilities may well exceed the cost of the belt weighing equipment by a magnitude of cost, even though they may be little used. Nevertheless, they must be available and of the correct standard and capacity, and suitably protected from the elements, otherwise testing may well be confined to dry, windless days!! Judicious selection of the location of the test facility and the size of the test load[s] will pay handsome rewards in terms of the ease with which the calibration and testing of the belt scales is conducted.

Consider the manner in which the test load is determined, in order to appreciate the above point.

The test load must be the larger of the following parameters:

1/ 2% of the hourly capacity

2/ The weight of material in one complete revolution at full capacity

3/ 200 scale intervals of the load totaliser

If we ignore 3/ for the moment, as it is rarely a constraint, and consider a capacity of say, 10 000 t/h and a belt length of 100 metres, then the test load will be 200 tonnes (2%).

If on the other hand, the belt length is 500 metres, then the test load will be well in excess of 200 tonnes, depending on the belt speed forcing a result from 2/. The regulations therefore punish the use of approved scales in long conveyors, and encourage, correctly so, the use of short conveyors for stamped scales.

Once the size of the minimum test load has been determined, it must be appreciated that many tests are conducted to fully test the scale over its entire operating range. For example, 2 pairs of tests are conducted at each of the following points on the load range:

20, 40, 60, 80 and 100%

The testing officer can therefore demand up to 20 tests at 200 tonnes each, which can occupy a great deal of time and many resources, thus mitigating against too frequent a testing regime.(see figure 3)

For this reason alone, the user of such weighing equipment has an intrinsic concern for, and interest in, the repeatability of the weighing equipment, and it thus behoves him to make sure that the supplier can actually deliver this single most important parameter of performance. Unfortunately, the regulations, whilst containing an intrinsic requirement for the scale to be repeatable, make no specific call for it to be so, relying on the energy and diligence of the TSO to maintain vigilence on the site in question. That is a pious hope in reality, and the user is really on his own in this regard, as it is quite impractical for the TSO to monitor the performance of a high capacity belt scale facility in the way he does for petrol stations and street traders.

One other aspect of terminal development that must be mentioned, is the problem of flow rates. As we saw previously, the question of flow rate is vital to the performance of belt weighing equipment, as like all instruments, non-linearity is ever-present and if the flow range varies widely with a fixed-speed conveyor, then there is a bottom limit of 20% which the regulations allow as the minimum.

At a typical ship-unloading facility, this can present severe operational problems which must be resolved if the law is to be complied with. The problem can be solved by a variety of means, such as:

A/ Storage capacity between the unloader and the scale

With this arrangement, the feed to the scale can be switched on/off as the amount of material in the storage bunker allows, such that the flow is always within the statutory limits of 100-20%.

B/ Continuous Unloaders

These devices ensure that the average flow rate through the ship is always well within the statutory limits, and the problem is largely avoided, except for clean-up operations which can still present problems.

C/ Variable speed conveyors

This is the elegant way to compensate for wide variations in flow rate without the use of either of the above two facilities. It is necessary to use a scale in the conveyor preceeding the trading scale, such that the flow rate signal from the first scale is used to control the speed of the conveyor in which the trading scale is located. In this way, the load on the second scale is kept within allowable limits irrespective of the actual flow rate, but very careful consideration must be given to the dynamics of the control scheme to guarantee that under all conditions, no flooding of the variable speed conveyor can occur. (see figure 4)

Finally, the terminal operator will be acutely concerned with availability and reliability of the weighing equipment, and this again emphasises the importance of repeatability - the ability to deliver accuracy over long periods of time, without the necessity to test and recalibrate. These last two actions are very involved in the case of belt weighing, as there is no easy way in which to determine if a scale which is functioning, i.e. is not broken, is actually measuring correctly, as the scale forms part of a complex system which must be carefully investigated as to overall repeatability. This is a skilled task and not one easily conducted either automatically or by the user himself.

For these reasons, it is becoming more common to have the supplier carry a continuing responsibility for the performance of the equipment he supplies, and within the environment of an approved Quality Assurance regime. Such a partnership can guarantee that the performance that is so necessary can be constantly available, thus enhancing the overall performance of the terminal and providing a tight control over the measurements of the materials moving through the terminal.

Recent developments in this area include the availability of automatic

Stock Accounting Packages at bulk terminals which maintain a careful watch over all movements into, through and out of the terminal, and provide Management reports on the performance of various parts of the plant. Such systems use a computer to interrogate all the belt scales in the plant and generate running and cumulative data on all weights, flow rates, durations etc., and can also alarm against illegal routings being selected. (see figure 5)

6/ CONCLUSION

The use of continuous weighing equipment at bulk terminals and process plants has assumed an increased significance in recent years due to the need for improved control and performance of those plants.

The use of recognised standards for specifying and contracting for weighing performance is essential if traceability of performance is to be achieved, as it must for overall quality assurance.

References:

(1) "Merchants of Grain" Dan Morgan

(2) "Bulk Solids Handling" April1991, J P Kelly

(3) Port Development International June 1986

(4) "Sampling & Weighing" Jan Merks

Appendix:

The International Regulations covering Automatic Belt weighers are designated RI 50, and are currently (1991) in the final stages of re-drafting for the purpose of under-pinning the EEC Directive on Automatics. When adopted, these will provide for common legislation within the EEC allowing free trade in such instruments, which hitherto have required National approval in each member state.

In addition, these Regulations will provide an International framework of common standards within which international trade can take place for the first time. Each weighment, on whatever continent, will be traeable to the appropriate National Standard of Certified weights, each of which is traceable to the International Standard.

The co-ordinating body for these matters is the Bureau of International Legal Metrology in Paris. National access can be gained via the Weights & Measures Authority in each country, or directly to the BIML.

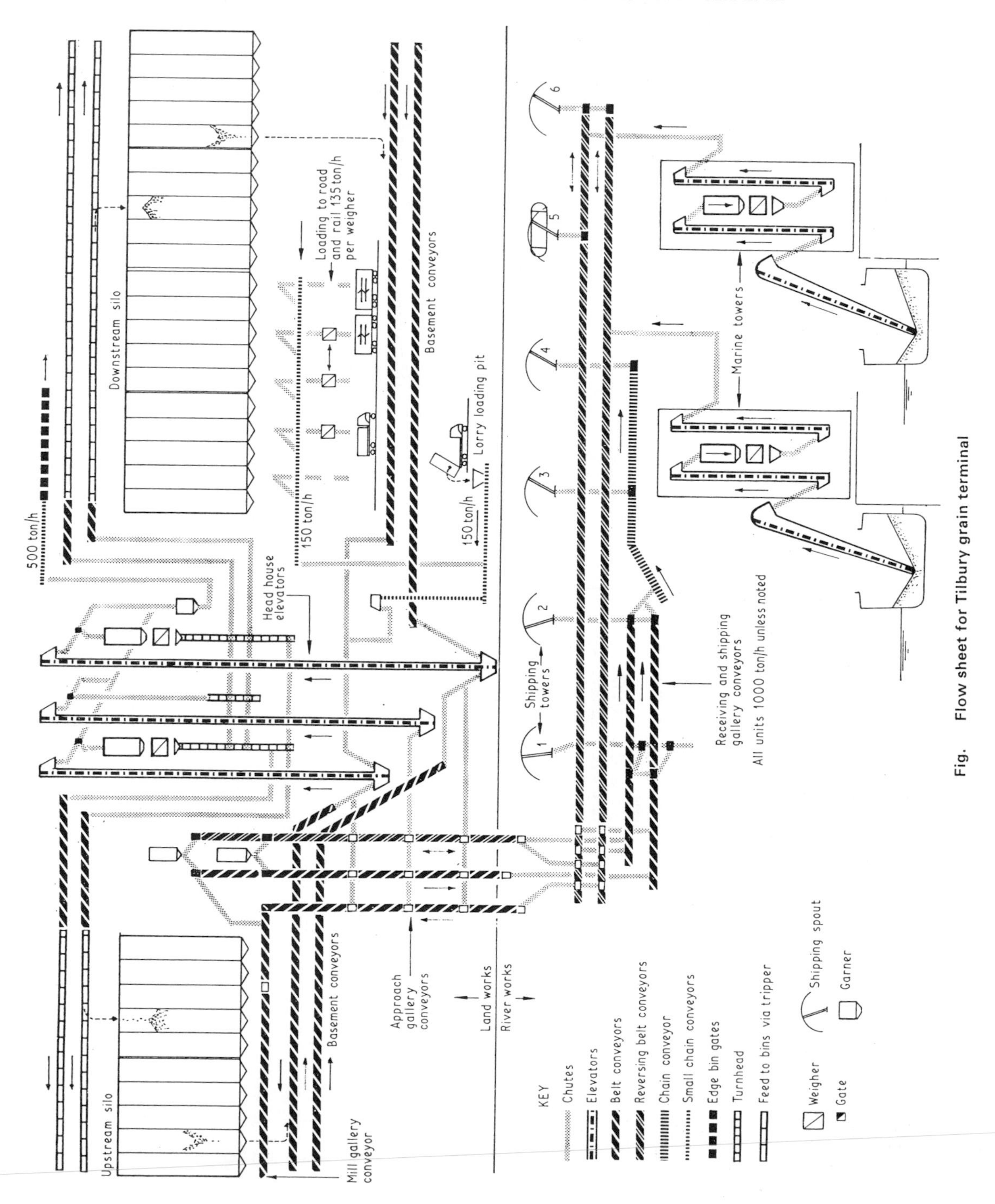

Fig. Flow sheet for Tilbury grain terminal

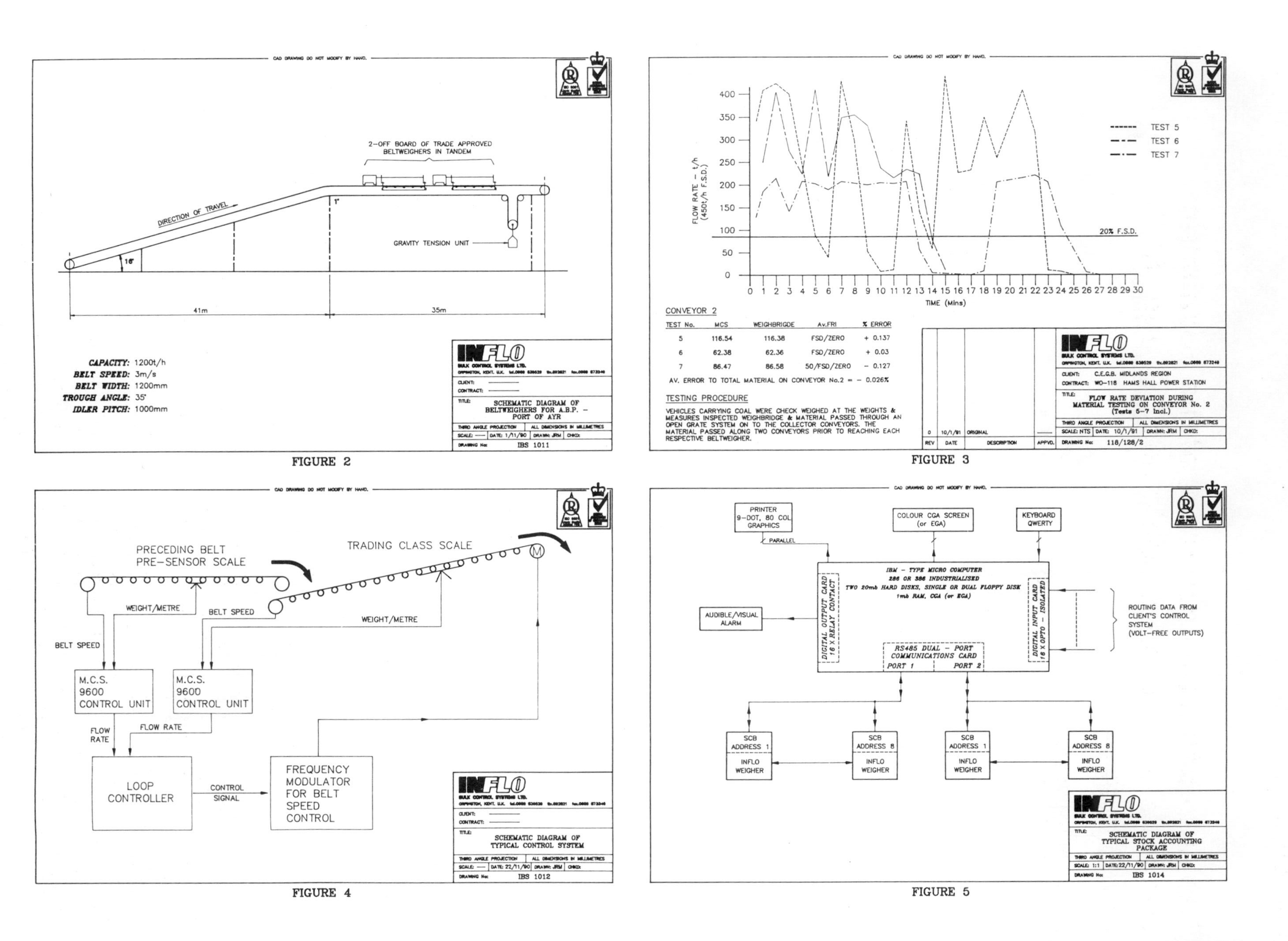

FIGURE 2

CONVEYOR 2

TEST No.	MCS	WEIGHBRIDGE	Av.FRI	% ERROR
5	116.54	116.38	FSD/ZERO	+ 0.137
6	62.38	62.36	FSD/ZERO	+ 0.03
7	86.47	86.58	50/FSD/ZERO	– 0.127

AV. ERROR TO TOTAL MATERIAL ON CONVEYOR No.2 = – 0.026%

TESTING PROCEDURE

VEHICLES CARRYING COAL WERE CHECK WEIGHED AT THE WEIGHTS & MEASURES INSPECTED WEIGHBRIDGE & MATERIAL PASSED THROUGH AN OPEN GRATE SYSTEM ON TO THE COLLECTOR CONVEYORS. THE MATERIAL PASSED ALONG TWO CONVEYORS PRIOR TO REACHING EACH RESPECTIVE BELTWEIGHER.

FIGURE 3

FIGURE 4

FIGURE 5

C418/051

The automated measurement of the volumetric contents of stockpiles contained in large 'A' frame facilities

R N BARNES, BSc(Eng)
The Wolfson Centre, Thames Polytechnic, London

Synopsis. Computer programs are used to model the shape of material stockpiles in large "A" frame storage buildings following typical charging and reclaim operations. The models are used to establish the optimum number and location of a series of depth measurement transducers that will yield an accurate on-going assessment of the volumetric contents of a typical configuration of such a stockpile under charging and reclaim operations.

1. Introduction.

Knowledge of the quantity of materials held in storage is important for the economic operation of production plant. When bulk solid material is contained in relatively small scale silos it is possible to measure the quantity of material in stock on a gravimetric basis by mounting the silo on load cells or by installing strain links on the steelwork supporting the silo. Such methods are capable of indicating the quantity of material to within one or two percent. However, when the material is contained in very large silos or in stockpiles it is impractical to use gravimetric methods of measurement. On the other hand estimates of the volumetric contents are possible and it is to practical solutions of this problem that this paper is addressed.

Clearly, the exact volumetric contents of a stockpile could be determined if the entire surface profile of the material were accurately known. This would entail making numerous measurements of the height of the material over the complete surface. This is clearly impractical, and indeed, unecessary as it is, in fact, possible to obtain a very good estimate of the volume from only a few spot height measurements. The main problem is to determine the number and position of the spot height measurements necessary to achieve a given measurement accuracy. For the purposes of this paper it is assumed that the spot height measurements could be obtained using ultrasonic transducers mounted at the appropriate points in the roof of the "A" frame building. There are several commercially available products with adequate power and resolution for this task and systems are available where up to 60 transducers may be multiplexed from a single signal processing console, which may also be linked through a serial communications port to a central computer for evaluating the volumetric contents, data logging and other tasks.

Methods of volumetric measurement of material contained in large cylindrical silos have been presented in previous papers (1) (2) where it has been shown that volumetric measurements, with an accuracy of the order of 2-3%, were possible using only 4 ultrasonic transducers. This paper extends the concepts of that work to the problem of the measurement of the contents of stockpiles contained in large "A" frame buildings.

2. Sample Problem.

The methods of stockpile measurement are presented in the context of a particular industrial "A" frame building shown in the diagrams of Fig.1. The building has a floor size of 110m (360ft) by 46m (150 ft) and the material may be loaded up to a height of 6m (20 ft) against the side and end walls. If the powder angle of repose of the material is 30 degrees, the maximum height of the stockpile along the centre line of the building will be 19m (63 ft), with a full capacity of about $5.6 \times 10^4\ m^3$. Material is charged onto the stockpile from a conveyor, having a trip feeder, positioned in the apex in the roof space. The material is reclaimed by 12 transverse underfloor conveyors each having 5 feeding hatches.

Based on a drained angle of repose of 30 degrees, the material left in the building following complete draw down with the reclaim system is $1.02 \times 10^4\ m^3$ giving a draw down efficiency of about 82%.

In order to examine the accuracy of volumetric contents measurement for the above situation, computer models have been developed and numerous simulations have been performed to gain an overall idea of what accuracy might be available with given numbers of depth measurement transducers.

3. Modelling Procedure.

The computer models operate by setting up a rectangular array of points over the base of the stockpile and the level of material at these points is then calculated. Suppose the building is initially empty and material is then added at a given point to a given depth. The level of material at every other position can be calculated with respect to the loading point using the angle of repose for the material. Further material can be added or removed from the same, or other, locations and calculations of the surface shape repeated. The volume under the surface can easily be found from the model and comparisons made with the estimates derived from the material height at selected locations where the transducers are to be positioned. Further details of the basic approach are given in (1).

4. Measurement Transducer Positions.

Two distinct transducer arrangements, utilising 12 and 24 transducers respectively have been considered as shown in the diagrams of Fig.2. An important general principle in the measurement technique is that each transducer takes a spot height measurement which should be the average height of the material for that area. The volume in that area is then the product of the area and the spot height and the total volume of the stockpile is the sum of all these individual volumes. Clearly, the more transducers that are used the smaller will be the area associated with each transducer, and the more accurate will be the overall estimate of the overall volumetric contents.

It is also important that the transducers should be at positions where the height of material is close to the average height for that area. Now the normal operations of loading and reclaim generally result in the material sloping from the building centre line to the side wall and the average height for the chosen areas, as shown in Fig.2 (a) and (b), will be halfway between the building centre line and side wall as shown in Fig.2(c). It will be seen that these choices give good results.

5 Test Results.

5.1 Test Stockpile Conditions.

A number of typical stockpile models have been generated to simulate typical operating conditions as summarised below in TABLE 1 and shown in the diagrams of Fig.3. These diagrams are screen dumps of the graphics display of the various stockpile conditions. The conditions simulated start with a full stockpile which is progressively emptied by operating the reclaim conveyors three at a time until complete drawdown is reached as shown in stockpile number 7. Two other cases are also tested.

TABLE 1. Stockpile Models

Stockpile No	Condition	Stockpile Volume	
		$m^3 \times 10^4$	($ft^3 \times 10^6$)
1	Full capacity.	5.62	(1.99)
2	As in 1 but with complete draw down on conveyors 10,11 & 12.	4.21	(1.49)
3	As in 2 but with partial draw down on conveyors 7,8 & 9.	3.67	(1.30)
4	As in 3 but with complete draw down on conveyors 7,8, & 9	2.88	(1.02)
5	As in 4 but with partial draw down on conveyors 4,5, & 6	2.32	(0.82)
6	As in 5 but with complete draw down on conveyors 4,5, & 6	1.64	(0.58)
7	Complete draw down on all conveyors	1.02	(0.36)
8	As in 4 but with stockpile partially recharged.	3.70	(1.31)
9	Partially filled stockpile from empty state.	2.20	(0.78)

5.2 Results for Measurement Simulation.

The various stockpile conditions described in TABLE 1 above, and illustrated in Fig.3, have been used to simulate the measurement of material quantity using the 12 and 24 transducer arrangements shown in Fig.2. The results are given in TABLE 2 below. The percentage error in the measurement is the difference between the measurement derived from the transducer array and the actual volume for the particular stockpile expressed relative to the full capacity of the "A" frame building, i.e. the volume for stockpile 1 in TABLE 1 above.

TABLE 2 Simulation Results

Stockpile No	Error %	
	12 transducer system	24 transducer system
1	1.99	1.56
2	3.75	0.17
3	5.57	-0.59
4	6.10	-1.47
5	5.58	-2.64
6	6.64	-3.21
7	5.58	-2.64
8	-0.65	-1.25
9	-0.61	-0.64

6.Conclusions.

It can be seen that the results when using 24 transducers give about twice the accuracy as when 12 transducers are used. It should be stressed that the errors given are "system errors" that arise from the particular disposition of the transducers. Any errors due to the transducers themselves will, of course, add to the errors given.

It is estimated that typical capital equipment costs for the measurement systems given above would be approximately £10k and £16k for the 12 and 24 transducer systems respectively with installation costs additional to these. The decision to install such a system in a storage facility would, of course, depend upon how important it was to know what quantity of material was held in stock and clearly this becomes more important with increasing unit value of the material under consideration.

The installation of transducers for either of the two schemes would require additional support brackets to facilitate the attachment of the transducers under the roof in the positions indicated in Fig.2. and in fact access to these positions would not be easy. It would be much more convenient if it were possible to mount the transducers along the centre line of the building as walkways are usually provided adjacent to the loading conveyor. This possibility was examined but because the transducers are positioned along the building centre line they measure the peak heights of the material. This means that an algorithm must be used to estimate the average value from the peak height but it was found, using 12 transducers positioned along the centre line, that errors as large as 20% occured with certain material configurations. These errors were so unacceptable it was not felt worthwhile to give detailed results. However, these tests did point to the necessity of carefully locating the transducers to positions where the material height was close to the average for the chosen area under consideration.

Finally, it may be stated that suprisingly accurate estimates of the volumetric contents of large storage buildings would seem to be possible using the methods discussed. However, it is felt that simulations should be performed to verify the likely accuracy for other configurations of building geometry, for different loading and reclaim arrangements, and also when different operational methods are to be used.

7. References.

(1) BARNES,R.N. The Measurements of the Volumetric Contents of Large Silos and Stockpiles. Proceedings of the Third International Conference on Bulk Materials Storage, Handling and Transportation. Newcastle Australia, June 1989. 295-306. (The Institution of Engineers, Australia.)

(2) BARNES,R.N. Optimum Positioning of Level Transducers for Silos Contents Measurements. Solidex 88 Conference, Harrogate, March 1988.

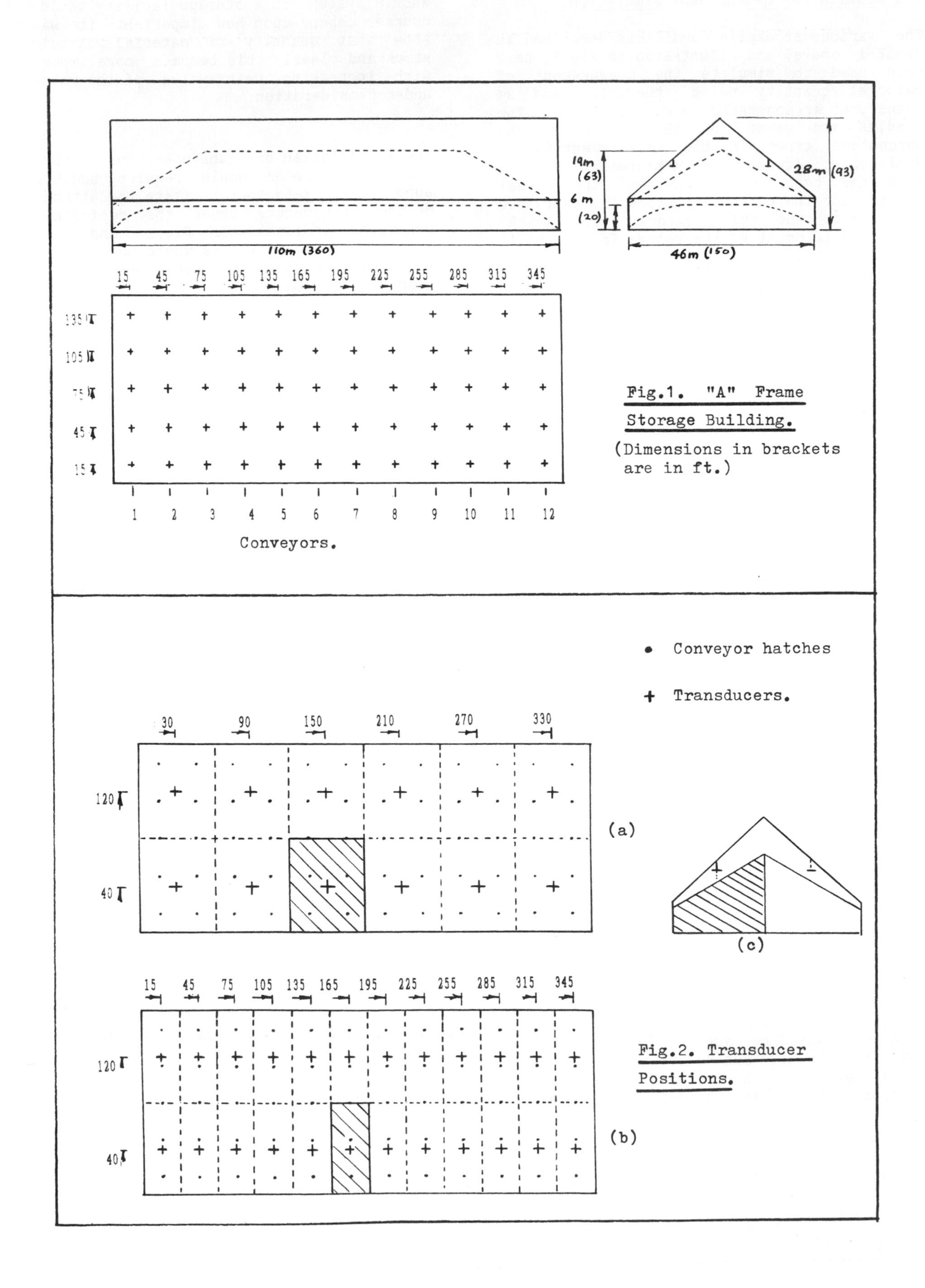

Fig.1. "A" Frame Storage Building.

(Dimensions in brackets are in ft.)

Fig.2. Transducer Positions.

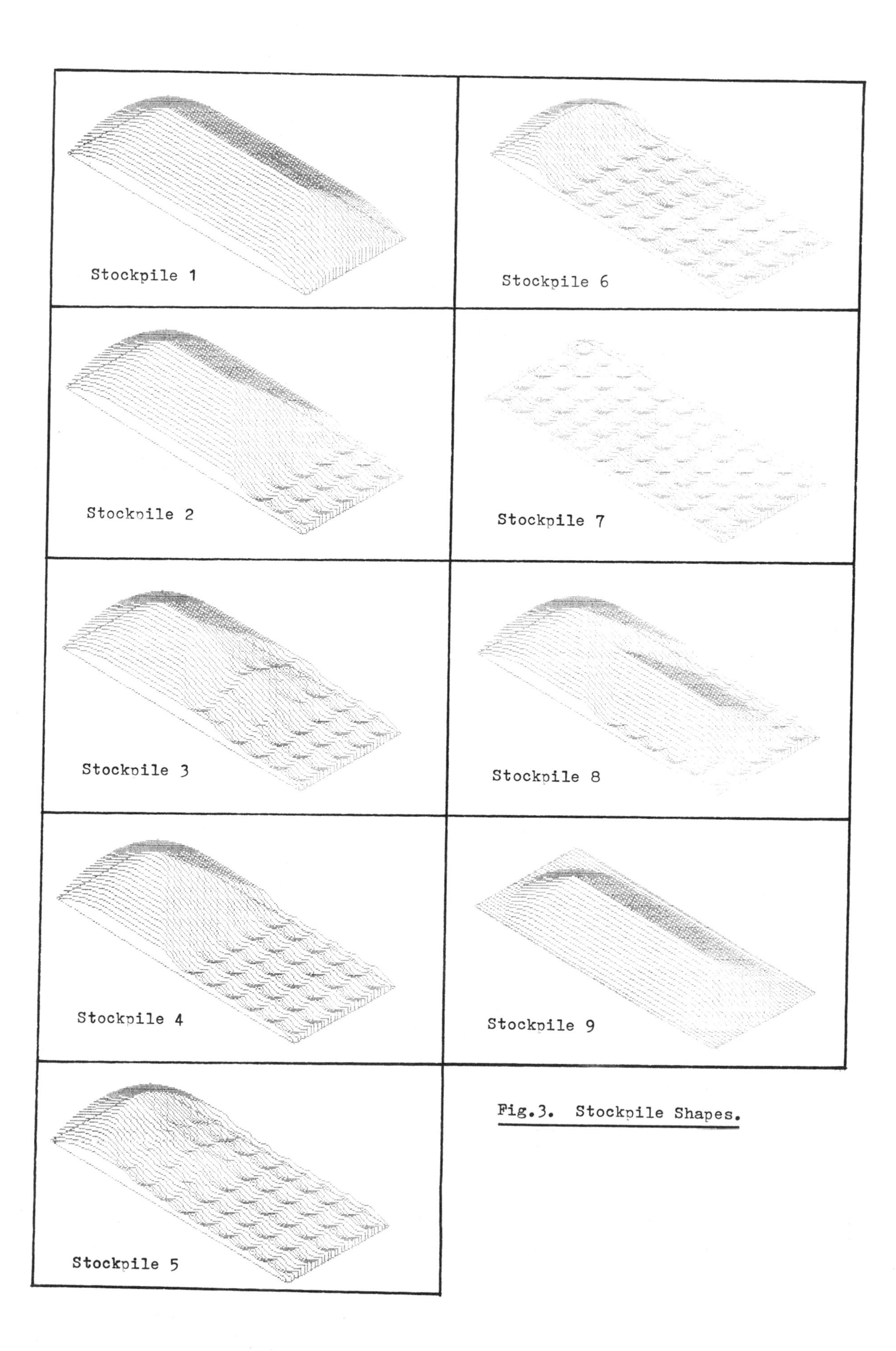

Fig.3. Stockpile Shapes.

C418/012

Conveyor transfer chute performance and design

P C ARNOLD, BE, PhD, CEng, MIMechE, FIEAust and G L HILL, BE
Department of Mechanical Engineering, University of Wollongong,
New South Wales, Australia

SYNOPSIS Some of the critical aspects of transfer chute design are examined, paying particular attention to such parameters as material trajectories and material flow properties. The merits of differing transfer chutes are discussed, as well as the concept of using small surge bins to overcome many of the problems found in existing transfer chutes.

NOTATION

g	Acceleration due to gravity
h_b	Material height on conveyor belt prior to discharge
R	Distance from centre of discharge pulley to material's centre of mass
v_a	Material exit velocity of flow-round zone
v_b	Conveyor belt velocity
v_p	Material entry velocity to flow-round zone calculated from the point of intersection of the trajectory of the centre of mass of the incoming stream with the plate
α_b	Conveyor belt inclination angle prior to discharge
α_d	Bulk solid stream discharge angle measured from the vertical
$\delta(\sigma_1)$	Angle of internal friction corresponding to a consolidation stress σ_1
σ_1	Major consolidation pressure or stress
σ_a	Adhesive stress
β	Angle of impact surface to the vertical
$\beta = 0$	for a vertical plate or surface
$\beta = +\beta$	for a plate angled towards the incoming material stream
$\beta = -\beta$	for a plate or surface angled away from the incoming material stream
γ	Specific weight of material
θ	Angle of velocity vector $\vec{v_p}$ to the horizontal
ϕ	Angle of wall friction
ϕ_{RW}	Rough wall friction angle
μ	Friction coefficient for material sliding on itself
δ	Effective angle of internal friction

$\mu = \sin\delta$ for material sliding on itself

1. INTRODUCTION

The transfer chute plays a vital role in the efficient and smooth operation of belt conveyor systems. A poorly designed chute will most certainly lead to a great number of problems as well as increase tremendously the cost of running the whole conveyor system. These costs can arise from lost production due to the down time required to repair badly damaged belts resulting directly from transfer point failings, or the extra cost of labour required to maintain the chute not only mechanically but to clear blockages and to clean up fugitive material that always seems to escape in great quantities from poorly thought out transfer areas.

More often than not the actual belt conveyor can be made to operate relatively efficiently and for this reason it is one of the most widely used methods for transporting bulk materials in the world today. Generally though it is the transfer area that is neglected the most, the design being left to the draughtsman with little or no experience of such a critical area. To compound the problem the transfer chute is often fabricated by one company and installed by another. All of these parties unaware of the requirements of the original conveyor designer and/or manufacturer.

Very few comprehensive works have been undertaken on the myriad of details that are required to produce a good transfer chute design [1, 2, 8, 9]; most designs are based on experience, the bulk of which has been learnt the hard way. This paper examines some of the more critical aspects in the design of transfer areas, with the emphasis on some of the points that are often overlooked. It also examines the use of small surge bins coupled with variable speed belt feeders to overcome many of the problems associated with the transfer area. This approach has been the focus of a research program carried out at the University of Wollongong Australia, to determine the suitability of such an arrangement as a transfer device, particularly for use in the case of high speed belt conveying where many of the problems are amplified.

2. DESIGN REQUIREMENTS FOR TRANSFER CHUTES

When designing a transfer chute particular care should be taken to ensure that the device will be able to carry out the required duty with the minimum of maintenance.

Following is a brief outline of the points required to be addressed in the design of a transfer area:

(a) Material flow rates

Careful consideration should be given to the style of transfer chute employed to handle the required flow rates of material. This aspect is discussed in more detail later in the paper, however, closely spaced skirt boards combined with high belt velocities and large flow rates often result in overloading the transfer area. This may be more prevalent particularly after the material has been deflected off flat impact plates or surfaces angled away from the incoming material stream as shown in figure 1, and cannot gain sufficient velocity in the direction of the receival belt to effect a smooth transfer.

(b) Transfer chute geometry

The chutework should be constructed such that it has a suitable geometry to enable it to handle the range of particle sizes and flow properties of the bulk solid, allowing it to flow freely within the chute without hanging or bridging on the walls or excessively wearing the belt.

Before designing any materials handling facility the material's properties and behaviour should be carefully assessed. Tests should be carried out to determine the frictional characteristics with a number of different low friction lining materials as well as the wear resistant linings intended for use inside the chute. This information is vital to determine the correct chute angles as well as one of the most critical aspects in chute design, the materials trajectory.

Careful attention should be paid to the dead area under the discharging stream as this is where a rain of fine material particles will tend to settle out. Due to the low normal contact stress these particles have with the surface on which they settle, a 'rough wall' friction condition can prevail [3]. This is shown by way of example in figure 2, stress levels below the point at which ϕ_{RW} crosses the line of ϕ result in the material preferring to shear across itself rather than slide along the chute surface. In this case the particles will adhere strongly to the chute tending to build up and block it. The variation for ϕ_{RW} versus normal contact stress σ_1 can be found from:

$$\phi_{RW} = \tan^{-1}(\sin \delta) = \tan^{-1}(\sin \delta(\sigma_1)) \qquad (1)$$

where: ϕ_{RW} = rough wall friction angle [°]

$\delta(\sigma_1)$ = angle of internal friction corresponding to a normal consolidation stress σ_1 [°]

Steps should be taken to eliminate the existence of any chute surfaces in these critical low contact stress areas or provide a means to promote flow down the chute. This may be achieved by such techniques as:

- keeping the bed depth thick,

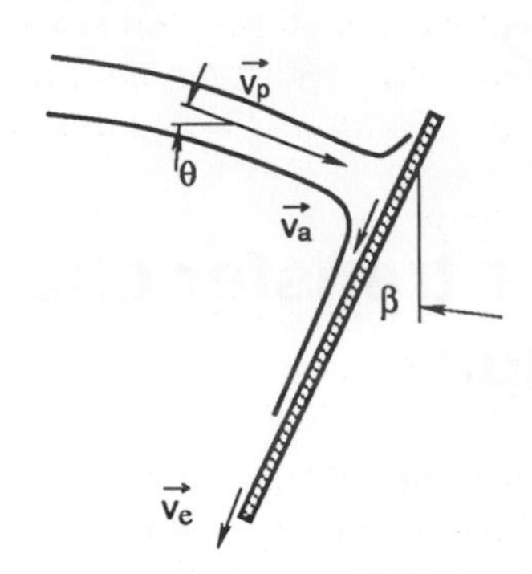

Fig 1 Cohesionless material behaviour on impact plates

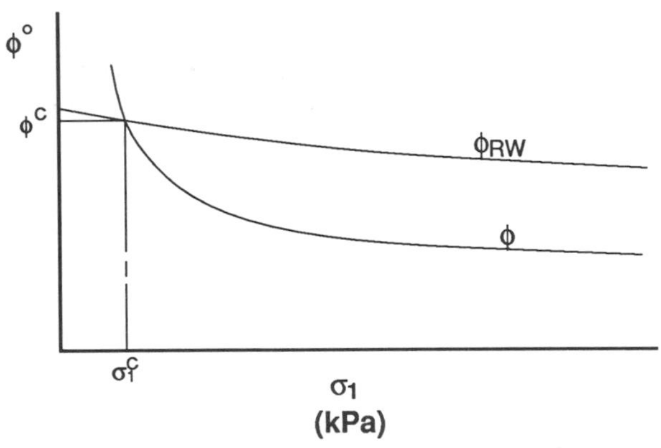

Fig 2 Typical ϕ_W and ϕ_{RW} versus σ_1 variations (after McLean [3])

- maintaining high material velocities,
- using low friction linings,
- employing mechanical methods such as air cannons or vibrators.

Particular care needs to be taken with dribble chutes used to collect belt cleaner scrapings, as such materials generally have adverse flow properties and hence are notoriously hard to shift.

(c) Material matches belt velocity

The area where most transfer chutes experience difficulties is in matching the material velocity with that of the receiving belt as closely as possible in order to obtain a smooth turbulence free transfer. When material is deposited onto the receiving belt without due consideration being given to the directions of both the impacting stream and the conveyor, the result is more likely to be a gross mismatch in the velocity of the two. The outcome is generally a large volume of virtually static material that must be accelerated away from the loading zone.

(d) Minimise material degradation

Most of the material degradation on a conveyor will occur at the transfer area. For fragile or brittle materials this can result in a large amount of unacceptable substandard product and the creation of large volumes of hard to control dust. The transfer area design should aim to minimise the degradation of material by impact or from large drop heights;

associated with this is minimising the amount of wear taking place inside the chute from the impinging or high velocity sliding of the material stream.

(e) Minimise spillage

Perhaps the most obvious sign of a badly performing transfer area is the amount of material that has managed to escape the chute boundaries. Some transfer points literally bury themselves in this material and hence great attention to detail should be paid in designing the skirtboard and other sealing areas of the chute.

(f) Low Maintenance

Wear is perhaps the biggest single contributor to transfer chute failings; although it can be reduced it is hard to eliminate altogether. To operate efficiently the transfer area must be easy to maintain with replaceable liners, large access hatches, including provision for a mono rail crane to remove and replace heavy parts.

The above points are but a few of the more important items that are often completely overlooked when designing transfer points. Most are common sense and should be considered as a minimum check list for chute design. The problems encountered in designing proper functioning transfer points increases as:

- the speed of the belt increases
- when the incoming and receiving belts are not in line
- when the receiving belt loading point is inclined.

On high velocity conveyors short acceleration belts are sometimes used to overcome part of the problem by bringing the material up to the velocity of the main conveyor before discharging it. This relieves the main belt of most of the wear problems associated with the free falling material stream leaving only a relatively cheap, short belt to be replaced. This however does not address the problems of material degradation, dusting, noise and containment of the falling material stream. One solution that has been proposed to counter these problems associated with both high and low speed belts is to use a mass flow bin combined with a belt feeder as a transfer point. This paper looks at the design of this type of transfer chute and the factors influencing its reliable operation.

3. MATERIAL TRAJECTORIES

One of the most critical aspects in transfer chute design is the knowledge of the material trajectory. This is a vital aspect of the design in both low and high speed belts as without this knowledge the chute could be rendered useless requiring costly and time consuming work later to rectify problems such as severe wear areas cased by the unexpected impingement of material on the chute walls. Computer modelling [4] of some of the various methods of trajectory prediction has been carried out combined with physically testing the full sized outputs against various low and high speed conveyors. This modelling has shown that methods relying on physical interaction between the belt and material such as [5,6] predict better trajectories than those that simply address the topic as a case of projectile motion of the material from the belt.

3.1 Belt conveyor trajectories

A number of methods exist in the literature to predict the material's trajectory [2, 5, 6, 7, 8, 9]. Some however will lead to large errors [8,9] discharging the material further than it would normally go for the case of high speed belts, and dropping the material short in the case of low speed belts. They do however posses elements in their derivation that can be usefully combined with methods [5, 6] such as material centre of mass calculations and cross sectional area of the material stream.

Trajectory prediction can basically be divided up into two parts:

- High speed belts, when the belt velocity is high enough that the material will traject at the point of tangency between the belt and head pulley, without wrapping around the head pulley. The condition for a high speed belt is satisfied if [5]:

$$\frac{v_b^2}{Rg} - \frac{\sigma_a}{\gamma h_b} \geq \cos \alpha_b \qquad (2)$$

where : σ_a = adhesion [kPa]
γ = specific weight [kNm^{-3}]
v_b = conveyor belt velocity [ms^{-1}]
h_b = material height on belt [m]
R = distance from centre of discharge pulley to material's centre of mass [m]
g = acceleration due to gravity [ms^{-2}]
α_b = conveyor belt inclination angle [°]

- low speed belts where the material wraps to some degree around the head pulley before discharging. The degree of wrap relies on many factors some of which are dependent on the bulk material's flow properties. Wrap will occur if [5] :

$$\frac{v_b^2}{Rg} - \frac{\sigma_a}{\gamma h_b} < \cos \alpha_d \qquad (3)$$

where: α_d = bulk solid stream discharge angle [°] measured from the vertical

Determining the wrap angle (α_d) and discharge velocity is of paramount importance in predicting an accurate trajectory. Methods that do not use at least the concept of material interaction on the belt due to friction such as [2, 7, 8, 9], cannot be relied upon to give worthwhile results when trying accurately to predict the material's trajectory particularly for sticky or cohesive materials.

3.2 Trajectories from impact plates

While not all chute designs utilise impact plates most rely on the material impacting a surface soon after discharging from the conveyor. Impact plates or 'crash boxes' are very commonly used in chutes that must deal with high velocity flows allowing the designer to reduce the length of the chute to a manageable size. Just as it is important to determine the trajectory of the material from the conveyor it is equally important to be able to predict where the material will go once it has departed the impact surface. To do this a value for the velocity of the stream must be found at the point it exits the flow-round zone as shown in figure 3.

For materials with little or no cohesion this exit velocity can be estimated from the consideration of momentum and frictional interaction between material and plate:

$$v_a = v_p [\sin(\theta + \beta) - \mu \cos(\theta + \beta)] \qquad (4)$$

where:

- v_a = material exit velocity of flow-round zone [ms^{-1}]
- v_p = material entry velocity to flow-round zone calculated from the point of intersection of the trajectory of the centre of mass of the incoming stream with the plate [ms^{-1}]
- θ = angle of velocity vector $\vec{v_p}$ to the horizontal [°]
- β = angle of plate to the vertical [°].
- β = 0 for a vertical plate or surface
- β = $+\beta$ for a plate angled towards the incoming material stream
- β = $-\beta$ for a plate or surface angled away from the incoming material stream
- μ = friction coefficient for material sliding on plate or impact surface
- μ = $\sin\delta$ for material sliding on itself

In order to make use of Eq. (4) the following condition should hold:

$$\sin(\theta + \beta) - \mu \cos(\theta + \beta) > 0 \qquad (5)$$

or,

$$(\theta + \beta) > \tan^{-1} \mu \qquad (6)$$

Experiments carried out on impact plates that meet the above condition using a material with a cohesive stress of 0.04 kPa on mild steel [10] have shown that Eq. (4) gives workable results.

For materials displaying higher amounts of cohesive stress or for geometries that do not meet the conditions set by Eq. (6) it is recommended that a multi step approximation procedure be used as set out in [11] by Korzen. This rather lengthy and moderately complicated method relies on the premise that the material will form a simple semi circular curved block that remains stationary on the impact surface and over which will flow the incoming stream. In practice it has been observed that the block of material takes a more complex partly spherical shape with the appearance of a rib in the centre which splits the material flow around either side of itself complicating the situation, this technique however provides a useful estimate which if combined with the sound practice of allowing field adjustment in the design will provide an effective solution.

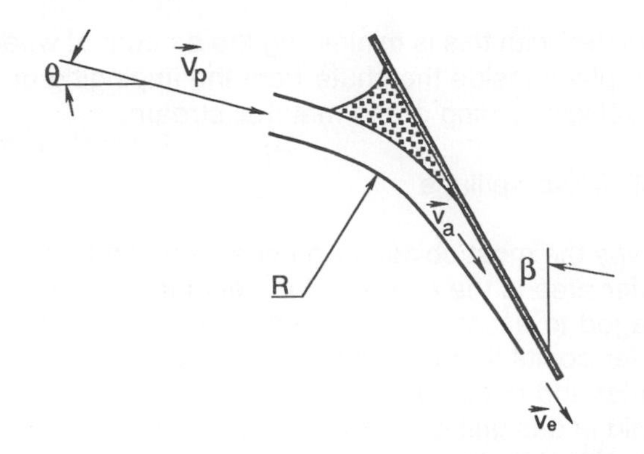

Fig 3 Cohesive material behaviour on impact plates

Plotting the trajectory from the impact surface should be done in the same way as it is for the belt conveyor once the final velocity has been found. This final velocity is however not necessarily the one found from Eq. (4) or the multi step procedure [11]; these velocities are only applicable at the end point of the flow-round zone and not at the exit edge of the impact surface where some increase in velocity can be expected due to the material rilling down the slope or sliding down the underside of the impact plate [2].

4. CHUTE TYPES

Many different designs of transfer chute have evolved in the past to suit a wide variety of conditions and materials. Most of these however can be categorised into four main groups by the way in which the material is dealt with after it trajects from the feed conveyor. Briefly these main types are:

4.1 'U' formed chutes

This title is meant to encompass all chutes that transfer material in a continuous sliding action, in which the material is taken from the discharge point to the receiving conveyor remaining in continuous contact with the chute work. The design of such chutes are well documented in works such as those by Roberts et al. [12]. They work well for free flowing low abrasive materials particularly where degradation is to be kept to a minimum.

4.2 Direct transfer

In this type of chute the material discharges from the feed conveyor and deposits onto the receiving conveyor, coming into contact with as little of the chute as possible.

The chute enclosure is used primarily as a means of dust containment and should be well sealed. To effectively seal the transfer area the following points should be noted:

- Material entry and exit points should be equipped with close fitting rubber curtains around both carry and return sides of the belt.
- The chute enclosure should be exhausted to a dust filter as all chutes should when handling extremely dusty or hazardous material.
- The material must not be loaded in the belt transition area.
- The belt should be well supported in the loading zone by closely spaced idlers and pads that ride under the belt in the area of the skirtboard seals.
- Skirtboards should be sufficiently long; it has been suggested [2] that 1m for every 1 ms^{-1} of belt speed be used. This minimum length is required to provide the material enough time to settle onto the belt before departing the transfer area.
- The skirt boards should be sealed with a labyrinth type of seal providing at least two adjustable lips. The first lip is constructed from a wear resistant material and rides just clear of the belt to take the extreme pressures off the secondary lip which does the actual sealing.

To aid in centralising the feed the onto the receival belt this type of chute should employ high angled troughing rollers in the loading zone. These rolls can be inclined up to 50° to provide good centralisation of the feed and after the loading zone the belt should be gradually coaxed back down to run at its normal troughing angle. The impact zone should be fitted with easily replaced wear liners in case stray material impacts on the chute walls through such factors as poor trajectory predictions, a change in the material properties or simply the fact that the material has to fall into a narrower area caused from the highly troughed belt.

Direct transfer has many advantages on 'slow running' belts where the material wraps on the head pulley before trajecting. Some of these advantages are:

- Sticky or cohesive materials can be more readily managed due to the limited amount of contact with stationary objects, though belt cleaner scrapings may require careful handling to incorporate them into the main stream of material.
- Little difficulty is experienced centrally loading the belt, promoting an even material profile on the receival belt, eliminating mistracking of the belt and spillage.
- Providing the material is made up of predominately fines with no large lumps, wear is kept to a minimum. The belt will experience some increased wear due to the direct impact of the falling material, but It must be kept in mind that the belt cover is made to take some abuse and this type of wear situation is preferable to the stresses imposed on the belt in crash stops or the sort of damage imposed on the belt edges from badly mistracking belts. Both of these situations are common occurrences resulting from poorly designed transfer areas.

Although the advantages of direct transfer are high, with some materials this method will not be possible. Fragile material would suffer high amounts of degradation in the drop to the receival conveyor and materials composed of extremely large sharp lumps will tend to damage the receival belt excessively and should be transferred by some other means.

4.3 Cascade transfer

One of the most common forms of cascade chute uses a number of 'crash boxes' or 'rock boxes' in which the incoming stream impinges on a layer of trapped dead material to absorb the impact and reduce the amount of degradation the material experiences. A typical cascade chute may utilise one or more 'crash boxes' to enable it to steer the material onto the receival belt.

The advantages in this style of chute lie in its ability to reduce degradation and to handle extremely heavy and very abrasive materials at high flow rates, where the costs associated with the continual inspection and replacement of impact plates, grizzly screens and chute liners may be prohibitive considering the life of these items in large throughput operations.

The drawbacks to the 'crash box' chute are:

- Poor performance with cohesive materials. The crash box has the habit of trapping enough material to start bridging the gap between itself and the head pulley of the discharging conveyor. This build up of material can readily block the chute or at least throw the trajectory of the falling material stream further out. The lateral spread will increase with the consequence that the material may start striking the wall of the chute causing extensive wear damage.
- Extremely high wear is generally experienced at the lip of the 'crash box' which will always be exposed to the high velocity material stream rilling down the slope before trajecting. This area should be constructed from the best affordable wear resisting material available and made easily replaceable. Access into the chute to change the lip should be made via large easily sealed doors to enable free movement of maintenance personnel.
- Difficulty in handling high velocity material streams. As the incoming material strikes a surface angled away from the discharging conveyor there will be quite a significant amount of material spreading laterally to the desired direction of flow. Most of the laterally spreading material will be made up from the coarse components of the material stream

due to the segregation effects of the conveyor belt [13]. It should be appreciated that the lateral spread of the material stream increases dramatically as the impact surface is tilted away from the incoming flow. Utilising a 'crash box' with high velocity conveyor belts will amplify the problem, as generally the material surface will be inclined between 20° and 50° away from the incoming material stream. In these situations it is advisable to check carefully the material's trajectory and the positioning of the 'crash box' to minimise the angle between the material stream and the stationary material surface. Reducing this angle will control to some degree the material spread, but allowance should be made in the sizing of the chute enclosure to accommodate the lateral flow of material preventing it from it impinging on the walls of the chute and aiming above all to incorporate it back into the main body of flow as smoothly as possible.

The other commonly used style of cascade chute utilizes impact plates to direct or steer the material flow onto the receival belt. With careful design this type of chute can be made quite flexible in its ability to control the material flow. Typical features that should be included into this chute are:

- Field adjustable impact plates. The impact plate should initially be positioned carefully with respect to the trajectory of the incoming material stream. Failure to use an accurate prediction of the material's trajectory in the early stages of design will often lead to a chute that requires extensive field work to rectify this basic design fault.

- Both upper and lower impact plates should be curved to steer the material and reduce the amount of lateral spread it experiences. The advantages to fitting a curved top plate, are that, compared to a flat plate, it allows the material to be coaxed more gradually to exit at the required angle, reducing lateral spread and to some extent degradation of the material due to impact. Figures 4 and 5 show the results of experiments carried out on impact plates [10] utilising washed black coal. In these experiments the plate's position relative to the discharge pulley remained unchanged except for its orientation, the curved plate was positioned as shown in figure 6. The flow rate from *one* of the material streams separating from either side of the main body of flow was recorded and averaged over ten readings to obtain the results. It can be seen from figure 4 for a constant belt velocity and hence contact velocity with the plate that the curved plate caused marginally less spreading as well as the added benefit of discharging the flow in a near vertical position enabling a more compact chute design to be utilised. Figure 5 indicates the high amount of material that is separated off the main body of flow from a surface that is inclined away from the incoming stream by as little as 10°, as well as the better performance of a curved plate over an inclined or vertical flat plate. It should be noted that in figure 5 comparisons should only be made between results at the same belt velocity. The plate

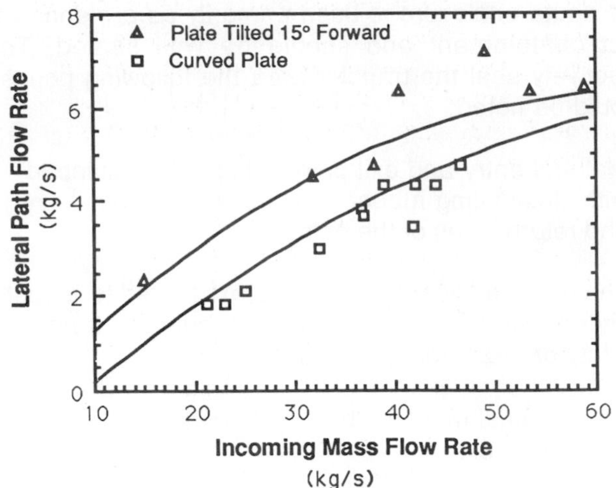

Fig 4 Lateral Path Flow Rate Vs Incoming Flow Rate, Belt Velocity 3.5 m/s

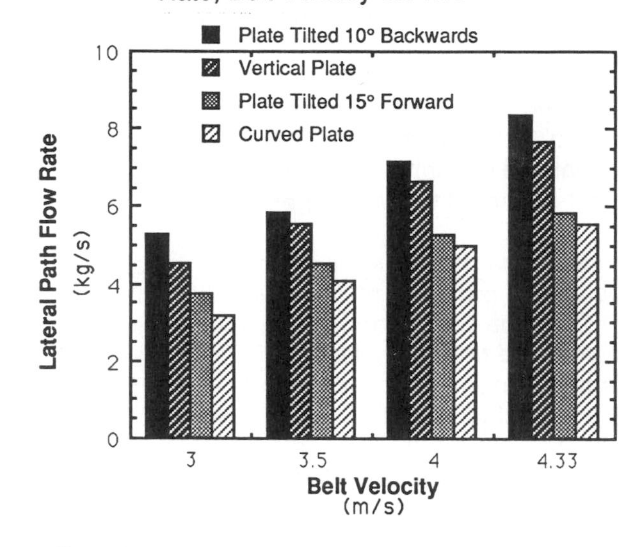

Fig 5 Lateral path flow vs belt velocity,incoming mass flow rate 38 kg/s

positions remained fixed and not varied to suit the differing contact angles due to increased belt velocities i.e. the higher the belt velocity the lower the contact angle and hence more lateral spread. Even though the lateral spreading of the material stream can be contained by the inclusion of winged type additions at the edges of the impact plate it is always in the designers interest to minimise the cause of this problem as close to the source as possible rather than trying tocorrect for it after its happened.

The lower plate should also be curved in order to steer the material onto the next belt depositing it there with a velocity that matches the belt as closely as possible in both magnitude and direction. Without the curve in the loading plate the material is deposited in a direction that does not coincide with the receival belt leading to turbulence, pooling and mistracking of the belt in the loading zone.

- Adjustable plates [1] should also be included to steer the lateral spread of the material stream back into the main flow at the loading end of the chute. Experiments [10] carried out on a large scale test rig have noted that lateral spread from material streams striking flat plates in the range of angles shown in figure 5 can be up to three times the width of the incoming stream at less than 1m from the point of impact, hence the need to gain control of these side streams before loading onto the next belt.

- Wear resistant liners in all areas subject to impingement of material. These liners should be made easily replaceable from the best quality material one can afford. Careful thought should be put into the design of the wear liners and their fixing systems to ensure that no ledges or bolt heads protrude into the material flow as these will cause hang ups and possible blockage of the chute in much the same way as they do inside bins and bunkers. Valley angles should be fitted with generous radii or fillets to help minimise the build up of fines.

4.4 Surge bin transfer chutes

The use of a small well designed mass flow bin as a transfer chute has been the subject of an intensive investigation as an alternative to the more traditional styles of chute described above.

The basic principle behind the surge bin transfer is to combine a mass flow bin topped by an adjustable curved impact plate with a variable speed belt feeder for discharge of the material. Bin design has reached a level of technology where reliable performance can be obtained for a given set of material properties. Mass flow can easily be achieved with some attention to detail and the bulk solid can be made to behave in a predictable manner whilst in the bin.

4.4.1 Details of experimental work with surge bins

Experiments [10] carried out using the combination of a mass flow bin of 5 m^3 capacity and belt feeder have shown that flow rates can be achieved from the surge bin that would make this system practical for use in many applications, particularly where high velocity belts are required to be fed at angles other than in-line.

Two different types of belts were used at varying belt feeder angles from 0° to 4° declination; these were:

- A rubber belt cleated in a chevron pattern, with the cleats in the reversed direction as shown in figure 7, as well as facing forward. The cleats were positioned at a 30° included angle and were 12 mm high.

- A smooth 2 ply vinyl belt.

The highest flow rates were obtained from the 4° declined feeder and a typical set of results is shown in figure 8. A maximum flow rate of 82 kgs^{-1} was obtained for the reverse cleated belt, corresponding to a belt feeder velocity of 3 ms^{-1}. Velocities greater than this were unobtainable due to a lack of power in the hydraulic drive system.

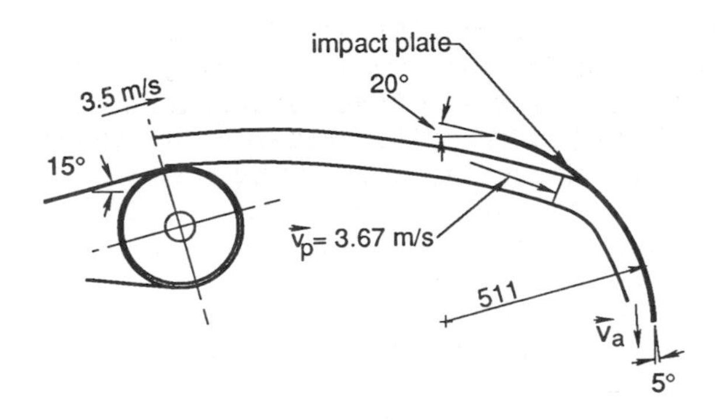

Fig 6 Curved impact plate positioning

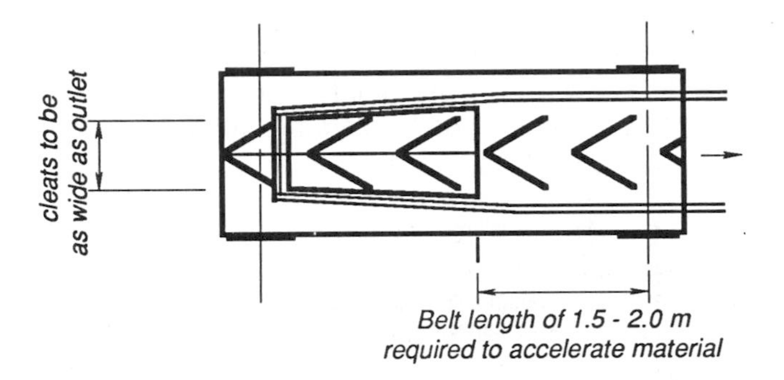

Fig 7 Cleated belt feeder showing cleats in reverse direction

The receival conveyor used in these experiments had a capacity of approximately 100 th^{-1} of black coal when running at 4 ms^{-1} while the bin/belt feeder could produce around 300th^{-1} at 3 ms^{-1}. Considering the aperture area which was little more than 200 mm x 280 mm this was quite a high flow rate.

The power requirements of the belt feeder were also closely monitored while running the experiments by recording the hydraulic pressure at the drive motor. Figure 9 shows the variation of running power consumption to belt velocity obtained for various cleated belt arrangements. The curve shows the agreement of the data to the theoretical model proposed by Reisner based on major consolidation stress (σ_1) [14], which has been modified to include the feeder resistances to motion as well as the taking into account the power required to accelerate the material up to the velocity of the belt feeder [15]. It is interesting to note the marginal power difference between the 0° and the 4° declinations, with the 4° declined belt discharging considerably more material than a level belt which used most of its power trying to shear the material out of the hopper opening and failed to reach belt speeds higher than 2.3 ms^{-1}. Although the Reisner method with feeder resistances is useful for calculating running power requirements, initial or start up power should be determined from a method such as that proposed by McLean and Arnold [16].

4.4.2 Features of the surge bin transfer chute

Some of the advantages of the surge bin as a transfer chute include:

- Reliable material flow in transfer area.

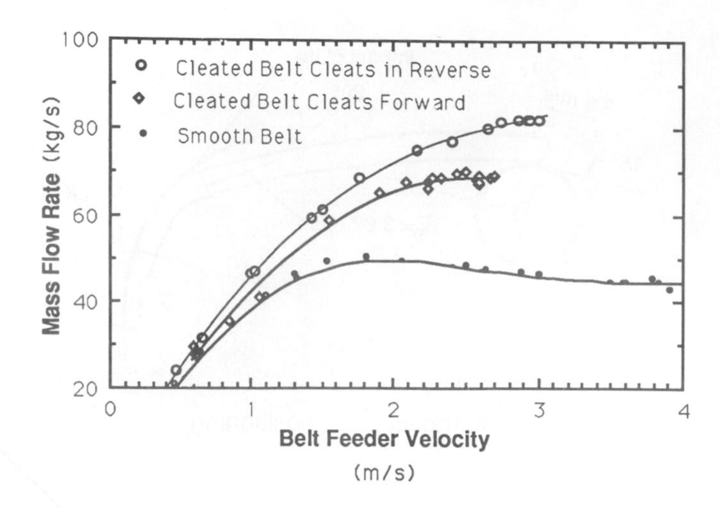

Fig 8 Mass flow rate vs belt feeder velocity, 4° declination, aperture area = 0.055 m^2

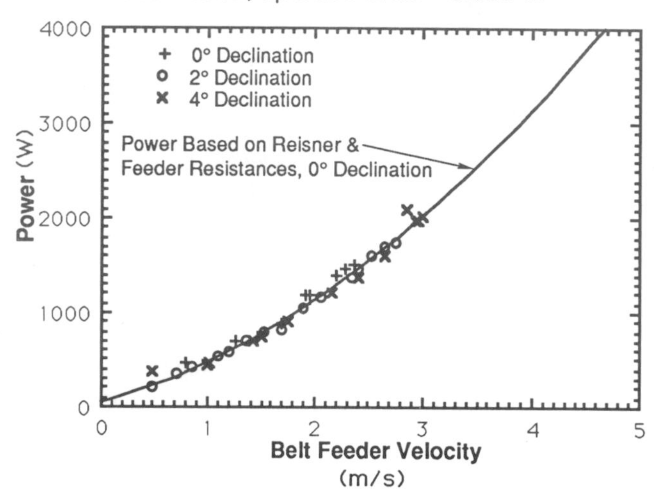

Fig 9 Power vs belt feeder velocity, cleats reversed aperture area = 0.055 m^2

- Material undergoes a minimal amount of free fall between belts, reducing breakage and the amount of dust generated.

- The material deposits onto the belt with less splashing and turbulence in the loading zone, stemming mainly from the close match of velocity vectors and the elimination of the lateral spreading caused by impact surfaces. Observations made during tests noted the lack of any turbulent mixing and the extremely smooth streamline flow in the loading zone.

- Direction of feed could be altered by rotating the surge bin/feeder to allow delivery to other belts running at any angle to the main conveyor with efficiencies equal to that of an in-line transfer. This could be accomplished on the fly with sufficient surge bin capacity and belt speed variation.

Although proving useful for materials displaying cohesion the surge bin transfer would have difficulty in coping with materials of large lump size if a mass flow outlet could not be designed to cope with such large sizes. Such a surge bin would also require careful design if the material was of a free flowing nature allowing it to flow uncontrollably onto the belt feeder when not running.

5. CONCLUDING REMARKS

The transfer point is one of the most problematic areas in the design of belt conveyor systems. It is here more than anywhere else along the length of the conveyor that a detailed knowledge of the material's properties are required to initiate a successful design. Perhaps one of the most basic factors in obtaining a good reliable transfer point is establishing the correct material trajectory from both the discharging conveyor as well any impact surfaces the material meets on the way to the receival belt. Once this aspect has been sorted out the choice of chute will be influenced heavily by such factors as the type of material to be handled, the lump size, abrasiveness, moisture content, cohesion/adhesion it possesses, frictional properties, belt velocity.

Even with extensive knowledge of the materials properties and trajectories, the transfer area will not perform well if the proper attention to detail has not been made. Areas such as wear liner fastenings, sealing, skirting and the allowance for easy access to carry out maintenance must all be carefully considered before the chute is built and erected and not after, as is often the case.

Surge bins as transfer chutes offer many advantages to the more traditional varieties of chutes outlined in sections 4.1 to 4.3. Work to date on a large scale model has shown they have application in areas requiring:

- high speed transfers;

- low material degradation;

- high efficiency loading transverse to the direction of belt travel;

- feeding of multiple belts, changing on the fly.

6. ACKNOWLEDGEMENTS

The authors wish to acknowledge the financial support provided by the National Energy Research Development and Demonstration Council under Project 1188 'Blockage Wear and Conveyor Direction Change of Conveyor Chutes'.

7. REFERENCES

1. SABINA, W.E., STAHURA, R.P., SWINDERMAN, R.T., Conveyor Transfer Stations Problems and Solutions. Martin Engineering Co 1984 67 pages.

2. M.H.E.A., Guide to the Design of Transfer Chutes and Chute Linings. The Mechanical Handling Engineers Association, 1989, 93 pages.

3. McLEAN, A.G., Wall Yield Locus to Friction Angle Variation Analysis and Implications. I.E. Aust. Third International Conf on Bulk Materials Storage Handling and Transportation, June,1989, Newcastle, NSW, pp 74 - 77.

4. ARNOLD, P.C., HILL, G.L., Predicting the Discharge Trajectory From Belt Conveyors. I.E. Aust. Third International Conf on Bulk Materials Storage Handling and Transportation, June,1989, Newcastle, NSW, pp 131 - 135

5. KORZEN, Z. Mechanics of Belt Conveyor Discharge Process as Affected by Air Drag. Bulk Solids Handling, Vol. 9, No 3, 1989, pp 289 - 297.

6. BOOTH, E.P.O., Trajectories From Conveyors Method of Calculating Them Corrected. Engineering and Mining Journal, Vol. 135, No 12, 1934, pp 552 - 554.

7. DUNLOP., Dunlop Conveyor Manual. Dunlop Industrial, 1982, pp 9.1 - 9.5.

8. M.H.E.A., Recommended Practice for Troughed Belt Conveyors. The Mechanical Handling Engineers Association, 1986, pp 33, 42 - 48.

9. C.E.M.A., Belt Conveyors for Bulk Materials. CBI Publishing Company, 2nd Edition, 1979, pp 54 - 57, 277 - 291.

10. ARNOLD, P.C., HILL, G.L., Design of Conveyor Chutes with Special Attention to Blockage, Wear and Conveyor Direction Change. End of Contract Report, NERDDP Proj. No. 1188, 1991,123 pages.

11. KORZEN, Z., The Dynamics Of Bulk Solids Flow on Impact Plates of Belt Conveyor Systems. Bulk Solids Handling Vol. 8, No. 6, 1988, pp 689 - 697.

12. ROBERTS, A.W., The Dynamics of Granular Material Flow Through Curved Chutes. Mechanical and Chemical Engineering Transactions, Institution of Engineers, Australia, Vol. MC3, N°.2, Nov 1967.

13. WATERS, A.G., MIKKA, R.A., Segregation of Fines in Lump Ore Due to Vibration on a Conveyor Belt. I.E. Aust. Third International Conf on Bulk Materials Storage Handling and Transportation, June,1989, Newcastle, NSW, pp 89 - 93.

14. REISNER, W., EISENHART ROTHE, M.V., Bins and Bunkers for Handling Materials. Trans. Tech Publ., 1971.

15. ARNOLD, P.C., ROBERTS, A.W., Estimation of Feeder Loads. Paper presented to Institution of Mechanical Engineers Seminar 'Recent Developments in Design and Operation of Bulk Handling Systems', London Sept, 1985.

16. McLEAN, A.G., ARNOLD, P.C., A Simplified Approach for the Evaluation of Feeder Loads for Mass- Flow Bins. J. Powder and Bulk Solids Technol. Vol. 3, N°.3, 1979, pp 25 - 28.

C418/057

Techniques to improve the efficiency of bulk solids conveyors

R P STAHURA
Martin Engineering Company, Neponset, Illinois, United States of America

SYNOPSIS: This paper will present an examine the causes and cures for common conveyor problems such as carryback, transfer point spillage, and belt wander and focus on ways to improve both the conveyors and the process used to engineer them. It will focus on new designs and techniques for inclusion in new material handling systems, and for use in retrofit improvements on existing conveyors.

INTRODUCTION

Like arteries in the human body, conveyors continuously supply materials to and carry finished products and waste away from vital on-going operations. And like arteries, when conveyors are running with less than maximum efficiency and productivity, it spells serious trouble for the system's overall health.

Inefficient conveyors suffer carryback, leading to the build-up of materials on idlers and other expensive components. They spill material at loading zones and transfer points, causing accumulations that can damage the belt, wear moving components, and increase both cleanup labor and maintenance expense. They wander or mistrack, spilling material and endangering personnel and conveyor components including the belt itself.

And through these inefficiencies, they multiply the operation's hidden costs. These added expenses are created in lost production, increased downtime for service and maintenance, and the cost of the replacement of prematurely worn equipment. In addition, the attitude that inefficiency is "good enough" is recognized elsewhere in the organization; as a result, other departments relax or operate at less than peak performance. Quality goes down and with it employee morale.

THE COSTS OF INEFFICIENT CONVEYORS

One key to efficiency is handling a product as few times as is necessary. That means conveyors should be loaded, track true to the unloading point, and then be unloaded, all without material escape or waste motion. Often an inefficient conveyor is signaled by accumulations of DURT--material that has escaped from the system and is now at large in the plant. This DURT adds costs to an operation that are uncounted and perhaps uncountable.

When material becomes fugitive and has to be gathered up and reintroduced into the system, this inefficiency increases the cost of production. Or if the fugitive material is lost and cannot be salvaged, it even more dramatically increases the inefficiency. The cost of producing this lost material raises the cost of the whole operation.

Critical components like idlers and pulleys and other conveyor subsystems are literally bombarded with fugitive materials--both fines and larger particles. The accumulations of material impair performance: seizing bearings, wearing rollers, and requiring increased maintenance or premature replacement. In addition to the cost of replacing these expensive components, there is the overhead of the additional downtime required to service or replace these systems.

Added to the visible costs--clean-up and equipment--there are hidden costs growing out of inefficient conveyors. A plant with efficient material handling is usually safer. There are fewer accidents which are costly in both personnel and in production. A study by the U.S. Bureau of Mines on accidents stated that a high percentage of them were conveyor related, and that for the most part the victims were in a dangerous situation because of accumulated "DURT" from the conveyor.

THE SOLUTION, IN 25 WORDS OR LESS

So, if inefficient conveyors are costly, why is it that the problems have been allowed to continue? The reason is the attention of those who can cause things to happen in plants is not focused in on the causes and solutions to conveyor inefficiency. But attention will soon be centered on fugitive materials, because the computer's ability to store data and spot profit leaks will expose these costs at the high price tag they really bear.

Now this is the point where I get to solve the problems of inefficient conveyors in 25 words or less. If you go away with nothing else from this presentation, remember this: the key to efficient conveyors is a stable operation. Improved stability--in belt support, in skirting, in belt tracking, in belt speed, in carryback removal--in short, in all things--will improve conveyor performance. I absolutely positively no excuses guarantee it.

Let's look at some ways to make conveyor systems more stable and, as a result, more efficient.

CONTROL OF BELT TRACKING

One way to improve belt efficiency is to make certain the belt is tracking true. A mistracking conveyor can spill material, and runs the risk of significant and expensive damage to the belt--damage that can shut an entire operation down. Nothing will more quickly demonstrate how conveyor efficiency affects the productivity of the entire operation than a plant shutdown due to the need to splice or replace a damaged belt.

To prevent belt mistracking, the first key is to get the conveyor line squared away. Return idlers should be at true right angles to the desired path. But under real conditions--allowing for variations in floor level, in operating temperatures, or occasionally, but more frequently than we'd like to admit, structural damage, like a forklift running into a conveyor stringer--there are problems with achieving this alignment. Consequently, each operation must make provisions to control belt wander.

Misalignment Switches

The first step in preventing belt damage due to mistracking is the installation of a series of wander or misalignment switches. These switches are installed near the outer limit of safe belt travel, and act to shut off the power to the conveyor in the event the belt moves past the safe limit. When the belt moves too far in either direction, it physically contacts the lever, pushing it over past its limit, and breaking the power circuit. The belt stops, and the operator must realign the belt. In many cases, he will need to walk the conveyor to manually reset the switch before operation can begin again. Of course, the frequent tripping of a belt misalignment switch is a signal that something more dramatic is wrong with the system. It is like the 'idiot' light in your car that shows a red light when the engine is too hot. You can restart the car or the conveyor, but it is really an alarm that there is a more serious problem.

Training Idlers

There are a number of systems available that work proactively against belt wander. These systems--most often installed on the return side of the conveyor--because that is where the belt is confined by hangers, or obstructions which can damage a wandering belt--are composed of some manner of bearing-mounted idler which senses and reacts to belt misalignment. They correct wander, without requiring the conveyor go 'down'.

One common system utilizes a pivot bearing mounted in the center of a idler framework, with vertical rollers or fixed pieces at the belt edge. When the belt moves too far in either direction, the belt hits the vertical roller, kicking the pivot bearing over so the belt is forced back into the middle. This type of training idler, while reasonably effective when new, poses several problems as wear and fugitive material impair its ability to function. The correcting action is caused by the belt literally slamming into one side or the other. When the correcting action takes place, it usually kicks over with such force that the belt is then slammed over to the far side of the structure. The belt is constantly in motion, back and forth between the two sides. In addition, the pivot bearing and its framework under the belt are exposed to fugitive material, which will impair its ability to pivot. The bearing can seize, preventing the idler from working, or--what is worse--freezing it into a 'misalignment' position, creating the problem it was installed to correct.

It is better to install a belt training system that is impervious to a build-up of material and that will correct the line of travel, rather than slapping the belt from one side to the other. A competing style of belt tracker uses a double row, spherical bearing mounted inside the center of the roller. The bearing both rotates and pivots simultaneously to provide support for the belt and angular displacement to correct misalignment. As soon as the belt starts to run off center, this tracker idler acts immediately to correct and re-center the belt. The belt is not required to crash to either side of the structure in order to actuate the belt training cycle. With the bearing sheltered inside the roller, it is free to function without material build-ups; it will not seize up in a position of misalignment.

THE TROUBLE WITH TRANSFER POINTS

Let's look at another place where instability is introduced into conveyor systems: the transfer point, the point on the conveyor where the belt is loaded with material, usually either from another conveyor or out of a storage container. In order to stabilize the belt travel and contain fugitive material, there are three distinct areas in a transfer point which must be addressed. In order of importance, they are: belt support structure, chute wear liner, and skirting.

It seems obvious that in order to have an efficient conveyor, the belt's line of travel should be as steady as possible. The conveyor should be run like a rope on a table, flat and true. But this is not typically the way conveyors are built. Even if designed properly, loading zones are often so poorly maintained they develop unwanted dynamic action, motion that destabilizes the belt, and hence reduces efficiency.

If the belt is allowed to sag or flex under the stress of loading, material will escape. (FIG. 1.) Airborne dust, fines, and even chunks of material will be carried by the dynamic forces of loading out of even the smallest gaps which open between the sealing system and the belt. As the belt moves up and down, material is released, to dirty the plant, to pile up on conveyor components, and depress efficiency.

FIG. 1. Insufficient belt support allows the belt to sag and material to escape

And worse, wedge-shaped pinch points will develop as the belt sags below the chutewall. Material is carried into these openings, and packed tighter and tighter by belt motion. This entrapped material, poses the risk of undue wear or dramatic damage.

Any belt sag--even one barely apparent to the naked eye--is enough to permit fines to start a grinding action, producing wear on rubber skirt and belt surface, and allow escape to the environment. Consequently, adequate support in the loading zone is required. For transfer points which see light loading, a simple solution may be to increase the number of idlers underneath the belt. This decreases the space between sets of rollers, and, consequently, reduces the potential for belt sag.

One limiting factor in close idler spacing, however, may be the distance required to lay an idler over on its side to allow it to be serviced. Track-mounted idlers, which slide into position and can be easily removed without requiring the raising of the belt or the movement of the idler base are one solution.

Although close idler spacing reduces belt sag, it cannot eliminate this condition in areas of heavy loading. Under severe loading conditions, an impact cradle should be used to provide complete support for the belt. (FIG. 2.) Constructed from multiple-layer elastomer bars, these cradles perform two functions: they absorb impact to prevent crushing damage to the belt and conveyor structure, and they serve to stabilize the line of belt travel, providing a surface that can be effectively sealed against the escape of material. The impact bars are fabricated with a slick top layer to allow the belt to slide smoothly across its surface with minimal friction. The lower layers of the impact bars should be composed of durable, spongy materials which absorb the energy of impacts to minimize belt damage.

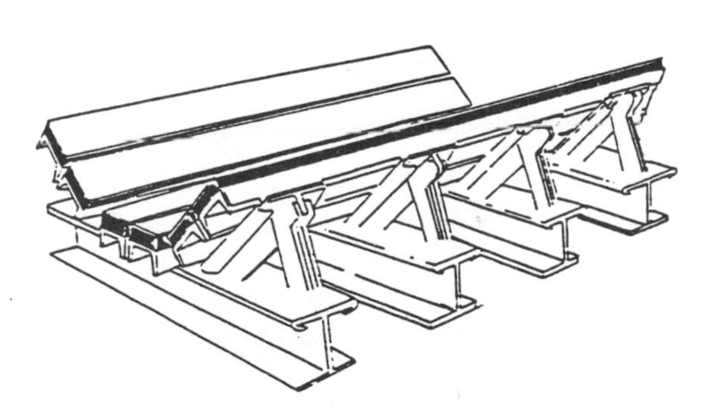

FIG. 2. Impact cradle installed under the belt provides in high stress loading areas.

A new technique for improving belt support and belt sealing in areas which see light loading calls for the use of slider bars installed as side rails under the loading zone. (FIG. 3.) Installed on either side of the belt, like the rails of a bed, these smooth surface bars ensure proper belt troughing and a stable belt to enhance sealing.

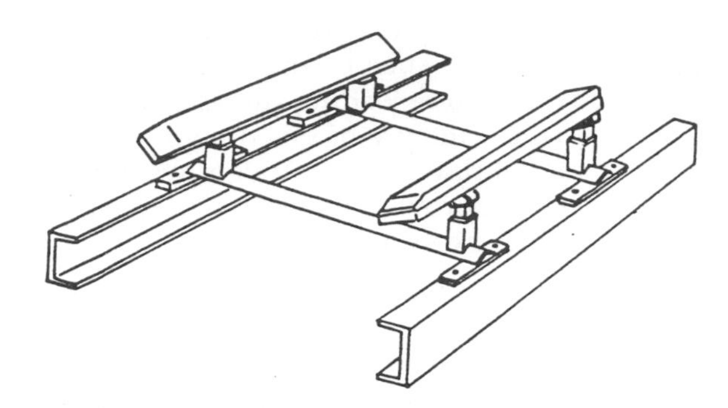

FIG. 3. 'Side rail' slider bar installation provides improved belt stability.

It is important that the belt be properly supported as far down the line as necessary. The length required in each transfer point depends on the individual characteristics of that conveyor system, including material handled, belt width, speed, and other factors. Belt support is like money: it is better to have a little extra than to fall a little short.

Good success has been obtained in installations with impact cradles under the loading point, followed immediately by sets of side rail slider bars. The impact cradle absorbs the energy of loading; the side rails--continued as far down the conveyor as required--maintain a sag-free, sealed surface.

The second item which must be considered in designing transfer points to minimize fugitive material is the use of a wear liner inside the chutewall. (FIG. 4.) Inside the impact area, deflector wear liner--replaceable steel plate bent to funnel material away from the belt edge--creates a dam to prevent undue wear on the rubber. The installation of wear liner prevents sideloading on the conveyor skirting, thereby improving the seal.

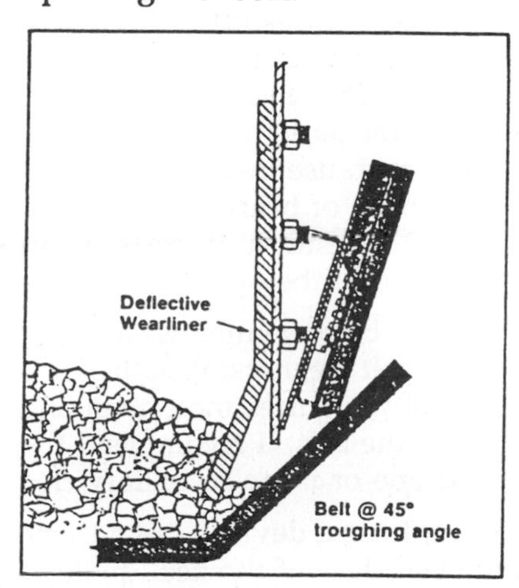

FIG. 4. Deflector wearliner inside the chute prevents stress on the skirt.

This wear liner can be fabricated from materials ranging from ceramic faced steel to UHMW polyethylene, depending on the specific application. Wear liner should be installed in slight, self-relieving taper, opening toward the transfer point's exit. This prevents material entrapment--and its risk of damage--between wear liner, chute wall, and the moving belt.

Deflector wear liner should be used in all circumstances where conditions permit. However, there are some conditions--such as small belts running near capacity, or where loading surges frequently tax the system--where the use of this design would constrict material flow. Here, straight wear liner would provide improved material placement without choking the system.

The final ingredient in designing a stable transfer point is the belt edge skirting system. There are a number of rubber skirting systems to provide sealing between the chute work and the belt currently on the market.

In my experience, it is best to use a skirting system that provides a continuous seal while using a design that allows easy adjustment and replacement of the wearable rubber. And the simpler the process of adjustment--the fewer fasteners to loosen, the fewer clamps to remove--the more it is likely that the required adjustments will actually take place. If the procedure is cumbersome or complicated, three bad things can happen. Adjustment may not happen at all. Or adjustment will be done only rarely, so material escapes under the worn rubber. Or, worse, the maintenance man or conveyor operator, feeling guilty for not making regular adjustments, will "compensate" and over-adjust the rubber, moving it too far down onto the belt at the risk of ripping out the entire section or damaging the belt. To prevent these problems, the maintenance procedure should be as free of fuss, tools, and downtime as possible.

CARRYBACK CONTROL

I'll now turn to a another subject important to belt stability: conveyor belt cleaning systems. Carryback is material that sticks to a belt past the unloading point and eventually drops off in piles beneath the conveyor, or worse, builds-up on idlers, pulleys, and other components. It is these accumulations on components that are particularly worrisome, for these build-ups can cause instability in tracking, or the premature failure of bearings, idlers, and pulleys. All of these result in decreased conveyor efficiency, which costs the operation money.

Carryback can best be prevented with well-designed, well-maintained belt cleaners, installed at the head pulley. It generally requires more than one device to properly clean the belt. I recommend the use of a pre-cleaner and one or more secondary cleaners.

As a pre-cleaner, I use a device I call a Doctor Blade located on the very face of the head pulley. The position of the blade in relation to the belt places it in a peeling position. (FIG. 5.)

With an angle of attack of less than 30 degrees, this blade cleans with very little tip pressure against the belt, resulting in low wear on the blade and the belt surface. If this angle of attack were greater (a "scraping" position), greater pressure would be required to hold it in position against the onslaught of material. Lower blade pressure also means the device's tensioning arrangement will allow relief when a mechanical splice moves past the cleaning edge.

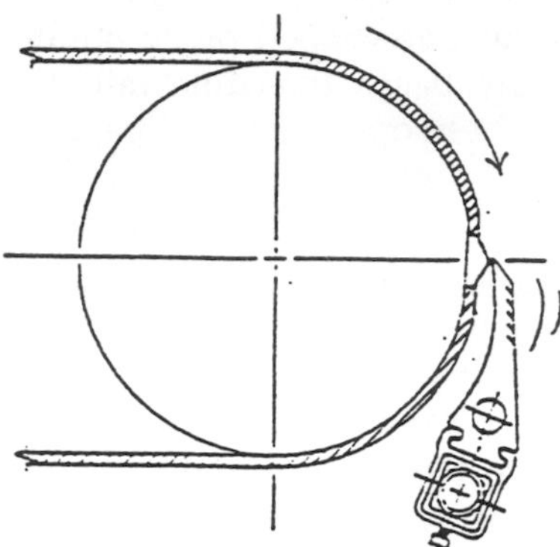

FIG. 5. A pre-cleaner in a peeling position provides effective carryback removal yet allows the passage of belt splices.

The Doctor Blade is designed to remove the majority of material clinging to the belt right at the face, so these fines can join the cargo discharged from the belt. It leaves a thin skim of fines to be removed by the secondary cleaner(s).

For the secondary cleaner, I recommend a design that incorporates multiple blades approximately six inches wide that span the belt. As this cleaner must be as efficient as possible, the blade must mate with the belt as flat as possible. There are irregularities on the surface of the belt that the cleaner must adjust to instantly. Separate blades that are individually suspended have the best potential to remain in contact as the belt passes underneath.

Again, the blade's angle of attack to the belt is an important consideration. Here, I recommend a scraping position, (FIG. 6.) as the peeling angle causes blades to wear to a razor edge.

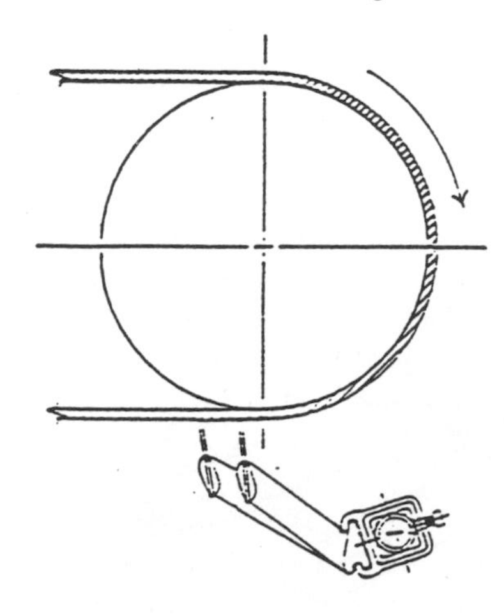

FIG. 6. The secondary cleaner is installed at a scraping angle to safely remove carryback fines.

This creates the risk that an unskilled or "in a hurry" adjustment might apply too much pressure for instant release from obstructions like mechanical splices. The end result could be damage to the belt, splice, or the cleaner.

The positioning of the final cleaner is critical. The closer you remove the carryback to the discharge point, the lower your potential for problems with fines build-up on the dribble chute. I recommend an installation so the blades contact the belt blade while it is still against the head pulley.

It is important to provide access to your belt cleaning devices. This will allow operations personnel to make inspections and provide appropriate service as needed, perhaps even several times a day on those occasions when the material is in its "worst case" condition. Access windows with easy to operate doors should be installed on both sides of the pulley and in line with the axle of the belt cleaners.

You'll want to perform these adjustments in the shortest possible time, to minimize conveyor downtime. It would be desirable that all adjustments and blade replacement be performed with minimal tools. For this reason, I recommend track-mounted belt cleaners which permit an operator--even if he is not a tool-carrying maintenance mechanic--to perform any necessary service.

DESIGNING FOR MAINTENANCE

Ease of maintenance is essential to the concept of having an efficient conveying system. Quick replacement of components that are worn or failing is a priority. For this reason, any component of the conveyor--whether it be impact bars, idlers, skirting, or belt cleaners-- should have features that make it easy for the maintenance people to perform a quick adjustment or replacement. To me, this suggests designs featuring track-mounted components--systems that slide in and out--to provide ease of both installation and replacement.

BELT MONITORS: THE WAVE OF THE FUTURE

Many modern conveyor belts are designed to provide performance at speeds of approximately 1000 feet (300 meters) a minute. That is a speed in the neighborhood of 11 miles per hour, or better than 18.2 kilometers per hour. Conditioned as we are by automobile speeds, 11 miles per hour (18 kilometers per hour) doesn't seem very fast. On the other hand, that speed translates as 17 feet (5 meters) per second, every second. That seems quite fast.

Now , in a car, you wouldn't think of driving--or at least of driving very far, or very fast--without your speedometer. In your car, your speedometer allows you to gauge when you will arrive at your destination, as well as clues as to whether you are traveling at a safe speed, and perhaps, what fuel consumption will be. Why settle for less information about a conveyor belt?

Belt speed monitors are now available to gauge the running speed of the belt. This information will tell operations personnel if the belt is keeping up with the demands of production--whether the belt is running too fast--thus increasing the chances for mistracking--or that the pulley is turning too fast. Belts that operate on steep declining slopes can run away due to the pull of gravity. Or there may be slippage, where the belt stalls or slows dramatically.

The belt monitor helps you to ensure consistent speed operation. After the parameters have been established, the belt speed monitor can alert you when the belt is not keeping up with production, or when production is not keeping up with the belt. These devices can be either electronic--with a sensor that notes the rotation of a pulse transducer affixed to a rotating pulley on the belt drive, or they can be mechanical, using a roller. The outputs can be remote from the location of the belt, allowing one control-room operator to make sure a number of belts are running at specification, without having to devote time to walking each belt. It allows the operator to become instantly aware of any problems, before it turns from a problem into a catastrophe.

As an example, if feeder equipment operates with normal loading patterns when the belt speed is dramatically reduced due to slippage or other problems, you can do serious damage to the belt or other conveyor components.

In addition to the mechanical and electronic monitoring systems mentioned above, there is now increased use of computerized systems for monitoring the efficiency of conveyor belts. Computer-based systems and sensors are now available to spot plugged chutes, belt tears and broken splices. In addition, sensors and computer software can also be used to monitor a conveyor's vital signs--its temperature, amp draw, vibration level, and other considerations--to spot a potential problem before it shuts down a line. The mining industry--with its miles of conveyors both above and below ground--has adopted these remote sensing systems around the world. In keeping with the statistical process control systems that are already in place in many industries, more sophisticated monitor systems will soon be available.

CONCLUSION

Conveyors can be made more stable, and better controlled, and hence more productive. That means the expenses for the improvements discussed here will be well worth it, because they will show themselves on the bottom line.

C418/004

The conveying technique of the 1990s. An enclosed belt conveyor which can run round sharp corners

R JOHANSSON
Scaniainventor Conveyor Sicon AB, Helsingborg, Sweden

SYNOPSIS The conveying technique of the next century will certainly be subject to much stricter environmental requirements than what has been accepted or tolerated sofar. As the new types of enclosed belt conveying systems are gaining acceptance, the conventional open type belt conveyor will be ruled out in many of its present applications.

In this paper I will describe one of these new systems where most of the environmental constraints of the conventional belt conveyor has been eliminated.

Cost comparisons with conventional belt conveyors include investment cost, maintenance costs and operating costs.

1. CONVENTIONAL BELT CONVEYOR TECHNIQUE

Conveyor belt systems are generally acknowledged as the most cost effective and reliable method of moving bulk materials on medium to long transport distances.

The advantages of belt conveyors compared to chain conveyors, screw conveyors and pneumatic conveyors are:

- low power requirement,
- gentle handling,
- long service life.

The disadvantages compared to other systems are:

- spillage of material and generation of dust particurlarly at the loading, discharge and transfer points,
- a limited degree of inclination,
- a very limited degree of horizontal curving.

1.1 Improvements

Throughout the years, the conventional belt conveyor technique has been the subject of intensive product development. Designers and operators have sought solutions that would help to overcome some of the most important constraints imposed by the belt conveyor, e.g.:

- Endless efforts have been spent on developing various systems for belt cleaning. In spite of all these efforts, most operators will agree that there is no equipment available today which is maintenance-free, reliable and which does not create other problems such as increased belt wear. Environmentally, belt conveyors are still considered a nuisance because of spillage and pollution, particularly at the transfer points.

- The restricted inclination of belt conveyors has called for improvements such as rubber cleats vulcanized to the belt. This in turn has further emphasized the problem of belt cleaning.

- Different designs of covered belt conveyors have contributed to a reduction in dust pollution but offer no solution to spillage at the transfer points or from the return part.

- Several solutions have been developed to allow belt conveyors to run in curves. These systems tend to be expensive, complicated and still only allow curves with very large radii.

1.2 Alternative designs

Several new concepts have been introduced which meet some of the criterias for enclosed handling.

- The load-carrying part of the belt can be covered, it can slide on a U-shaped duct or inside a pipe instead of being supported by idlers.

  Hereby the material and the dusty air will be enclosed.

  However, the spillage from the return part still remains.

- A sandwich belt arrangement can improve the maximum inclination of the conveyor but again it does not solve the problem of spillage.

- The conveyor belt can be formed into a pipe which makes both the load-carrying and the return part enclosed. The ability to negotiate curves is improved but it still requires considerable radii.

The common conveying element which has been the subject for all the above development and improvement is still the old, conventional, flat conveyor belt, used in various forms of troughed or tubular sections.

To overcome the fundamental problems of the belt conveyor there was a need for innovative thinking which was not constrained by the present technique, in particular with regard to the design of the belt.

2. THE SICON CONVEYING SYSTEM

The unique part of the SICON system is the conveyor belt itself. Although it is different in its design, it still provides the same functions as a conventional conveyor belt.

The three basic functions are:

* To support the cargo

* To provide a track for the idlers and pulleys

* To take up the tension from the drive machinery

Whereas a conventional conveyor belt has a uniform construction over the whole width of the belt, the SICON belt is designed with the aim of optimizing each of the above functions.

The result is a belt where the track for the idlers and the longitudinal reinforcements are concentrated to the edges of the belt and the intermediate part of the belt has only one function, that of supporting the cargo. *See figure 1.*

Hereby it has been possible to use steel wire ropes for the longitudinal reinforcing and the profiles at the edges of the belt are designed and reinforced in the same manner as for normal V-belts. This arrangement has made it possible to design a conveyor system with some exceptional features.

2.1 Enclosed handling

A cross-section of the conveyor shows the two steel cord reinforced rubber profiles vulcanized to the load-carrying belting. The profiles serve as tracks for the angled support roller and vertical guide roller and, with steel cord reinforcing, they absorb the tension of the conveyor drive system. The pearshape belt is manufactured from a quality of rubber which has very flexible, elastic and wear resistant properties. This makes the belt capable of negotiating very sharp corners. *See figure 2.*

Except for loading and discharge, the belt is at all time enclosing the material.

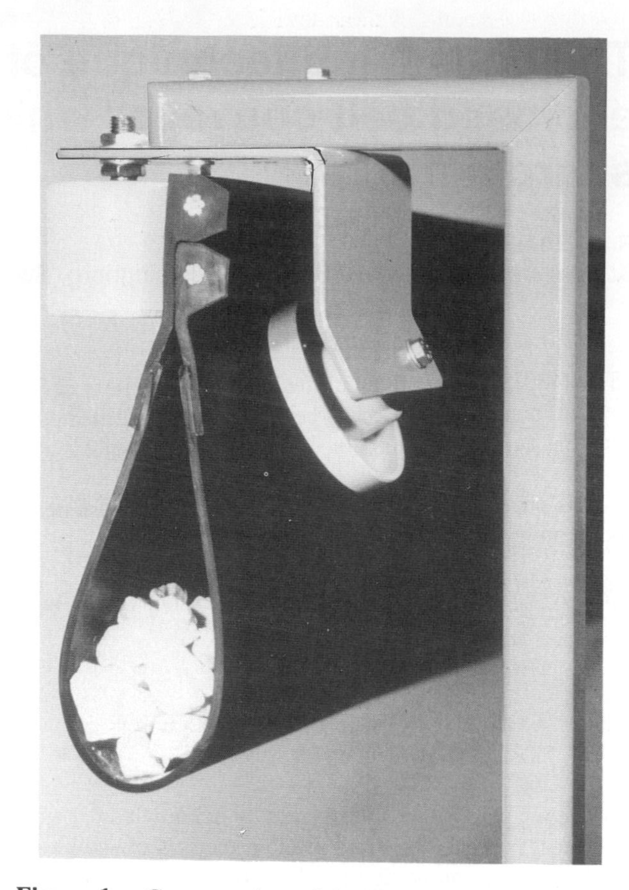

Figure 1 Cross-section of the SICON belt

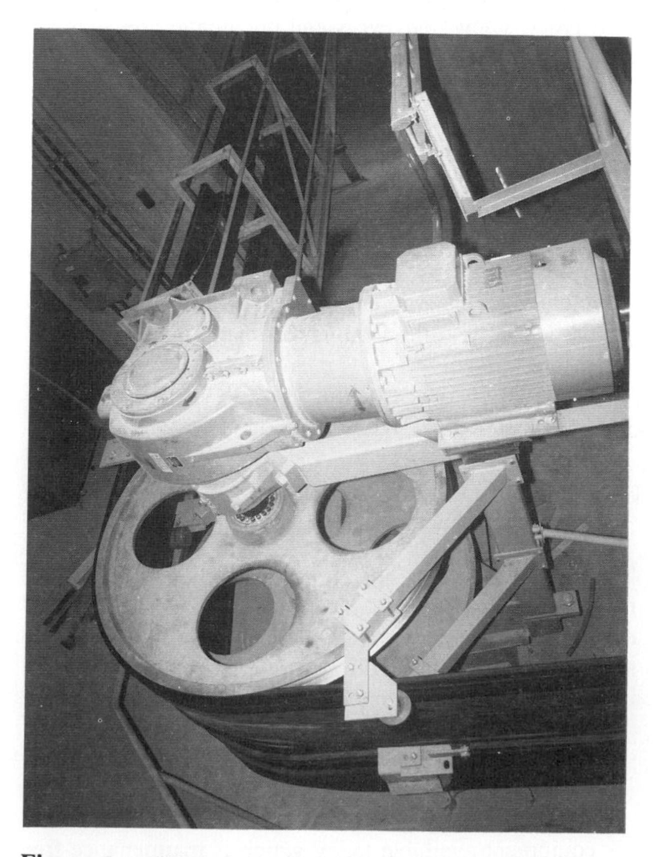

Figure 2 90° corner with intermediate drive unit

Also the return part of the conveyor is closed in the same manner. The result is a totally enclosed conveyor which transports bulk materials from the loading point to the discharge point, without transfers, spillage or pollution.

The environmental advantages when handling dusty, sometimes even toxic materials, are quite obvious but it also leads to direct cost savings in reduced maintenance and cleaning, caused by dust and spillage.

2.2 Ability to negotiate sharp corners

Thanks to its unique design, the SICON conveyor can negotiate very sharp corners. The minimum radius of the deflection sheaves depends on the size of the belt.

Data	SICON 100	SICON 1000
Handling rate m^3/hour	10-100	100-400
Belt speed m/sec	1-5	2-5
Min. curve radius m	0,4	0,75
Max. lump size mm	40	70

2.3 Steep inclination and two-way conveying

The conveyor can run in a steep inclination (max. 20-35° depending on the transported material).

The return part does not have to follow the load route and therefore the conveyor can be arranged as a closed loop with maximum flexibility in plant lay-out. It is also possible to transport material in the return part simultaneously, thus making it a true two-way conveyor.

2.4 Multiple drive units - low belt tension

Due to the very good traction on the V-shaped profiles, it has been possible to install multiple drives. Drive units can be fitted at any curve where the belt wrapping is minimum 90°. *Figure 2*. This will reduce the maximum belt tension for long conveyors.

The energy consumption is the same as for conventional belt conveyors, thus considerably lower than for screw or chain conveyors and a fraction of the power required for corresponding pneumatic conveyors.

2.5 Smooth loading and acceleration even at high belt speeds

Loading of the belt can take place anywhere along the conveyor. In order to ensure a controlled flow rate of the transported material, it is recommended to use either a vibrating feeder, screw feeder, rotary vane feeder or similar.

At the loading stations, the conveyor belt is opened into a U-shape with sufficient opening for the loading chute.

As all the belt tension is taken up by the two reinforced profiles, the payload-carrying belting is completely slack and can therefore, in the best possible way, absorb the impact of the loaded material. The result is very efficient and quick acceleration of the material, even at comparatively high belt speeds. *See figure 3.*

For a conventional belt conveyor, the material falls onto a belt which is in tension and therefore acts like a trampoline making the material bounce and start rolling before it finally comes to a rest on the belt.

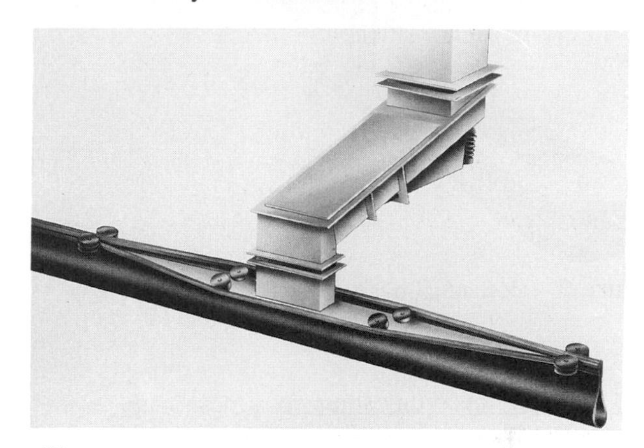

Figure 3 Loading point with a vibrating feeder

2.6 Multiple discharge methods

Vertical discharge
The belt opens out by a gradual transition from its pear-shape into a flat, vertical orientation. After discharge, the return part again closes to its original shape. *Figure 4.*

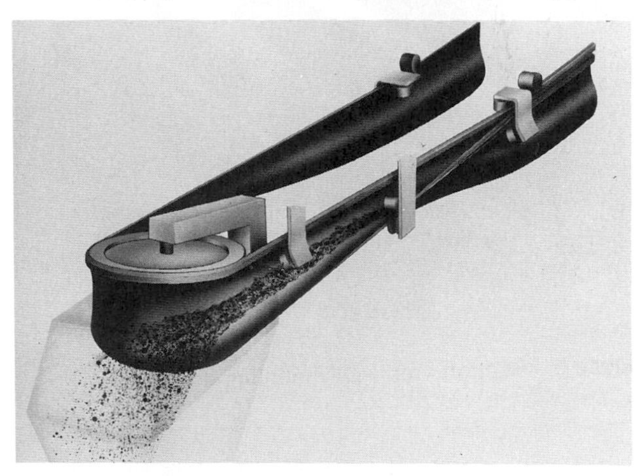

Figure 4 Vertical discharge

Horizontal discharge
Here the belt opens out in a horizontal orientation (similar to conventional belt conveyors). After discharging the belt remains in its flat position while it turns forward again to get the dirty side facing upwards. Any spillage from the discharge point is gathered on the lower part of the belt before it closes again to the pear-shape.

The receiving shute can be fitted with a gate which will direct the material either onto the next mode of transportation or back into the belt for further transport to the next discharge station. *See figure 5.*

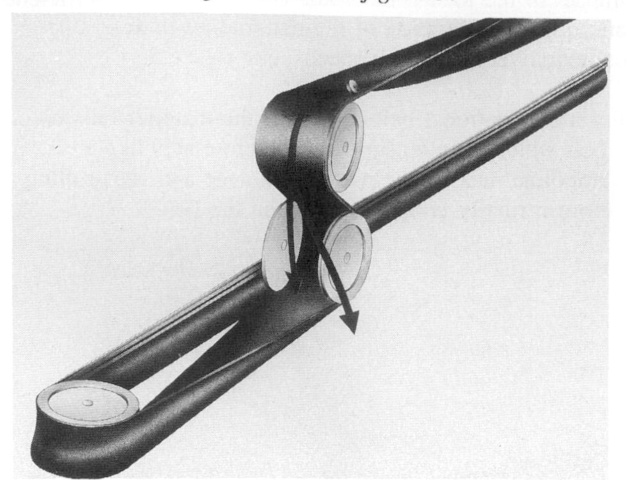

Figure 5 Horizontal discharge

2.7 Simple supporting structure

The SICON conveyor requires less space than a conventional belt conveyor and the supports are simple.

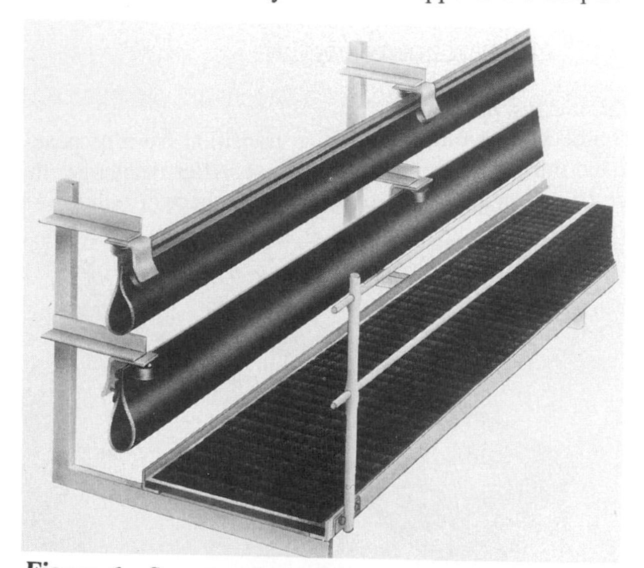

Figure 6 Supports for wall mounting

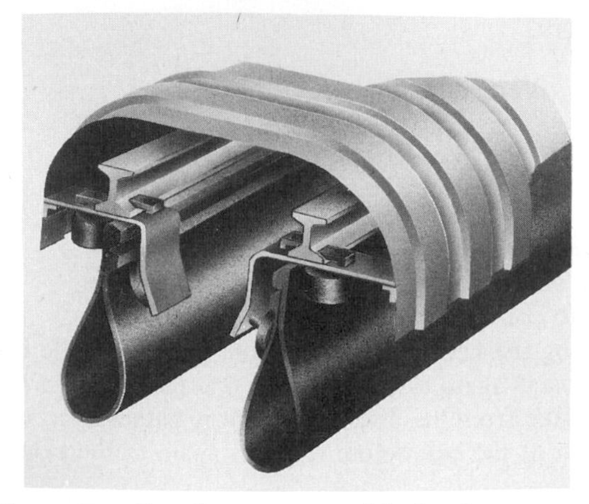

Figure 7 Out-door support with simple cover

Figure 8 Covered conveyor bridge

2.8 Capacity range

Sofar the conveyor has been manufactured in two basic sizes which are related to the size of the profiles (SICON 100 and SICON 1000). Each size of profile can be used for different widths of belt and, by operation at different speeds, the present two sizes cover a handling rate from 10 to 400 m^3/hour. (*See section 2.2*)

The longest belt in operation has a conveying distance of 400 m and the highest capacity sofar is 250 t/hour.

The largest conveyor on order has a handling rate of 360 m^3/hour and a transport distance of 640 m.

3. OPERATING EXPERIENCE

3.1 Reference installations for SICON 1000

The first SICON conveyor was installed by Cementa AB at their main production plant in Slite, Sweden. This conveyor has been in operation since April 1988.

Cementa uses the SICON belt to transport 250 t/hour of sand and limestone from the stockpile up to the buffer hoppers at the mill. Two feeder conveyors from under the stockpile load the material into the SICON belt which at the other end discharges onto a distributing conveyor above the buffer hoppers.

Earlier, the same transport was handled by four conventional belt conveyors connected via transfer chutes. The new conveyor takes the material all the way, without any transfers and the related dust and spillage.

The order was placed in August 1987 and the installation was commissioned in April 1988.

As a result of the very good performance of this first conveyor, Cementa placed an order for a second conveyor of similar size for handling of hot cement clinker. This has called for a conveyor belt of special quality, approved for handling material with a temperature of max. 90°C.

The handling capacity is 150 t/hour with possibility to increase to 250 tons/hour and the total length is 220 m. This conveyor runs round three 90° curves making it possible to do the complete transport in one single conveyor. Conventional belt conveyors would have needed three transfers with corresponding spillage and costs.

The second conveyor was commissioned in June 1989 and with suitable dust extraction at the loading and discharge points it has been possible to make the total system dust-free although the product (cement clinker) is very dusty.

3.2 Reference installation for SICON 100

The first conveyor of the SICON size 100 was ordered by the Swedish company Höganäs AB for installation at their new production unit at Höganäs. This conveyor handles iron powder at a rate of 30 t/hour.

The total transport distance is 280 m and the conveyor has 10 loading points and two alternative discharge points. Strict environmental requirements have influenced the design of loading and discharge points, all of which are connected to a dust extraction system. Hereby the transport is free from any dust emission and spillage and the material is protected from moisture.

The conveyor was commissioned in January 1990 and in spite of being the first of the size 100, only minor modifications had to be made after the trial period.

Since then, a series of conveyors of the size 100 has been commissioned.

The longest conveyor (AVEBE TAK, Holland) has a conveying distance of 400 m.

Materials range from fine, dry powders to sticky and lumpy materials.

4. APPLICATIONS

The SICON conveyor has created a great interest from many different sectors of the industry and for handling of a wide range of products. The following type of products require special attention regarding spillage and pollution.

* *Building Industry*
  Sand, Lime-stone, Clay, Cement, Gypsum, etc.

* *Power Generation*
  Coal, Wood-chips, Peat, Bottom Ash, Fly Ash, Lime, Gypsum, etc.

* *Paper Industry*
  Wood-chips, Paper Pulp, Kaolin, Chemicals, Waste Paper, etc.

* *Steel and Metal Industry*
  Ore, Coal, Slag, Metal Concentrates, Alumina, Foundry Sand, etc.

* *Mining Industry*
  Ore, Minerals, etc.

* *Food Industry*
  Flour, Sugar, Vegetables, Fish, Frozen Food, Cereals, Residues, Waste, etc.

* *Feedstuff Industry*
  Grain, Soya Meal, Fish Meal, Tapioca, Pellets, Pet-food, etc.

* *Chemical Industry*
  Minerals, Chemicals, Ready-made Products in granular of powdery form, etc.

5. COST COMPARISONS

5.1 Investment

5.1.1 The SICON belt itself is <u>more</u> expensive than corresponding conventional belts. The reason for the higher cost is the fact that the belt is built up from many components and that the production volume is still comparatively low.

5.1.2 The idlers are fewer and <u>less</u> expensive than for conventional belt conveyors.

5.1.3 The head and tail ends are <u>more</u> expensive for the SICON system as the diameter of pulleys are larger.

5.1.4 The cost for introducing a curve in the SICON system is <u>considerably less</u> than the corresponding cost for a transfer point for conventional belt conveyors. Particularly if the comparison includes the cost for a transfer house, dust filter, servicing platforms and the loss of elevating height at all the transfer points.

5.1.5 The supporting structure is in most cases <u>simpler and less expensive</u> for the SICON system. Also, the flexibility of the conveyor allows a lay-out where one can make use of existing structures. It could be an advantage to suspend the conveyor from the walls of a building, or to follow an existing pipe line rather than building a new structure along a straight line between loading and discharge points.

The above cost comparisons indicate clearly that the SICON conveyor is not very competitive for a short conveying distance along a straight line and for a material which is not creating any environmental problems.

However, if the route calls for one or more transfers over a longer distance and the nature of the materials handled is such that it requires a cover or enclosure, the SICON system will be competitive even comparing the first costs (investment).

5.2 Maintenance on the equipment

5.2.1 Belt wear

Wear on conventional conveyor belts is caused by:

- Impact and acceleration of material
- Misalignment of idlers and pulleys
- Dirt accumulation on idlers and pulleys
- Belt scrapers
- Bad tracking/Off-centre loading (worn edges)

The SICON system differs from conventional belt conveyors in the following respect:

- Impact is absorbed by the elasticity of the rubber and the acceleration is improved correspondingly. *See section 2.5.*

- The idler assemblies are self aligning with little need for adjustments.

- Dirt stays inside the conveyor and idlers operate in a clean environment.

- No need for belt scrapers.

- Off-centre loading is not possible as material always rests in the centre of the belt.

In view of the above, we foresee less wear on the SICON conveyor than on conventional belts. The operational experience supports the assumption but the number of operating hours is still too low to confirm the expected life.

5.2.2 Wear or malfunction of idlers

The main reasons for replacement/maintenance on conventional idlers are:

- Idlers stop due to dirt or moisture penetrating into the bearings.

- Idlers stop due to accumulation of dirt on the outside.

By eliminating dirt outside the conveyor belt these problems will not be applicable to the SICON conveyor.

5.2.3 Service and maintenance on shutes, drives, pulleys and control gear

Here we can not see any major difference between conventional systems and the SICON conveyor except for one specific item which is common for all closed belts.

All closed conveyors are sensitive for overfilling.

Measures have to be taken to control the feed rate and in addition it is recommendable to introduce a overfilling sensor after each loading point to stop the belt if a certain level of material is exceeded.
Such sensors must be checked regularly to ensure a safe function.

5.2.4 Service on belt scrapers and dust filters

Belt scrapers are not required on the SICON conveyor and the dust extraction is limited to the loading and discharge points.

5.3 Other operating costs related to dust and spillage

In a survey made in 1987 by the Bulk Materials Handling Committee of the Institution of Mechanical Engineers with a view to quantify the cost of dust, mess and spillage associated with bulk handling operations, the conclusion was that these costs are in the range of £ 200 million for the UK industry as a whole.

The report, presented by Mr. H.N. Wilkinson, Dr. H. Wright and Professor Dr. A.R. Reed, specified the costs related to:

- Loss of bulk material
- Spillage clearance
- Special costs, attributable to the type of industry and its environment but all related to pollution and spillage.

Although there was a great variation between the individual types of industries, the average costs could serve as an indication of the magnitude of such costs:

	Average (Pence/ton)	Range (Pence/ton)
Loss of material	47.0	200 - 5
Spillage clearance	6.9	49.7 - 1.3
Extra maintenance	4.7	33.1 - 0
Special costs	10.7	33.1 - 3.8
Total	69.3	

In a similar study carried out in 1986 in Sweden by Mr. O. Öberg at the Royal Institute of Technology, the corresponding costs were (convertion 1 £ = SEK 10.50):

	Average (Pence/ton)
Loss of material	42.0
Spillage clearance and extra maintenance	16.5
Special costs	14.7
Total	73.2

The Swedish study was concentrated to material spillage at belt conveyors and covered 40 plants handling various types of materials.

It is certainly possible to reduce the above costs by the introduction of the better design of the transfer points, improved belt scrapers, dust extraction, etc. but by changing to enclosed handling the cause of all the above costs are eliminated.

6. RESTRICTIONS

At this stage of development, the SICON conveyor has the following limitations:

- The handling rate is limited to approximately 400 m^3/hour.

- Max. lump size is 70 mm.

- It is not possible to incorporate a travelling tripper (travelling discharge point).

7. CONCLUSIONS

Every type of conveyor has its own, limited area where it best services its purpose.

For belt conveyors, chain conveyors, screw conveyors, bucket elevators, pneumatic conveyors, etc. these areas are quite well defined and often related to the type of material which is handled.

By introducing a new type of conveyor which combines certain features from several of the above conveyors, a new area of applications is created. An area where one can combine e.g. the power requirement and the gentle handling of a belt conveyor, the enclosed handling associated with screw or chain conveyors, and the flexibility in lay-out of the pneumatic conveyor.

This area also covers an increasing range of bulk materials that require a special conveying technique.

- Chemicals and minerals tend to become more refined, more concentrated, more toxic and more expensive.

- Large scale production plants in the food and pharmaceutical industry require continuous conveying within the plant in a closed and hygenic manner.

- The use of air and water as transport media (pneumatic and slurry conveying) will be restricted due to pollution problems.

The choice of Bulk Materials Handling Equipment, towards the year 2000, will be governed; on the one hand by demands, set by the authorities for a better working environment for the employees and on the other hand by demands from the industry for improved materials handling systems utilizing less labour and allowing a high degree of automation in order to become more cost effective.

REFERENCES

1. WILKINSON, H.N., REED, A.R. and WRIGHT, H. The Cost to U.K. Industry of Dust, Mess and Spillage in Bulk Materials Handling Plant.

   Bulk Solids Handling, volume 9, number 1, February 1989, pages 93-97.

2. ÖBERG, O. MATERIAL SPILLAGE AT BELT CONVEYORS.

   Royal Institute of Technology, Sweden.

C418/036

Mechanical behaviour of wheat under storage stress histories

J Y OOI, BE, PhD and J M ROTTER, MA, PhD, FIEAust
Department of Civil Engineering, University of Edinburgh, Scotland

SYNOPSIS Recent developments in numerical techniques for predicting overpressures in silos during discharge suggest that discharge pressures may be strongly affected by the complex stress-strain response of the bulk solid. The traditional view, that the response of a bulk solid can be simply described by its internal and wall friction angles together with its flow function, is giving way to descriptions which involve the dilation of the material during flow. For these more complex models, much more knowledge of bulk solid mechanical behaviour is needed. This paper describes a series of tests on wheat, in which the stress history of the sample was arranged to follow that of wheat placed in a silo, buried beneath material placed after it, and then expanded vertically as the discharge took place. The stress-strain response of the material was followed closely. The tests are part of a test programme undertaken to study the complete deformation and strength properties of dry bulk solids. The paper presents results for only one material, but some general conclusions can be drawn for other commonly stored products.

1 INTRODUCTION

Most recent attempts to predict overpressures in silos during discharge using numerical modelling and some careful experimental observations on silo wall pressures suggest that wall pressures may be strongly affected by the detailed stress-strain response of the stored solid. However, the stress-strain behaviour of dry bulk solids stored in silos has not been widely investigated to date.

Traditional tests on bulk solids generally only give the angle of internal friction and are mainly concerned with the flow of the solids. The commonest test is conducted using a shear box (1), which is adequate to the task but not capable of giving more general information on the properties of a bulk solid. As a result, it is difficult to correlate unexpected behaviour found in tests with properties found using this apparatus.

There are strong indications that material properties other than failure properties affect the pressures exerted on silo walls. In particular, there is much evidence that measures of the volumetric response, such as the effective Poisson's ratio, shear modulus and dilation angle may have a very significant bearing both on pressures in silos and on bulk solids flow (2). The response of the bulk solid under a variety of loading conditions must be explored thoroughly if more powerful flow and pressure prediction techniques are to be developed.

This paper describes the results of a series of tests on wheat conducted using a triaxial testing apparatus which has been modified to work with interstitial air and at the very low stress levels encountered in silos. Considerable care was taken to determine volumetric changes in the specimen, as well as deviatoric strains. The repeatability of the results was carefully verified. Some interesting features of the mechanical properties of wheat are described.

Several investigations have used a triaxial apparatus to measure the properties of grains (3-14), but only last three references appear to have examined features other than the deviator stress-axial strain response. Some other researchers have studied the one-dimensional axial response with radial strains fully prevented (15-18).

2 EXPERIMENT

A schematic diagram of the experimental rig is shown in Fig. 1. The sample was enclosed in a rubber membrane and sealed against the top and bottom platens. A large sample (100 mm diameter x 200 mm height) was used to improve the precision of the test at low stress levels. Instead of using water in an enclosing cell to confine the sample, the confining pressure was obtained by reducing the air pressure inside the sample. This arrangement has several advantages: small but accurately controlled deviator stresses can be used, water is kept away from the hygroscopic dry sample and free access to the outside of the sample permits direct dimensional measurements. The rig

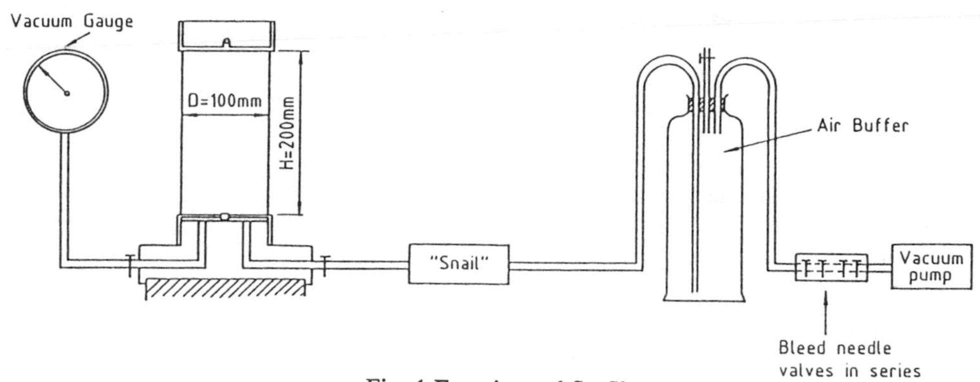

Fig. 1 Experimental Set-Up

was housed in a temperature controlled room to minimise thermal volumetric effects. The development of this modified triaxial apparatus has been described in detail elsewhere (12,13).

Confining isotropic pressure was achieved by opening or closing needle valves (Fig. 1). Anisotropic loading and unloading were implemented by manually adjusting the confining isotropic pressure continuously to maintain a constant ratio of axial to radial stress whilst the samples were axially strained at a slow rate. The required radial stress was determined from the nominal axial stress, which was continuously displayed by the data acquisition unit.

Deviator loads were applied under displacement control. In calculating the stresses in the sample, corrections were made for the stresses in the enclosing rubber membrane. Considerable care was taken to ensure the repeatability of tests at low stresses (12,13).

During discharge, the material expands under a regime of reducing vertical stresses, but often exerts pressures larger than the filling values on the silo walls. Dilation of the material may occur during shearing failure. The volumetric behaviour of the solid can therefore be expected to play an important role in the loads exerted on the wall. With this in mind, special attention was paid to measurements of volumetric strains throughout the experiments. A detailed discussion of the technique can be found elsewhere (13).

3 TEST PROGRAMME ON WHEAT

A test programme was undertaken on wheat to investigate its mechanical behaviour and properties. Wheat was chosen as a common, well-behaved granular solid. The stress-strain response during both isotropic and anisotropic compression as well as during deviator loading and unloading were examined in detail. Special attention was paid to the response during initial deviator (axial) loading. Several factors which are likely to influence the induced wall pressures were investigated, including the stress history, stress level dependency (barotropy) and grain orientation.

Since wheat has elongated grains, it displays anisotropic behaviour in a mass and the orientation of the grains affects the properties (6,19). Thus it is important to define, if possible, the dominant orientation of the grains in a sample. The effect of dominant grain orientation on the stress-strain behaviour was explored by using tests with three different grain packing patterns. Tests with different orientations are described elsewhere (13,20).

The 'axisymmetric' grain samples described here were prepared by allowing the grains to fall from a funnel placed about 150 mm above the grain surface. These samples simulated grain in a centrally filled silo. Under these conditions, the major axis of each wheat grain usually lies tangential to the cone of the instantaneous surface, leading to systematically axisymmetric inclined grain axis orientations.

A description of this extensive experimental study, together with many interesting observations, is given by Ooi (13). The anisotropic loading and unloading of two samples with a principal stress ratio of one half is presented here. The stress history used in these samples followed that of grain stored in a silo. Deviatoric extension tests on similar anisotropically compressed samples with the axisymmetric grain orientation are also described.

4 CONSTANT STRESS RATIO COMPRESSION

During the filling of a silo, both the vertical and lateral stresses increase gradually. Theoretical predictions of pressures in silos and hoppers usually assume a constant ratio of wall pressure to mean vertical stress, namely the lateral pressure ratio, k. In this section, the results of tests with a ratio of the principal stresses of one half (k=0.50) are presented (these tests are termed anisotropic compression). This constant stress ratio compression is close to the stress path during filling of a silo.

4.1 Volumetric Response

The relationship between the specific volume and the mean pressure p ($p=(\sigma_a+2\sigma_r)/3$) during this test is shown in Fig. 2. The specific volume is the volume which contains a unit volume of grains. The result for a similar sample under isotropic compression (k=1) is also plotted for comparison.

The volumetric strain under anisotropic compression (k=0.5) was found to be very similar to the corresponding isotropic value at the same mean pressure. At low pressures the specific volume decreased rapidly with pressure. At higher pressures, the sample became denser and the volume changed less rapidly.

In each test, the pressure was first increased to a certain value (initial loading), then decreased almost to zero (unloading), then increased again to a higher value (reloading) and decreased again. These cycles were repeated several times. After reloading beyond the maximum pressure which had been achieved earlier, reloading displayed the same response as initial loading, so the pattern of behaviour can be divided into the initial loading or virgin line, and the unloading-reloading hysteresis loops.

All stress cycles gave similar unloading-reloading hysteresis loops. On unloading, the samples were observed to swell only slightly, but the rate of swelling rose as the mean pressure decreased. The reloading lines were found to be almost straight. The reloading bulk modulus had a mean value of 5.8 MPa with a coefficient of variation of 24%. Rather surprisingly, the reloading bulk modulus appears to be higher when unloading has occurred from a lower peak pressure.

It was noted that the curved unloading paths (Fig. 2) appear linear when the mean pressure p is plotted on a natural logarithmic scale (13). This behaviour has also been observed in soil testing (21,22). Further testing is needed to verify and explore the barotropy (stress dependency) of the unloading and reloading stiffness parameters.

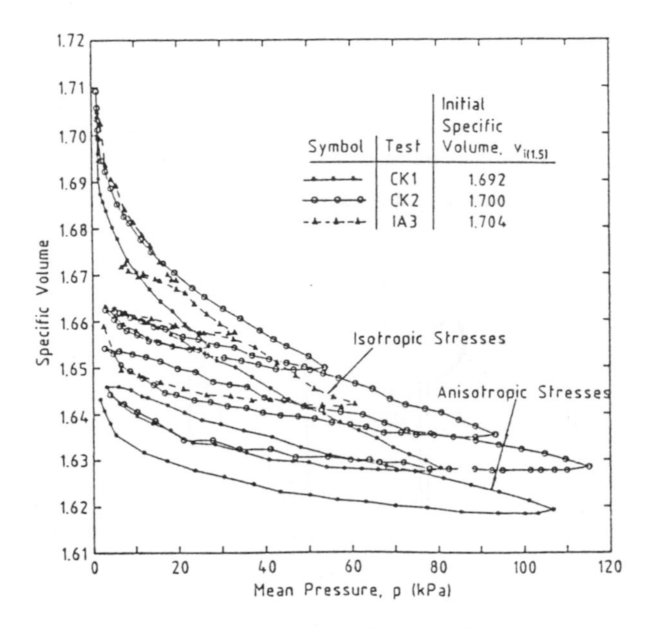

Fig. 2 Anisotropic Compression

4.2 Anisotropy

The anisotropy of the strains developing in the test was examined by plotting the volumetric strain against the axial strain as shown in Fig. 3. Despite the similar changes in volumetric strain for isotropic (k=1) and anisotropic (k=0.5) compression reported above, the radial and axial strains were very different. The mean radial strain under anisotropic loading was consistently smaller than that in the isotropic test. This was expected and was a result of the lower radial stress for a particular value of mean pressure p. The volumetric strain changed almost linearly with the axial strain, except at very low mean pressures (less than 10kPa).

At very low pressures, comparatively large axial strains developed quickly (Fig. 3). This was because each sample was first set up with a small and nearly isotropic pressure (zero state for strain measurement), and axial loads were next applied to bring the stress ratio down to 0.5.

The mean slope of this relationship (Fig. 3) for pressures greater than 10kPa during virgin compression was estimated to be 0.98. This is very close to the condition of zero lateral strain which corresponds to the condition commonly found in silos.

Although the results above show that the grain behaves quite anisotropically, it is often assumed to be an isotropic material. An experimental value for Poisson's ratio, assuming isotropic behaviour, can be deduced from this test. With the radial to axial stress ratio k of 0.50, the isotropic Poisson's ratio ν is given by

$$\nu = \frac{1-2m}{3-2m} \tag{1}$$

in which m is the ratio of radial to axial strain (m=0.98). From Eq. 1, a constant isotropic Poisson's ratio of 0.34 may be deduced for the virgin anisotropic compression. This is not very different from the commonly assumed value of 0.3 (23-25).

The strain response during unloading and reloading is closer to isotropic. There is an almost linear relation between the volumetric strain and the axial strain. The average unloading slope is 2.3 and the average reloading slope is 1.8. A stress cycle slope of about 2 implies that the 'isotropic' Poisson's ratio during unloading and reloading is close to zero. However, the tests should really be interpreted using orthotropic or anisotropic elasticity theory. An orthotropic interpretation of the test results was presented elsewhere (13).

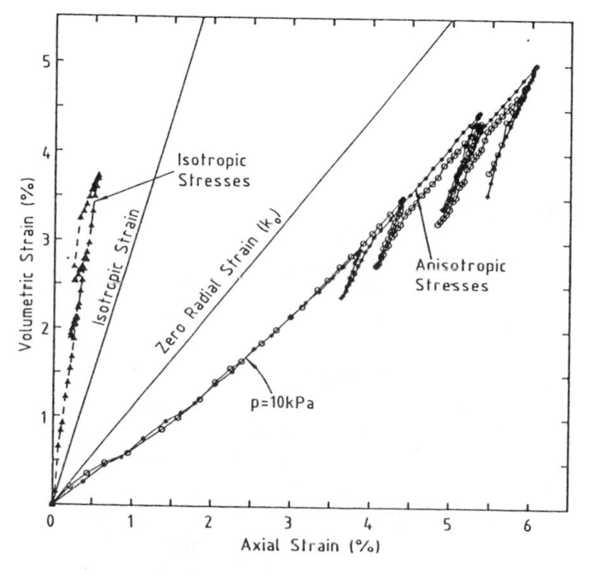

Fig. 3 Strain Anisotropy

5 DEVIATORIC EXTENSION

During the discharge of a silo, the bulk solid expands vertically under a reducing vertical stress regime. To simulate the discharge condition, several samples were subjected to deviatoric extension. These tests were termed the KE series. After compression under a stress regime with k=0.5, the radial stress was held constant and the axial stress steadily reduced (axial extension). This allows the study of a stress history which is closer to that involved in the discharge of a silo. The samples were prepared in a systematic manner and had similar initial specific volumes.

5.1 Deviator Stress Response

The deviator stress response under axial extension, normalised using the initial mean pressure p_i, is presented in Fig. 4. In this test, the major principal stress (most compressive) coincides with the axial direction initially (since k=0.5), but it progressively decreases until an isotropic stress state is reached (k=1). After that, the confining lateral pressure σ_r becomes both the major and the intermediate principal stresses (k>1).

The stress-strain curves indicate a very stiff response at very small axial extensions (Fig. 4). Thus the axial stress must drop very much before the solid extends very much at all. The response was found to be much stiffer than in other tests where the samples were deviatorically compressed instead of extended (13). This may be because the initial stress path lies well inside the yield surface (22). The deviator stress continued to increase to the end of the test at about 6% tensile axial strain (Fig. 4).

The tangent and secant moduli were deduced from the stress-strain plot and were found to increase as the initial mean pressure increases (13). For the range of initial mean pressures p_i of 14 to 109kPa used in this test series, the tangent modulus (slope of the stress-strain plot at any instant) was found to reduce from an initial value of 20-80MPa (at very very small axial strains) to 0.2-2MPa at 1% axial strain (13).

5.2 Failure Properties

The failure of a granular mass can usually be adequately represented by the Mohr-Coloumb failure criterion (21). Under deviatoric extension, the criterion can be expressed as

$$(\sigma_r-\sigma_a)_f = (\sigma_r+\sigma_a)_f\sin\phi + 2c\cos\phi \tag{2}$$

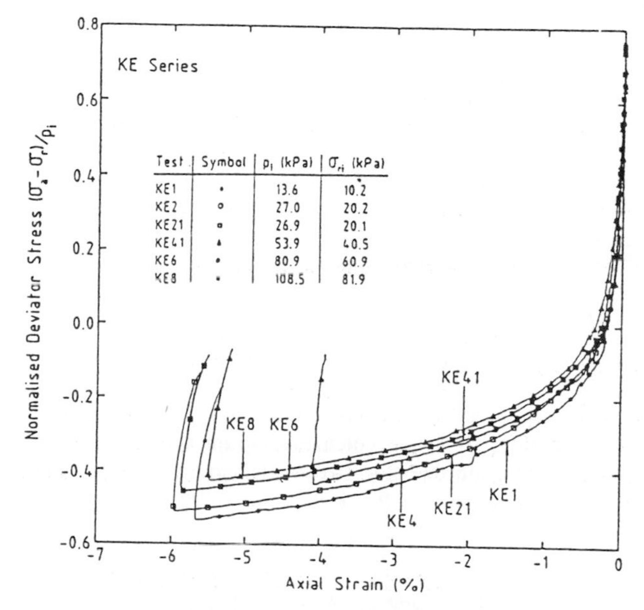

Fig. 4 Deviator Stress Response

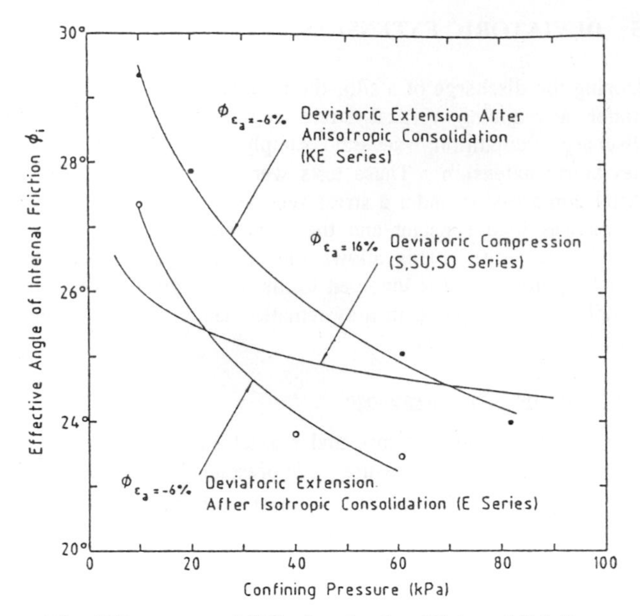

Fig. 5 Barotropy of Effective Angle of Internal Friction

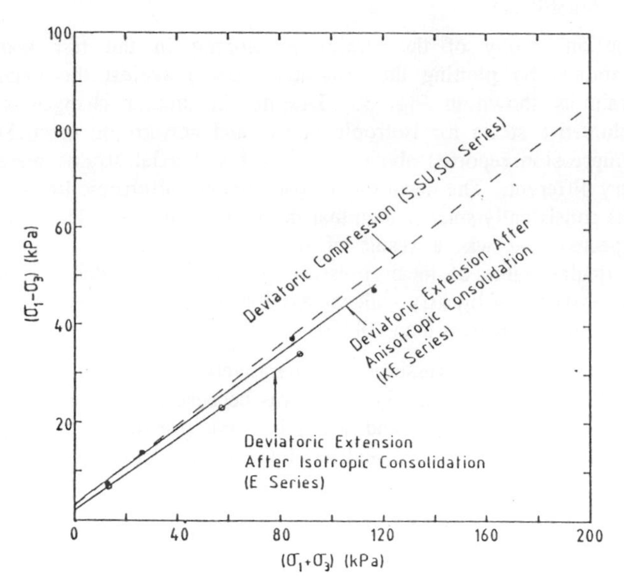

Fig. 6 Failure Stress States

in which σ_a and σ_r are the vertical and the radial stresses, and c and ϕ are material constants termed the cohesion and the internal friction angle.

In bulk solids handling, materials are almost always treated as cohesionless (c=0). The effective angle of internal friction ϕ_e can be calculated from Eq. 2 assuming zero cohesion.

Since the tests had to be stopped before a peak or limiting deviator stress could be reached, the friction angle was estimated as the value at -6% axial strain. The effective angle of internal friction (Fig. 5) may be seen to reduce with increasing confining pressure, especially when the confining pressure is small. Similar behaviour is found using a Jenike shear cell.

The wheat was next treated as a typical c - ϕ material (having both internal friction and cohesion) to see whether the higher friction angle at low confining pressures could be represented by the presence of cohesion. The deviator stress (σ_r-σ_a) is plotted against the sum of the major and minor principal stresses ($\sigma_r+\sigma_a$) in Fig. 6. The cohesion and the friction angle can be deduced from the slope and the intercept of the line (Eq. 2). The plot shows that the failure properties of the wheat can be described very well by the Mohr-Coulomb failure criterion with a friction angle ϕ=22.8° and a cohesion c=1.7kPa for the range of mean pressures tested (14 to 109kPa). The good fit suggests that the variations of the friction angle with confining pressure assuming zero cohesion (Fig. 5) can mostly be attributed to the presence of a small cohesion.

The failure data for two other test series are also shown in Figs 5 and 6 for comparison. The sample behaviour in these tests is not reported here. The friction angle is seen to be slightly larger than the value for the extension tests of isotropically consolidated samples (E test series) but slightly smaller than those derived from deviatoric compression tests (S, SU, SO test series).

5.3 Volumetric Response

The volumetric response is indicated in the specific volume-axial strain plot (Fig. 7). The volumetric strain appears to be less repeatable than the deviator stress-axial strain relation, and it fluctuates much more with axial strain, partly as a result of small errors in the tiny radial deformations measured at small axial strains.

In general, all samples showed a similar response. Up to a tensile axial strain of about 3%, the volume decreased.

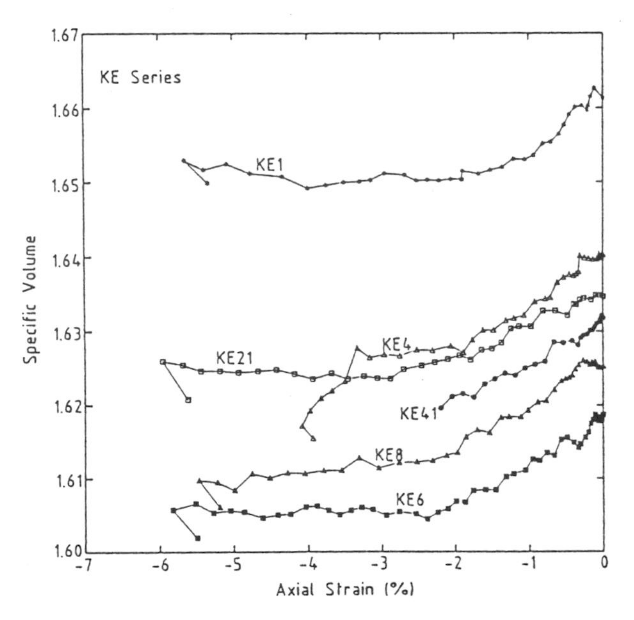

Fig. 7 Volumetric Response

Thereafter, little or no further volume change occurred. Towards the end of the test at about -6% axial strain, some samples showed a slight increase in volume. This pattern is similar to that found for isotropically consolidated samples subjected to deviatoric extension (series SE; (13)).

These measurements, showing that the volume decreases as the sample is extended, are not easy to explain. At very small axial strains, with stress states far from shearing failure, the response might have been expected to be pseudo-elastic, possibly following Hooke's law locally. This is not the case here, since the compressive volume changes suggest a Poisson's ratio greater than 0.5. Such a value is commonly regarded as unrealistic, though it is possible to develop simple inter-particle mechanism models for anisotropic solids which reproduce it. It should be noted that the grains are axisymmetrically inclined to the horizontal at a small angle and the vertical stress is decreasing. It is therefore possible that a mechanism may have developed where the grains could easily slip radially inwards. This could lead to greater compressive radial strains and a net decrease in volume. Further work is required to develop a satisfactory understanding of the behaviour observed here.

6 DISCUSSION

The tests reported above were undertaken to explore the strains developing in wheat under a stress history corresponding to that occurring in a vertical silo section. The chief findings may be described as follows.

When the material is placed in the silo, quite large reductions in volume occur at low stress levels, but the material becomes much stiffer as the stress level rises (Fig. 2). This indicates that significant consolidation, leading to slip against the wall, can be induced by quite small amounts of later filling.

If the material is poured into a conical heap and subjected to linearly increasing stresses with a lateral stress ratio of around k=0.5, very little radial straining occurs (Fig. 3). In the silo, the radial strain is effectively resisted by the wall, and one may therefore expect a lateral stress ratio of aound k=0.5 to develop in this wheat.

When the same material is subject to uniform isotropic stress, the strains in different directions are far from equal (Fig. 3). This shows that materials with elongated particles will behave very anisotropically in a silo. Numerical methods which attempt to predict silo pressures must account for this response. A distinction is needed between bulk solids which are free of special orientation features (such as cement or sand) and those which have particles which pack in a special manner (most grains, mica, plastic film, plastic pellets, some coals etc).

When discharge begins, and the material begins to move down the silo, axial extension of the solid might have been expected, with either constant or increasing radial stress (overpressures). However, most existing theories (Walker, Walters, Jenike) suggest very little change in the vertical stress. These tests show (Fig. 4) that only exceptionally small axial strains and tiny volumetric or radial strains (Fig. 7) are likely to occur unless the radial stress is changed. Thus, unless a geometric imperfection is encountered on the silo wall, overpressures are not likely to be generated in the vertical section of a silo in mass flow. The switch theory of overpressures (26,27) thus seems difficult to justify from careful material testing.

Most existing theories of solids flow and silo pressures are based on the failure properties of the solid. Some earlier tests on wheat (6) suggested that the failure properties might be very dependent on the stress history to failure, so that a compression test could not adequately represent the extension failure which might be expected in a silo. The present tests do not reproduce this finding (Fig. 6), but suggest that the failure properties of samples with the same grain orientation are similar for both tests.

7 SUMMARY

Some of the findings of an extensive experimental investigation into the mechanical behaviour of wheat at low stresses have been presented. The tests described here have investigated the stress-strain behaviour under a stress history which wheat in a mass flow silo might experience during filling and discharge of the vertical section. Several significant findings have been made in relation to the numerical modelling of silo pressures, the potential for switch pressures in parallel silos during discharge, the anisotropic character of typical solids and the stress path dependence of failure properties. These have been outlined in the discussion section.

REFERENCES

1) Jenike, A.W., Elsey, P.J. and Woolley, R.H. "Flow Properties of Bulk Solids", Proc., ASTM, 1960, 60, pp 1168-1181.

2) Ooi, J.Y. and Rotter, J.M. "Wall Pressures in Squat Steel Silos from Finite Element Analysis", *Computers and Structures*, 1990, 37, No.4, pp 361-374.

3) Hartlen, J., Karlsson, L.-A. and Norten, P. "Spannmals mekaniska egenskaper", (The Mechanical Properties of Grain) (in Swedish), 1979, Rpt No. 1, Dept of Soil Mechanics, Lund Inst. of Tech..

4) Hartlen, J. "The Strength Parameters of Grain", European Symposium on Particle Tech., June 3-5, 1980, Amsterdam, Holland.

5) Hartlen, J. and Johannesson, L.-E. "Hallfasthet hos spannmal", (The Strength of Grain) (in Swedish), 1981, Rpt No. 11, Dept of Soil Mechanics, Lund Inst. of Tech. and Swedish Geotechnical Inst., Linkoping, Sweden.

6) Hartlen, J., Nielsen, J., Ljunggren, L., Martensson, G. and Wigram, S. "The Wall Pressure in Large Grain Silos", Swedish Council for Building Research, Stockholm, 1984, Document D2:1984.

7) Smith, D.L.O. and Lohnes, R.A. "Frictional and Stress-Strain Characteristics of Selected Agricultural Grains as Indicated by Triaxial Testing", Proc., Powder and Bulk Solids Conference, May 12-14, 1981, Chicago, Illinois.

8) Smith, D.L.O. and Lohnes, R.A. "Frictional Properties and Stress-Strain Relationships for Use in the Finite Element Analysis of Grain Silos", *Jnl of Powder and Bulk Solids Tech.*, 1982, 6, No. 3, pp 4-10.

9) Smith, D.L.O. and Lohnes, R.A. "Bulk Strength and Stress-Strain Behaviour of Four Agricultural Grains", *Jnl of Powder and Bulk Solids Tech.*, 1983, 7, No. 4, pp 1-17.

10) Bokhoven, W.H. and Lohnes, R.A. "Preconsolidation Effects on Flow Characteristics of Soybean Meal", *ASAE*, Paper No. 86-4075 Summer Meeting, 29 June-2 July, 1986, California.

11) Bokhoven, W.H. and Lohnes, R.A. "Strength and Stress-Strain Characteristics of Soybean Meal as Determined by Triaxial Testing", Proc., 11th Annual Powder and Bulk Solids Conference, 1986, Rosemont, Illinois, pp 49-57.

12) Ooi, J.Y., Rotter, J.M. and Hull, T.S. "Triaxial Testing of Dry Granular Solids at Very Low Stresses", Proc., 11th Australasian Conference on the Mechanics of Structures and Materials, Auckland, New Zealand, August, 1988, pp 379-384.

13) Ooi, J.Y. "The Mechanical Behaviour of Bulk Solids and the Prediction of Wall Pressures", PhD Thesis, 1990, University of Sydney.

14) Kolymbas, D. "Stress Strain Behaviour of Granular Media", Proc., 3rd Int. Conf. on Bulk Materials Storage Handling and Transptn, I.E. Aust., Newcastle, June, 1989, pp 141-149.

15) Clower, R.E., Ross, I.J. and White, G.M. "Properties of Compressible Granular Materials as related to Forces in Bulk Storage Structures", Trans., *ASAE*, 1973, 16, No. 3, pp 478-481.

16) Thompson, S.A. and Ross, I.J. "Thermal Stresses in Steel Grain Bins using the Tangent Modulus of Grain", Paper No. 82-4069, *ASAE*, 1982, Summer Meeting, Madison, Wisconsin.

17) Bishara, A.G., Ayoub, S.F. and Mahdy, A.S. "Static Pressures in Concrete Circular Silos Storing Granular Materials", *ACI Jnl*, 1983, 80, No. 3, pp 210-216.

18) Zachary, L.W. and Lohnes, R.A. "A Confined Compression Test for Bulk Solids", Proc., 13th Powder and Bulk Solids Conf., Rosemont, Illinois, 9-12 May, 1988, pp 483-491.

19) Nielsen, J. "Load Distribution in Silos Influenced by Anisotropic Grain Behaviour", Proc., Int. Conf. on Bulk Materials Storage, Handling and Transptn, I.E.Aust., Newcastle, Aug, 1983, pp 226-30.

20) Ooi, J.Y. and Rotter, J.M. "Volumetric and Deviatoric Behaviour of Wheat under Triaxial Testing", Proc., 2nd European Symp. on the Stress and Strain Behaviour of Particulate Solids- Silo Stresses, Prague, Aug, 1990, 10 pp.

21) Lambe, T.W. and Whitman, R.V. *Soil Mechanics*, 1969, Wiley, New York.

22) Atkinson, J.H. and Bransby, P.L. *The Mechanics of Soils: An Introduction to Critical State Soil Mechanics*, 1978, McGraw-Hill, London.

23) Arnold, P.C., McLean, A.G. and Roberts, A.W. "Bulk Solids: Storage, Flow and Handling", Tunra Bulk Solids Handling Research Associates, 1980, 2nd edn, Uni. of Newcastle, Aust.

24) Gorenc, B.E., Hogan, T.J. and Rotter J.M. (eds) *Guidelines for the Assessment of Loads on Bulk Solids Containers*, I.E. Aust., 1986.

25) SAA , *Loads on Bulk Solids Containers*, Draft Aust. Santard, Santards Assoc. of Aust., Sydney, 1990.

26) Walters, J.K. "A Theoretical Analysis of Stresses in Silos with Vertical Walls", *Chemical Engg Sci.*, 1973, 28, No. 1, pp 13-21.

27) Jenike, A.W., Johanson, J.R. and Carson, J.W. "Bin Loads-Part 2: Concepts; Part 3: Mass Flow Bins", *Jnl of Engg for Industry*, ASME, 1973, 95, Series B, No. 1, pp 1-12.

C418/039

Confidence intervals in mass flow hopper design

O J SCOTT, BE, ME, MIEAust
Department of Mechanical Engineering, University of Newcastle, New South Wales, Australia

SYNOPSIS

In the determination of the bulk handling characteristics of granular products, random variations in laboratory test data have been traditionally accepted and catered for within the analysis procedures. Recent investigations by the author suggest that the procedures adopted may lead to inadequate estimations of critical hopper dimensions when the laboratory data is analysed.

This paper reviews the current procedures and show how an analysis of the distribution in the laboratory test results can be used to estimate upper and lower confidence intervals in the critical hopper geometry.

1. INTRODUCTION

The procedures for determining the geometry of bulk materials storage bins and bunkers has been widely discussed since the pioneering work of Jenike and Johansen [1-3]. While a number of alternative techniques have been investigated in the intervening years [4], shear cell tests have become widely accepted as the principal means for determining the required test data on which the designs are based. In the course of a project to investigate a quality control procedure for the handleability of coal supplied to power stations, the uncertainty of the Jenike procedure for determination of critical hopper geometry was questioned. As the Jenike procedure had become a norm for the prediction of hopper geometries any alternative abbreviated procedure needed to produce results that compared favorably to the standard shear test results. While practical experience showed the critical opening dimension determined from the standard Jenike shear test were adequate, the uncertainty in the prediction was not widely known. Normal practice includes a factor of safety of 20 - 25% in the the hopper outlet dimension and an increase of 3° in the inclination of mass flow axisymmetric hopper walls [5,6]. However for contractual purposes any quality control test procedure would need to be well defined, reliable, repeatable and not subject to unacceptable influence by random or systematic errors. Large uncertainties would either restrict the acceptance limits and exclude potentially acceptable coal or if higher acceptance limits were set, lead to flow interruptions from coal that was deemed acceptable.

Singhal and Hogg [7] had used extensive test replications to support variations in the analytical procedures used with the Jenike test and to suggest that by replication of all shear tests, the critical arching stress in a mass flow hopper could be estimated to within ±10%. This work was conducted on three samples of a laboratory prepared coal with maximum particle sizes of 1.18, 0.6 and 0.212mm at unspecified moisture contents. With normal Jenike test procedure using materials screened to -4mm and at as-supplied moisture contents up to near saturation levels it was necessary to assess the uncertainty in results from materials at their production conditions.

2. NOTATION

B = critical arching dimension
α = flow channel half angle
δ = effective angle of internal friction
ϕ_w = boundary friction angle
ϕ_i = internal angle of friction
σ = normal stress
σ_1 = major consolidation stress
σ_c = unconfined yield strength

σ_p = preshear normal stress
σ_s = normal stress at shear
τ = shear stress
τ_p = shear stress at preshear
τ_s = shear stress at failure (shear point)
τ'_s = prorated shear stress at failure
ρ_b = bulk density

3. CRITICAL ARCHING DIMENSION

To many engineers the concept of "flowability" refers to the ability of a material to arch or form ratholes within the storage system, though perhaps to these should be added the ability to adhere to chute walls. The more generally recognised of these obstructions to flow result from a high value of the cohesive strength of the bulk material. In the design of a bin, bunker or stockpile, it is important to ensure that a cohesive arch will not form, either at the hopper outlet or further up within the bin, nor that a stable rathole will develop. Generally, cohesive bulk solids gain strength with increasing consolidation pressure, with many materials tending towards a limiting value. For this reason, the critical location for an arch is normally at or near the outlet where the flow consolidating stress in the hopper is lowest and the unconfined yield strength of the material is also the lowest. Thus the relationship between the unconfined yield strength and the major consolidation stress commonly referred to as the Flow Function (FF) is an important relationship in defining the critical dimension at which the coal will arch across an outlet in a particular hopper.

In functional form, the critical arching dimension is given by

$$B = f_1 (\delta, \alpha, \phi, \sigma_1, \sigma_c, \rho) \qquad \text{...........(1)}$$

Referring to Figure 1, the critical opening dimension is obtained for the condition $\sigma_c = \sigma_1$ defined by the intersection point of the Flow Function and Flow Factor ff which represents the stress condition set up in the arch. Flow factors for various geometries are readily available from the literature [3,5,6].

Thus to define the critical opening dimension measurements need to be undertaken to define the Flow Function, the effective angle of internal friction, the boundary friction angle and the bulk density. All of these measurements will introduce some uncertainty in the prediction of the critical opening dimension.

4. FLOW FUNCTION

Of the available test methods to determine the flow function, the linear direct shear tester of the type developed by Jenike [1-3] is probably the most widely used; it allows simple though indirect measurement of the relevant bulk solid parameters, including the boundary friction. The tester is equipped with a shear cell of circular cross section. It comprises a base, which is located on the machine, a shear ring, which rests on top of the base, and a cover which has a shear force loading bracket attached to it as illustrated in Figure 2. The shear cell is mounted in the test machine and shear strain is applied by means of the loading stem, the sample of material contained in the cell being subjected to normal pressure through the applied load; the shear force (and hence stress) is measured by means of a load transducer and is recorded continuously with time. Generally, the cell used has an inside diameter of 95 mm giving the cell a cross-sectional area of 7.14×10^{-3} m^2. For high consolidation stress conditions, it is usual to use a cell of 65 mm diameter having a cross-sectional area of 3.18×10^{-3} m^2. The applied strain rate is approximately 2.5 mm/min.

A description of the procedures used is available in various publications [1-6], though in particular the report by the Working Party on the Mechanics of the European Federation of Chemical Engineering [8] gives a definitive and detailed procedure. As indicated in Figure 2, the shear ring is initially offset, the objective being that during the shear consolidation phase of the test, consolidation to the critical density or voidage state should be completed by the time the shear ring has moved to a position of concentricity with the base. In view of the limited travel for this condition to be reached, it is necessary to commence each test with a pre-consolidation phase. The pre-consolidation phase involves the overfilling of the assembly, the application of a prescribed consolidation force and at the same time, a twisting top is given a pre-determined number of oscillatory twists through an amplitude of approximately ±45° [8]. On completion of pre-consolidation, the mould ring is removed and the sample screeded level with the top of the shear ring to remove excess material. The cell is then set-up as indicated in Figure 2 for the consolidation procedure to achieve the optimum voidage in the shear zone. This is then followed by the shear test at a reduced normal load. A large number of replicate samples are produced at each consolidation state and sheared under a variety of shear normal loads to enable a family of yield loci to be produced. One disadvantage of the shear test is that the unconfined yield strength at a specific principal consolidation stress is not directly measured but has to be determined by applying Mohr's theory to the yield loci. To produce a relationship between the bulk strength over the range of consolidation states that may apply within the handling chain requires a large number of samples to be tested. Reference 7 recommends that between six and ten individual shear tests be conducted at each of at least four consolidations stress states. This requires between 24 and 40 individual samples being prepared and tested to determine the Flow Function.

5. TEST PROCEDURE

Over the past fifteen years a very large number of Jenike tests have been performed in the Bulk Materials Handling Laboratories at the University of Newcastle on over 500 different bulk materials at various moisture contents. For design purposes the repeatability and reliability of the test results has been assessed intuitively by the laboratory technician and a reasonable factor of safety allowed in the design. Materials that showed little deviation from straight line yield loci and whose effective angle of internal friction and kinematic angle of internal friction displayed acceptable trends had limited repetitive tests conducted. Materials which showed considerable scatter in their results, particularly those at high levels of moisture or with significant proportions of cohesive components such as clay, would be subject to considerable repetitive testing to establish acceptable results. In all cases however the end result of testing would be a single determination of the Flow Function and other parameters for each material condition. This would lead to a single value for the minimum arching dimension, half hopper angle or critical piping dimension from which a recommendation on acceptable bin geometry could be made.

However as suggested, for contractual purposes any test procedure needs to be well defined, reliable, repeatable and not subject to unacceptable influence by random or systematic errors. The final product of a quality control test procedure needs to have statistically supportable uncertainties available as well as a mean value. It could also be argued that for design purposes knowledge of the uncertainty in the hopper dimensions could result in more acceptable designs with the risks involved in using undersized dimensions more adequately recognised.

In the early 1980's a Working Party on the Mechanics of Particulate Solids of the European Federation of Chemical Engineers was established to assess the reproducibility and reliability of measurements of the flow properties of particulate solids. The work of this committee resulted in the release of Reference 7. In the initial stages of their study, samples of a well defined powder were supplied to various laboratories throughout the world for comparative tests. These showed a wide disparity in results and supported the need for a definitive procedure for the Jenike test. While recognising the desirability of using a precisely defined and controlled monosized fine cohesive powder for comparative assessment, the variation in results if a widely sized moist bulk material had been used is not known.

One of the difficulties of the Jenike procedure is in obtaining replicate test samples for a single value of the preshear normal stress (σ_p) which have been conditioned to a consistent repeatable critical density or voidage state. For the procedure outlined in [8] a minimum of between six and ten replicate samples are required for each determination of a yield locus. Standard practice is to accept that a shear stress at preshear (σ_s) which has reached a steady state when the split cell is approximately concentric indicates that a sample is neither under-consolidated nor over-consolidated as shown in Figure 3. This condition is reached by judicious selection of the preconsolidation load and the number of twists used during the preconsolidation phase. Reference [8] requires that for correctly consolidated samples the steady state preshear shear stress does not deviate by more than ±5% from the average steady state value but notes that for some particulate solids this tolerance may not be achievable. Practical experience suggests that materials with a large size distribution subject to moisture variation may show a much wider spread in the steady state shear stress.

Ideally all samples would be conditioned by the application of the preconsolidation load and the number of preconditioning twists in such a way as to ensure that the shear consolidation force was identical. However because of the recognition that experimental variations occur and must be accepted within a range of ±5% or greater of the average for many materials, an empirical procedure referred to as prorating adjusts the shear stress at failure τ_S according to the relativity of the individual consolidation shear to the average preshear shear stress $\bar{\tau}_p$ according to the relationship

$$\tau_s' = \frac{\tau_s \times \bar{\tau}_p}{\tau_p}$$

In the analysis it is assumed that for replicate tests undertaken on a single sample consolidated to a consistent voidage, the preshear shear stress and the shear stress during failure have a Normal or Gaussian distribution about their mean. This assumption, shown in Figure 4 and verified by Singhall and Hogg [7], allows upper and lower limits for a given confidence to be established for both the singular value of the preshear stress and a linear interpolation of the yield loci.

The uncertainty in the preshear shear stress and its impact on the prorating of shear stress values must be accommodated in the analysis of the raw data. If a 90% confidence interval is assumed acceptable for engineering design purposes, then the individual raw shear stress values should be prorated not only to the mean, but also to the upper and lower 90% confidence limits. These prorated shear stress data can be analysed to predict upper, mean and lower yield loci at each level of consolidation normal stress. As each failure shear stress value in itself also has an uncertainty in its measurement independent of the initial preshear shear stress then each yield locus must be recognised to have an individual uncertainty. If a Mohr-Coulomb behaviour is assumed in the bulk material then the yield locus is of a linear form. Standard linear regression techniques can be adopted to determine an upper and lower bound confidence interval on each prorated yield loci as shown in Figure 5. By using the lower bound yield loci and the lower bound preshear shear and normal stress combinations, a lower bound unconfined yield strength / major consolidation pressure data set can be obtained for each yield locus to plot a lower bound Flow Function. Similarly the use of the mean and upper bound data set can be used to compute a mean and upper bound Flow Function. The best fit for the flow function and a limiting envelope within which the flow function can be assumed to lie with 90% confidence i.e. a measure of the reliability of the Jenike test procedure, is shown in Figure 6.

The techniques also allows for an upper and lower bound to be determined for the effective angle of internal friction as shown in Figures 5 and 6, and to determine a confidence in the values for the angle of internal friction.

6. TEST RESULTS

Using the procedures outlined an analysis has been carried out on a number of samples of bulk materials. By way of example Figures 7, 8 and 9 show the results for two coal samples at their as supplied or used moisture contents.

Figure 7 shows the results for a washed clean coal (source AA) with a moisture content of 12.6% which is at or near its maximum moisture content. This material was regarded as well behaved by the laboratory technician and would normally need little replication of tests. In its initial form the eight yield loci data points fit acceptably to a straight line yield locus and would in normal circumstances be regarded as most reliable. Yet on a confidence interval analysis the data lead to a significant variation in the computed Flow Function as is shown in Figure 7.

Figures 8 and 9 show the results for a ROM coal (source BB) at two moistures of 5% and 17%. The lower moisture content represents the air dried state for this coal which is near the as supplied moisture while the higher moisture content is equivalent to near the saturation limit for this coal corresponding to when the coal is stockpiled for long periods. This coal has a high proportion of ash in the form of clay. At the lower moisture content the scatter in the boundary conditions is similar to those for the washed clean coal of Figure 7, however at the higher moisture content the degree of uncertainty in the calculated parameters is considerably higher. This indicates that as the moisture content and thus the difficulty in handling the coal increases so does the magnitude of the confidence intervals

Other bulk materials such as copper ores, bauxite etc subject to the influence of moisture when similarly tested have shown data that was comparable to that given in Figures 7 to 9.

7. CRITICAL ARCHING DIMENSIONS

One of the major uses of the Flow Function (FF) is to determine a critical outlet dimension to ensure that a bulk material will not arch at the outlet of a hopper. While the Flow Function is the major characteristic used in this calculation another significant variable is the bulk density (ρ). The wall friction angle (ϕ_w) and the effective angle of internal friction (δ) through their influence on the stresses at the outlet have a secondary influence on the calculation of the critical arching dimension. To isolate the influence of the confidence intervals of

Table 1. Variation in Critical Arching Diameters and Effective Angle of Internal Friction

	Moisture %	B_c - metres Upper	Mean	Lower	δ - degrees Upper	Mean	Lower
Washed Clean Coal AA	12.6	0.90	0.45	0.15	60	54	48
ROM Coal Source BB	5.0	0.20	0.0	0.0	42	38	35
ROM Coal Source BB	17.0	1.30	0.75	0.0	75	64	54

the cohesive strength measurements on determining the arching dimension, calculations have been made for the two coals detailed here assuming a conical half hopper angle of 20° and a constant bulk density of 800kg/m^3. The results of these calculations using the mean, upper and lower bounds of the FF and δ are shown in Table 1.

Thus while the mean Flow Function for say the Washed Clean Coal suggests a critical arching diameter of 0.45 m within a conical hopper of 70° inclination, using the upper and lower 90% confidence interval on the Flow Function suggests that the arching diameter lies somewhere between 0.15 and 0.90 metres. Traditionally, a tolerance of plus 20 to 25% was added to the computed critical arching diameter when recommending a suitable hopper outlet diameter. Analysis would suggest that confidence intervals in excess of this could commonly occur. It is also important to note that the results suggest that some materials may have a significant chance that they will flow through substantially smaller outlets than those predicted if mechanical arching did not occur to obstruct the flow.

Table 1 also shows the variation in the effective angle of internal friction (δ) that can be computed from the yield loci data. While this has no major effect on the computed arching dimensions it is significant when using the bulk material properties to predict the loads that the material imposes on the walls of the bin or bunker. For the ROM Coal from source BB where the moisture varied between 5 and 17% the effective angle of internal friction could lie, with a confidence of 90%, anywhere between 35 and 75°.

8. BULK DENSITY TESTS

The other measured parameter that can have a substantial influence on the calculation of a critical arching dimension is the bulk density. While it has often been standard practice within the industry to define a single value for the bulk density of a bulk material the variation in bulk density with consolidation stress has long been used when calculating the critical arching dimension.

Analysis of the influence in variations in this parameter has not been finalised at the date of preparation of this paper however two important aspects must be noted. Figure 10 shows the variation in measured bulk density for three replications of the ROM coal source BB at 17% moisture content where it can be seen that substantial variations can occur in the measured result. There will be a noticeable effect of using an upper and lower bound value for the bulk density in the prediction of a outlet dimension as it is directly proportional to the inverse of the bulk density. The second feature is that there are often considerable changes in bulk density as the moisture content changes. Figure 11 shows the results for a ROM steaming coal from a third source which displays a significant decrease in the bulk density with increasing moisture followed by a substantial increase at higher moisture contents.

9. WALL FRICTION TESTS

The third major test data used in the prediction of hopper geometries is the wall or boundary friction characteristics of bulk solids. Reference [8] details the procedures for this test. While statistical analysis is incomplete it is readily apparent that significant variations do occur in its measurement. It is common while conducting wall friction tests for the shear stress to fluctuate widely under the application of a constant normal stress. Variations also exist between static and kinematic wall friction. As an example Figure 12 shows the variation that occurs when testing the friction characteristics of the ROM coal (BB) at 17% moisture on a rusted 250 Grade Structural Steel surface. Both the maximum stress and the minimum stress at each level of normal pressure are shown for each of three replications of the test. An analysis of this data would lead to significant variations in the wall friction angle at a 90% confidence interval.

Figure 13 shows the variation that exists between the mean values of the wall yield loci for the ROM Coal (BB) at the 5% and 17% moisture contents. These differences are substantial and if confidence intervals were added the variation between a lower bound value at the low moisture and an upper bound value at the high moisture could be quite substantial.

The results as shown in Figures 12 and 13 suggest that under some circumstances the wall friction angle may differ significantly from that predicted from a single test of the wall yield locus. This has particular importance in the prediction of the loads exerted by a bulk solid on the walls of a bin or silo. Hoppers that have been designed for funnel flow may through a variation in material moisture content switch into mass flow with a consequent change in the load distribution and magnitude. It is important that the limits on these variations be known in advance.

10. CONCLUSION

The major result of the study to date has been to develop a procedure which can be used to define confidence intervals in the measured flow functions for the bulk strength of granular materials. These confidence intervals may be significantly greater than equivalent tolerances used in the past to recommend hopper opening dimensions. By using the procedures outlined, a prediction of the critical hopper geometry can be calculated with tolerances at various levels of confidence. These can then form the basis of an engineering decision on the final design outlet dimension, hopper inclination and design loads.

11. REFERENCES.

1. Jenike, A.W., "Gravity Flow of Bulk Solids", Bul. 108, Utah Engng. Exper. Station, University of Utah, 1961.
2. Johanson, J.R., "Stress and Velocity Fields in Gravity Flow of Bulk Solids", Bul. 116, Utah Engng. Exper. Station, University of Utah, 1962.
3. Jenike, A.W., "Storage and Flow of Solids", Bul. 123, Utah Engng. Expt. Station, University of Utah, November 1964.
4. Bradfield, B. & Lohnes, R.A. "Strength Testing of Powders and Bulk Solids: An Overview", Proc. Powder and Bulk Solids Conf., Chicago, May 15-18, 1989
5. Arnold, P.C., McLean, A. & Roberts, A.W. "Bulk Solids: Storage, Flow and Handling". TUNRA Publication, 1979, 1980, 1982.
6. Roberts, A.W. "Modern Concepts in the Design and Engineering of Bulk Solids Handling Systems". TUNRA Publication, 1988.
7. Singhal, I.K. and Hogg, R. "Evaluation of Shear Cell Test Procedures for the Characterisation of Powder Flow", 2nd Int. Conf on Bulk Materials Storage handling and Transportation, Wollongong, Australia, July 86.
8. "Standard Shear Testing Techniques for Particulate Solids using the Jenike Shear Cell", ISBN 0 85295 232 5. Published by The Institution of Chemical Engineers, European Federation of Chemical Engineers, 1989.

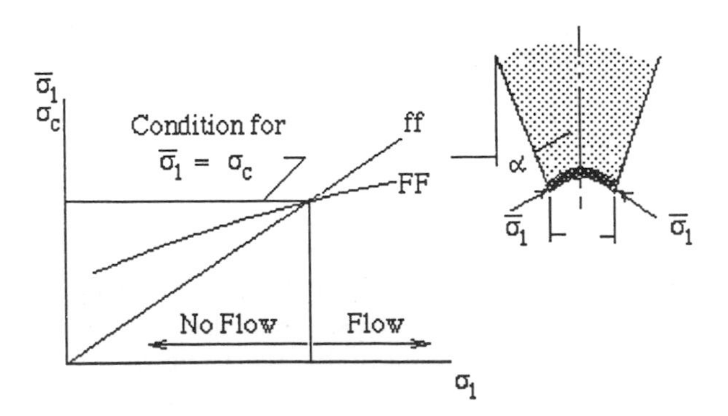

Figure 1. Flow/No Flow Condition for Mass-Flow Design

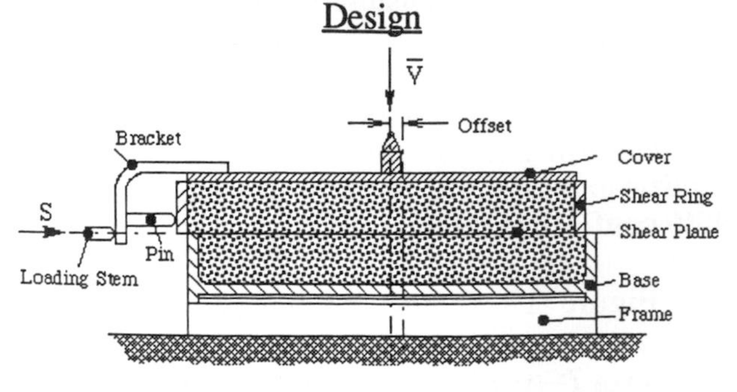

Figure 2 Jenike Type Shear Cell

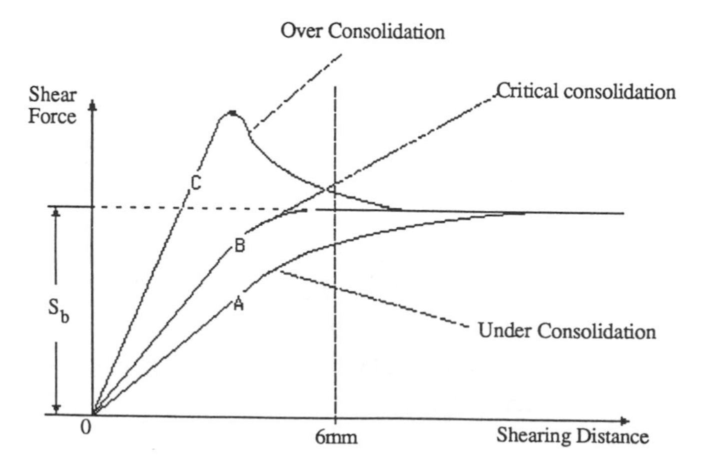

Figure 3. Shear Force at Consolidation

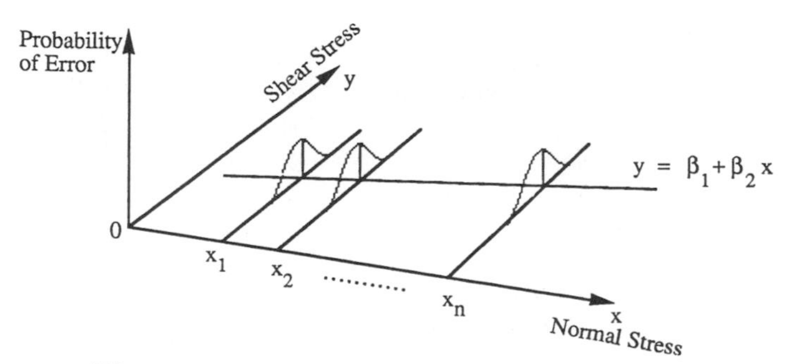

Figure 4 . Gaussian Distribution about Mean

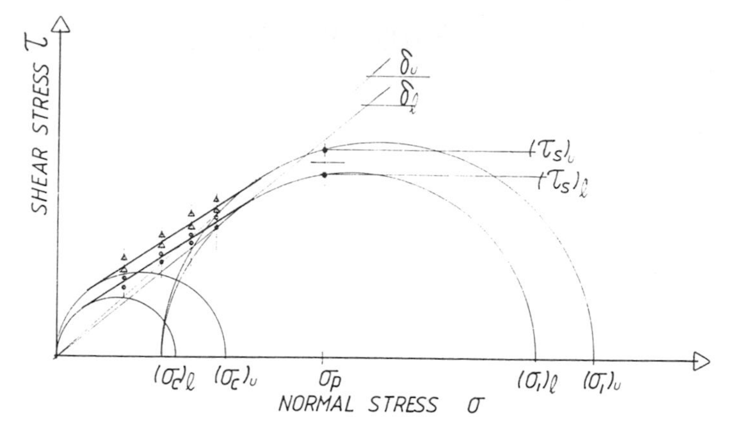

Figure 5. Limit Yield Loci

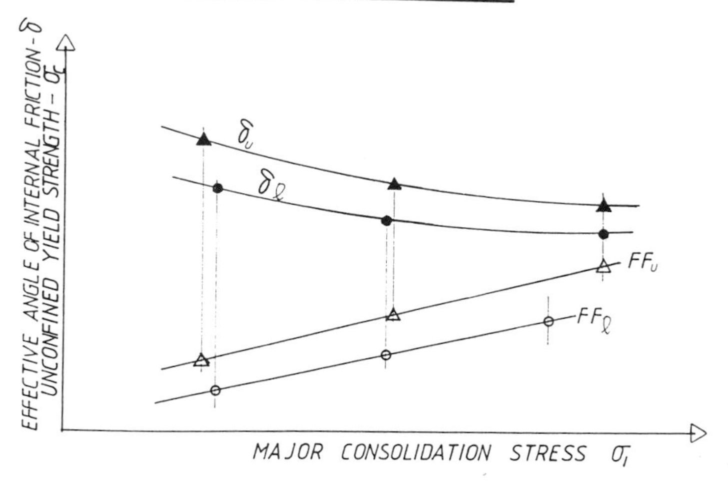

Figure 6. Upper and Lower Limit Flow Function and Effective Angle of Internal Friction

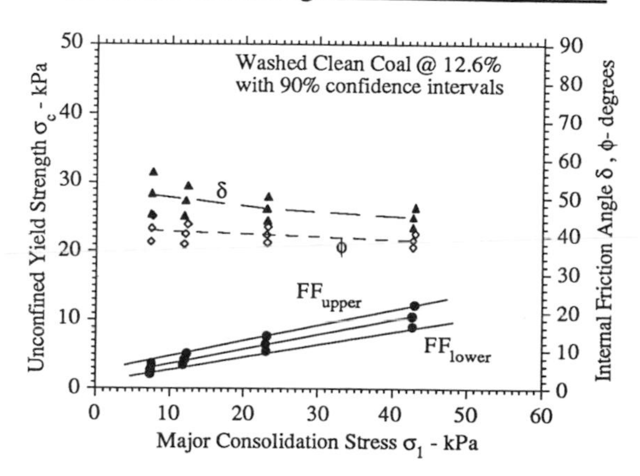

Figure 7 Jenike Test Results for a Washed Coal at 12.6% moisture

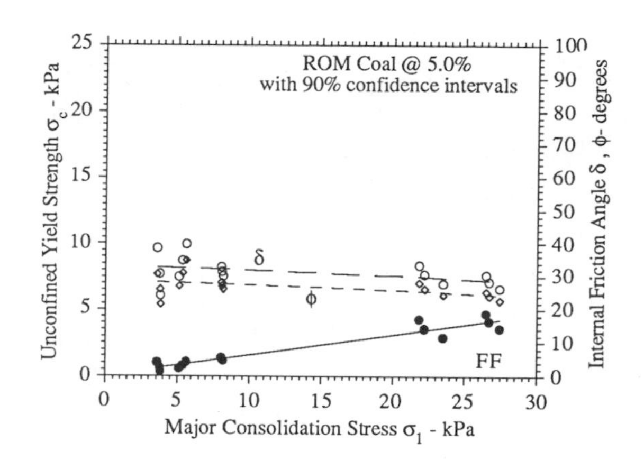

Figure 8 Jenike Test results for ROM Coal at 5.0% moisture

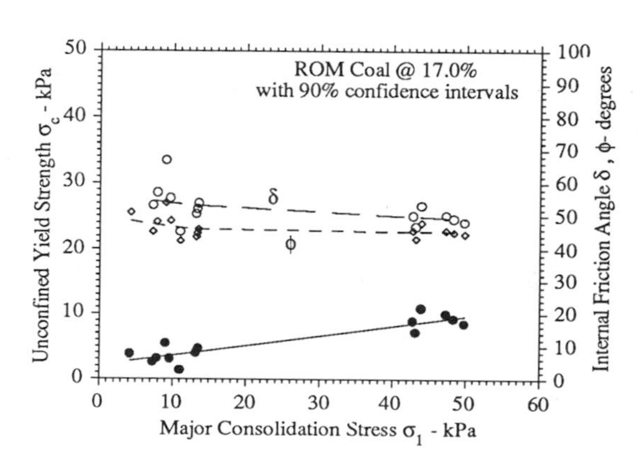

Figure 9 Jenike Test results for ROM Coal at 17.0% moisture

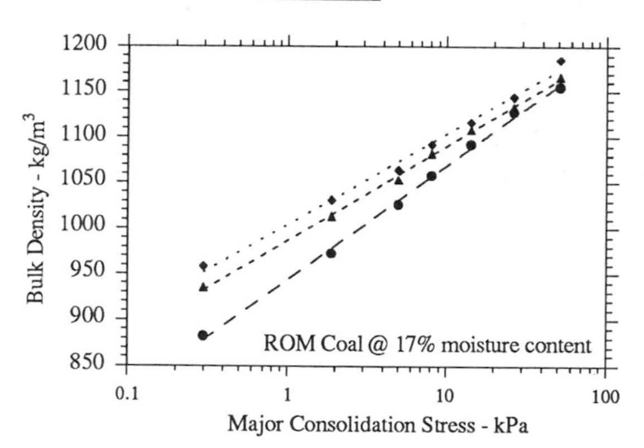

Figure 10. Variation in Bulk Density with Three Replications

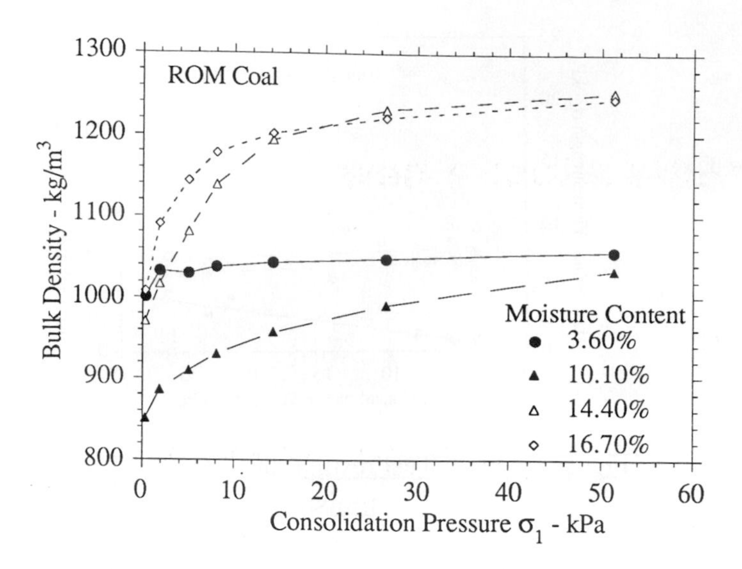

Figure 11. Variation in Bulk Density with Conolidation Pressure and Moisture Content

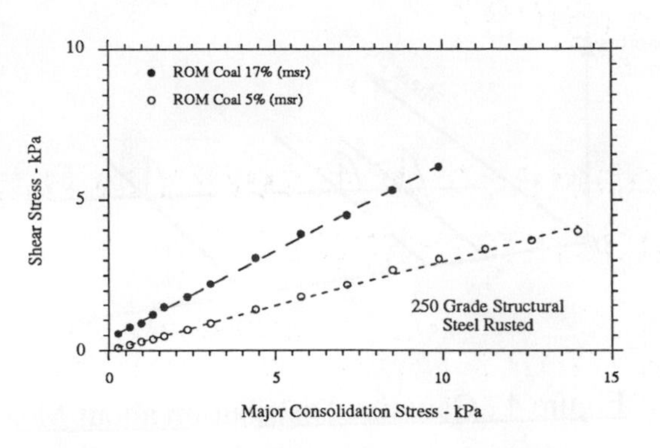

Figure 13. Variation in Wall Yield Loci with Moisture Content

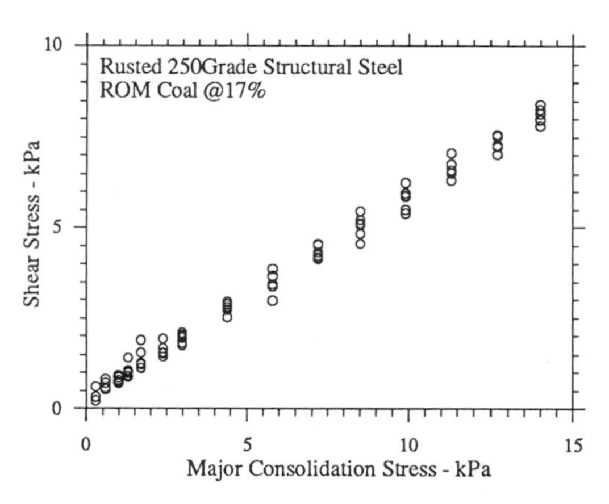

Figure 12. Wall Yield Loci Variation for Mild Steel Rusted

C418/037

Measured pressures in full-scale silos: a new understanding

J Y OOI, BE, PhD and J M ROTTER, MA, PhD, FIEAust
Department of Civil Engineering, University of Edinburgh, Scotland

SYNOPSIS Many experiments have been conducted over the last two decades to study the pressures which occur in silos when they are filled and discharged. Unfortunately, many of the existing tests on silos were conducted with an inadequate appreciation of what was needed, especially in terms of the pattern of pressure developing on the silo wall. This paper describes an analysis of test results from a full scale working silo in Sweden, paying special attention to the patterns of wall pressure. Several different kinds of experiment were conducted, and these show that very unsymmetrical but repeatable pressure regimes exist in the silo, even when it is most carefully symmetrically filled and discharged. The difference between filling and discharge is seen to be less significant than the circumferential variation. The results have far-reaching implications for the design of silo structures and for standards which specify silo design loads.

1 INTRODUCTION

In the last twenty years, many experiments have been conducted to explore the pressures exerted by granular solids on the walls of silos. Unfortunately, these experiments are very difficult and very expensive, so a limited number of observations have almost always been made. In addition, unsatisfactory measuring instruments were sometimes used, which led to incomplete or unreliable test results.

It has been widely assumed that the pressure at a particular level in the silo is the same at all points around the circumference. It was generally assumed that the highest pressure was the most damaging to the structure, and it was not recognised that a pressure which varied between 20kPa and 40kPa around the circumference might be much more dangerous to the structure than a uniform pressure of 60kPa. The importance of circumferentially varying pressures was not recognised until quite recently (e.g. 1,2), so it is natural that silo pressure experiments were designed without special attention to this feature.

This paper describes an analysis of a series of very carefully conducted experiments on a full scale silo in service in Sweden. From these experiments, estimates of the pressure variations around the silo circumference can be made. The analysis of the pressures shows that the same unsymmetrical pattern apppeared repeatedly in many experiments, irrespective of some changes in the filling and discharge arrangements. This paper attempts to describe some of these repeated patterns, and to indicate their significance for silo design.

If the variations around the circumference are ignored, as in most previous tests, the mean pressure pattern is found to be quite close to the Janssen distribution. This is found to be true both for initial filling and during discharge. The result shows that previous experiments have missed a vital characteristic feature of silo pressures by ignoring these circumferential variations. The difference between filling and discharge is seen to be less significant than the circumferential variation.

The silo was more fully instrumented than those used in most other tests, and pressure cells which had been developed with much special knowledge of the measuring difficulties were employed. Tests were conducted using wheat and barley, but only the barley tests are discussed here due to space restrictions. Here, the wall pressures from tests with different methods of filling and discharge are analysed statistically. The results have far-reaching implications for the design of silo structures and for standards for silo design loads.

2 EXPERIMENTS

The experiments were conducted in Sweden on a prototype grain silo in service at Karpalund. Summaries of the many experiments are reported elsewhere (3-11). The experiments led to many new but rather complex discoveries (7,8,11-13).

The silo was a reinforced concrete structure, 7m in diameter and 46m high, the corner cell of a block of six (Fig. 1). Pressure cells were placed at seven levels on four equally spaced vertical generators, with an additional eight cells at the level 17.5m above the bottom, where the eccentric discharge arrangement was expected to lead to high pressures (Fig. 1). The cells were mounted with special knowledge and particular care learned in previous studies (14).

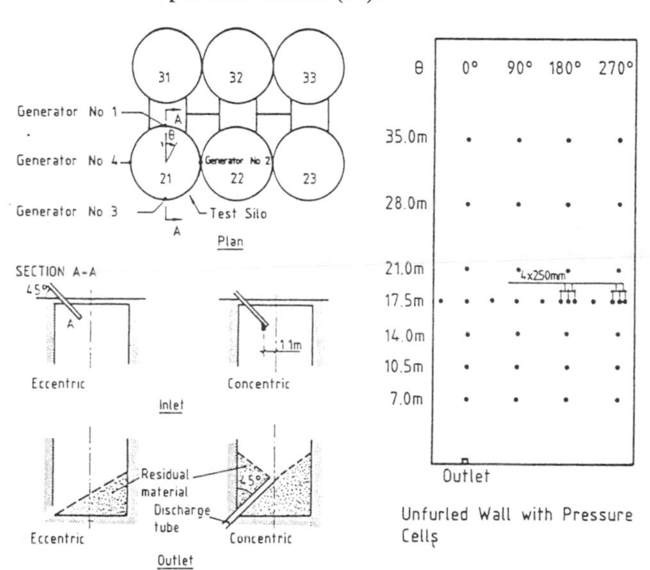

Eccentricities of filling and discharge were investigated by using different inlet and outlet arrangements (Fig. 1). Three different methods of filling were used. In the 'eccentric' filling arrangement, the inlet chute was inclined at 45°. The grain thus entered the silo with a significant horizontal velocity away from Generator 1, and impacted on the far wall (Generator 3) at about 22m above the base. The 'concentric' filling arrangement had a vertical extension tube added to the inclined chute, resulting in a 1.1m eccentricity from the central axis (Fig. 1). The falling grain trajectory was slightly inclined towards Generator 1. The 'distributed' filling was achieved by 'raining' the grains down from a special rotating device (15).

Two different outlet arrangements were used. The 'eccentric' discharge arrangement had the outlet at the base of the wall, giving an eccentricity equal to the full radius (Fig. 1). In the 'concentric' discharge arrangement, a tube was used to remove grain from the central axis of the silo (Fig. 1). The methods of filling and discharge for each test analysed here are shown in Table 1.

The measured pressures were recorded as a function of time. Filling continued over two days, followed by a storing period. Discharge was carried out over a further two days. The records studied here concern the end of the filling process together with the storing period, termed 'storing' and the first two hours of discharge, termed 'discharge'. These were the periods when the pressures were generally highest and most variable.

The internal surface of the silo wall was accurately surveyed (8,16) and revealed extensive bulges of up to 30 mm. As this is almost 1% of the silo radius, the wall imperfections may have affected the silo pressures significantly.

3 REGRESSION ANALYSIS

The readings of a given pressure cell were taken as sample observations of a stochastic population. The readings of the four equally spaced pressure cells at a given level were subjected to a harmonic regression analysis, giving the mean pressure over the period, three systematically unsymmetrical harmonic components, and the standard deviation of the remainder, which constitutes the randomly varying component.

The circumferential coordinate origin was taken as Generator 1 (Fig. 1) and the pressure assessed as

$$p = p_o + p_1\cos\theta + p_2\cos2\theta + p_1^*\sin\theta + p_r \qquad (1)$$

in which θ is the circumferential coordinate and p_r is the component of the observed cell pressure which varies randomly with time.

The term p_o is the mean pressure at the level, which would normally be expected to correspond to the Janssen (17) pressure. The significance of the different terms of this analysis have been described elsewhere (13). The silo geometry, the filling and the discharge arrangements are all situated symmetrically with respect to the plane through Generators 1 and 3. As a result, it is natural to expect that the pressures measured on Generators 2 and 4 would be the same, except for their random stochastic variations p_r.

Unfortunately, there were only four cells at all levels but one in the silo. As a result, the pressure variation within each quadrant of the circumference cannot be determined. By choosing the lowest harmonic terms in the regression analysis, as in Eq. 1, rapid variations in pressure around the circumference are ignored. This low harmonic assumption has

Table 1 Test Specifications

Series	Test	Date	Material	Method of Filling	Method of Discharge	Number of Readings		Mean COV (%)	
						Filling	Discharge	Filling	Discharge
BEE	BEE1	07.05.79	Barley	Eccentric	Eccentric	.8	86	4.2	25.4
	BEE2	02.05.79	Barley	Eccentric	Eccentric	20	38	4.3	25.2
	BEE3	13.05.79	Barley	Eccentric	Eccentric	4	83	9.6	18.8
	BEE4	23.04.79	Barley	Eccentric	Eccentric	10	24	2.9	23.6
	BEE5	28.05.79	Barley	Eccentric	Eccentric	10	119	3.6	24.9
	BEE6	28.04.80	Barley	Eccentric	Eccentric	18	119	4.9	18.0
BEC	BEC1	18.06.79	Barley	Eccentric	Concentric	24	121	4.8	24.9
	BEC2	12.06.79	Barley	Eccentric	Concentric	23	113	6.0	23.6
BDE	BDE1	23.03.82	Barley	Distributed	Eccentric	3	115	1.9	7.9
	BDE2	30.03.82	Barley	Distributed	Eccentric	17	121	0.1	10.6
BCE	BCE1	30.03.80	Barley	Concentric	Eccentric	9	121	1.5	10.1
	BCE2	07.04.80	Barley	Concentric	Eccentric	15	107	2.5	11.4
	BCE3	14.04.80	Barley	Concentric	Eccentric	15	120	0.4	9.7
	BCE4	13.04.82	Barley	Concentric	Eccentric	4	125	0.3	12.6

been shown to be unjustified (1,18). However, it is not possible to deduce more components of the circumferential pressure variation until experiments are conducted with many more installed pressure cells around the circumference.

4 PRESSURES DURING STORING

The pressure readings during storing were subjected to the harmonic statistical analysis described above. Each harmonic term of each test series is found to follow a repeated pattern. These harmonic terms are not be presented here because of space limitation, but many results may be found elsewhere (13,18). The systematic pattern of pressure down each of the four generators was reassembled using Eq. 1. The pressures after filling for the tests with eccentric filling are shown in Fig. 2 and those for concentric and distributed filling are shown in Fig. 3.

The natural expectation is that, after filling, the pattern down each generator (a) should be the same and (b) should be reasonably close to Janssen's theory (1895). Neither of these expectations is borne out by the observed pressures. The pressures shown in Figs 2 and 3 are not measured values at any instant, but the mean value at each point over an extended period. Each plotted point in the Figs 2 and 3 typically represents the results of 13 independent observations, so departures from an expected pattern cannot be simply ignored as a scatter in the readings. The difference between the instantaneously measured pressures and the systematic pattern is indicated by the coefficient of variation from the statistical analysis (Table 1). The mean coefficient of variation is of the order of 3 to 5%. This low value indicates that the systematic pressure patterns describe almost all the readings very well.

Even though there is an apparent plane of symmetry through the silo inlet and outlet, the pressure profiles down Generators 2 and 4 are quite different (Figs 2 and 3). The pressures differ from Janssen pressures quite significantly (the grain surface is at 43m). In particular, high pressures were consistently recorded at the 21m level on Generator 2 and low pressures at the 17.5m level on Generator 4. This pattern is found irrespective of the method of filling. The pattern is thus influenced by some existing physical features of the prototype silo. Since the bulk solid compresses under its own weight during the filling process, it is subjected to the changing shape and size of the silo cross-section. It therefore seems likely that unsymmetrical imperfections in the silo wall may strongly influence the pressures.

It may be noted that distributed filling (tests BDE, Fig. 3) almost always gave smaller pressures on all four generators than

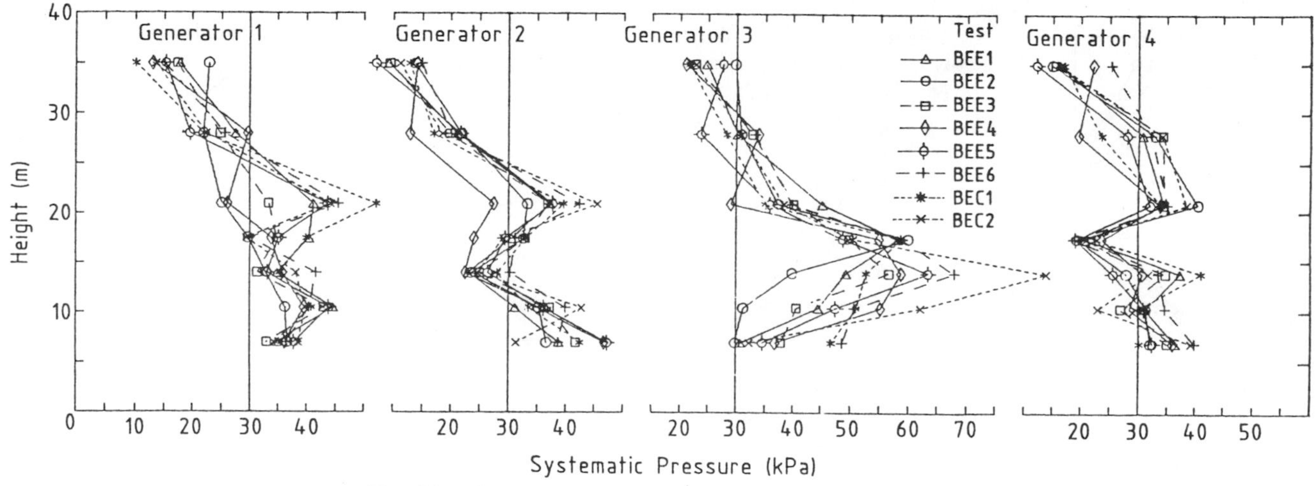

Fig. 2 Profiles of Systematic Pressure After Eccentric Filling

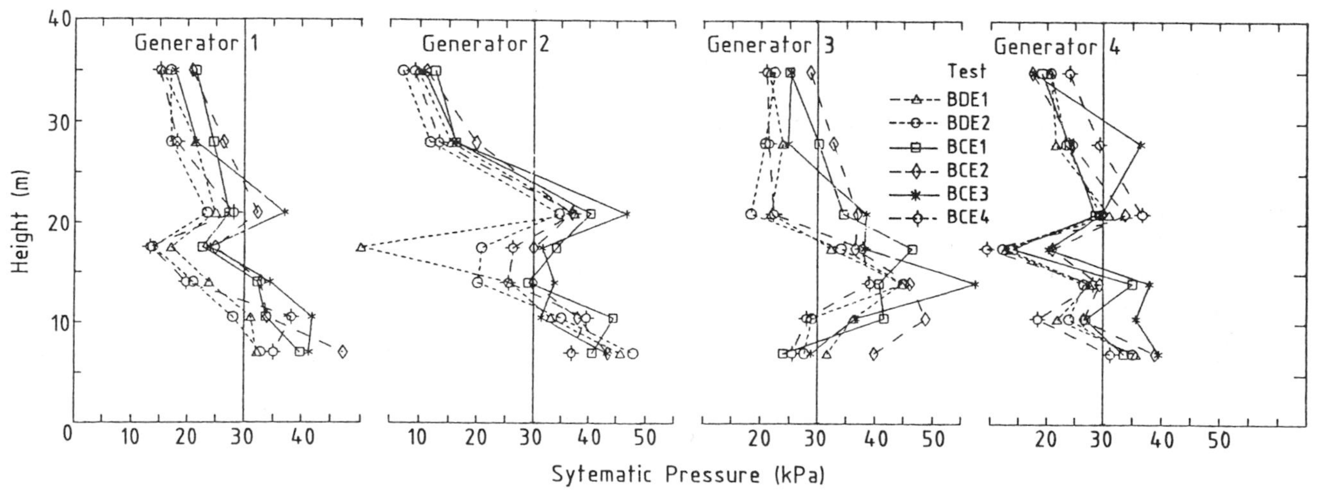

Fig. 3 Profiles of Systematic Pressure After Distributed and Concentric Filling

eccentric and concentric filling (tests BEE, BEC, Fig. 2 and tests BCE, Fig.3), though the pattern did not change. This is contrary to what one might expect, since distributed filling led to the highest observed bulk weight density (γ=8.3 kN/m^3, compared with 7.9 and 8.0 kN/m^3 for eccentric and concentric filling respectively). Since the mechanical behaviour of granular solids is strongly influenced by the orientation of the grains, it seems likely that distributed filling, producing a horizontal surface, caused a horizontal orientation in the grains as they landed, leading to a smaller lateral pressure ratio k.

Since grains were discharged into and out of the silo in the plane through Generators 1 and 3, the pressures on Generators 1 and 3 may be affected most by the eccentricities of filling and discharge. Down Generators 1 and 3, concentric and distributed filling gave quite similar systematic patterns of pressure (Fig. 3). Down Generator 1, the repeated pressure profile has low pressures at the 17.5m and 14m levels, whereas larger pressures are seen on Generator 3 at these levels.

The natural supposition of most researchers is that the unsymmetrical pattern of filling pressures must have arisen from the eccentricity of filling (Fig. 1). Thus, an experiment involving distributed filling should have been substantially symmetrical. However, this was not borne out by the experiments (tests BDE in Fig. 3), so the highly unsymmetrical patterns must be caused by existing features of the structure such as imperfections in the wall or grain-structure interaction associated with adjacent silo cells.

The pressure profiles down Generators 1 and 3 for eccentric filling are quite different from those resulting from the other two filling methods (contrast Figs 2 and 3). The difference appears to be caused only by the eccentricity of filling. A detailed examination of the differences suggests that eccentric filling affects only the pressures below the wall impact level (at about 22 m), resulting in increased pressures there on Generators 1 and 3.

5 PRESSURES DURING DISCHARGE

The systematic pressure profiles for the experiments in which eccentric or concentric discharge was used are shown in Figs 4 and 5. Since discharge began after filling was completed, the pressures during discharge were affected by both the filling and the discharge processes.

5.1 After Eccentric Filling

The pressures during concentric and eccentric discharge (after eccentric filling) are shown in Fig. 4. For both methods of discharge, the resulting pressure profiles down all generators are not distinguishable within the scatter of the experimental data except near the base of Generator 1 where the eccentric outlet was located. The greatly reduced pressures there for the eccentrically discharged tests seem naturally attributable to the flow zone near the outlet. The similarities of the systematic pressure patterns elsewhere indicate that the flow pattern is probably similar for both cases. This is consistent with the semi-mass flow pattern observed to have developed in both cases (10).

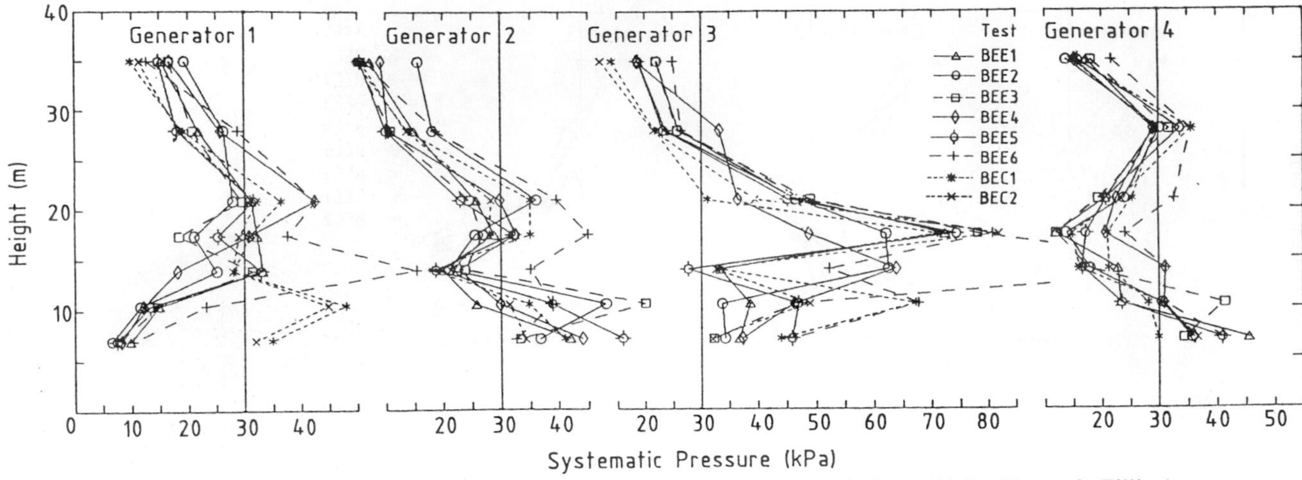

Fig. 4 Profiles of Systematic Pressure During Eccentric and Concentric Discharge (After Eccentric Filling)

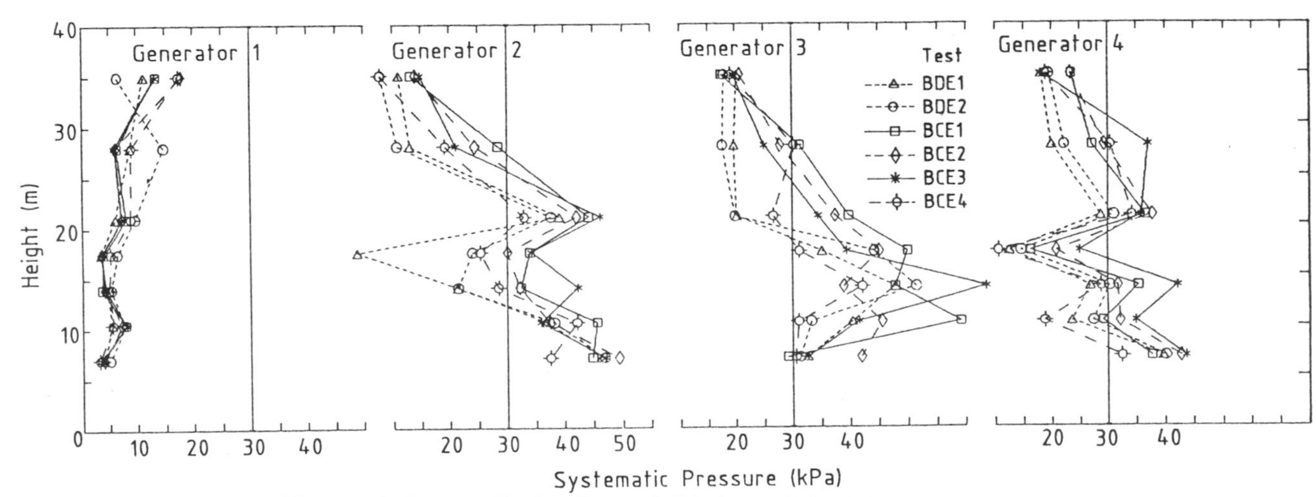

Fig. 5 Profiles of Systematic Pressure During Eccentric Discharge (After Distributed and Concentric Filling)

The systematic pressures under eccentric and concentric discharge (Fig. 4) can be compared with the initial pressures after eccentric filling (Fig. 2). It may be noted that the pattern during discharge generally follows the storing pressure pattern quite closely, though the pressures have reduced near the outlet on Generator 1 and increased at the 17.5m level on Generator 3. These changes are directly attributable to the flow pattern. A high variability is seen in the pressure at the 14m level. One test (BEE3) gave a very high systematic pressure of 125kPa at the 14m level. This high pressure point was sustained for a long period of time so it cannot be ignored. It may also be noted that a large peak systematic pressure of 84kPa was seen at the same level in test BEC2 during storing (Fig. 2). It is not clear whether these are in some way related.

5.2 After distributed and concentric filling

The systematic pattern for eccentrically discharged tests (after distributed and concentric filling) are shown in Fig. 5. The pressure patterns on all generators except Generator 1 closely follow the patterns developed during storing. However, down Generator 1, a completely new pattern was established as the systematic pressures reduced drastically to about 10kPa at the top and about 5kPa at the bottom.

This greatly reduced pressure field down the whole of Generator 1 indicates that a flow channel probably formed adjacent to the wall above the outlet. Since the pressures were substantially unchanged down the other generators, the flow channel was probably confined to a width much smaller than the diameter of the silo. This is again consistent with the internal flow pattern observed to have developed in these tests (10).

6 PRESSURE DISTRIBUTION

An approximation to the systematic unsymmetrical pattern of pressures on the silo wall may be obtained by taking the mean value of each harmonic coefficient at each level (for each test series), and recombining these harmonic terms to produce the mean systematic repeated unsymmetrical pressure distribution. When this is done using Eq. 1, the pattern of mean pressure on the unfurled wall may be illustrated as shown in Fig. 6. For space reasons, only the mean distributions for eccentric filling (Fig. 6a) and eccentric discharge following eccentric filling (Fig. 6b) are shown here (tests BEE: see Table 1).

The results are less easy to comprehend than might be supposed, because they represent a two-dimensional surface of pressures on a three-dimensional silo wall. However, they show the complete picture of how the pressure varied on the whole silo wall.

7 DISCUSSION OF STORING AND DISCHARGE PRESSURE PATTERNS

The similarities in the flow pressures for both concentrically and eccentrically discharged tests (Fig. 4) indicate that the flow pattern and the stress field developing in the silo are not as dependent on the eccentricity of discharge as one might expect, but are strongly influenced by the method of filling. For the eccentrically discharged tests with concentric or distributed filling (Fig. 5), the flow pattern and the flow pressures are similar regardless of whether the silo is subject to concentric or

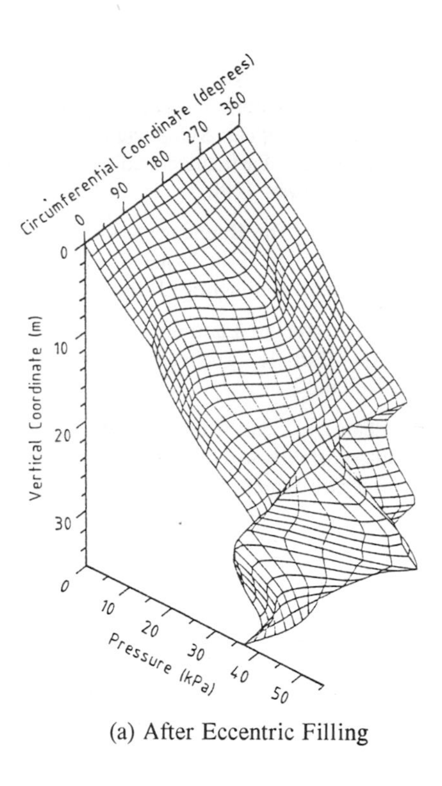

(a) After Eccentric Filling

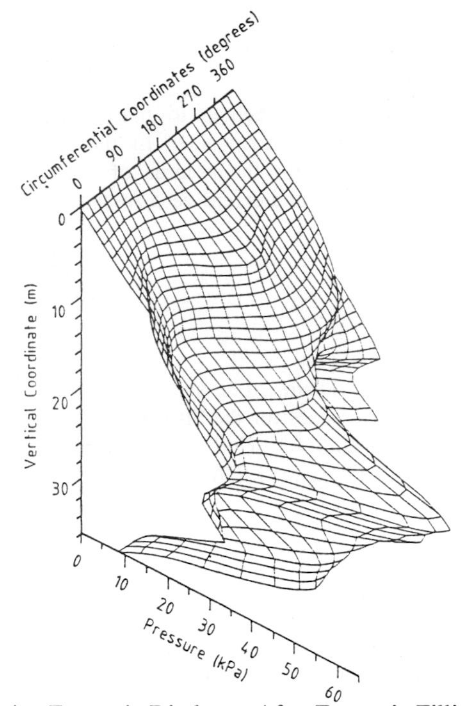

(b) During Eccentric Discharge After Eccentric Filling

Fig. 6 Mean Pressure Distribution

distributed filling. Thus, eccentric discharge does not always lead to a special pressure distribution.

These observations suggest that the flow pattern is strongly dependent on the filling process. This has been reported before (10). It has been recognised for some time now that the flow pattern strongly affects the pressures during discharge.

It is likely that different filling methods gave rise to different packing and orientation of the individual grains, which then resulted in different bulk mechanical behaviour. The grain packing and orientation also significantly affect the bulk density. Since barley is known to display highly anisotropic behaviour resulting from the dominant grain orientation (8), the above may not be surprising.

The pressures after filling and during storing are quite remarkably unsymmetrical. This lack of symmetry may be attributed to the presence of wall imperfections and to the grain inhomogeneity resulting from the filling process. The pattern has been repeated consistently in many separate tests so the phenomena cannot be safely ignored in design, even if the quantitative relationship between the high pressures and causative factors has not yet been established.

Depending on the methods of filling and discharge, the pressures during discharge may change around the effective transition, or where the flow channel forms near the wall. Since the pressure pattern in most of the silo is quite similar under both filling and discharge conditions, it may be satisfactory to use factored initial pressures, together with additional patch pressures, to represent flow pressures.

Unsymmetrical pressures of this kind revealed here lead to bending of the wall in reinforced concrete silos and to high compressive stresses in steel silos. Suggestions that unsymmetrical pressures should be recognised in design have been made for some time, chiefly couched in terms of the need for wall bending strength (2,6,8,19,20). However, it is not clear that the same bending strength should be required in silos of all sizes, all aspect ratios, all construction techniques and for storing all solids. Ideally the silo should be directly designed for unsymmetrical pressures, rather than for a certain bending strength which may or may not relate well to the real requirements of a particular installation.

The statistical treatment presented in this paper may be used to derive more realistic, rigorously based pressure distributions for silo design. The coefficient of variation of the random component (Table 1) gives a measure of the extent to which the mean pressure may be exceeded. The design pressure may be defined as a characteristic value, being the mean plus one or more standard deviations, depending on the probability of exceedence for which the silo is to be designed. This procedure follows that used in codifying other loading conditions for structures (wind, snow, earthquake etc).

8 CONCLUSIONS

A statistical analysis of many careful experiments on a prototype tall concrete silo storing barley has been presented. These experiments included distributed, concentric and eccentric filling, storage and both concentric and eccentric discharge.

The results have shown that quite large systematic unsymmetrical pressures occur during filling and storage, as well as during discharge. The pressure profile may deviate significantly from Janssen pressures even though the mean distribution may be quite close to the Janssen prediction.

The systematic unsymmetrical pressure patterns are not explicable solely in terms of the eccentricity of filling and discharge, and many must be attributed to very small aberrations in the silo geometry.

Because the high pressures observed in these full scale tests have been shown to be related to anisotropic properties in the stored solids, it is clear that more extensive investigations must be undertaken into the mechanical behaviour of granular solids stored in silos. A better understanding of the mechanical behaviour of bulk solids would allow the effects of grain anisotropy and grain inhomogeneity on wall pressures to be explored and quantified. Designs could then account properly for the characteristics of the materials to be stored in the silo.

ACKNOWLEDGEMENTS

The authors are deeply indebted to their colleagues Jorgen Nielsen and Jorgen Munch-Andersen of the Danish Building Research Station (SBI) for the records of these experiments, many useful discussions and extensive correspondence.

REFERENCES

1) Rotter, J.M., Pham, L. and Nielsen, J. "On the Specification of Loads for the Structural Design of Bins and Silos", Proc., 2nd Int. Conf. on Bulk Materials Storage Handling and Transptn, I.E.Aust., Wollongong, July, 1986, pp 241-247.

2) Schmidt, K.H. and Stiglat, K,. "Anmerkungen zur Bemessungslast von Silos", *Beton und Stahlbetonbau*, 1987, 9, pp 239-242.

3) Nielsen, J. and Askegaard, V. "Scale Errors in Model Tests on Granular Media with special reference to Silo Models", *Powder Technology*, 1977, 1.

4) Nielsen, J. and Kristiansen, N.O. "Pressure Measurements on a Silo in Karpalund" (in Danish) Nordic Group for Silo Research, 1979, Rpt. No.5, Technical Uni. of Denmark, Dept. of Structural Engg.

5) Nielsen, J. and Kristiansen, N.O. "Related Measurements of Pressure Conditions in Full Scale Barley Silo and in a Model Silo", Proc., Int. Conf. on Design of Silos for Strength and Flow, 1980, Univ. of Lancaster, Sep.

6) Nielsen, J. and Andersen, E.Y. "Loads in Grain Silos", *Bygningsstatiske Meddelelser*, 1982, 53, No.4, pp 123-135.

7) Nielsen, J. "Load Distribution in Silos Influenced by Anisotropic Grain Behaviour", Proc., Int. Conf. on Bulk Materials Storage, Handling and Transptn, I.E.Aust., Newcastle, August, 1983, pp 226-30.

8) Hartlen, J., Nielsen, J., Ljunggren, L., Martensson, G. and Wigram, S. "The Wall Pressure in Large Grain Silos", Swedish Council for Building Research, Stockholm, 1984, Document D2:1984.

9) Munch-Andersen, J. and Nielsen, J. "Size Effects in Large Grain Silos", *Bulk Solids Handling*, 1986, 6, pp 885-889.

10) Munch-Andersen, J. and Nielsen, J. "Pressures in Slender Grain Silos: Measurements in Three Silos of Different Sizes", Proc., 2nd European Symposium on the Stress and Strain Behaviour of Particulate Solids- Silo Stresses, CHISA, Prague, August, 1990.

11) Munch-Andersen, J. "The Boundary Layer in Rough Silos", Proc. 2nd Int. Conf. on Bulk Matls Storage Handling and Transptn, I.E.Aust., Wollongong, July, 1986, pp 160-163.

12) Hartlen, J. "The Strength Parameters of Grain", European Symposium on Particle Technology, June 3-5, 1980, Amsterdam, Holland.

13) Ooi, J.Y., Pham, L. and Rotter, J.M. "Systematic and Random Features of Measured Pressures on Full Scale Silo Walls", *Engineering Structures*, 1990, 12, No.2, pp 74-87.

14) Askegaard, V., Bergholdt, M. and Nielsen, J. "Problems in Connection with Pressure Cell Measurements in Silos" (in English), *Bygningsstatiske Meddeselser*, 1971, 42, No.2, pp 33-74.

15) Wigram, S. "Report on Test of Using a Distributing Device When Filling a Grain Silo Bin", 2nd Int. Conf. Design of Silos for Strength and Flow. Stratford-upon-Avon, UK, Nov, 1983.

16) Nielsen, J. "Opmaling af Silo i Karpalund" (in Danish), Nordic Group for Silo Research, 1979, Rpt. No.4, Technical Uni. of Denmark, Dept. of Structural Engg., 21pp.

17) Janssen, H.A. "Versuche uber Getreidedruck in Silozellen", *Zeitschrift des Vereines Deutscher Ingenieure*, 1895, 39, No.35, pp 1045-1049.

18) Ooi, J.Y. and Rotter, J.M. "Unsymmetrical Features of Measured Pressures on Prototype Grain Silos", Proc., 2nd European Symposium on the Stress and Strain Behaviour of Particulate Solids- Silo Stresses, Prague, August 1990, 10 pp.

19) Peter, J. "Design of the Central Cone Silo with Special Regard to Eccentric Discharge", Proc., 2nd Int. Conf. on the Design of Silos for Strength and Flow, Stratford upon Avon, Nov, 1983, pp 526-541.

20) Safarian, S.S. and Harris, E.C. "Empirical Method for Computing Bending Moments in Circular Silo Walls due to Asymmetric Flow", 1989 Meeting of ACI Committee 313, Atlanta, GA, Feb, 1989.

C418/023

A field study on the dust emission from a coal storage and handling site

E L M VRINS, PhD
Buro Blauw, Wageningen, The Netherlands

SYNOPSIS During one year a field study has been performed to the windblown dust emission from a coal storage and handling site. The effects of meteorological conditions, handling activities and dust reduction measures were investigated in order of priority: wind direction, rainfall, watering of the stock piles, windspeed and daily variations. In the winter period wind erosion from stock piles was the main dust source. In the summer period the piles were watered and the wind erosion largely reduced. Traffic movements were supposed to be the main dust source in the summer.

1 INTRODUCTION

For the last 8 years an extended research program on coal was supported by the Dutch Government. A field study on windblow dust from coal stock piles was carried out within this research-program. During one year, coarse dust was sampled at a location about 500 metres from a coal storage site. The aim of the study was to provide a better insight in the dust emission of the storage site and the relevant emission factors.

Nuisance as a result of dust deposition in urban areas is regarded as the main environmental impact of windblown coal dust. The dust consists of particles with diameters varying from 0.1 µm to several hundreds of µm. The particle size plays an important role in the dispersion of the dust. The largest particles are deposited near the stock piles and contaminate only the site itself. The smallest particles are widely dispersed and contribute only little to the total airborne dust concentration. The particles causing the nuisance are small enough to be blown away over some distance and large enough to deposit, before being dispersed to insignificant concentrations. The size fraction that should be considered, depends on meteorological and local conditions. Roughly speaking, the particle dia-meters are between 10 and 100 µm.

It may be clear, that nuisance caused by coal dust is a local problem. This statement can also be reversed: when coal dust is present it is likely to come from a nearby source. Tracing the source of coarse dust particles is easier than that of the fine particles, which may originate from a source far away. For tracing the source accurate data of the wind direction are necessary. Because of the fast wind direction variations a high time resolution is required for the dust concentration data.

These considerations lead to the following requirements for the dust sampling device, wich should be used in this field study:

- Adequate for sampling particles in the size range of 10 to 100 µm.
- A high time resolution.

Most commercially availabe dust samplers are designed for sampling the fine dust fraction (< 10 µm), the so-called health related fraction because of its inhalibility. Devices, which sample the total airborne dust, mostly require a sampling time of 24 hours. This is too long for making accurate correlations between wind direction and dust concentration. Therefore, a new sampling device has been developed for this study: the Course Dust Recorder.

2 EXPERIMENTAL SET UP

The Coarse Dust Recorder (Fig. 1) consists of a tube, through wich air is sucked by four fans. It is kept in the wind direction by a windvane. Inside the tube a cassette (Fig 2) is placed. This cassette contains a 12 m long closed strip, which is moved stepwise. Every 18 minutes a 20 mm long piece of the strip is exposed to the airflow and the coarse dust particles are collected by inertial impaction. After one week the complete strip is loaded with dust particles and the cassette is replaced by another one. In the laboratory the cassette is automatically analysed by an image analysis system. From the area coverage of each piece of strip the coarse dust concentration is calculated. The relation between area coverage and dust concentration depends on the sort of dust and the size distribution. In this research project this relation was experimentally established by optical and gravimetrical analysis of flat plates, covered with the relevant sorts of dust. Only dark coloured particles were taken into account. So sand and seasalt particles, which were not of interest for this study, were not detected by the optical system. The Coarse Dust Recorder was calibrated under atmospheric conditions with a monodisperse aerosol (di-ethyl-hexyl-sebacate) generated by spinning top aerosol generators. The Aerosol Tunnel sampler (Hofschreuder and Vrins 1986) was used as a reference. As the air flow inside the tube is established by fans, the windspeed may influence the air flow and thereby the sampling rate. Thus the windspeed has to be known for a correct calculation of the dust concentration. The sampling velocity also influences the sampling efficiency.

The magnitude of this influence depends on the square root of the variations in sampling velocity and the steepness of the sampling efficiency curve as a function of aerodynamic particle diameter. The 50 % cut-off diameter was assessed to be 19 µm. Due to the windspeed in the range of 1 to 10 m/s a shift in sampling efficiency of at the most 5 % can be expected.

The storage site under investigation is located at the "Maasvlakte", near "Hoek van Holland". It stretches almost from west to east for about 1 km and has a width of about 200 m. Ship loading and unloading occurred along the entire south and west side. The main roads are all east-west orientated. The measuring site was north of the piles, almost halfway, at a distance of 500 m. The western half of the site is covered with iron ore.

The data presented in this paper were collected in the period from January 1989 until August 1989. This study is restricted to coal dust, although iron ore will clearly add a large amount of dust at the measuring site. However, coal is the subject of the research program. The data collected on iron ore emissions may be processed later.

3 RESULTS

The effects of meteorological conditions, handling activities and dust reduction measures on the dust concentration at the measuring site are investigated in order of priority. The following order is passed through:

Wind direction. The wind direction determines the direction of the dust source and thus is the first emission factor to be analysed. It shows the relative contribution of the different sources to the total dust exposure at the sampling site. In the following only the data with the wind coming from the coal storage site are processed.

Rainfall. Rainfall may have an effect on the dust emission, but certainly has a dramatic effect on the dust dispersion by washing out the airborne dust almost completely, before it reaches the measuring site, thus masking the dust emission. For the investigation of the other dust emission factors the data with rainfall are skipped.

Watering of the stock piles. The stock piles are watered in order to reduce wind erosion. The watering system is only in use in the summer, while frost may damage the system in winter time. For further evaluation of the data the data set is split in the winter period (no watering) and the summer period (watering).

Windspeed. Windspeed may have an effect as emission factor. However, it also influences the dust dispersion. Calculations showed, that in the given situation for the particle size fraction under consideration a slight decrease of the concentration-emission ratio is expected with increasing windspeed. This decrease is in first approach neglected, because the effect of windspeed on the dust emission is expected to be more dramatic.

Daily variations. The handling activities are continuous for 24 hours a day. An evaluation of the daily course of the dust concentration may reveal special information on the dust sources. Special attention deserves the fact, that spraying of the roads occurs only in the mornings on weekdays.

From a period of 7 weeks the median hourly dust concentration was calculated for all wind directions (Fig. 3). The storage site is in the direction from 140 to 220 degrees. Its contribution to the dust concentration is quite obvious. The coal storage site stretches from 140 to 180 degrees. From 180 tot 220 degrees the iron ore is stored. In the following only the data which relate to the coal storage are evaluated. The data with rainfall are skipped.

Fig. 4 shows all measured hourly concentrations in the period January 1989 to August 1989 with the wind coming from the coal storage site. The highest concentrations were measured in the winter period (up to 700 µg/m3), when the stock piles were not watered. In the following the winter and summer period are avaluated separately.

Subsequently the effect of windspeed is evaluated (Fig. 5). The effect of the windspeed is obvious in the winter period. With increasing windspeed higher dust concentrations occur. It shows wind erosion as the main dust emission factor. The figure for the summer period looks different. There are no data with high windspeeds. Therefore it is difficult to make a good comparison with the winter period. However, the relatively high concentrations with low windspeeds is remarkable.

Fig. 6 shows the daily course of the coarse dust concentration on weekdays. In the winter period, no daily trend can be seen. In the summer period, it is striking, that all high concentrations occur in the night time. This might be explained by the fact that watering of the roads occurs only in the mornings, while verhicle movements take place all day long. Dust emission by vehicle movements does not depend on windspeed. This is in agreement with the fact that no relationship was found between dust concentration and windspeed in the summer period. No daily trend in dust concentration is found in the winter period. Firstly, the large wind erosion may hide the emission due to vehicle movements. Secondly, the drying up of the roads will be much faster in the warm summer then in the cold winter. Thus watering the roads only in the morning may be sufficient in the winter, whereas repeating the watering in the evening is recommended in the summer.

4. CONCLUSIONS

In this study an effort has been made to describe the dust emission from a coal storage site using dust concentration measurements, meteorological data and general information about the handling activities and dust reduction measures at the storage site. In most cases this information is readily available. It is shown that it is possible to identify special features of dust sources and evaluate dust emission reduction measures. However, it is required, that the coarse dust particles are sampled representatively and that sampling occurs with a high time resolution. Most currently available sampling devices do not fit these requirements. Therefore the Course Dust Recorder was developed, which turned out to be adequate for these measurements.

The main conclusions are:

- High dust concentrations at the measuring site were due to the dust emission from the coal and iron ore storage site.
- Highest coal dust concentrations in winter time were due to wind erosion from the unwatered storage piles.
- Highest coal dust concentrations in summer were due to vehicle movements.

REFERENCES

(1) VRINS, E. The Coarse Dust Recorder. AEROSOLS, Science, Industry, Health and Environment (eds. Masuda S and Takahashi K). Proc. 3rd Int. Aerosol Conf. pp 970-973.

(2) HOFSCHREUDER, P., VRINS, E. (1986). The Aerosol Tunnel Sampler: A total airborne dust sampler. Proceedings of the 2nd Int. Aerosol Conf. Berlin, 491-191.

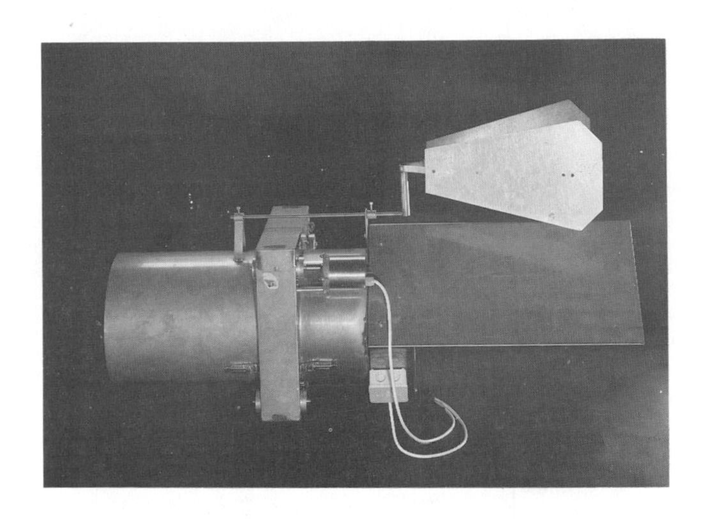

Fig. 1 The Coarse Dust Recorder

Fig. 2 The Cassette

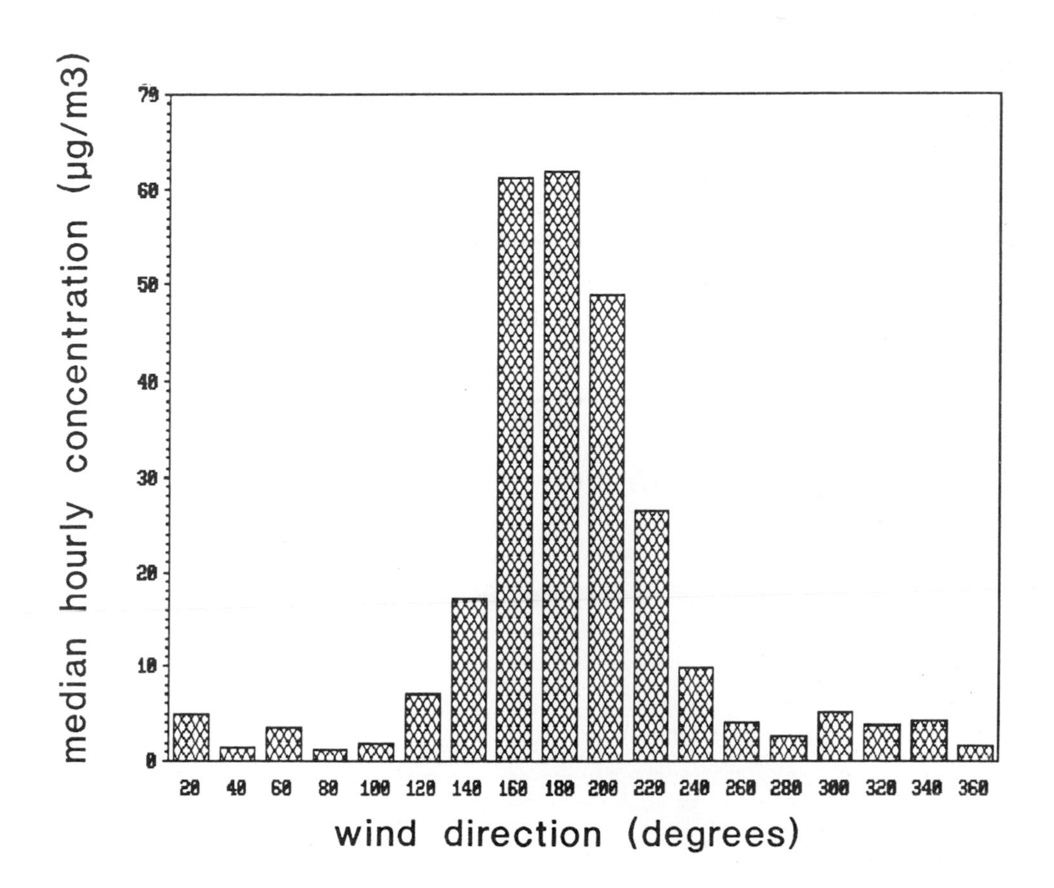

Fig. 3 The hourly dust concentration versus wind direction.

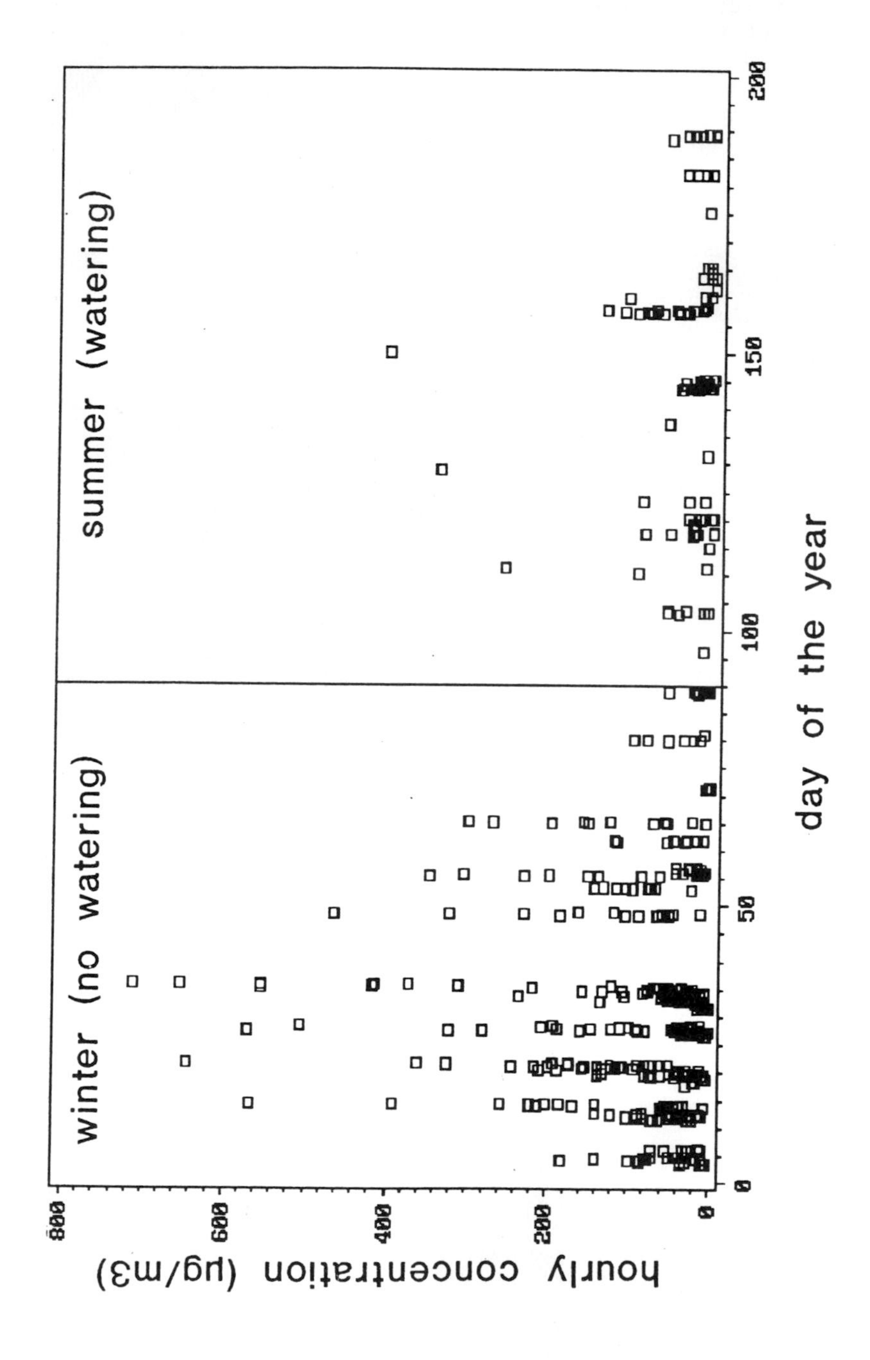

Fig. 4 Hourly dust concentrations during period from January 1989 to August 1989.

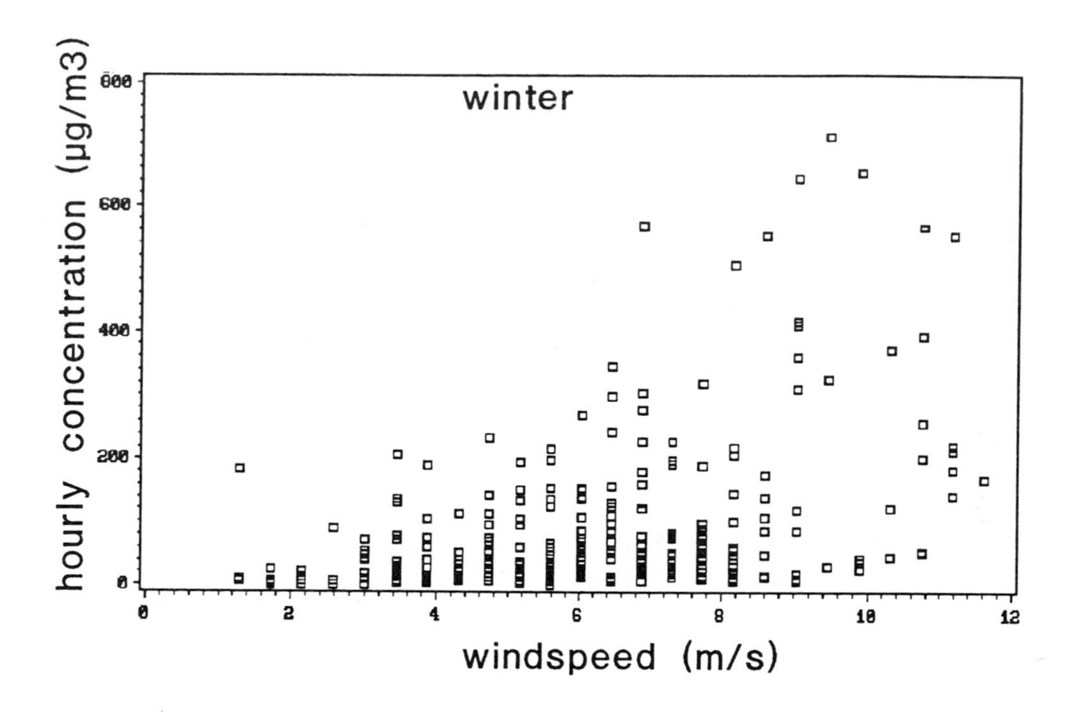

Fig. 5a Hourly dust concentrations versus windspeed in the winter period.

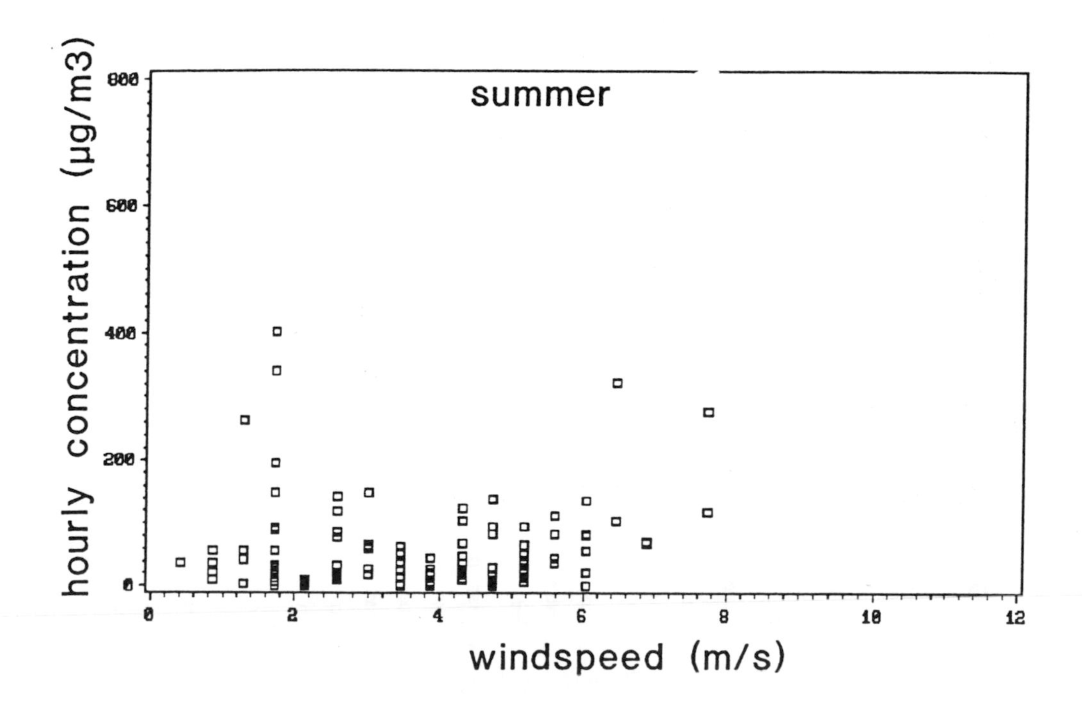

Fig. 5b Hourly dust concentrations versus windspeed in the summer period.

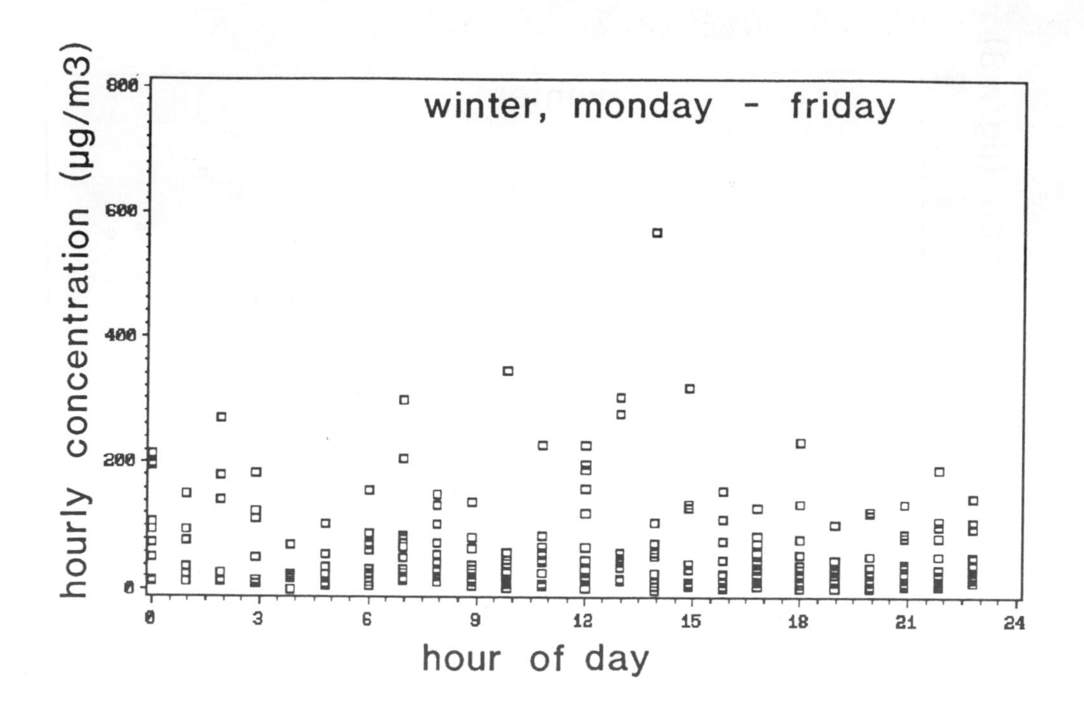

Fig. 6a Hourly dust concentrations versus hour of the day in the winter period.

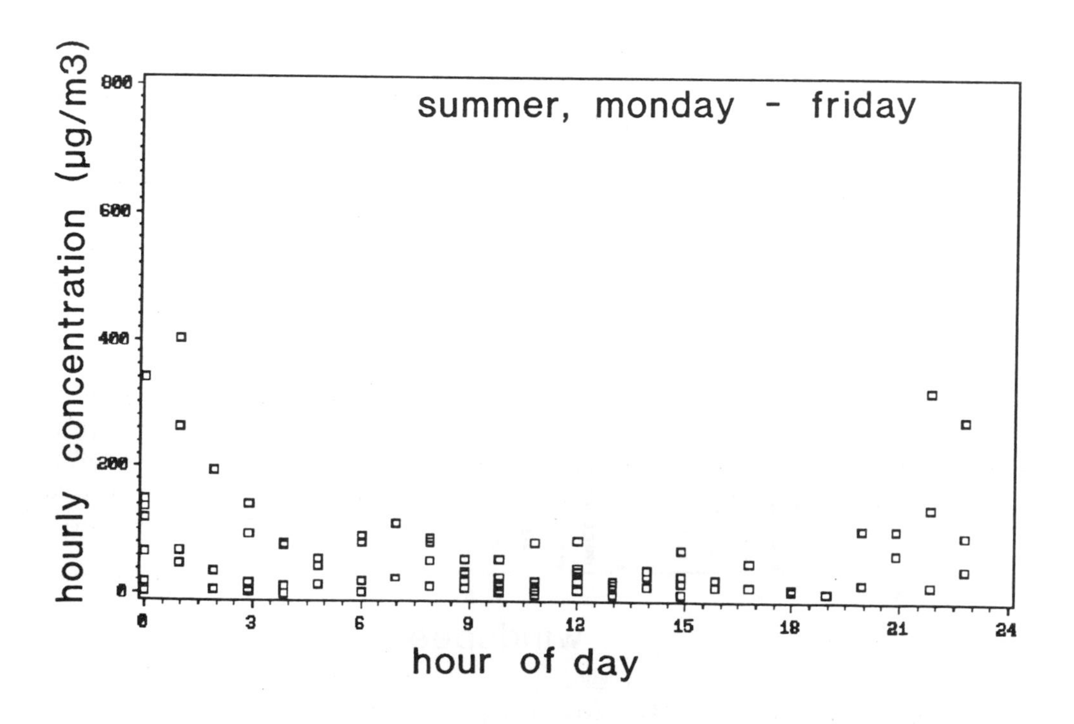

Fig. 6b Hourly dust concentrations versus hour of the day in the summer period.

Microfoam – innovative dust control

H C COLE
Deter Company Incorporated, Burgin, Kentucky, United States of America
F W PARRETT, BSc, PhD, CChem, FRSC
Deter UK, London

SYNOPSIS The need to reduce dust emissions during bulk material handling is a continuing problem. Deter Microfoam[R] offers a different approach in dust suppression which is considerably more effective and uses much less water than normal wet suppression. Microfoam[R] is applied where it can be mixed with bulk materials to ensure suppression of fugitive dust particles in with the bulk materials.

1. INTRODUCTION.

Many raw and manufactured materials produce dust during handling or dry processing. These airborne particulates can cause environmental, health or even explosion problems. In the U.K. the recent Control of Substances Hazardous the Health (COSHH) legislation has now set limits for even nuisance dusts which should not be exceeded (10 mg/m^3 for total dust, 5 mg/m^3 for respirable dust). Respirable dust is made up of particles less than 10 microns, and it is these very small particles which can reach the lungs. With this additional legislation setting these limits the need for more innovative approaches to dust control is needed. The traditional approaches to dust control - mechanical enclosure and extraction will continue to play an important role in reducing dust emissions, but the innovation provided by Deter Microfoam[R] is the best suppression system for reaching the new limits imposed by environmental legislation.

2. FOAM GENERATION.

The foam which is generated by traditional venturi systems (e.g. fire fighting foam) is not suitable for dust suppression. Such foams are essentially large (5mm) bubble foam. For dust suppression the requirement is for small bubbles (less than 100 microns). Microfoam[R] is a small bubble pressure foam and has proved to be extremely effective in dust suppression.

3. THE MICROFOAM GENERATOR

For the consistent production of good quality foam it is essential to mix the three components, water, compressed air, and foam agent accurately at all times. The Deter Microfoam Generator patented by Deter Co. Inc. does this and converts this mix into a small bubble, high energy foam, 50–100 microns in diameter. The foam can be pumped through small bore pipes (25–50mm) to special foam nozzles where it is discharged for mixing with the material to be suppressed.

4. SUGGESTED ACTION OF MICROFOAM[R]

The very small bubbles in Microfoam[R] make it extremely stable. Hence it can be piped under pressures of up to 3.5 bar (50 psi) to the required discharge point. If the foam comes into contact with a larger particle its surface will be wetted will but the foam will be substatially unaffected. However it is suggested that dust particles, which are comparable in size to Microfoam bubbles will penetrate the bubbles. This breaks the bubble and "wets" the dust particle, which will adhere to the moistened surface of the larger particles. Thus by using a small bubble foam it becomes possible to selectively coat the small particles (dust) in a bulk material with water without saturating everything, such as the larger particles, and the surrounding machinery as can happen with water sprays. Typically a Microfoam[R] application will often require 60 litres (2 cu.ft.) of foam per tonne of dusty bulk material to achieve dust suppression. This amount Microfoam[R] contains about 2 litres of water, so dust suppression can be achieved using only 2 litres of water per tonne for many materials. This water will contain 1–2% of Deter Foam Agent, which is formulated to be non–toxic and biodegradable.

5. USE OF MICROFOAM[R]

Whereas simple water sprays are used by spraying onto dusty material Microfoam[R] must be mixed with the dusty material. Therefore careful consideration of the handling and process system for each bulk material must be examined to find the best place to introduce the foam e.g.

Flowing the foam into crushers along with raw material

Injecting the foam into the free falling material at transfer points

Injecting foam into screw conveyors

Injecting foam into pug–mill mixers

Applying foam to materials as they are dumped from trucks

Whilst processes such as crushing can be the reason a material becomes dusty, subsequent handling means that transfer points are also major dust producers. In the free fall period the air movement will remove "fines" from the bulk material. Foam should be introduced into the material as early as possible in the handling since one treatment will often keep the material dust suppressed through all subsequent handling operations.

6. SOME RECENT MICROFOAM INSTALLATIONS

a. Quarry operations. To illustrate the typical lay out of a Microfoam^R installation Diagram 1. shows how foam is applied at a belt transfer point. This system is being used at a limestone quarry in Canada. The photographs in this quarry show high dust generation without Microfoam (Fig.1) and the cleaner situation three minutes after the foam application commenced (Fig.2).

b. Screw conveyor. A screw conveyor does introduce a mixing action and can be a very useful point to introduce foam into a material. A small brown coal handling operation (2–3 tph) was extremely dusty. When the conveyor was operating levels at the output tip measured by HFM1 dust meter were 3 mg/m^3 and rose rapidly to above 10 mg/m^3. When Microfoam^R was injected into the screw conveyor the dust levels measured at the output rapidly fell to below 0.5 mg/m^3 and operators could safely stand around the output area.

In another Screw conveyor application in the USA, the "fume" from a steel furnace which is heavily contaminated with heavy metals (hazardous) are collected by a precipitator and stored in a hopper which feeds a relatively short screw conveyor. Microfoam^R injected into the screw conveyor ensures that the fume is dust free when discharged into a 20–tonne trailer for overland transport to a hazardous disposal tip. The treated material can be dumped from the trailer into the disposal pit under high wind conditions without being blown away.

When a screw conveyor is used in this way foam injection points should be below the maximum material level in the conveyor tube, and it is possible to enhance the mixing action by welding small bars 180° apart between the flights of the screw.

c. Stockpiles. Wind erosion load–in and load–out on stockpile operations can cause dust emissions. The effectiveness of Microfoam^R treated material in minimising such losses was illustrated by wind tunnel tests on untreated and treated material which showed that in the wind tunnel moisture additions of 0.1 percent as Microfoam were more effective at coagulating fugitive particles than similar applications of water.

7. CONCLUSION

The development of small bubble pressure foam has provided another option for engineers concerned with control of dust from bulk material handling and processing. It will not replace the existing mechanical enclosure systems or water based spray systems, but is a true innovation in tackling a very old problem.

Fig. 1. Limestone Quarry Operation
with no dust suppression

Fig. 2. Limestone Quarry Operation
3 minutes after Microfoam application
commenced

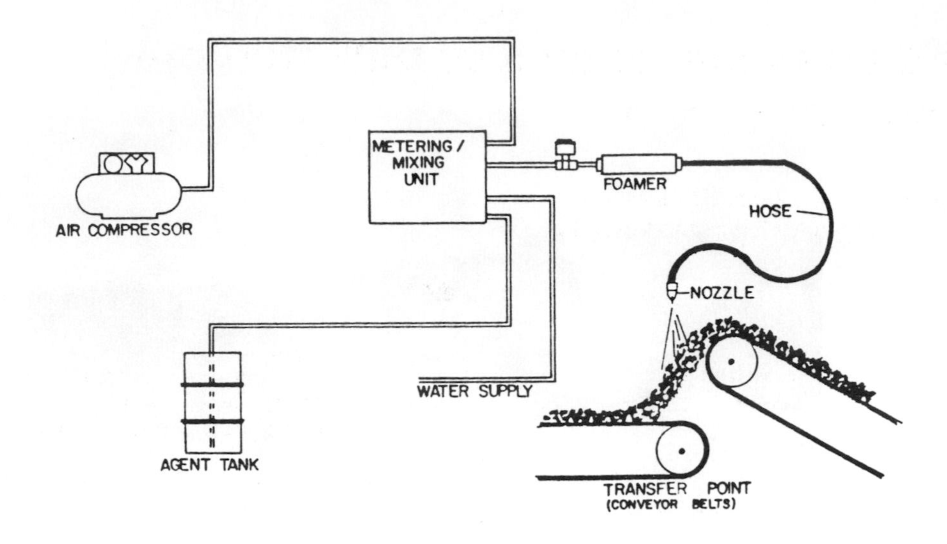

Diagram 1. Deter Microfoam. Installation on a belt conveyor system

C418/006

Blending and train loading of bulk iron ore: a case study

R ATTWOOD, BTech, FIEAust, FAusIMM
Kinhill Engineers Pty Limited, Adelaide, Australia

SYNOPSIS At the recently opened iron ore mine at Iron Duke in South Australia, the selective production of mined material in conjunction with a multi-stockpile blending programme guarantee an acceptable product for the blast furnaces. This paper reviews the methods used to stockpile, reclaim and load the ore into railway wagons, and the way in which knowledge of the ore's bulk flow properties was applied to the design of the holding bins. Lessons learned from difficulties faced during the design and commissioning of the plant are also reviewed.

1 INTRODUCTION

For over forty years iron ore from the Middleback Ranges in South Australia has been used in the blast-furnaces of the city of Whyalla in South Australia. The ore deposits occur in pockets in the ranges, and a number of these have now been exhausted.

The Iron Knob deposit, located at the extreme northern tip of the Middleback Ranges, remains the main source of iron ore for the Whyalla blast-furnaces.

The Iron Duke deposit lies at the extreme southern tip of the Middleback Ranges, about 33 km from the now worked out Iron Baron facility and 50 km from the BHP Long Products Division steelmaking facility at Whyalla.

The long-term security of the supply of raw material to Whyalla necessitated the development of this ore body as a replacement for the Iron Baron material. The challenge was to reduce the variation in the quality of the supplied material from both old and new sources, as measured by the standard deviation of ore contaminants in the fluxed pellets delivered to the blast-furnace.

The Iron Baron deposit had been worked for over twenty years during which time the problem of contaminant variability was tackled by a number of techniques, mostly of a hit or miss nature. By 1986 a manual blending system was in operation in which crushed ore was trucked to a laydown stockpile area where it was dumped into layers. This crude first blend was refined by using front-end loaders to recover the material by taking vertical cuts through the stockpile. These activities resulted in a halving of the standard deviation of contaminant elements in the ore from Iron Baron.

The development of the Iron Duke deposit presented both the opportunity and the challenge to introduce further blending improvements through the mechanization of a second blending step. The new system has shown a further reduction of over 30% in the standard deviation for key contaminants, (particularly SiO_2) during the first year of operation of the mine. Further improvements are expected as mining proceeds and the surface ore, which is particularly variable, is exhausted.

The process of achieving the desired control over contaminant variability of the ore is measured at six stages in the ore preparation cycle:

- in situ ore body;
- in situ pellet ore in the ore body;
- pellet ore in the pre-crusher stockpile;
- pellet ore in the post-crusher blended ore stockpile;
- pellet ore in the train wagons;
- pellet ore after final blending with Iron Knob ore and formed into fluxed pellets.

The changes in variability at each of the six stages are shown on Figures 5.1 to 5.6, and shows that the standard deviation of ore contaminants is reduced from 5.41 to 0.123.

The key activities in the process of providing uniformly blended ore from the Iron Duke Mine, to acceptable standard deviation limits for contaminants are the:

- identification and selective mining of over thirty different ore types (classes) of ore at the depot;

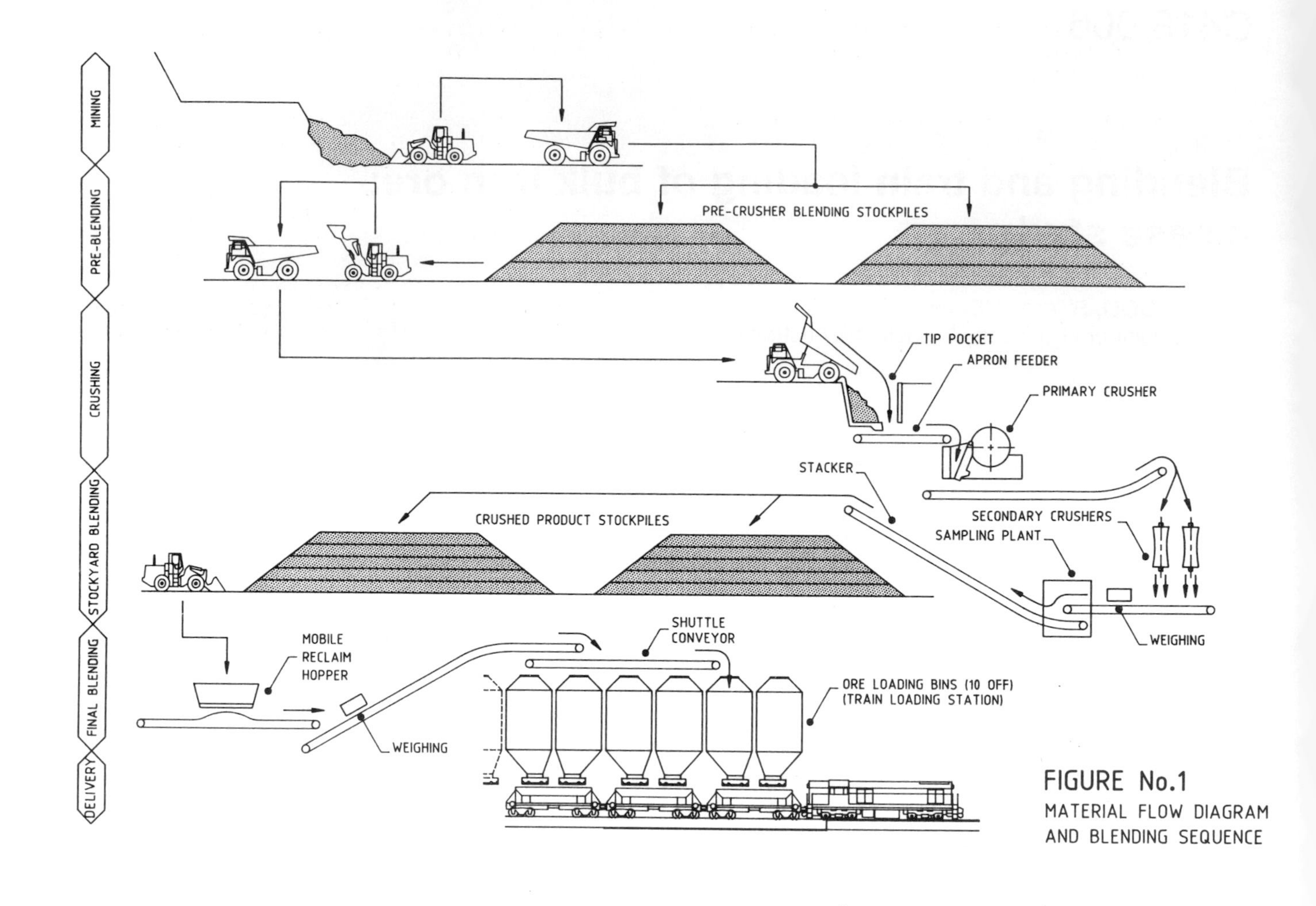

FIGURE No.1
MATERIAL FLOW DIAGRAM
AND BLENDING SEQUENCE

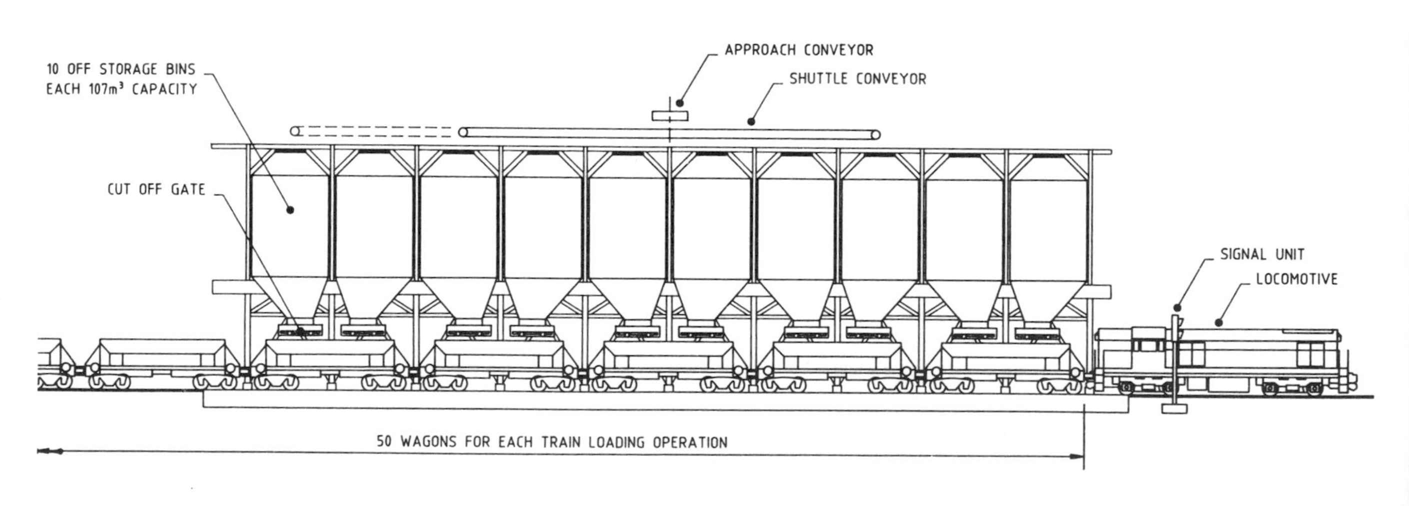

FIGURE No.3
TRAIN LOADING STATION
GENERAL ARRANGEMENT

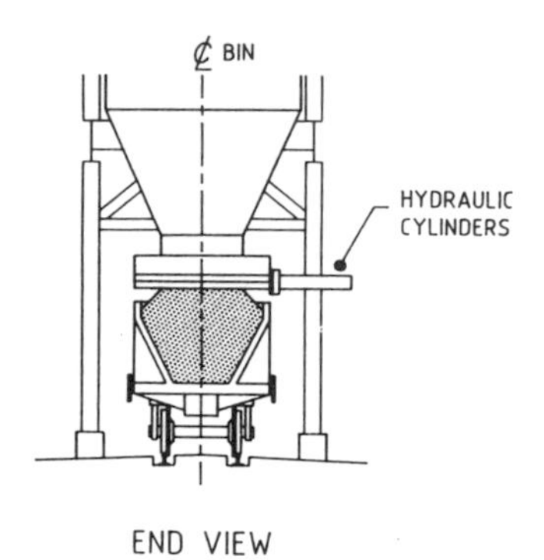

FIGURE No.4
BIN / WAGON OFFSET

- pre-blending of these products through a pre-crushing material laydown and recovery operation;
- application of a programmable mechanized windrow stacking operation to the product stockpiles;
- use of a technique to reclaim ore from the product stockpiles.

The total blending system combines elements of mechanized plant with the use of wheeled mine vehicles for a capital and operating cost-effective solution.

2 MATERIAL FLOW AND BLENDING SEQUENCE (Figure 1)

Prior to mining the ore body is extensively drilled and sampled. The recovery of ore is then programmed for pre-crusher blending in stockpiles. These stockpiles are formed by successive layering of selected product grades delivered from the mines by haul trucks (Figure 2a). Wheel loaders recover this material by reclaiming across the ends of the stockpiles, taking a vertical slice of the layered product grades.

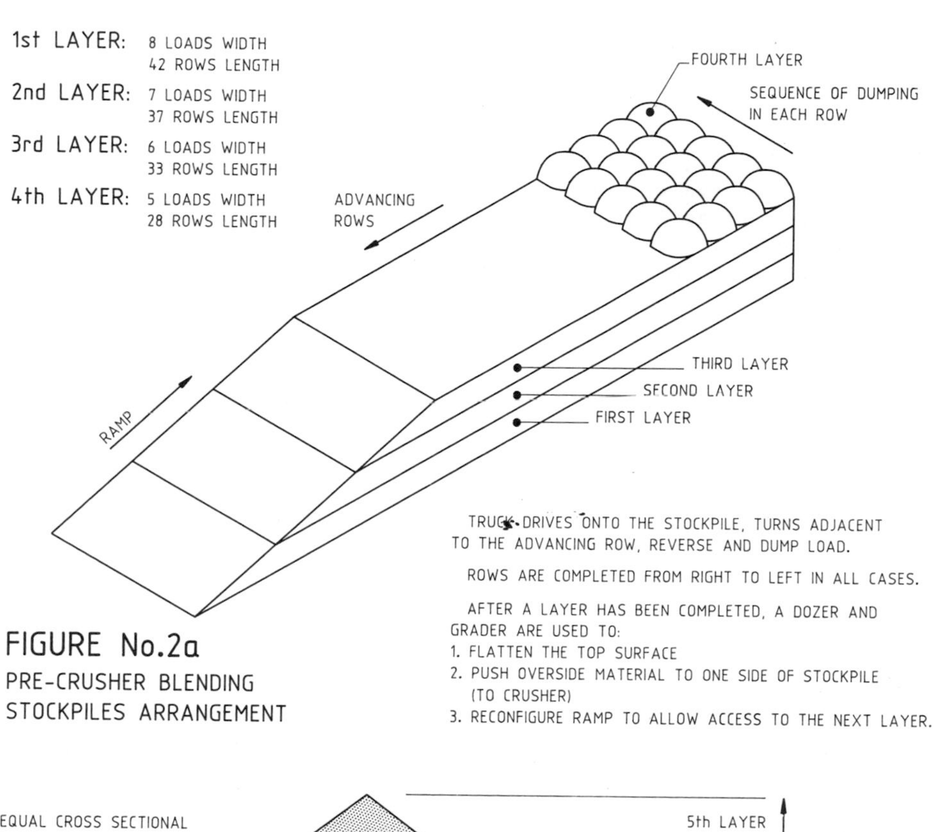

1st LAYER: 8 LOADS WIDTH
42 ROWS LENGTH
2nd LAYER: 7 LOADS WIDTH
37 ROWS LENGTH
3rd LAYER: 6 LOADS WIDTH
33 ROWS LENGTH
4th LAYER: 5 LOADS WIDTH
28 ROWS LENGTH
ADVANCING ROWS
FOURTH LAYER
SEQUENCE OF DUMPING IN EACH ROW
THIRD LAYER
SECOND LAYER
FIRST LAYER
RAMP
TRUCK DRIVES ONTO THE STOCKPILE, TURNS ADJACENT TO THE ADVANCING ROW, REVERSE AND DUMP LOAD.
ROWS ARE COMPLETED FROM RIGHT TO LEFT IN ALL CASES.
AFTER A LAYER HAS BEEN COMPLETED, A DOZER AND GRADER ARE USED TO:
1. FLATTEN THE TOP SURFACE
2. PUSH OVERSIDE MATERIAL TO ONE SIDE OF STOCKPILE (TO CRUSHER)
3. RECONFIGURE RAMP TO ALLOW ACCESS TO THE NEXT LAYER.

FIGURE No.2a
PRE-CRUSHER BLENDING STOCKPILES ARRANGEMENT

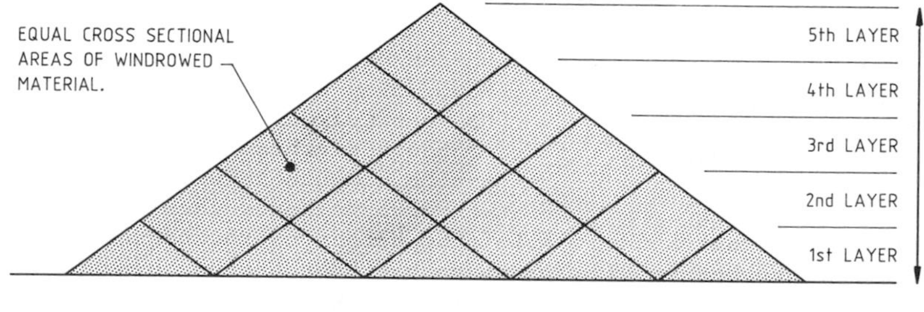

FIGURE No.2b
WINDROW PATTERN

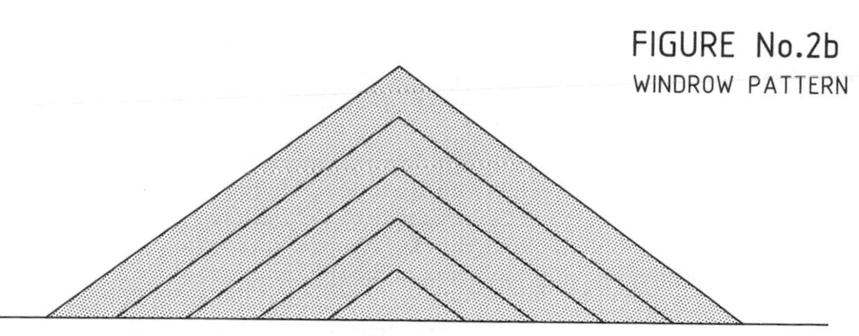

FIGURE No.2c
CHEVRON PATTERN

No size distribution sample has been taken of the material in the pre-crusher stockpiles, and there is little to be gained by attempting this exercise. Material is campaign mined from ore body zones and laid down in the pre-crusher stockpiles as products from the defined zones. Since ores from different zones break in different ways during the blasting process, a painstaking analysis of the size distribution from any one zone would have little meaning in the overall picture. On-site observations of a large number of tipping operations at the primary crusher indicate that a typical 85 tonne load of ore contains up to six 1 m^3 lumps and 50% of the remaining ore is less than 200 mm. Of the -200 mm material some loads contain over 50% of this -200 mm material as 'fines' of less than -75 mm. There is no attempt to control feed size to the crusher other than by the blasting process. The procedure of campaign mining from specific ore zone types reduces the impact of segregation of the large lumps at the pre-crusher stockpile area since all zone types generate approximately the same percentage of very large lumps. Some short duration surges of unblended material will pass through the primary and secondary crushing process, but the impact of this disappears during the blended stockpiling process where successive, small cross-sectional areas, strips of material are laid down in a windrow pattern (Figure 2b).

The material passes through a primary and secondary crusher without screening or bypass operations, with the product reduced from -1000 mm to -75 mm lump size. There is a high proportion of fines in the product at this stage. After passing through sampling and weighing operations, the ore is stockpiled by the stacker, which is capable of laying down up to 60 windrows in each 50 000 tonne stockpile.

The windrow pattern laydown of material allows for continuous modification of the quality of the input material by selecting from the various ore zone types to achieve the total blended quality requirement. A programmable travelling, slewing, luffing stacker is used to form the windrow stockpile. This has the advantage over a chevron formed stockpile (Figure 2c) of reducing significantly the effect of segregation of material in the stockpile. (Ref. Weddig (1), Gerstal (2) and (3).)

Once a 50 000 tonne stockpile has been completely formed, then the overall composition of the material contained in the stockpile cannot be changed. The act and method of reclamation of material from the stockpile and its subsequent distribution to the ten circular bins and loading into the rail wagons can be used to improve the homogeneity of the material.

The method used in the Iron Duke project of taking vertical slices through the ends of the stockpile (i.e. across the windrowed strips) and the subsequent distribution of the material into the rail bins by a shuttling conveyor contributes to the homogenization process.

Each 3000 tonne trainload of ore is railed to the Whyalla stockyards where it is further crushed and blended with ore from the Iron Knob deposit before being pelletized and dispatched to the blast furnace in 1500 tonne consignments.

3 TRAIN-LOADING BINS (Figure 3)

The use of overhead bins for loading iron ore into wagons has a number of advantages for this project, as follows:

- It enables the quantity of material for a full train-load to be organized in advance of the arrival of the train, eliminating the potential for delay as a consequence of equipment breakdowns.
- The train-loading operation can be undertaken using a minimal labour force (train-driver and bin gate operator only).
- Separation of specified blended material from different stockpiles between train-loads can be achieved.

These advantages, together with the relative economy of capital costs compared with alternative systems, were key aspects in the choice of this system for the project. However, the designers of the plant were faced with a number of engineering problems.

Of particular concern were the limitations on the geometric configuration of the bottom cones of the bins resulting from the need to apply the system to the use of existing rolling-stock (wagons). This had the unfortunate consequence of giving the bottom cone geometry of the bins an offset discharge from the centre line of the bins (Figure 4).

The bins are designed for funnel flow to minimize the potential for wear on the vertical walls. However, it was necessary for the designers to ensure that the critical opening dimensions were such that a stable rat-hole could not form and that the angle of the slopes of the bottom cone was sufficient to enable complete emptying of the bins. It was recognized that the potential for wear on the bottom cone was significant, and both this region and a small proportion of the vertical wall were provided with wear plates.

Assistance to the designers was provided by The University of Newcastle Research Associates (TUNRA) in New South Wales. While sample material from the mine was not available at the start of the design process, TUNRA considered the performance of a number of iron ores from different locations against the geometric constraints set by the designers. TUNRA found that, with the exception of one sample of material that had a moisture content in excess of 11%, the proposed discharge opening dimensions were sufficient to ensure that a stable rat-hole would not form, and that a bottom cone angle of not less than 45° would be appropriate.

The prospect that any of the ore from Iron Duke would have a moisture content in excess of 11% seemed (and still seems) extremely unlikely, with the specified moisture content being 2.5-3.0%. With this information, the designers prepared the designs of the bins for capacity, strength and life. The final bin geometry was as follows:

- 10 bins each of 4260 mm internal diameter with a volumetric capacity of 107 m^3;
- cut-off gate dimensions of 1600 mm x 1000 mm, with the gate centre-line offset 600 mm from the centre-line of the bins;
- asymmetric bottom cone with a shallow side angle to the horizontal of 45°, and a steep side angle to the horizontal of 69°.

The design material specification was as follows:

Iron ore	Bulk density of 3.1 t/m^3/Max
Moisture content	2.5-3.0%
Angle of repose	35°
% fines	47% (-6 mm) by mass
Effective internal angle of friction	63°
Angle of friction between ore and vertical wall	25°

4 CUT-OFF GATES

At the time of tender acceptance for the project it was proposed that clamshell gates would be used to cut off the material from the choke-filled rail wagons. Some doubts were raised about the ability of the clamshells to cut through the static heap of material, and proposals were obtained for alternative solutions.

A hydraulically-operated horizontal flat-plate was selected, and this has proven to be very satisfactory in operation. Initially there was a tendency for material to be pushed over the side of the wagon by the advancing edge of the cut-off plate; however, a simple extension to the lower lip of the cut-off pocket eliminated this problem.

5 TRAIN-LOADING (Figure 4)

The ore train consists of fifty wagons, each of 60 tonnes nominal capacity. It is a specified requirement that train-loading, including spotting of the wagons under the bins, be completed in one hour. The system is designed to allow thirty-five minutes for material discharge and twenty-five minutes for the spotting operation.

Each of the ten ore bins contains sufficient volumetric material to fill five wagons, hence each train requires a total of ten indexing operations involving the spotting of wagons under the bin. The spotting operation is achieved by a series of marker lights along the rail track indicating the locomotive position for each operation. The simplicity of the system and the skill of the operators results in a degree of accuracy in locating wagons under the bins that is appropriate for the installation.

6 THE RAIL BINS IN OPERATION

The project was officially opened by Brian Loton, Managing Director of BHP, on 27 February 1990. On that day, to the intense interest of a large group of engineering and non-engineering observers, iron ore was placed into rail wagons in the manner intended by the designers.

Up to the end of 1990 approximately 200 train-loading operations were undertaken, and the production rate for the mine met its targets. The bins have been the subject of modifications aimed at overcoming some spillage from the cut-off gate operation and to eliminate a propensity for four of the ten bins to retain some material on the shallow side of the discharge cone.

No stable rat-holes have formed in the bins, and the material in the bins discharges at a rate that comfortably meets the specified requirement for complete train-loading in one hour (including wagon spotting time). Depending on the bulk density of the product, this is equivalent to a rate of over 5000 tonnes of iron ore per hour through the cut-off gates.

The retention of a small quantity of material in four of the ten bins has been a cause for concern. The bins are all identical in design, and there are no significant measurable differences between the bins that completely empty every time and those that do not. Persistent attempts to cause the lumpy material to segregate in the bins to bias that material to the shallow cone side of the bin have greatly improved, but not eliminated, the problem.

A small quantity of infill using sand-filled epoxy is being applied to the critical transition point between the shallow side of the bottom cone and the vertical side of the wall of the bin. There is no proposal to provide vibrators, air blasters or any other flow enhancers to the bins, and the designers envisage that the infill material and the improved polished surface of the bottom cone wear plates will enable the performance of the bins to meet all expectations.

7 COMMENTS

From the design point of view, the predicted result for the performance of the bins has been good, with small but interesting problems occurring during the commissioning stage with respect to the hang-up of material in the bins. It is useful to note that the tendency of material passing into these bins to segregate was used to largely overcome the hang-up problem without recourse to any further activity. The constructor, designers and the owner were anxious not to use any vibrating systems as flow enhancers on the bins.

The difficulty the designers faced through sample material from the mine not being available prior to the design being undertaken was not as important as imagined at the time, the results from other iron ores proving sufficiently close to those for the Iron Duke deposit to make the design angles close to optimum. It was of some surprise to the designers that the bulk density of the mined material turned out to be significantly lower than was specified and that this seemed to have little effect on the flow characteristics of the material.

The use of overhead bins with choke cut-off gates was a cost-effective alternative to a large overhead bunker or an on-ground bunker with a large single hopper, which are used widely throughout the coal industry. The system is simple to operate, requiring minimal care by the train-driver and with some margin for error in the spotting of wagons under the bins. Spillage is minimal and, while fugitive iron ore dust does escape during the loading operation, it is unlikely that this is in sufficient amounts to cause any long-term problems at the site.

The total operation, involving pre-blending before crushing, windrow stockpiling after crushing, end-on recovery of material to the train-loading station and final blending of material through the bins to the wagons has shown a reduction of one-third in the standard deviation of contaminant variation in the product from Iron Duke mine. It has been estimated by the Whyalla steelmaker that the cost of loss of production of steel as a result of eratic blast-furance performance originating from contaminant variability would be in the order of A$5 million a year should the improvement shown by the Iron Duke system not have been attained.

8 ACKNOWLEDGEMENTS

This paper has been prepared with the co-operation of the Design Department and Mine Planning Group of BHP Long Products Division, Whyalla, and acknowledges that the success of the design owes a great deal to the accurate prediction of flow characteristics by The University of Newcastle Research Associates test laboratory.

9 REFERENCES

1. Gerstel, A.W. 1977, Bed Blending Theory, Chapter 24 of Stacking, Blending, Reclaiming, edited by R.H. Woehlbier Trans Tech Publications, Clausthal-Zellerfeld, W-Germany.

2. Weddig, H.J., 1977, Prehomogenisation of Raw Materials, Chapter 33 of Reference 1.

3. Gerstel, A.W. 1982 - A Model for Judging Bulk Materials Preparation in Blending Piles - Bulk Materials, Storage, Blending & Sampling, Proceedings of the First Australian International Conference, 1982, Sydney.

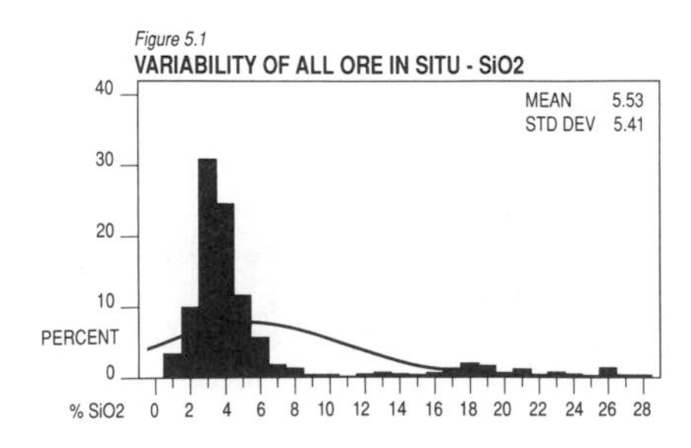

Figure 5.1
VARIABILITY OF ALL ORE IN SITU - SiO2

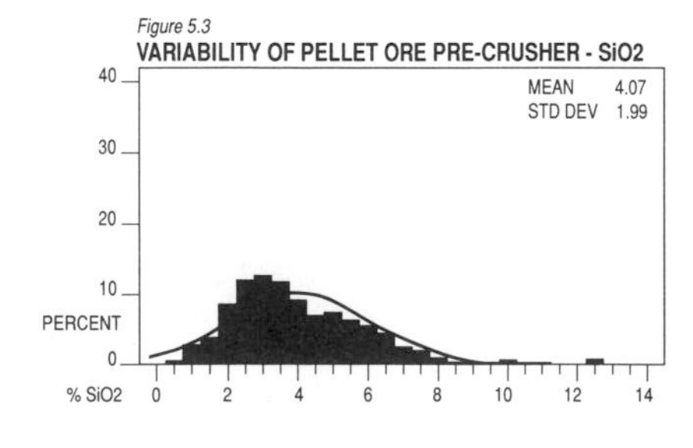

Figure 5.3
VARIABILITY OF PELLET ORE PRE-CRUSHER - SiO2

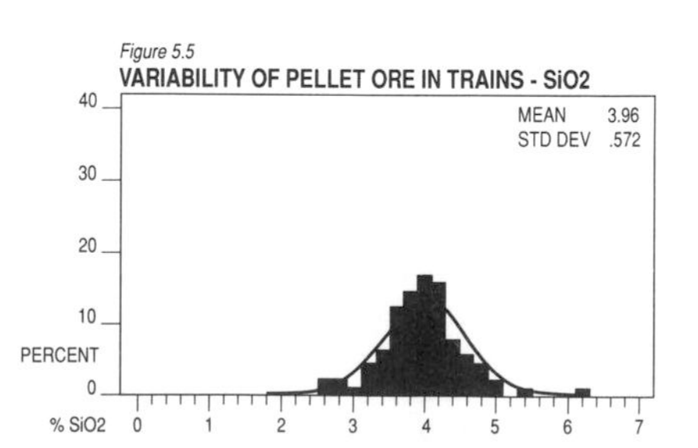

Figure 5.5
VARIABILITY OF PELLET ORE IN TRAINS - SiO2

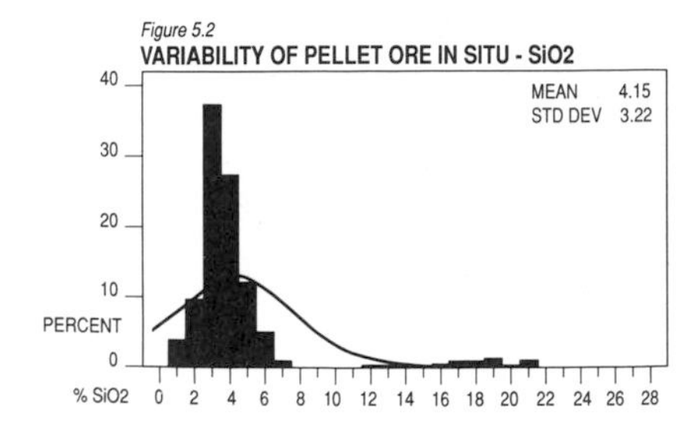

Figure 5.2
VARIABILITY OF PELLET ORE IN SITU - SiO2

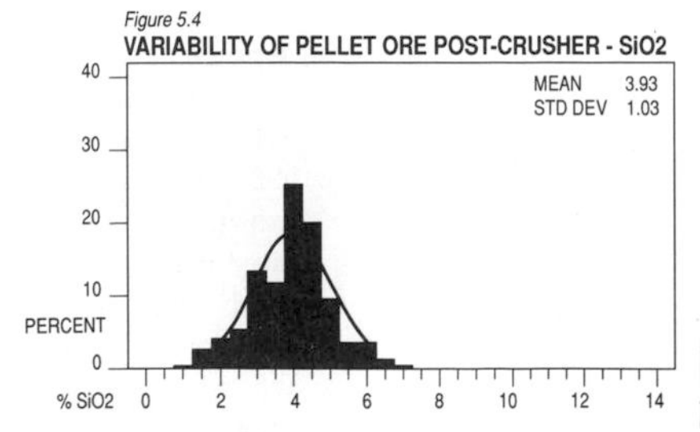

Figure 5.4
VARIABILITY OF PELLET ORE POST-CRUSHER - SiO2

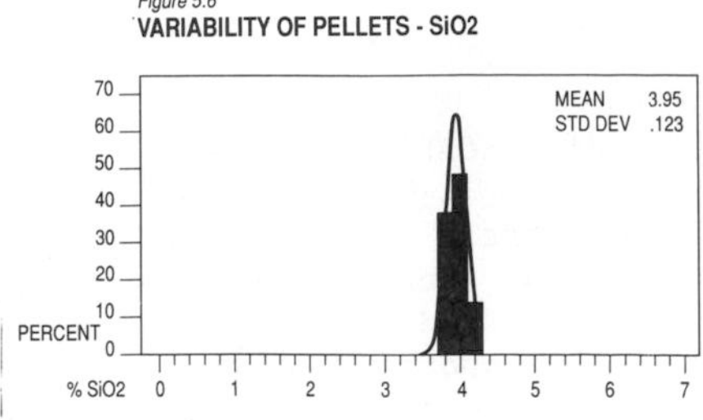

Figure 5.6
VARIABILITY OF PELLETS - SiO2

C418/018

Computer simulation of ballistic separation

C HOGUE, BSc and D E NEWLAND, MA,ScD, FEng, FIMechE, FIEE, FIOA, FIMA, MSEE, MASME, MASA
Department of Engineering, University of Cambridge

Synopsis *The separation of particulate materials on a vibrating sieve is being studied by a new computational method. Movement of individual particles is represented by Newtonian mechanics but, to allow acceptable computation times, the sieve is partitioned into separate sections. Particle movement between partitions is controlled by a probabilistic law. This computational strategy appears to offer significant advantages for modelling ballistic separation.*

1 Introduction

The separation of granular materials under vibration is used widely for grading or for improving product quality. Although vibrating screens have been in use for many years in industry, many problems arise for which only temporary or very costly solutions can be found. The grading of cylindrical particles, for example, is a recurring problem, since conventional sieves cannot prevent oversized cylinders from falling vertically through the sieve apertures. Optimal sieve performance is also of major concern.

There is extremely limited knowledge on the fundamental mechanisms of segregation and sieving for granular particles of different physical properties. The large number of parameters involved has led to mainly empirical analyses of specific combinations of sieves and materials. Recently attempts have been made to model the behaviour of granular materials in a more general context and these are summarized below.

2 Relevant previous work

The process of thin-layer screening has been shown empirically to produce optimal separation with good product quality and is being used increasingly in industry (see Mandrella [9] and Beenken [1]). In thin-layer screening, individual particles are observed to follow ballistic trajectories. The increasing availability of computing power has therefore led to the development of discrete models (as opposed to continuum models) for describing the behaviour of granular materials in sieving applications.

The difficulty with discrete models is (1) the amount of time required for their computation and (2) the difficulty of modelling inter-particle contact correctly. An example of a discrete model is Cundall's two-dimensional distinct element method (DEM) [4], which was originally conceived to model the progressive failure of rock slopes using a relatively small number of large elements. The model has now been applied by several authors [3, 11, 5] to systems containing hundreds of particles.

In the DEM, the particles are modelled as polygons and a key feature is the contact detection algorithm. Only neighbouring particles are checked for contact. The motion of the particles is updated according to Newton's second law, solved numerically by a simple central-difference algorithm. A force-displacement law is used to find contact forces from displacements.

Haff and Werner [6] have modelled one large and 30 small, solid inelastic disks moving in two dimensions. The normal force between particles is taken to be a function of their overlap at the contact point. The shear force is assumed to be simply related to the normal force through a coefficient of friction. Even for such a primitive model, only small-scale simulations are possible because of the time required for computation.

A number of modifications of the DEM algorithm for polygonal particles have been made. Ting and Corkum [11] have applied the algorithm to a segregation problem involving a 24-sided rotating drum containing 700 disks. Walton [12] and Campbell and Brennen [2] have studied the flow of granular particles down an inclined chute. An advantage of the DEM algorithm is its possible extension to three dimensions, as shown by Ghaboussi and Barbosa [5]. Although a large number of particles may be simulated, the computational requirements of these simulations are enormous.

Rosato *et al.* [10] have proposed a pseudo-random approach using the Monte-Carlo method of statistical mechanics to simulate segregation under conditions of vibration for a system of about 100 disks. Although the concept of the model is promising because computing time is saved by not having to numerically solve the equations of motion, it is not clear how to extend the model to a realistic practical situation. So far, no model has been applied to the problem of particle separation on an inclined vibrating screen. This is because these models have not yet reached a degree of computational efficiency which allows them to be used for a practical sieving application. In the next section, we describe our new model which can simulate both the segregation and the sieving process. Its objective is to reduce the computational requirements as much as possible, without sacrificing satisfactory modelling of the dynamics of particle contact.

3 Partition Model

The new model proposed here is based on the result that, in a granular material, individual particle behaviour is influenced primarily by collisions with particles in its neighbourhood. To exploit this feature, the (inclined) vibrating surface carrying the granular particles is divided into several partitions along its length. Each partition defines a "region of interest" within which particle behaviour is monitored. However, the size of the partitions is so chosen that only one partition at any given time needs to be under scrutiny. Which partition is "active" in this way is determined by the location of a "tracer particle" (for example the only black particle); the active partition is the one containing this tracer. Its trajectory is followed from partition to partition. The number of partitions varies according to the length of the sieve and the particle size distribution. An important feature of this model is that the equations of motion are only integrated for the particles contained within the active partition.

In order to simulate the motion of all the particles within the active partition, classical dynamical equations are used. When any particle leaves the (active) partition through one of its side boundaries, a new particle is introduced pseudo-randomly at the opposite boundary. If an undersized particle falls through an aperture of the sieve, it is not replaced.

The general algorithm of the partition model is given in table 1. For the current version of the model, the sieve is harmonically vibrated perpendicular to the screen surface.

```
Set simulation parameters
REPEAT
  Enter new partition
  Set initial configuration
  REPEAT
     FOR all particles within partition
       Compute contact forces
       Integrate equations of motion
       Check for particle crossing side
         boundaries in order to introduce
         new particle pseudo-randomly
       Check for particle being sieved,
          and remove
     Increment time
  UNTIL tracer has been removed or has
        passed to the next partition
  IF tracer has been removed GOTO END
UNTIL tracer has left the last partition
END
```

Table 1: Partition Model Algorithm.

Sieving is introduced to the model by allowing undersized particles to fall through apertures at the bottom of the sieve. The initial configuration of the first partition is set randomly according to a specified particle distribution. For subsequent partitions, the initial state is taken to be the final configuration of the previous partition.

The main features of the partition model algorithm, which are (i) the computation of the contact forces, (ii) the particles' motion, (iii) the flow across the boundaries, including the sieving process and (iv) the initial configurations, are described in more detail in the later sections.

4 Assumptions

Our model has the following main simplifying assumptions:

i. only smooth, spherical particles are considered;

ii. motion is two-dimensional with the centre of all particles moving in one plane;

iii. the angular rotation of the particles is ignored;

iv. inter-particle contact ends when the distance between the centres of each pair of interacting particles exceeds the sum of their undeformed radii (because of damping, actual contact may end sooner);

v. when a particle passes through the wall of a partition, it is reintroduced at the opposite wall but at a randomly-chosen height;

vi. when the strategy for introducing particles into a partition fails because another particle blocks the planned introduction, the randomly chosen entry position is reselected according to rules which avoid two particles having to occupy over-lapping positions;

vii. when the tracer particle passes through a wall, the initial state of particles in the newly-active partition is taken to be the same as the concluding state in the previously active partition.

The details of assumptions (v) to (vii) are discussed in detail where they arise in the text. Our intention is to refine all the assumptions and validate them where possible as the research continues.

5 Contact forces

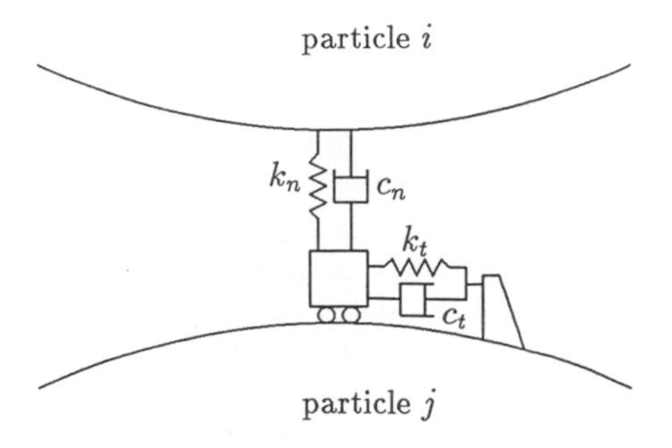

Figure 1: Simplified dynamical model of the inter-particle forces.

Let particle i have a normal stiffness k_i and normal damping c_i. At an impact between particles i and j, the resultant inter-particle normal stiffness and damping k_n and c_n are defined as

$$1/k_n = 1/k_i + 1/k_j \quad (1)$$
$$1/c_n = 1/c_i + 1/c_j \quad (2)$$

The value of k_i for each type of particle may be obtained experimentally from deflection measurements under given loads. The value of the normal (viscous) damping c_i may be obtained from a corresponding value of the coefficient of restitution by choosing c_i to give approximatively the same loss of energy at impact when two identical particles collide.

According to Hertzian contact theory, the ratio of the tangential stiffness k_t to the normal stiffness k_n may vary between 2/3 and 1. In their calculations of discrete particle behaviour, Cundall and Strack [4] showed that the results were not sensitive to this ratio and we have assumed that $k_t/k_n = 1$.

For typical values of mass, inertia and stiffness, the shear proportional damping factor $\beta_t = c_t/k_t$ is calculated by Corkum and Ting [3] to be approximately equal to the normal proportional damping factor $\beta_n = c_n/k_n$. Consequently, c_t may also be taken equal to c_n.

6 Inter-particle forces

Equations (3) through (6) describe the inter-particle contact forces for the model in figure 1. The axes used and the definitions of reference unit vectors are shown in figure 2. The x-axis is horizontal whereas the X-axis is parallel to the screen. Thus the y-axis is vertical and the $Y-$axis is perpendicular to the screen. The geometrical centre of particle i is defined by position vector $\mathbf{r}_i$ (not shown) with respect to the fixed origin O. The forces calculated are those exerted by particle j on the adjacent particle i; equal and opposite forces react back on particle j from particle i.

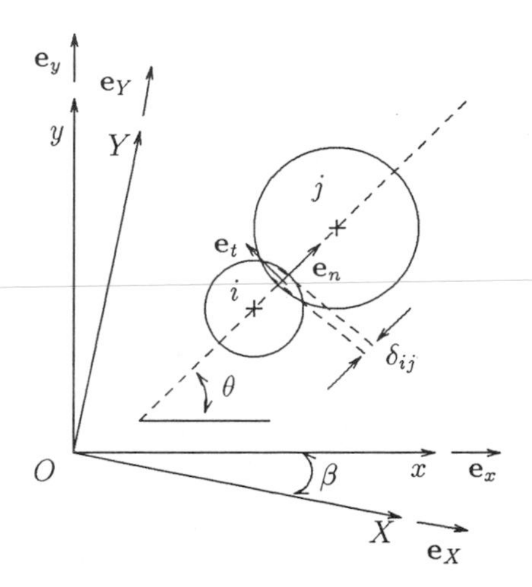

Figure 2: Coordinate systems, angle of impact and normal deflection. Note that the x-axis is horizontal and the X-axis is parallel to the screen.

The inter-particle normal stiffness force (K_n) is proportional to the normal deflection at the point of contact, whereas the normal damping force (C_n) is assumed to be proportional to the relative velocity between the centres of the particles in contact. The tangential (shear) stiffness and damping forces, K_t and C_t, are functions of the relative tangential velocity of the particles at their point of contact.

The inter-particle normal stiffness and damping forces may thus be written as

$$K_n = \delta_{ij} k_n \quad (3)$$
$$C_n = (\dot{\mathbf{r}}_i - \dot{\mathbf{r}}_j) \cdot \mathbf{e}_n \, c_n \quad (4)$$

The inter-particle tangential stiffness force is given by

$$K_t = \int_0^{\Delta t} (\dot{\mathbf{r}}_i - \dot{\mathbf{r}}_j) \cdot \mathbf{e}_t \, k_t \, dt$$
$$\approx (\dot{\mathbf{r}}_i - \dot{\mathbf{r}}_j) \cdot \mathbf{e}_t \, k_t \, \Delta t \quad (5)$$

assuming the tangential velocities are not altered by the normal impact, and where Δt is the duration of the contact. The tangential damping force is given by

$$C_t = (\dot{\mathbf{r}}_i - \dot{\mathbf{r}}_j) \cdot \mathbf{e}_t \, c_t \quad (6)$$

At every contact, we assume that there exists a limiting tangential resistance to sliding T_{max}, defined as

$$T_{max} = \mu_{ij} (K_n + C_n) \quad (7)$$

where μ_{ij} is the coefficient of friction between particles i and j. If the magnitude of the shear force (K_t+C_t) exceeds T_{max}, (K_t+C_t) is set to T_{max} and sliding occurs.

We also assume that the surface velocities deriving from angular rotation are small compared with the tangential components of the linear velocities. This has allowed the angular rotation of the particles to be ignored. It alters the detailed trajectories of individual particles (unless the particles are very small compared with their mean-free path) and we shall examine the errors that result in future research. However, for the present, we want to simplify the particle dynamics as far as possible in order to concentrate on how best to represent the movement of large groups of particles.

The instantaneous inter-particle forces are assumed to be $(K_n + C_n)$ in the normal direction and $(K_t + C_t)$ tangentially, where K_n, C_n, K_t and C_t are defined as above, and subject to the limit (7). For this particle model, these forces are resolved along the reference directions X, Y so that the resultant inter-particle force applied on particle i by particle j can be written as

$$F_{ij}^X = -[K_n + C_n] \cos(\theta + \beta) + [K_t + C_t] \sin(\theta + \beta) \quad (8)$$
$$F_{ij}^Y = -[K_n + C_n] \sin(\theta + \beta) - [K_t + C_t] \cos(\theta + \beta) \quad (9)$$

These equations apply only for the duration of the contact between particles i and j. To determine when this contact ends, it is assumed that the particles remain in contact as long as the distance separating them does not exceed the sum of their radii. That is, the inter-particle deflection is assumed small enough to justify neglecting the case where the particles separate before having regained their initial shape.

7 Screen forces

The first partition is chosen to start immediately after the loading area of the sieve as we expect that there is little to be gained by including the loading area in the simulation. This assumption is based on the fact that the length of the loading surface is small compared to the length of the screen. Therefore, in a two-dimensional model of a vibrating sieve, no wall effect, except for the screen, needs to be considered.

The normal force imparted on the ball by the harmonically vibrating screen can be written as:

$$F_s^Y = k_i\delta_s + c_i(\dot{z} - \dot{Y}_i) \qquad (10)$$

for $\delta_s > 0$; F_s^Y is zero otherwise. $\dot{Y}_i$ is the velocity of particle i along the Y-axis. The deflection δ_s of the particle at the contact point and the motion of the screen (which is normal to its plane) are described respectively by equations 11 and 12:

$$\delta_s = z - (\mathbf{r}_i \cdot \mathbf{e}_Y - R_i) \qquad (11)$$

$$z(t) = z_o \sin \omega t \qquad (12)$$

where R_i is the radius of particle i.

The tangential force between the screen and particle i is calculated in the same manner as the tangential interparticle forces, subject to the maximum magnitude of the tangential force defined by

$$F_{s_{max}}^X = \mu_{si} F_s^Y \qquad (13)$$

where μ_{si} is the friction coefficient between the screen and the surface of particle i.

8 Equations of motion

The equations of motion of particle i in the X and Y coordinates are given by

$$m_i\ddot{X}_i = F_s^X + F_{ij}^X + m_i g \sin\beta \qquad (14)$$

$$m_i\ddot{Y}_i = F_s^Y + F_{ij}^Y - m_i g \cos\beta \qquad (15)$$

where $\ddot{X}_i$ and $\ddot{Y}_i$ are the X and Y coordinates of the acceleration of particle i, m_i is its mass, β is the angle of inclination of the sieve, and the forces F_{ij}^X, F_{ij}^Y, F_s^X and F_s^Y are as defined above. Note that these forces are generally zero except when particle i overlaps with a particle j or with the screen.

The above equations are solved using a second-order Runge-Kutta algorithm with varying step size. The accuracy check is performed by integrating over the same region twice, first by using a single step, then by using two half steps. The step size is adjusted to ensure a maximum error of 1% at each step. A fourth-order Runge-Kutta algorithm has also been used but it was found that the increase in accuracy was not worth the increase in computation time, since a high level of accuracy is not essential in such a semi-probabilistic system.

9 Boundary conditions

A particle is considered to have left the active partition when its centre of mass has crossed one of the side boundaries of the partition or, in the case of an undersized particle, it has passed through the sieve. (This sieving process will be discussed in detail in the next section.) When a particle has left the partition by a side boundary, it is replaced by another particle introduced at the opposite boundary. A trial position for the height at which the particle is introduced is determined pseudo-randomly according to a constant distribution centered around the centre of mass of that type of particle, so that the segregation phenomenon which affects the vertical distribution of particles within the bed is taken into account. The trial position will be rejected if the particle's overlap with an existing particle is greater than the accepted tolerance (which is currently set to be 5% of the sum of the two particles undeflected radii). If such is the case, a new trial position is generated and is tested again for excess overlap. After three trials, the position is adjusted by moving the entering particle to the nearest position where its overlap lies within tolerance.

This variability in the mean value of the probability distribution of the height of reintroduced particles ensures that the proportion of undersized particles is increasingly greater in the lower layers of the bed as the partition is closer to the end of the sieve. The velocity vector of the newly introduced particle is chosen to have the same value as had the previous particle when it left the partition.

The above scheme is not applied to the tracer particle when it enters a new partition. The incoming height is the height at which the tracer left the previous partition, but if there is excessive overlap with the other particles this overlap is ignored. The resulting motion soon separates strongly overlapping particles. This strategy is chosen because it is assumed that the number of partitions is small so that an overlap with a particle close to the same position is not likely to occur often enough to justify the extra computing requirements of a more complex strategy.

A more elaborate scheme in which the rate of incoming particles will be "independent" of the rate of outgoing particles is currently being studied.

10 Sieving

When an undersized particle percolates to the bottom layer of the bed of materials and is within a small distance from the screen, and its velocity vector points in the direction of one of the apertures (see figure 3), the force on the particle exerted by the screen is set to zero. At the next integration step, the particle will simply "fall" through the aperture and when its height is below that of the sieve, it is deleted from the data structure.

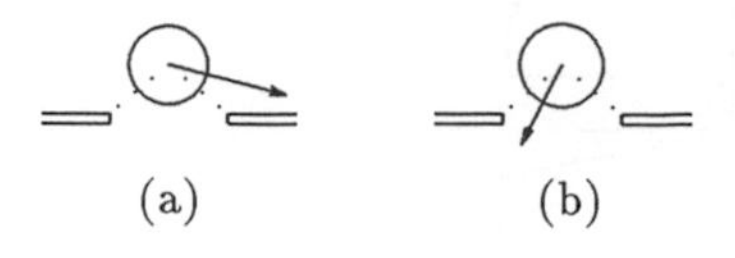

Figure 3: The condition on the velocity direction for the sieving algorithm: (a) particle will bounce off the sieve, (b) particle will go through the aperture.

11 Initial Configurations

Each time a tracer leaves a partition, a suitable initial configuration must be found for the new active partition. Several methods for generating initial configurations have been tested, including a Monte-Carlo method. The main disadvantage of such methods is the considerable computing requirement. Since the minimization of the computational effort is a major goal of the partition model, such methods were discarded.

For the first partition, a simple algorithm has been adopted. It consists of generating rows of non-overlapping particles of various sizes according to the chosen particle distribution and the desired depth. These are left to "settle" under gravity for several time steps. After this initialization period the sieve starts to vibrate and a tracer is introduced on the top layer of the bed of materials.

The initialization of the other partitions consists of taking the last configuration of the previous partition as the initial state for the new one. As a result of the sieving process, the proportion of fines decreases at the beginning of each new partition as the tracer approaches the end of the sieve.

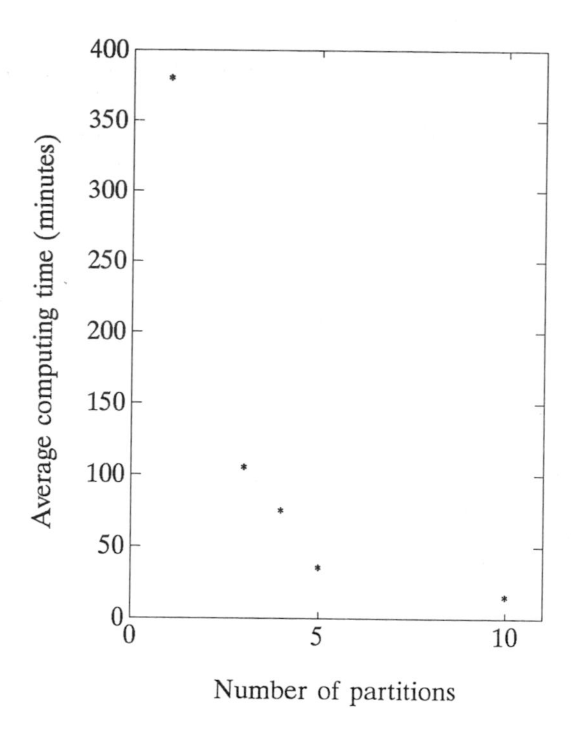

Figure 4: Typical relationship between the number of partitions and the required computing time.

12 Simulation results

A simulation is completed when the tracer leaves the sieve, either through one of the mesh's apertures or by passing through the boundary of the last partition.

Simulations were performed on a binary mixture of particles, on a 1 m long sieve, inclined at 5° with the horizontal and vibrating harmonically at 10Hz (normal to the sieve) with an amplitude of 4mm peak-to-peak. The diameters of the undersized and oversized particles were 10 and 20mm respectively. The screen apertures had a dimension of 19mm and were regularly spaced a distance 3.3mm apart (22.3mm between centres).

These simulations were repeated for different numbers of partitions. Figure 4 shows the relationship between the average time of the simulations and the number of partitions used. On an HP 800 series workstation, the model using 5 partitions proved to be about 10 times faster than the single partition one. This number increased to 25 when 10 partitions were used. The optimal number of partitions will be a compromise between minimizing computing time and minimizing the errors introduced.

Although for research studies very long computation times are possible, for example more than a week of c.p.u. time, we are interested in practical design calculations where an hour of computing time may be regarded as completely unacceptable.

For the parameters given above, it was decided that 5 partitions was a good comprise between numerical efficiency and accuracy. In figure 5, snapshots of the simulation show the tracer particle making its way through the different partitions. Note that a partition freezes to its last configuration once the tracer has left it.

Figure 6 shows successive snapshots of partition 3 in the above simulation, to illustrate the behaviour of particles within a single partition.

In order to examine the validity of the equations of motion, predictions for the motion of particles in a rectangular walled container, vibrated parallel to the vertical walls, have been compared with experimental tests. These experiments were performed on a vibrating perspex rig, containing smooth plastic spheres of one size restrained to move in two dimensions. Their trajectories were obtained by digitizing and processing video images taken during the motion. Apparently good correspondence has been obtained between theory and experiment [7].

More comprehensive tests of the full computer model on an experimental sieve rigged with measurement equipment will be performed shortly. The proportion of fines at the end of the sieve and the rate of sieving, which can be easily computed, will be used to assess the sieve's performance using different particulate media.

13 Conclusion and further work

The intention of the work described here is to develop a computationally attractive model for ballistic separation and sieving. The model described in this paper approaches this goal by partitioning the field of motion into several separate sections. Motion in each section or partition is simulated separately but the continuing motion of an identified particle (the tracer) can be followed as it passes from section to section.

In order to make the computation tractable, at present it has been limited to simulating the motion of spherical visco-elastic particles moving in one plane. This two-dimensional model allows the effect of different size particles, different vibratory conditions at the screen, different bed depths and different particle masses, stiffnesses, damping and friction coefficients to be studied. Also the effect of differing angles of inclination to the horizontal of the sieve can be examined. All these features are of great importance if such a computational model is to be

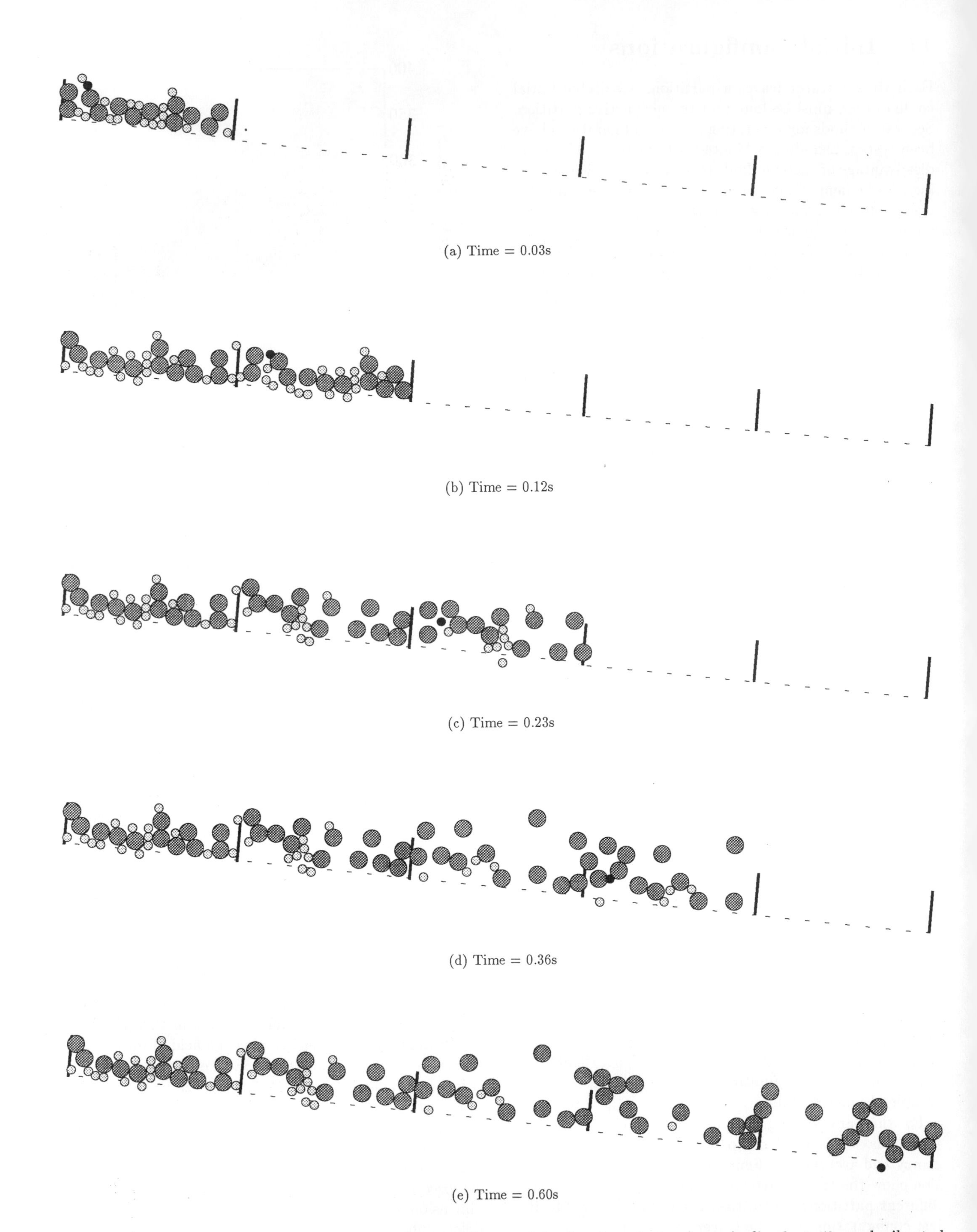

(a) Time = 0.03s

(b) Time = 0.12s

(c) Time = 0.23s

(d) Time = 0.36s

(e) Time = 0.60s

Figure 5: Snapshots of a simulation using the partition model: the sieve is 1m long, inclined at 5°, and vibrated harmonically at an amplitude of 4mm peak-to-peak and a frequency of 10Hz (normal to screen); the screen apertures' size is 19.0mm and they are 3.3mm apart; the screen coefficient of friction is 0.3; the particle diameters are 10 and 20mm, with masses of 0.01 and 0.02kg, stiffnesses and damping of 10 000 N/m and 200 Ns/m for the undersized, 5000 N/m and 100 Ns/m for the oversized and friction coefficients of 0.5 and 0.3 respectively.

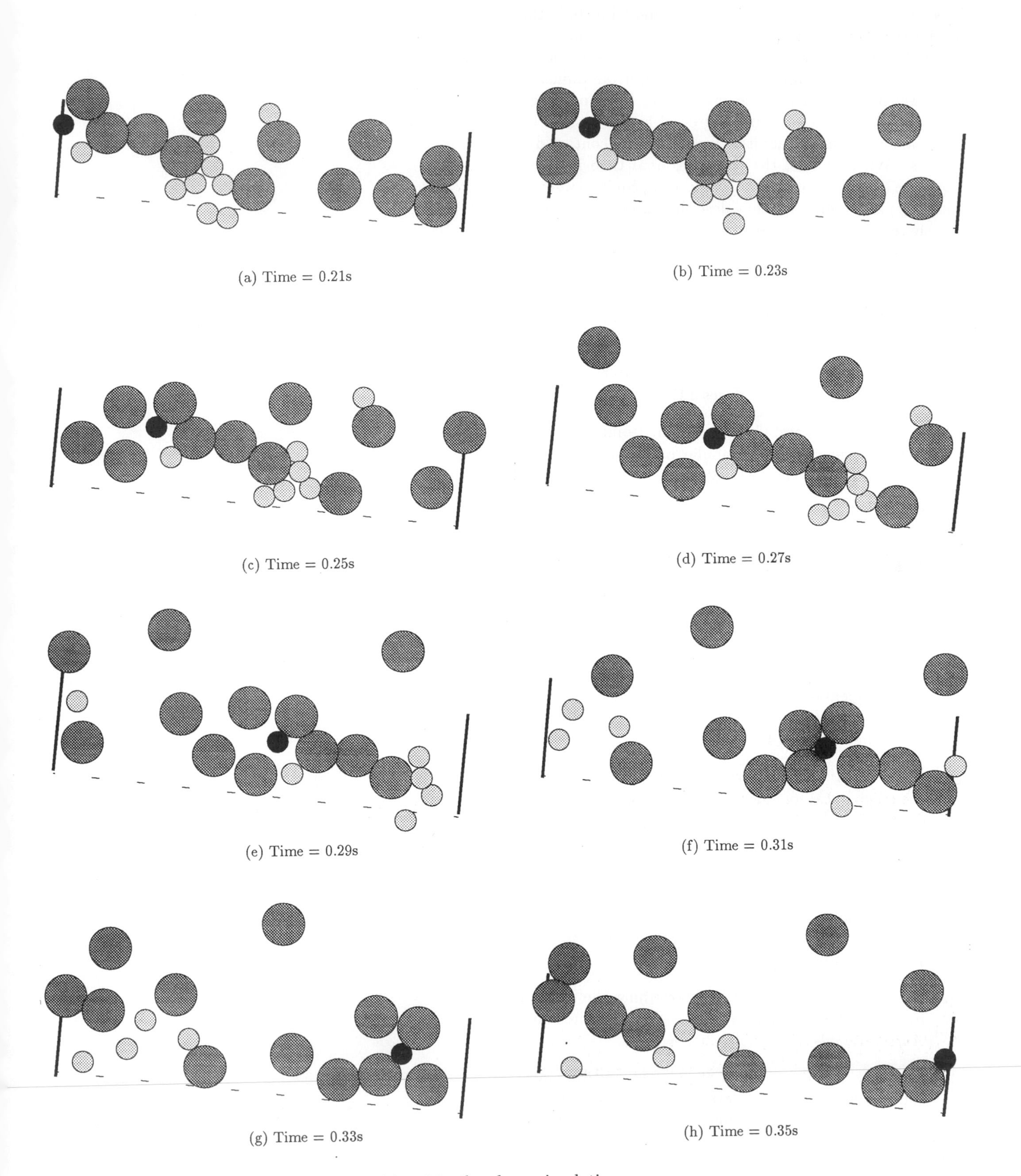

(a) Time = 0.21s

(b) Time = 0.23s

(c) Time = 0.25s

(d) Time = 0.27s

(e) Time = 0.29s

(f) Time = 0.31s

(g) Time = 0.33s

(h) Time = 0.35s

Figure 6: Zoom on several states of partition 3 in the above simulation.

used directly in industry for the design and operation of vibrating sieves.

Next steps in the refinement of this model include the development of equations of motion which permit non-spherical particles to be modelled and the extension of the simulation to three dimensions. Because of the difficulty of establishing valid particle interaction laws for rough non-spherical particles, some degree of averaging will be necessary using a probabilistic model for the contact dynamics. Without this, the computational time required would be excessive for existing computing equipment even with the partitioning strategy described above. Nevertheless there seems to be good scope for further development of the model described here.

References

[1] W. Beenken. Method for the modelling of the screening process on circular vibratory screens. *Aufbereitungs-Technik*, pages 147–156, 1990.

[2] C. S. Campbell and C. E. Brennen. Computer simulation of shear flows of granular material. In Jenkins and Satake [8], pages 313–326.

[3] B.T. Corkum and M.J. Ting. The discrete element method in geotechnical engineering. University of Toronto. Internal Report, 1986.

[4] P. A. Cundall and O. D. L. Strack. A discrete numerical model for granular assemblies. *Géotechnique*, 29(1):47–65, 1979.

[5] J. Ghaboussi and R. Barbosa. Three-dimensional discrete element method for granular materials. *International Journal for Numerical and Analytical Methods in Geomechanics*, 14:451–472, 1990.

[6] P. K. Haff and B. T. Werner. Computer simulation of the mechanical sorting of grains. *Powder Technology*, 48:239–245, 1986.

[7] C. Hogue. Ballistic separation. PhD thesis, University of Cambridge. To be published, 1992.

[8] J. T. Jenkins and M. Satake, editors. *Mechanics of Granular Materials: New Models and Constitutive Relations.* Elsevier Science, Amsterdam, 1983.

[9] H Mandrella. Criteria for determining the sizes of machines for fine and medium particle screening. *Aufbereitungs-Technik*, pages 3–8, 1984.

[10] A. Rosato, F. Prinz, K. J. Standburg, and R. Swenson. Monte-Carlo simulation of particulate matter segregation. *Powder Technology*, 49:59–69, 1986.

[11] J. M. Ting and B. T. Corkum. Discrete element models in geotechnical engineering. In *Proceedings of 3rd International Conference on Computing in Civil Engineering*, Vancouver, Canada, August 1988.

[12] O. R. Walton. Calculations of shear flow. In Jenkins and Satake [8], pages 327–338.

C418/059

Dust, mess and spillage – an intractable or solvable problem?

H N WILKINSON, BSc(Eng), CEng, MIMechE, FIMatM
Stockton-on-Tees, Cleveland

SYNOPSIS A major problem with bulk materials handling is dust, mess and spillage. This causes many problems such as loss of material, extra maintenance, reduced plant life and availability and a poor working environment. These problems are estimated to cost UK industry about £200 million p.a. Various solutions are available involving containment, better design, dust suppression etc. which need to be made more widely known and applied. Greater encouragement needs to be given to promulgation of the best available technology and to research, development and innovation.

1. INTRODUCTION

When the I.Mech.E's Process Industries Division formed a Bulk Materials Handling Committee in 1983 one of the first tasks it undertook was to send out a questionnaire to identify the main problem areas in the handling, storage, physical processing and transport of bulk materials.

The most serious problems, in order of severity, were found to be;

(a) Dust, mess and spillage.
(b) No flow.
(c) High maintenance.
(d) Low reliability.
(e) Equipment unable to meet performance expectations.

The first of these problems, owing to its severity and universality, became the subject of a further study, as described in Section 3.

2 PROBLEMS CAUSED BY DUST, MESS AND SPILLAGE

2.1 Loss of bulk material

This occurs when spillage cannot be re-cycled or, in the case of fine dry materials, the loss is in the form of windborne dust. Such windage losses occur from open stockpiles, from transfer operations (e.g. grabbing, discharge from a boom stacker) and when material is churned-up by mobile equipment.

2.2 Cost of spillage clearance

Whether this is done manually or with the aid of machines (e.g. pneumatic pipeline) it can be expensive in manpower. Additionally the plant may have to be taken out of productive use while spillage clearance is in progress.

2.3 Extra maintenance

The presence of spillage etc. often causes equipment to wear out more quickly (e.g. through dirt in bearings), especially if the spillage material is abrasive or corrosive.

2.4 Reduced plant life

This follows from 2.3, especially if the necessary maintenance is neglected.

2.5 Reduced plant availability

This is a further likely outcome from 2.3. Spillage may also cause plant to mal-function (e.g. conveyor belt wear etc.). With some types of equipment (e.g. grabbing cranes) the plant may have to be stopped in windy weather owing to the excessive dust nuisance. In extreme cases it has been known for a plant to be out of action for a long time owing to spillage (e.g. a belt conveyor gantry collapsing under the weight of accumulated spillage.)

2.6 Contamination

With a plant producing high quality material the spillage may cause contamination of other products and may itself have to be reprocessed before it can be re-cycled.

2.7 Low staff morale

A dirty plant is unpleasant to work in. Thus maintenance work is likely to be less well done, the best staff may want to work in more attractive surroundings, leading to generally run-down conditions.

2.8 Low grade technology

Following on from 2.7 above, a dirty plant is inhospitable to sophisticated measurement and control equipment, leading to heavy dependence on manual operation and supervision.

2.9 Poor environment

This applies particularly where dusty materials are involved. In windy conditions clouds of dust are likely to be deposited on surrounding property. This is especially damaging if the material is in any way injurious.

2.10 Health hazard

This follows on from 2.9 above and may apply both to plant personnel and people living/working in the plant vicinity.

2.11 Dangerous working conditions

There is a general danger of personnel slipping/stumbling on walkways where spillage has accumulated. There is a special hazard when potentially explosive materials become airborne.

3. THE COST OF DUST, MESS AND SPILLAGE

In view of the formidable list of problems attributable to spillage etc. one might ask why greater efforts have not been made to deal with the matter. An obvious reason is that industry is not aware exactly how much spillage etc. is costing. Accordingly, the I.Mech.E. commissioned a study into this very subject and in 1988 a report was produced by Wilkinson, Reed and Wright(1).

The main finding of the study was that dust mess and spillage cost UK industry at least £200 million p.a. This is apart from costs that are difficult to quantify such as the general depressant effect and the environmental aspects mentioned previously.

The cost of dust, mess and spillage was reduced to the following formula:

1% loss of product + 22 pence per tonne of throughput

In the I.Mech.E. study, visits were made to eight UK plants as follows;

- coal burning power station
- china clay producer
- lead/zinc ore smelter
- flour mill
- cement works
- fertilizer manufacturer
- aluminium ore smelter
- coke works (for domestic coke)

These were considered to be reasonably representative of industries handling bulk materials on a large scale. The main areas where costs attributable to spillage etc. could be quantified were loss of material, spillage clearance and extra maintenance. Other cost areas could probably have been quantified if an in-depth study over a long period of time had been undertaken.

4. SOLUTIONS TO THE PROBLEM

4.1 Eliminate handling, storage, transport

This might be achieved by receiving bulk materials on a JIT (= Just in Time) basis, by using alternative materials (e.g. oil in place of coal as a fuel) but in most instances scope for this approach is very limited.

4.2 Containment

This can take many forms:

(a) Use of silos or covered stockpiles in place of open stockpiles.
(b) Use of a totally enclosed form of ship discharger (e.g. Siwertell) instead of a conventional grab.
(c) Employing a totally enclosed form of conveyor (e.g. 'Cleanveyor', airslide, en-masse, pneumatic pipeline, slurry pipeline in place of a conventional belt conveyor.
(d) Fitting shrouds around loading point (e.g. hoppers into which grabs or lorries discharge) or by loading wagons, lorries etc. through sealed spouts.
(e) Fitting dust extraction equipment at situations where airborne dust might arise (e.g. at belt transfers, over vibrating screens etc.).

4.3 Better design

This can be achieved in various ways;

(a) Taking into consideration the physical properties of the bulk materials being handled (e.g. to ensure the material flows properly so that blockages - a potent cause of spillage - do not occur). N.B. Guidance on measuring the physical properties of bulk materials is given in a BMHB Guide (2) and by Svarovsky (3).
(b) Designing on a more generous basis with respect to capacity.
(c) Using better construction material (e.g. abrasion/corrosion resistant).
(d) Employing proprietary peripheral equipment (e.g. skirt seals, tracking idlers, belt cleaners) where appropriate.
(e) Providing better access for cleaning-up spillage.

4.5 Dust suppression

This covers both containment (section 4.2) but also the use of water sprays, foam etc. In some situations the use of anti-degradation devices to reduce the generation of fines will be appropriate.

5. ACTION

The following are some of the ways in which improvement is likely to be made.

5.1 Cost effectiveness

Although the previously mentioned sum of £200 million p.a. (section 3) is a large one and justifies expenditure on a collaborative basis to reduce dust, mess and spillage, it does not necessarily justify heavy extra capital expenditure on a particular plant. For a typical plant handling 1mt. p.a. of bulk material costing £40/t the total savings from zero spillage etc. amount to £620 000 p.a. using the previously given formula. On the other hand, the initial capital cost of such a plant might be around £100 million - not much scope here for excessive lavishness in design! Thus the emphasis in any appropriate action must be in getting value for money, in particular making the right choice of handling system in the first place.

5.2 Better awareness

There is a considerable amount of information relevant to the dust, mess and spillage problem that needs to be made more widely available. The I.Mech.E. has already made some progress in this direction by organising seminars on 'Belt Conveyor Transfer Points' (4) 'Getting Value for Money in Bulk Handling Plants' (5) and 'Solutions to Industrial Dust, Mess and Spillage Problems' (6).

However, one difficulty is that many of the people in industry who need to be made aware of the solutions to the various problems are not members of the I.Mech.E. or indeed of any other technical institution. So the question that remains unresolved is who should pay for the organising of seminars, writing of publications etc. and, not least, ensuring that knowledge of such seminars and publications gets to the right people.

A further problem is that much of the best information on dealing with dust, mess and spillage problems is in the form of research reports that are not readily understood by industry. The need is for such information to be put in simple, practical terms but again, who pays for this to be done?

For completeness it should be added that the BMHB (British Materials Handling Board) has published a guide to the handling of dusty materials in ports (7) while the whole subject of bulk handling is covered in a textbook by Woodcock and Mason (8).

5.3 Equipment assesment

One of the problems facing plant and equipment designers and operators is knowing the exact capability of a piece of equipment with respect to the bulk material(s) being handled. Sometimes the people involved do not have specialized knowledge of bulk handling. Knowing the physical properties of the bulk material(s) involved is a considerable help and has been referred to previously (section 4.3). What is also needed are test facilities where equipment can be tested under real-life conditions. To some extent equipment manufacturers already do such testwork but in most cases facilities are limited to small throughputs. At present the best that can be hoped for is that equipment is tested and developed while installed on a production plant, with all the associated problems. Ideally an independent test facility is required but, as with 'Better Awareness' (section 5.2), who is going to pay for it?

Such a test facility is needed, not only to assess the capability of existing equipment but also to encourage development and innovation.

5.4 Research, development and innovation

It is a moot point whether basic research, development of existing equipment or innovation (i.e., introduction of new concepts) is likely to be the most profitable way forward. Ideally the need is for progress on all three fronts. As previously stated (section 5.3), facilities for testing and assessing new ideas would be beneficial.

Regarding research, the need is to encourage research into bulk handling topics and ensure that such research gets a fair share of the available funding. Topics for research are given in a report recently commissioned by the DTI (Dept. of Trade and Industry) and produced by Michael Neale and Associates (9). An earlier report commissioned by the DTI (10) gave suggestions for product development needs.

6 CONCLUSIONS

Dust, mess and spillage is still the biggest problem with bulk materials handling. In many cases solutions are available, particularly when dry dusty rather than wet cohesive bulk solids are being handled. In order to improve matters greater effort needs to be made in getting the best available technology more widely known and applied and in encouraging research, development and innovation. Other factors that will encourage more serious attention being given to the problem are pressure for a better and safer environment and a better appreciation of the costs involved with a badly designed / maintained/operated plant.

REFERENCES

(1) WILKINSON, H.N., REED, A.R. and WRIGHT, H.

The real cost to UK industry of dust, mess and spillage in bulk materials handling plants.
I.Mech.E. report dated June 1988.

(2) Bulk solids physical property test guide, 1983, British Materials Handling Board, Ascot, Berks.

(3) SVAROVSKY, L. Powder testing guide, 1987, Elsevier Applied Science.

(4) Belt conveyor transfer points, London, Oct. 1985, I.Mech.E seminar.

(5) Getting value for money in bulk handling plants, London, Feb. 1989, I.Mech. seminar.

(6) Solutions to industrial dust, mess and spillage problems, London, April 1989, I.Mech.E. seminar.

(7) SCHOFIELD, C. and SHILLITO, D. Guide to the handling of dusty materials in ports, 1983.

(8) WOODCOK, C.R. and MASON, J.S. Bulk solids handling, 1987, Blackie and Son Ltd.

(9) A survey of mechanical handling equipment to discover new product opportunities and relevant subjects for basic research work, 1988. Report SRS 368 by Michael Neale and Assoc. Ltd. for the Dept. of Trade and Industry.

(10) DAVIES, B. et al. The product development needs of the UK mechanical handling industry, 1979. Report RB/ME/79/30 for the Dept. of Trade and Industry.

C418/040

Preliminary results of a new dust deposition monitor

E J VAN ZUYLEN, PhD, L A M RAMAEKERS, PhD and T C J VAN DER WEIDEN, PhD
Ecofys, Utrecht, The Netherlands

SYNOPSIS A special apparatus (the EDDM) has been developed to measure the deposition of coarse dust in relation with the nuisance it causes. At short time intervals the additional amount of deposited dust is determined by measuring the decrease of the reflection of an opaline glass sample in a sample holder with a NACA profile. The results of measurements near a coal- and ore-handling area showed that relatively short periods (lasting some hours) with a high deposition rate were responsible for most of the deposited dust. The EDDM showed great accuracy in determining the time course of deposition and the direction of the source of the dust.

NOTATION

EAC	Effective Area Coverage, expressed in a percentage of a surface
R_q	Reflection quotient
h	typical height of the sample holder
μ	Viscosity of the air
ρ_{DEHS}	Specific density of DEHS
d_{ae}	Aerodynamic diameter of a particle
v_a	Velocity of air flow
η	Sampling efficiency
Stk_p	Stokes number for a particle "p" near the sample holder
$Stk_{p,50}$	Value of Stokes number where sampling efficiency is 50%

1 INTRODUCTION

People living in the neighbourhood of bulk materials transhipment areas often experience nuisance caused by the deposition of (coarse) dust. It is difficult to quantify this nuisance and to evaluate measures taken to prevent the emission of dust.
Most dust measuring techniques are not specifically aimed at coarse dust but, on the contrary, at fine dust. Also, nuisance is related to the accumulation of dust in relatively short periods, whereas existing deposition measuring techniques do not give information about short time intervals. They are not directionally sensitive and results are only available after laboratory analysis. Furthermore, nuisance is related to the pollution of surfaces, but deposition measurement techniques generally determine the mass of deposited dust instead of, for instance, the resulting decrease of reflectance. Finally, impaction is supposed to contribute significantly to the pollution of objects, whereas deposition measurement techniques are mainly aimed at sedimentation only.

Ecofys research and consultancy has developed an apparatus called the EDDM, the Ecofys Dust Deposition Monitor (see Figure 2), to meet at least some of the above mentioned requirements.
A specially designed sample holder collects coarse dust. Instantaneous information about the deposited dust and the direction of the source is collected every two hours with an on-line technique to measure the reflection of the sample. The results are collected together with the actual wind direction.
In this manuscript, the properties of the apparatus will be briefly described.
Further, the results of experiments near a coal- and ore-handling area will be described. These experiments were carried out to test the practicable usefulness of the EDDM, its capability to determine the source of the dust and to compare the results of the EDDM with coarse dust concentration measurements.

2 THE EDDM

2.1 Demands for a device to measure dust deposition

Problems with existing deposition measuring techniques, which are mentioned in the Introduction, can be translated into demands for a new technique.
- In the first place, complaints related to nuisance caused by dust are often connected to the accumulation velocity and they develope in a short time [1]. Therefore, short sampling intervals are wanted. Furthermore, when a relationship to the source and to meteorological conditions is wanted, a high time resolution is necessary. However, existing techniques to measure the deposition of (coarse) dust usually use long time intervals (a month or at least a week) to determine the amount of deposited dust. These long intervals are necessary because relatively

insensitive gravimetric techniques are used for the analysis, combined with low sampling efficiencies.
- In the second place, results should be instantaneously available if appropriate dust prevention measures are to be taken. However, nowadays laboratory analysis is generally necessary afterwards, to obtain results for existing techniques.
- With respect to pollution of surfaces, impaction and sedimentation (and turbulent deposition) are important deposition processes [2]. Which is the most important depends on the meteorological conditions, particle size distribution and the receptor among other things [3]. At high wind velocities, impaction gains weight as a cause of pollution, whereas the strength of coarse dust sources, for instance related to wind erosion, also increases. However, current deposition measurements are mainly aimed at sedimentation only. Therefore, existing deposition measurement techniques are mainly useful for the evaluation of long term effects and the determination of the overall distribution of dust.
- Techniques for instantaneous measurement of dust concentration generally suck in air. They are sometimes able to give on-line information about short time intervals (e.g. Beta-monitor). This apparatus as well as commonly used high volume sampling assemblies such as PM10, are in general not specifically sensitive to coarse dust (> 10μm), which is supposed to be the most significant cause of pollution in the surroundings of dust sources. Finally, the relationship between measured concentration and nuisance is poorly known.

Summarising, a new nuisance related dust deposition measuring technique should mainly be aimed at impaction. The shape of the receptor can not be uniquely defined because no receptor is representative for all objects. Furthermore, the technique should sample coarse dust with short sampling intervals and with short response times. It should not suck in air nor measure concentration. The analysis technique should be related to the reflection or to the EAC of samples.

2.2 Description of the EDDM

Dust is collected on an opaline glass sample with an adhesive coating to prevent particle bounce. The sample holder has an especially developed aerodynamic profile (see Figure 1) so as to disturb the airflow as little as possible. This shape was chosen on theoretical grounds and after comparison with a more simple profile of the same dimensions, which consisted of a flat plate with rounded edges. The deposition of dust particles on the NACA sample holder was higher and much more homogeneous than on the simple alternative.

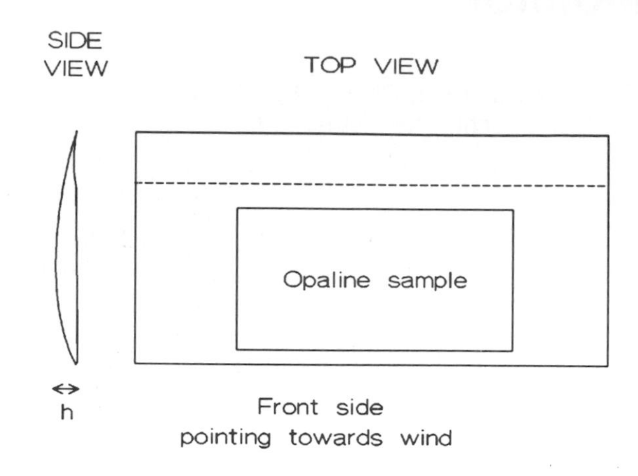

Figure 1 The sample holder (l · w · h = 150 · 60 · 6 mm). Area below dotted line is flattened slightly to hold the sample (l · w = 75 · 42 mm).

In the sampling position, the sample holder is located 2.3 m above the ground level (see Figure 2). The sample holder is tilted at an angle between 5° and 15°, depending on the (expected) deposition rate (see also Chapter 5).
The direction of the wind is continuously measured and is used to position the sample holder perpendicular to the airflow and to determine the direction of the source of the dust. After an adjustable time interval (e.g. one hour), the sample is retracted into the apparatus housing and analysed. If rain is detected by the mounted rain detector, the sample is also retracted into the apparatus to prevent that previously sampled particles are washed away. The housing of the apparatus consists of an aluminium frame covered with synthetic plates. The dimensions of the housing are 0.4m · 0.5m · 1.2m.
The quantity of the dust on the sample is determined by an optical technique. 40 Sensors, which consist of pairs of LED's and phototransistors, measure the reflection of infrared light by the sample. These values are compared to the preceding measurement. The resulting quotient is mentioned the Reflection quotient (R_q). The maximal load of the sample is reached when about 10% of the surface is covered with particles. The chance that new particles bounce off due to collision with other particles, which were caught in an earlier stage, increases above acceptable values. Further, adhering particles may partly cover each other. The relation–

Figure 2 Picture of the EDDM at the location of the experiments.

ship between R_q and quantity of collected dust might then be changed. So, sampling efficiency as well as analysis results can be affected when the sample is overloaded. By adjusting the angle of the sample holder or the length of the period before sample exchange, overloading can be prevented so that a 10% coverage of the sample surface should not occur in practice.
A built-in control and measuring computer controls the movements which are executed with great precision by stepping motors. The computer also takes care of the initial data analysis and storage. Temperature correction for each sensor and averaging over all sensors takes place instantaneously.
The overall measuring accuracy for the determination of the EAC was determined in earlier experiments to be better than 0.1%. The wind direction is averaged during the sampling time. The time course of the EAC as well as the direction of the wind (that is the direction of the source of the collected dust) are stored or can be retrieved immediately.

2.3 Sampling theory

Dust is collected on the sample holder by various processes. Impaction, which is supposed to play an important role, can be described in the following way.
The behaviour of a particle in a viscous airflow is described as a function of the Stokes number. This parameter is defined as follows:

$$Stk_p = \frac{\rho_{DEHS} * d_{ae}^2 * v_a}{18 * \mu * h} \qquad [1]$$

with

ρ_{DEHS} = $0.964 \cdot 10^3$ kg/m^3
μ = $1.815 \cdot 10^{-5}$ Ns/m^2
h = $6.0 \cdot 10^{-3}$ m (the sample is continously adjusted perpendicular to the wind direction; see also Figure 1)

and in the experiments:
d_{ae} in a range between 15 μm and 40 μm
v_a in a range between 0.8 m/s and 5.5 m/s

This is a valid description for particles with a Reynolds number smaller than 0.1, whereas errors are slight up to a value of 1. For particles near the sample holder this corresponds to a aerodynamic diameter (d_{ae}) smaller than about 60 μm.
When the airflow is obstructed, the particle is dragged along to a certain extent. Some particles are caught by the sample holder. The sampling efficiency is also given as a function of the Stokes number. It is defined as the ratio between two numbers (or two masses) of particles. On the one hand this concerns the particles sampled. This is divided by the number (or mass) of particles in the undisturbed flow in an area with the size of the sample surface projected perpendicular to the flow. The sampling efficiency can be described by the following empirical formula which is based upon reference [4]:

$$\eta = \left(\frac{Stk_p}{Stk_p + Stk_{p,25}} \right)^2 \qquad [2]$$

or

$$\eta = \left(\frac{Stk_p}{Stk_p + (\sqrt{2} - 1) * Stk_{p,50}} \right)^2 \qquad [3]$$

The value $Stk_{p,50} = Stk_{p,25}/(\sqrt{2} - 1)$ gives the value of the Stokes number at which 50% of the particles is sampled.
This description describes the impaction portion of the deposition process. It implicitly states that the angle at which the sample holder is positioned does not influence the sampling efficiency, because the angle is not incorporated in this equation. This assumption will be discussed later in view of the results. However, because the projected surface of the sample holder, perpendicular to the airflow, does depend on this angle, the amount of sampled dust will also depend on the angle and increase with an increased angle.

3 METHODS

3.1 Determination of properties of the sample holder and analysis technique

Experiments were conducted to determine the sampling efficiency of the sample holder and the relationship between R_q and the quantity of sampled dust.

Sampling efficiency of sample holder
In a wind tunnel experiment the sampling efficiency of the sample holder was determined for various wind velocities and for various particle sizes. The organic fluid DEHS (Di–Ethyl–Hexa–Sebacaat) was used in a spinning–top generator to generate mono–disperse spherical particles with diameters between 15 and 40 μm. These droplets do not bounce off so no adhesive coating was used. The efficiency was determined by simultaneously measuring the amount of DEHS sampled on the EDDM sample in a fixed period together with the concentration of DEHS in the wind tunnel. This concentration was measured by isokinetically sucking air through two filters in small filter holders which were mounted above and below the sample holder. The quantity of DEHS on both the samples and the filters was analysed by gas chromatography.

The reflection quotient: R_q
The R_q that results from the EDDM analysis of a specific sample after it has been exposed, can be compared to the fraction of the sample surface that is covered by dust particles. This fraction is estimated by microscopic analysis for a number of test samples covered with coal dust and for samples from the experiments at the coal– and ore–terminal. Video techniques are used to discriminate automatically between pixels with a grey–scale value above and below a particular threshold. In this way the percentage of pixels that is occupied by particles is determined. Test samples have been covered with coal dust, to a range of levels which appeared to be between 0% and 1.4%. The samples from the experiments showed levels between 0% and 5%. For coal dust particles in fact the Effective Area Coverage (EAC) is determined in this way because of the very low reflection of coal. For other materials with a lighter colour the microscope estimate of the EAC is less accurate because of the rather arbitrary choice of the threshold. The EDDM does not suffer from this problem because it does not use a threshold technique (see also Results).

3.2 Description of the experiments at a coal– and ore–handling terminal

Experiments were conducted to investigate the practical behaviour of the EDDM. During the experiments, the EDDM was placed west–south–west to a coal– and ore–terminal about 2 km away, near Europoort, Rotterdam. There were no major obstacles present that could obstruct a free flow of air to the EDDM. The sample holder was fixed at 15° to the horizontal. Every two hours the R_q was automatically determined.
During 10 weeks (October–December 1989) the EDDM measured the deposition of dust. Every week the samples were exchanged and stored for further analysis. Every two hours, or every time rain was detected, the samples were analysed. The total R_q compared to the measurement of the reflection of the clean sample at the start of the week as well as the R_q compared to the previous measurement were stored in the battery RAM of the computer. The averaged wind direction in the sampling period and the starting and stopping times of rain periods were also stored. The relationship between the wind direction and the R_q should give an insight into the direction of the source of the sampled dust.

During a period of 6 weeks the concentration of coarse dust was also determined every 2 hours with the help of the Coarse Dust Recorder (CDR; [5]). The results were used to interpret the results of the EDDM. The data of this apparatus are obtained by laboratory analysis of a strip by means of optical analysis. Parts of this strip have been exposed consecutively during the experiments to an airflow that is sucked through a tube at 5 m/s (independent of the wind velocity). The cut–off diameter, above which more than 50% of the particles are sampled, for this apparatus is 19 μm.

The resulting EAC of the strip is uniformly related to the coarse dust concentration [5]. For coal dust, an increase of the EAC in one hour of 1% corresponds to a concentration of 6.6 $\mu g/m^3$.

4 RESULTS

4.1 Properties of the sample holder

Sampling efficiency
In Figure 3 the results concerning the sampling efficiency of the sample holder are shown. The relative sampling efficiency is displayed as a function of the Stokes number which is given in Formula [1]. For a selection of the datapoints the error in the Stk_p as well as in the resulting sampling efficiency is indicated with error bars. These errors result when the accuracy of the various parts of the procedure are taken into account. The result of a fit to a function with the shape described in Formula [2], [3] is also given.

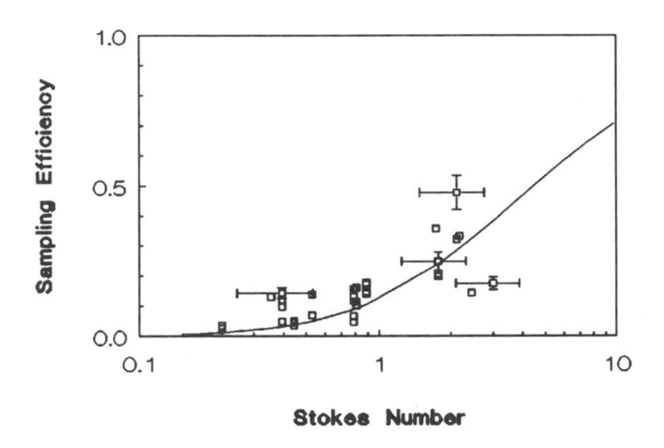

Figure 3 Sampling efficiency as a function of the Stokes Number. Solid line gives model fit.

Statistical analysis showed that the value of the Stokes number at which 50% ($Stk_{p,50}$) of the particles is sampled is 4.5 ± 0.4. This value in fact results from an extrapolation of the results because no values of the Stokes number were available above 3.0. It is difficult to extend the used technique to larger particles or Stokes numbers. These data, which resulted from experiments with "wet" and spherical droplets, can be translated to (coal-) particles using the aerodynamic shape factor [6], which corrects the behaviour of particles according to their irregular shape, and assuming that the adhesive coating on the opaline sample prevents particle bounce. For coal dust (specific weight: ρ = 1450 kg/m^3; aerodynamic shape factor: K = 1.18) it corresponds to a 50% cut-off equivalent diameter that ranges from:

(27 ± 1) μm at a wind velocity of 10 m/s to (38 ± 2) μm at a wind velocity of 5 m/s and (60 ± 3) μm at a wind velocity of 2 m/s.
The main conclusion of these experiments is that indeed predominantly "coarse" dust is sampled.

Dependence on the angle of the sample holder
From statistical analysis it appeared that the angle between the horizontal and the sample holder does not influence the sampling efficiency. The resulting cut-off diameter was not significantly different (α = 0.05) for the three investigated angles. However, this conclusion is based on a rather small number of datapoints.
Furthermore, when the angle is varied, the amount of sampled particles does vary according to the sinus of the angle, despite this conclusion. This follows from the fact that a variation of the angle corresponds to a variation in the flow rate through the projected surface of the sample holder perpendicular to the flow (see definition of the sampling efficiency in § 2.3). These findings confirm the implicit notion mentioned in the description of the EDDM.

Apart from these calibration experiments, sample holders have been located for 6 months in the surroundings of the same terminal mentioned earlier. These experiments are not described in full detail here. The sample holders were placed at an angle of 5° and 15°, respectively, with the horizontal. Daily samples did show amounts of dust that on average differed by the sine of the angles. A lineair regression showed that on average 2.8 ± 0.1 times more dust was collected on the 15° sample than on the 5° sample. This value does not differ significantly (α = 0.05) from the ratio between the projections of the sample perpendicular to the airflow (sin15°/sin5° = 2.97). This finding again confirms the above mentioned notion. It also indicates that impaction is the dominant sampling principle, since sedimentation should have resulted in a ratio of about 1.0 (= cos15°/cos5°) between the two sets of results. A simple model comparison between impaction (according to the calibration) at v_a = 10 m/s and sedimentation for particles of 40 μm (according to their speed of fall) gives a relative contribution to the EAC of about 4% for sedimentation.

4.2 Properties of the analysis technique

Samples obtained under test conditions as well as during the practical experiments near the transhipment area have been optically analysed. The resulting estimate of the EAC is compared to the R_q in Figure 4. From this figure it can be seen that the EAC of the test samples (indicated with black dots) covered with (black) coal dust is linearly related to the R_q. A lineair regression analysis on this data set resulted in:

$$R = a.EAC + b \quad [4]$$

with

$$a = -(5.1 \pm 0.4)$$
$$b = 0.999 \pm 0.006$$

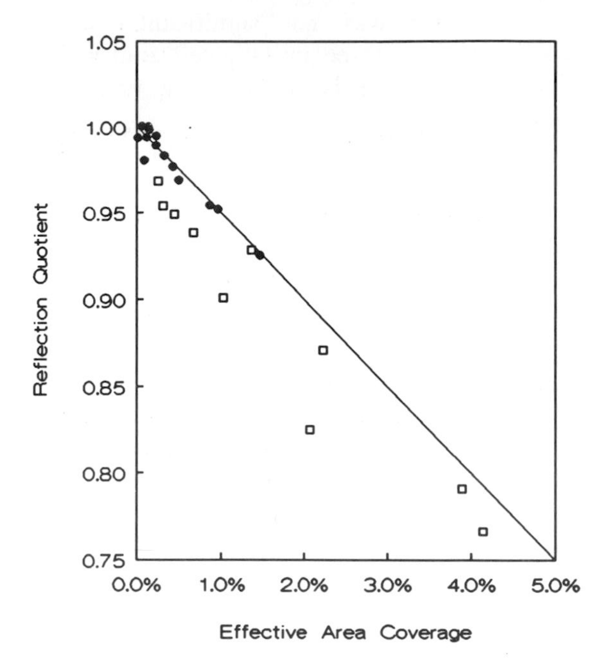

Figure 4 EAC (optical technique) versus R_q (EDDM). The line is a lineair fit to coal-dust test samples (black dots). Squares resulted from outdoor experiments.

It is remarkable that for these samples the percentage of decrease in R_q is about 5 times greater than the corresponding increase of the EAC. A possible explanation is found in the diffraction of emitted light by the disturbance of the sample coating, caused by the adhered particles.

The decrease in the reflection R_q for the samples gathered in the experiments at the coal– and ore-handling terminal (squares in Figure 4) is relatively even greater compared to the corresponding EAC. This difference can be explained by assuming that the various kinds of dust that were sampled in these cases have different effects in the two analysis techniques. The ore particles with light colours will have some influence on the R_q, whereas they are ignored in the optical analysis when their grey-scale value is below the threshold (see Methods). For the interpretation of the experimental results the values of R_q have been translated in values of EAC using the relation obtained for coal-dust.

4.3 Experiments at a coal– and ore-handling area

Time course of deposition

An example of the results of measurements at the coal– and ore-handling terminal is given in Figure 5a. The sample holder was fixed at an angle of 15°. The EAC is estimated using relation [4].

The time course of the EAC is shown in two ways in this figure. The solid line displays the change in EAC compared to the first measurement at the start of the week when the sample was clean. The datapoints show the EAC when each measurement is compared to the preceding measurement. It is obvious that in this example one period which lasted about 5 hours was responsible for about half of the total amount of deposited dust. Typically, for most of the weeks, a limited number of periods is responsible for a major amount of deposited dust.

At the end of each week the samples were replaced and stored. In Figure 4 the total R_q's for all samples are shown as a function of the results of the subsequent optical analysis of the EAC of the EDDM samples. The total R_q after one week sampling ranged from 0.97 (corresponding to the lowest deposition rate) to 0.76.

Figure 5b shows the coarse dust concentration in the same week. This figure resulted from measurements with the Coarse Dust Recorder which was situated next to the EDDM for 6 weeks. This concentration is expressed in arbitrary units. These units are derived from the averaged increase in the Effective Area Coverage of the exposed strip per two hours as determined by optical analysis. It is expressed in %/hour. The highest coarse dust concentration (16%, which is found on the 29th of october) corresponds to about 100 $\mu g/m^3$, if we assume that only coal dust was sampled.

The solid line in Figure 5b is only given for comparison with the results of the EDDM. It represents the summation of the two-hourly measurements of the concentration. From this figure it can be seen that there is a high correlation between the two types of measurements. However, a high concentration did not automatically lead to a high deposition and vice versa. The increase in the deposition on the 28th of October can hardly be retrieved in the concentration trace. This is probably because the wind velocity and the particle size distribution, among other things, determines the exact relationship between measured deposition and concentration.

Dependence of source direction

In Figure 6 the relationship between the change in EAC is displayed as a function of the averaged direction of the wind during the sampling period. The 3 datapoints with the highest EAC all resulted from a period when the wind was from the west-south-west. This direction corresponds to the direc-

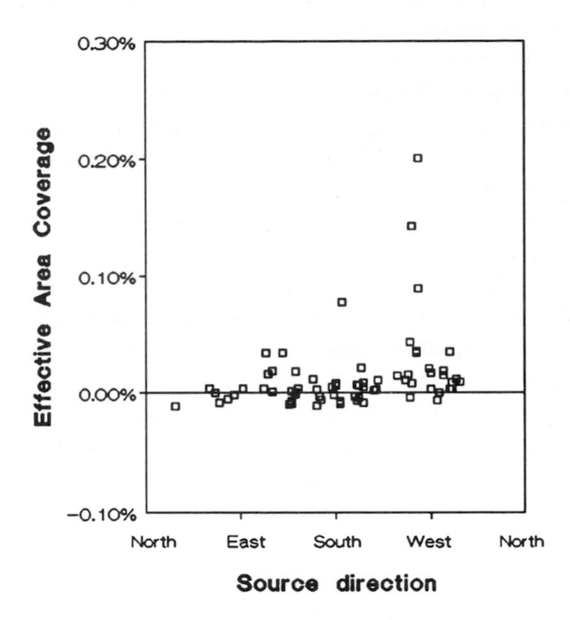

Figure 6 Two–hour changes in EAC as a function of the source direction. The wind direction corresponding to three datapoints with the highest value is the direction of the terminal.

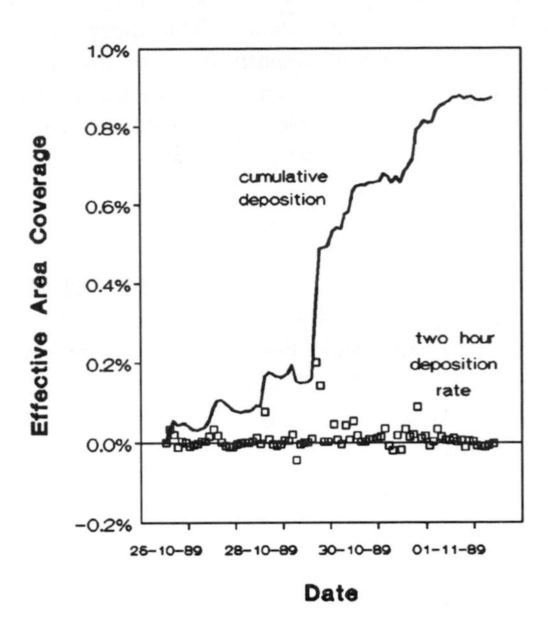

Figure 5a Results from the EDDM. Squares indicate the change in EAC per two hours. The solid line gives the total deposition during one week in October.

5 COMMENTS ON RESULTS

About a number of results described in this manuscript additional remarks can be made.

In the relationship between EAC as determined with an optical threshold technique and the values for R_q from the EDDM a difference was observed between the two sets of data (see Figure 4). Light coloured particles were supposed to account for these differences. The EDDM does signal these particles whereas the optical technique ignores particles with a grey–scale value below a particular threshold. In this respect it is worth mentioning that nuisance is not only related to black particles although they might be predominant. Additional research is needed to determine the sensitivity to various kinds of dust.

The EDDM cannot discriminate directly between different sorts of dust coming from different sources. However, in combination with the directional information of the EDDM (see figure 6) and geographical data about the studied area it might be possible to distinguish various sources.

The range of R_q for the samples that were used during one week at the transhipment area was between 0.97 and 0.76. This corresponded to a maximum EAC of about 5%. This is lower than the maximum acceptable value of 10% EAC (see Chapter 2). By changing the angle of the sample

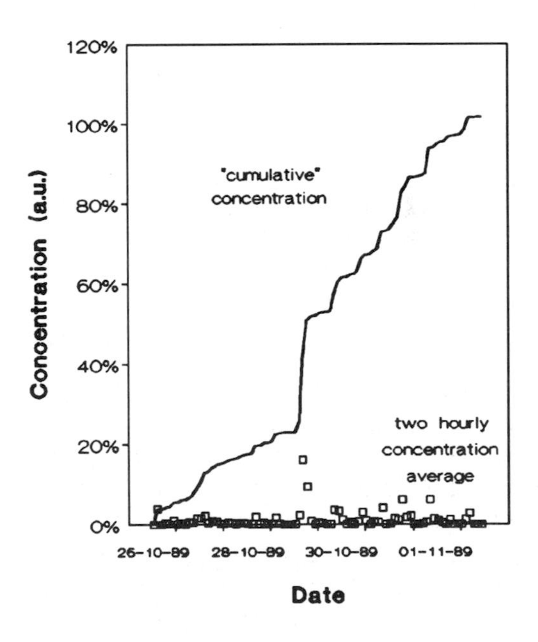

Figure 5b Results of measurements with the CDR for the same period as in figure 5a. Averaged (two hour) concentration and "cumulative" concentration are given in arbitrary units (see text).

tion of the transhipment area. More than 75% of the cases with a change of the EAC larger than 0.1% resulted from periods having this wind direction. The direction of the fourth highest datapoint did not correspond to high values during the other weeks.

holder, the rate with which R_q changes can be adapted to a particular location. The angle will be a compromise between sensitivity and the risk of overloading. The optimal sensitivity of the EDDM is reached when the tilt angle is 15° because then most dust is collected. On the other hand total coverage of the sample surface should not exceed about 10% before the sample is exchanged after one week.

The distance between the source and the EDDM was rather large in these experiments (about 2 km). It is often said [7] that pollution in the surroundings of dust sources is limited to distances less than 1 km. However, periods with a high load of dust could be easily distinguished.

In Figure 5a a (relatively small) decrease in the EAC can be observed in a number of periods. This might result from inaccuracy of the EDDM analysis. However, this inaccuracy is less than 0.1% EAC (see description EDDM). Furthermore, the fact that the two hourly results show only little variation during the indicated periods is in contradiction with this explanation. Another explanation might be that particles are blown away from the sample. Also, rain might have washed away a number of particles. However, this last explanation is not plausible. The sample holder is automatically retracted into the housing of the apparatus when rain is detected.

6 CONCLUSIONS

The EDDM has proven to give valuable information about the deposition of coarse airborne dust particles. A number of arguments can be summed up to support this conclusion.

- The deposition rate can be determined with high frequency.
- Because the EAC of surfaces of objects is supposed to be strongly related to the nuisance experienced by people, it is favourable that the Reflection quotient R_q is directly related to the EAC of the samples.
- The overall measuring accuracy (better than 0.1% EAC) appears to be more than sufficient to determine periods with an increased dust deposition rate.
- It seems important to confine nuisance-related measurements to coarse dust. Fine dust contributes only slightly to pollution of surfaces and therefore might only obscure results. The EDDM has a 50% aerodynamic cut-off diameter ranging from 35 μm at a wind velocity of 5 m/s to 25 μm at 10 m/s and therefore it appears to be selective enough. The dependence of the cut-off diameter might seem to be a drawback when representative sampling of dust is required. However, at high wind speeds smaller particles will also play an increasing role in the pollution of other (arbitrary) surfaces.

The above mentioned properties make the EDDM a useful apparatus for research on nuisance-related dust deposition.

Furthermore, the possibility to determine the direction of dust sources may be used for source detection. It is worth remarking that coarse dust is only spread over a relatively short distance from the source. Therefore, in many cases it will be possible to identify the source with the help of the EDDM. In this respect it is a drawback that only weekly samples can be analysed regarding the kind of dust unless a higher sample exchange rate is employed.

Finally, the short on-line response time of the EDDM may be used to signal situations with high loads of dust and to trigger measures to prevent the emission of dust at the sources.

The actual relationship between the results of the EDDM and nuisance caused to people needs further research. It was not possible to study this relationship in the experiments described because there are no inhabitants in the area where the experiments were conducted. However, planned experiments in another region might give an indication in the near future. In these experiments people will be asked to give an opinion concerning their evaluation of the environment, two times a day during two months. The concentration and deposition will be measured simultaneously using an EDDM and a CDR. Further, the (weekly) pollution of a number of surfaces will be determined and meteorological data will be gathered.

If results of these experiments give satisfactory results the EDDM might be suitable to "measure" nuisance, which in the end is its purpose. Then, eventually legislation might be based on its measuring values.

ACKNOWLEDGEMENTS

This research was supported by the Netherlands Agency for Energy and the Environment (NOVEM), within the framework of the Dutch National Program on Coal Research.

REFERENCES

[1] SEHMEL, G.A. Particle and gas dry deposition: A review. Atm. Environment, 1980, 14, pp. 983-1011.

[2] HAYNIE, F.H. Theoretical model of soiling of surfaces by airborne particles. In: S.D. Lee and T. Schneider (eds): Aerosols: Research, Risk Assessment and Control Strategies, 1986 pp. 933-949, Lewis Publishers, Inc., Chelsea, Michigan, USA.

[3] SCHOFIELD, C. and SHILLITO, D., Guide to the handling of dusty materials in ports, impact, prevention and control, 1983, pp. 74–87, British Materials Handling Board, Berkshire, UK.

[4] FUCHS, A. The mechanics of aerosols, 1964. Translation edited by Davies, C.N. Pergamon Press, Oxford.

[5] VRINS, E. The Coarse Dust Recorder. Proc. Third Int. Aerosol Conf. In: AEROSOLS, Science, Industry, Health and the Environment, 1990, Eds. Masuda, S. and Takahashi, K., 626–629.

[6] STOBER, W. Dynamic shape factors of non–spherical aerosol particles. In: Assessment of airborne particles Ed. Mercer, T.T.

[7] GILETTE, D.A., BLIFFORD, I.H. and FRYREAR, D.W. The influence of wind velocity on the size distribution of aerosols generated by wind erosion of soils, J. Geophys. Research, 1974, 79, pp. 4068–4075.

C418/003

Dust controlled loading and stockpiling of dry bulk materials

S W DOWNING
BLS Engineering Limited, Princes Risborough, Buckinghamshire

SYNOPSIS: This paper provides some useful information and comparisons for those people involved the movement and storage of dry bulk materials; who have identified the need for more effective methods of dust control than have been available previously, during outloading and stockpiling.

DUST IS THE PROBLEM

The present green revolution and all its ramifications is having a marked effect on industry. Existing standards are being tightened, new regulations are being introduced (COSHH for example) and there is a greater awareness of, and a feeling of responsibility for, the environment in general.

No longer is it acceptable to let pollution blow or flow away. The people next door or downstream don't want it and won't have it. The result is that new techniques are being developed and adopted for areas of dust nuisance that were formerly considered unimportant or for which no practical solutions were available.

CONTAINMENT

This paper deals specifically with the control of dust during the discharge of bulk materials from one receptacle to another and the first and most important consideration is containment.

As long as airborne dust is contained within an enclosed area, its nuisance or hazard potential is drastically reduced. With the use of conventional dust extraction systems, it can be drawn away from the risk area and collected safely for disposal or re-processing.

So it is the means of containment which hold the key to the problem. In the case of retractable loading spouts, the principle remains the same but the method of use and application will vary. The most common applications are discussed below.

TANKER AND BIN LOADING

The same system may be used for the various manifestations of what is probably the most common and sometimes the most troublesome of loading situations. Often involving high loading rates to achieve quick vehicle turn-round times, large volumes of dust laden air are displaced and must be contained and extracted effectively. The retractable spout can do this without reducing the loading rate or making the operation more complex.

First, a brief description of the spout. (See Fig.1). It has a flanged inlet which connects to the silo or conveyor outlet, usually below a feed control valve of some kind. This inlet leads into a feed chute which may be a flexible bellows, a straight telescope or, more practical, an articulated chute made up of a series of overlapping open-ended cones. The latter type has the advantage of flexibility which the telescope does not, and smooth contact surfaces which the bellows does not. So it allows the product to fall freely through into the receiver, with the minimum of disturbance or pick-up.

Around the inlet is the plenum chamber which includes an off-take spigot for connection to the extraction system. To the bottom of the plenum is fitted the outer sleeve or bellows which is made from a strong, flexible synthetic material and to which are fitted a number of spacing rings. These maintain the concentricity of the sleeve when it is under negative pressure.

At the base of the sleeve is the locating cone, an inverted, open ended cone which rests in the circular hatch opening in the tanker or bin.

Through the centre of the locating cone protrudes the lower part of the feed chute and between the two is an annulus through which the extraction air must pass on its way back through the spout to the dust collection system.

Once the dust laden air has passed through the lower annulus, it is then drawn back through the larger annulus area formed by the outer flexible sleeve and the feed chute. As the area of this is greater than the lower annulus, the velocity will drop, allowing larger dust particles to fall back through the spout. The remaining dust will be carried by the air stream, back through the outlet spigot at the top of the spout and through the ducting system to the collector. The subject of extraction volumes, velocities and overall system design, will be dealt with in greater detail later.

In addition to the component parts described above will be the main housing, gearbox and integral winch system, in the case of electrically operated units; the handwinch and pulley arrangement for manual operation; or the two pneumatic cylinders which replace the wire lifting ropes in the case of air operated systems. (See Fig.2).

The selection of the type of motive power for raising and lowering the spout will depend upon a number of factors. The first and most obvious is the cost. The manually operated unit with a simple handwinch, with wire rope and pulley systems, is less expensive than the pneumatic unit which is, in its turn, cheaper than the electric winch system.

Therefore the first consideration may be one of economy. However, there are other factors which may affect choice and these should be considered carefully if the loading spout is to perform its function correctly.

The electric winch and pneumatic systems offer potential for a higher degree of automatic control, especially where the spout is the final link in a complex process.

The electric winch, with the use of appropriate limit switches, will ensure that the spout follows the vehicle down as it settles on its suspension during loading. Of course, it relies on its own weight to do this, while the pneumatic system can continue to apply downward pressure throughout the loading cycle to ensure positive contact between spout and receiver.

While the pneumatic cylinders are fitted with knuckle joints at both upper and lower clevis, they do not have quite the degree of tolerance to misalignment as either of the wire rope systems. In addition they require rather more headroom and are not suitable for long vertical travel. About one metre is a reasonable maximum although this can be exceeded in certain cases.

If the product being loaded is hazardous, it may be necessary to provide an even more positive connection during loading. This can be achieved either by modifying the locating cone so that it can be attached to the tanker by means of two or more of the hatch clamps, or more quickly, by fitting an inflatable collar at the base of the spout. The spout is lowered until the collar assembly enters the hatch. The collar, which has a strong rubber membrane around the outside, is then inflated until it grips the outside perimeter of the hatch opening.

With this option, it is important to confirm that there is not too great a variation in the hatch diameters of the vehicles to be loaded as the collar can only inflate within a certain diameter range which must be laid down at the time of manufacture.

Another option available is the shut-off cone. This is a device by which the spout can be closed at the bottom end when it is withdrawn from the receiver. While it has some advantages it should never be used as a feed control device.

This function, however it is performed, should always be above the spout inlet. Otherwise, the spout may be closed at the bottom while still full of product. The damage resulting from retracting the spout in this condition, with the resulting compression of the product within, can be considerable.

The loading spout can be fitted with a sensing probe which will provide a signal when the receiver is full. This can be important where the loading bay does not have a weighbridge or other means of capacity control. Probes of different types are available and should be selected according to the type and characteristics of the product being loaded.

In the same way the materials of construction of the spout must also be selected bearing in mind not only the product but also the location and site conditions. For this reason it is important to provide full details of the product, its temperature range, its corrosive or abrasive qualities and other relevant information, together with accurate details of site conditions such as maximum and minimum temperatures, humidity levels, maximum wind speeds, etc.

It is now time to consider the dust extraction system in detail. The air volume to be extracted will depend on several points. The first is the required loading rate and bulk density of the product. This will dictate the volume of air which is displaced over a given period. It is important to make an accurate prediction of the loading rate as it is often under-estimated with the result that the extraction system is unable to cope. It is advisable therefore to estimate the air volume using the maximum possible loading rate.

It is also important to consider any other source of air which may become entrained with the product. For example, where the feed is coming from an airslide, a cyclone separator or a silo with a fluidizing system. Any extra air volume must be taken into account.

Having said that, the next point to consider is the proposed or actual configuration of the extraction system. It is quite possible that the air volume required to meet the displacement will not be sufficient to achieve an adequate conveying velocity back to the filter, especially if the dust laden air has to travel to the top of the product silo. Care should be taken to size the fan correctly, taking into account the ductwork diameter, the overall length and the number and type of bends required.

It is important to place a balancing damper in the extraction line adjacent to the spout and, in addition, an airbleed valve. The purpose of this valve is to ensure that even if the system volume has to be greater than is required for the displacement, the extra air can be drawn from outside the spout, thus avoiding pulling back too much product into the filter. It is also useful for balancing the system during the commissioning stage, once the loading rate has been confirmed.

If two or more spouts are to be served by the same extraction system, the ductwork must be designed to ensure equal distribution between the spouts or alternatively a shut-off damper may be fitted to each spout so that it can be isolated when not in use.

However the system is arranged, remember that the extraction system must have a suitable purging time after loading has finished to ensure that the ductwork is free of product.

OPEN LOADING AND STOCKPILING

The same general principles apply in these cases as for the loading of enclosed vehicles or containers. The exception is that containment is more difficult to achieve. In the first instance, the spout for open loading does not have a locating cone as previously described. Instead it is fitted with a flexible skirt. This is kept in contact with the product pile and provides the means of containing the dust so that it can be prevented from escaping to atmosphere. (See Fig.3).

Secondly, the options for raising and lowering the spout are not quite as numerous. Pneumatic operation is unlikely to be practical due to the lengths of travel involved. Hand or electric winches are equally effective, except that the manual option does not offer the advantage of the electric system when adopting the automatic operation described below, neither is it practical in the case of the larger size spouts.

In operation the open loading spout does not remain in a stationary vertical position. By monitoring the height of the product pile contained within the flexible skirt, the spout is raised incrementally by the winch system, releasing product from within the skirt and providing space for more of the falling product to expand and settle.

It is important that the skirt is in contact with the product pile throughout the loading operation. If this is not the case, dust may be released and will be impossible to recapture. In the loading of open vehicles it is tempting, once the first pile has been formed, to move the vehicle forward gently, building a ridge along the centre line and allowing the product to fall down the open face. The result is that dust entrained with the product will be carried away from the open side of the skirt and released to atmosphere. The same problem applies if loading a ship or stockpile with a slewing conveyor.

The correct method is to build the first pile, then stop the feed, reposition the spout and build the next pile. Once a series of piles has been formed, the areas between can be infilled until a sufficient quantity has been loaded. In the case of a vehicle, it may be better to have several spouts loading either simultaneously or one after another from a common conveyor. In this way the vehicle need not be moved during the loading cycle.

Where a slewing conveyor is loading a ship or stockpile, it may be possible to drag the skirt across the pile without losing contact to any great extent. There will inevitably be some release of dust, but this may be acceptable from an economic standpoint, bearing in mind the time lost each time the feed is stopped.

The design of skirts can vary according to the product and the loading technique to be employed. The solid, curtain-like skirt with a weight such as a chain stitched into the bottom is the most common, but a slotted, hula-type skirt, made up of a series of overlapping strips, may be more satisfactory if the spout must be moved across the product pile during loading. In some cases, a combination of the two may be necessary.

Where the product is very abrasive or contains large particles, the skirt may be made from a heavy duty rubber rather than the lighter, synthetic material.

Extraction air volumes will be calculated differently in open loading as containment is not so positive. The final figures will be calculated using a combination of factors including the loading rate and density of the product, the type and diameter of the skirt, the length of fall and the size of the inner annulus. In addition, empirical information based on similar applications will be taken into account.

On the subject of materials of construction mentioned earlier, it is worth stating that there is sometimes a tendency to overdesign the spout. This may come from a genuine desire to prolong the life of the equipment in a harsh and arduous environment, but at the same time, the questions of weight and cost should be kept firmly in mind.

The typical ship-loading or stock-piling spout is suspended from the outer extremity of a conveyor boom and will add substantially to the structure. It is useful to remember that, so long as the spout is in a vertical attitude during loading, there is relatively little attrition on the feed chute, provided that the product has been properly directed through a suitable headchute into the spout inlet. This inlet can be made from heavy guage, wear resistant material, as it is not likely to increase the overall weight of the spout by a significant amount. However, if the feed chute (telescopic or articulated) is too heavy, it not only adds its own weight to the total, but also, because of the increased torque, it affects the size of gearbox and motor and consequently the whole of the winch system. The knock-on effect of this can render a project uneconomic.

Experience has shown that a correctly designed feed-chute shows no appreciable evidence of wear even after five years handling an abrasive product such as petroleum coke.

ALTERNATIVES TO FREE FALL

One problem which may occur is a product which can be damaged and degraded if allowed to fall considerable distances, and a number of possible solutions have been put forward.

One is to fit a series of baffles within the spout to slow down the product's fall. These may be effective, but they are difficult to design and maintain and they may also reduce the amount by which the spout can be retracted.

Another method is to allow the product to fill the spout almost to the top and then to raise the spout by increments, as previously described, but in this case taking the signal from the top of the spout rather than the bottom. This will not protect the initial flow of product, but will prevent further damage once the spout is full.

The disadvantage is the increased weight created by the column of product contained within the spout, the extra strength of construction required to contain it and the additional size of the winch system capable of lifting it.

SUMMARY

In summing up, it must be said that retractable loading spouts, while they have made considerable progress in design and efficiency over the past ten years or so, can benefit from continuing development. There is still a great deal of experience to be gained by co-operation between the manufacturers and the end user.

With so wide a range of loading applications it is impossible to provide standard solutions in each case. However, a great many problems have been solved and development will continue as long as the need is there. It is important to remember that the loading spout is often the last item to be considered in what may be a highly complex system.

If it is to produce the best results, it must be considered as an integral part of the process. If the dust extraction system is not adequate, no loading spout, however good, will fully make up for the deficiency. Likewise, a loading spout which is undersized or incorrectly specified, will detract from the efficiency of the system of which it is a part.

Consultation with the manufacturers at the design stage is the best way to ensure that the equipment supplied will produce the best possible results at the point of loading, where it matters most.

- 0 -

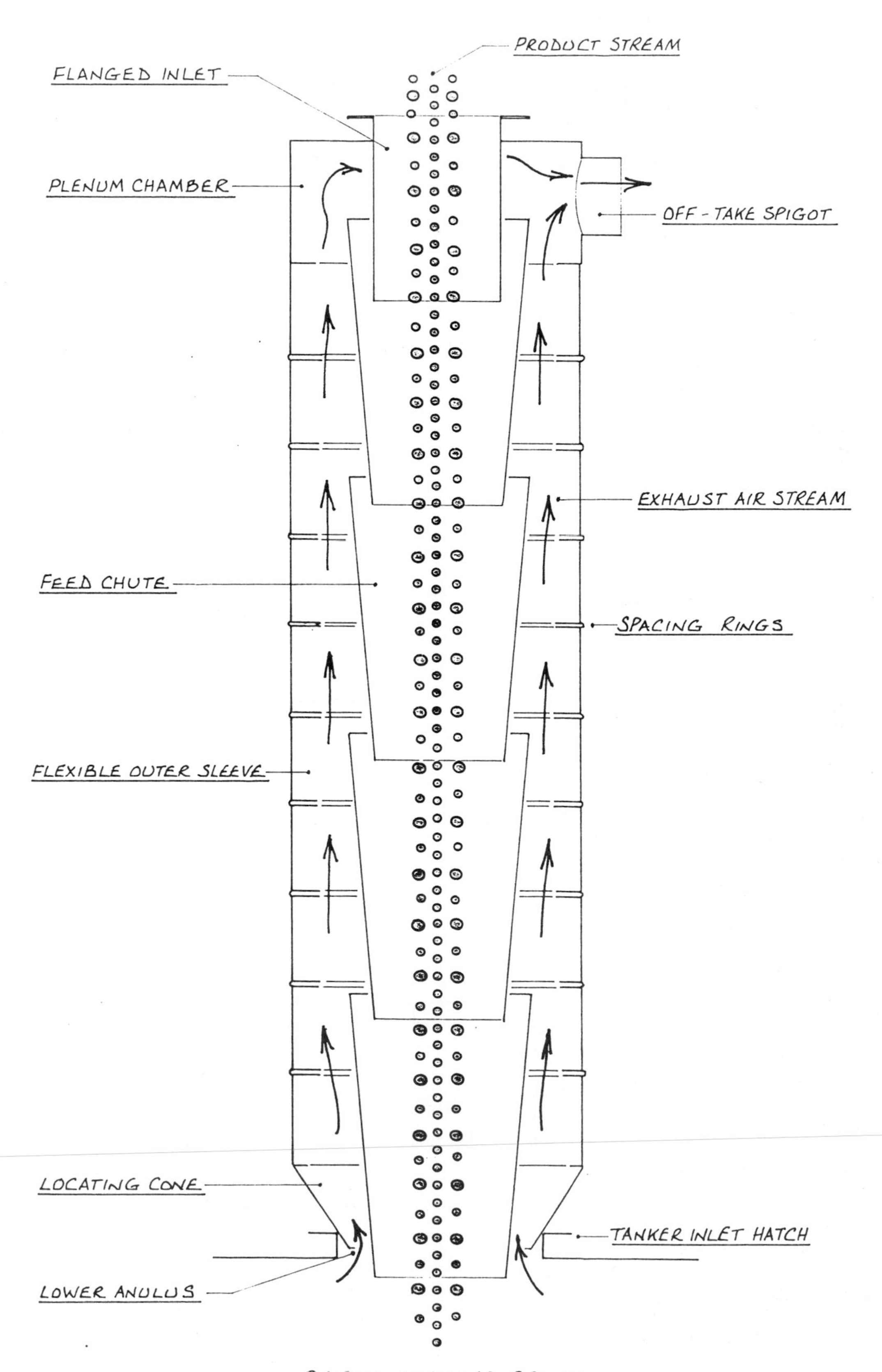

BASIC LOADING SPOUT

FIG. 1

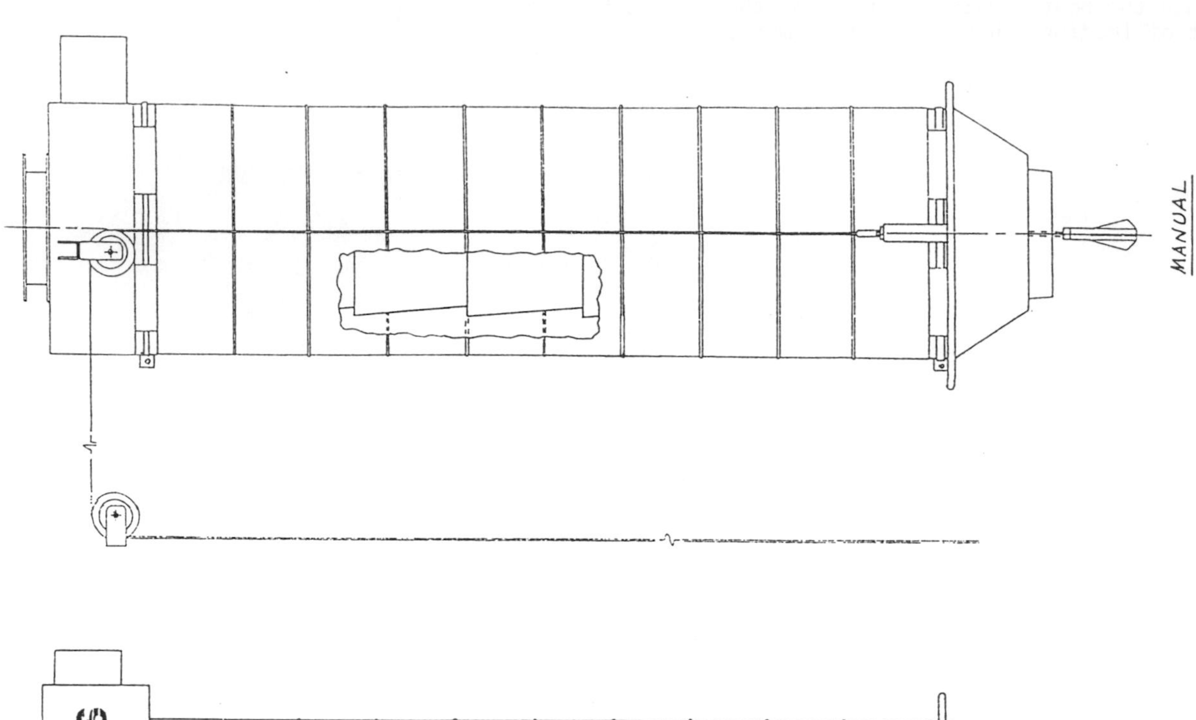
MANUAL

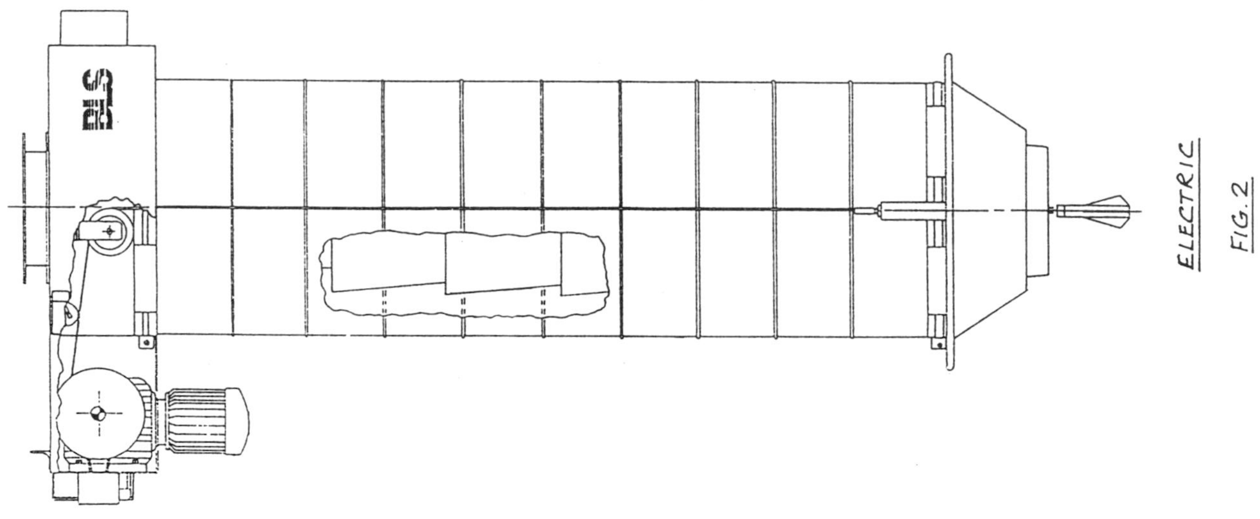
ELECTRIC

FIG. 2

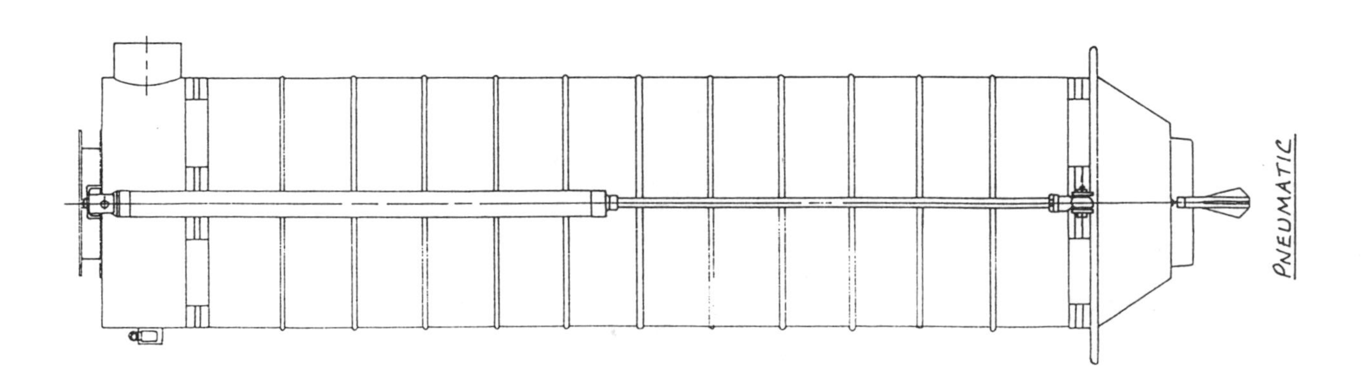
PNEUMATIC

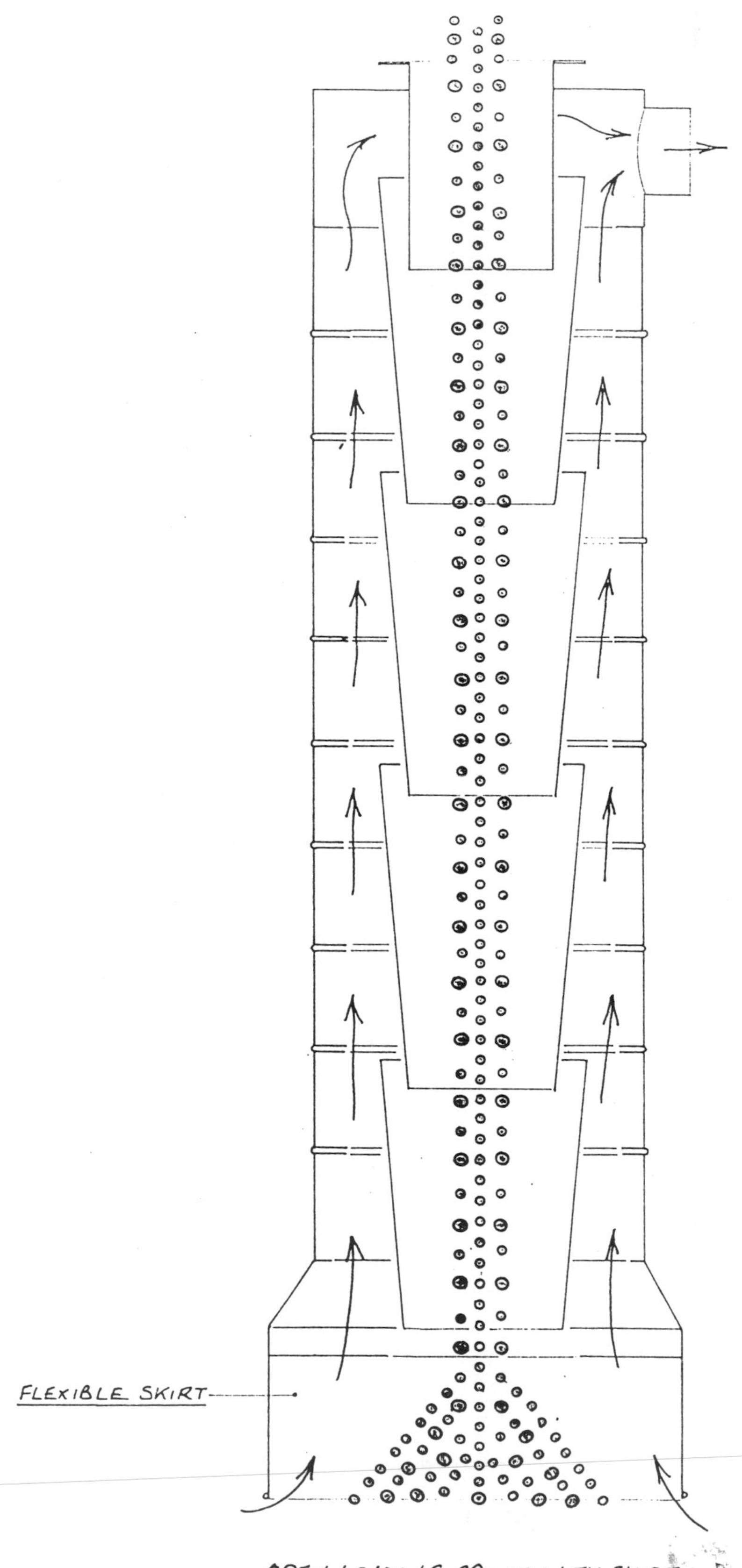

OPEN LOADING SPOUT WITH SKIRT

FIG. 3

C418/056

Techniques of in-bin blending

J W CARSON, PhD, MASME, MAIChE, MASCE and T A ROYAL, MS, MASME
Jenike and Johanson Incorporated, North Billerica, Massachusetts, United States of America

SYNOPSIS Storage bins and silos can be used effectively as in-bin blenders. Requirements for such devices are stated first. Then two of the more common types of in-bin blenders (multi-tube and BINSERT®) are described. Characteristics and limitations of each are listed along with examples of applications.

1 INTRODUCTION

Quality control is more important in today's competitive global economy than ever before. A useful technique to improve quality in many industries involves blending of bulk solids. This can be helpful with both the raw material stream which enters a process as well as the output stream from a process.

Blending can be done on either a continuous or batch basis. Continuous blending is often used to dampen out upsets either going into or coming from a process, while batch blending is used for close, well-defined quality control.

A variety of types of blenders is available on the market. Some blenders are purely mechanical in operation, e.g. employing ribbon flights or impellers. Others depend upon rotation of the blender to cause blending, such as single or twin vee-shaped shells or drum-type blenders. The blending action in some fine powder blenders is achieved by fluidizing the material. Blending can also be accomplished by metering individual streams of material onto a common collecting conveyor.

Static in-bin blenders operate differently than any of the other types of blenders listed above. An in-bin blender consists of basically a storage bin or silo which doubles as a blender. Two of the more common types of in-bin blenders are the multi-tube (e.g., Phillips or Young type of blender) and the BINSERT® blender. Characteristics and limitations of each will be described in this paper.

2 REQUIREMENTS FOR A GOOD IN-BIN BLENDER

There are a number of requirements for a good blender, such as:

- No "dead" regions (i.e., mass flow design).
- Large velocity gradients throughout blender. Thus, for example, there should be a large differential between the time it takes a particle in the fastest flowing region to exit the blender compared to a particle in the slowest flowing region. In addition, particles in the fastest flowing region should start to discharge as soon as possible after they enter the blender.
- Minimum need for recirculation of the contents of the blender.
- Blending uniformity which is independent of the blender's fill or discharge rate.
- Ability to blend a wide variety of materials (e.g., fine and coarse particles, free-flowing and cohesive materials).
- Blending uniformity which is independent of the level of material in the blender.
- Ability to cause segregating materials to remain blended as they discharge from the blender.
- Cost effective.

3 MULTI-TUBE BLENDERS

There are a number of different types of multi-tube blenders on the market. The most common in N. America is the so-called Phillips blender which is based upon a patent originally issued to Robert Goins [U.S. Patent #3,216,629].

Numerous follow-on patents have been issued, one of the more recent of which is U. S. Patent #4,472,064. Two other types of multi-tube blenders include the so-called Young blender [U.S. Patent #3,583,681] and the Fuller blender [U.S. Patent #4,569,596].

Since the Phillips-type blender is probably the most common, we will concentrate our attention on its characteristics.

The Phillips blender consists of a number of internal vertical tubes, each of which is partitioned into typically three tubes. Each partition contains a series of holes located at different heights along its length. The design of the holes is such that material flows into only the topmost hole in each partition. As the material level in the blender drops, the next lower hole in each partition of the tube becomes the active hole. These holes are arranged in a pattern such that no two active holes are at the same elevation. Blending is further enhanced by adjusting the velocities in the individual tubes through adjustments at the bottom of the blender where the tubes terminate.

The Phillips blender can be used as a continuous blender but its performance is improved dramatically if it is operated on a batch basis, i.e., the material level in the blender goes from essentially full to empty. In this mode of operation, this blender provides excellent blend uniformity in a single pass with materials which meet certain requirements as described below. Further recirculation does some good in achieving a better blend, but it is generally better to discharge the blender completely and then refill it.

The only exception to this involves the material first loaded into the blender and the last material discharged from the blender. Recirculating some material back to the top before discharging from the blender lessens problems at the beginning and end of discharge caused by the volume at the bottom below the lowest blend tube opening.

The Phillips blender is widely used for blending of plastic pellets which are similar in size and shape but may vary in color, chemical composition, melt index, etc. Such materials are well-suited for use in this blender because of their uniform particle size, low angles of internal friction, and because they are free-flowing.

Some problem applications for a Phillips-type blender are as follows:

- Achieving a uniform blend of a small amount of additive.
- Blending of materials that contain a wide range of particle sizes. A higher concentration of fines typically discharges last because fines sift through the coarse particles and segregate while flowing toward a tube opening.
- Blending of materials that have high angles of internal friction. Such materials often experience flow problems in the tubes and steep flow channels outside the tubes. The result is limited blending because a higher concentration of particles from the top surface exit first.
- Blending of cohesive (i.e., non-free flowing) materials. The tube openings are much too small to permit even slightly cohesive materials to flow in them reliably without experiencing pluggages.
- Achieving consistent and reliable flow of fine powders through the tubes while preventing flooding. Because of brief retention times in the tubes, fine materials easily aerate and flood.
- Plugging of tubes which goes undetected.

Earlier designs of the Phillips-type blender involved the use of an inverted conical insert located in the hopper section. The purpose of this insert is primarily to expand the material's flow channel to the hopper walls, much like is done in some bin designs to approach a mass flow pattern in a bin that would otherwise provide funnel flow. With some materials this causes erratic flow in the blender because of slopes which form below the insert at an angle of repose which is inherently unstable. This problem was corrected in U. S. Patent #4,472,064 by connecting a tail cone to the bottom of the insert thereby forcing particles to flow in a narrow, defined annular flow channel.

The blending uniformity of a Phillips-type blender for a range of materials is difficult to predict. Fairly large-scale model testing is required to anticipate performance for a given material.

4 BINSERT® BLENDER

The design of a BINSERT® blender is described in U. S. Patent 4,286,883 and various foreign patents [e.g. U.K. Patent #2,056,296]. It consists of a hopper within a hopper, both of which are usually conical in shape. Particles flow through the inner hopper as well as through the annulus between the inner and outer hoppers. By varying the relative position of these two hoppers as well as the configuration of the outlet geometry, it is possible to achieve between a 5:1 and 10:1 velocity differential between particles in the inner hopper compared to particles in the outer annular region.

An anti-segregation ring and distributor are often used at the top of the blender. They distribute incoming material uniformly on the top surface to prevent material from sliding down an angle of repose and segregating by the sifting of fines through coarse.

Some of the advantages of a BINSERT® blender over other types of in-bin blenders are:

- Easy to clean since all of the internal parts are exposed and easily accessible.
- Blends cohesive (i.e., non-free flowing) materials since the outlet can be sized as large as necessary for flow.
- Blends materials with high angles of internal friction.
- Blends materials which are highly segregating.
- Requires low headroom since the walls of the outer hopper can be made relatively shallow.
- Does not require mechanical moving parts other than perhaps a feeder (e.g., belt, screw or rotary valve used at the outlet to control the discharge rate).
- Can often be retrofitted to an existing storage bin causing it to act as an in-bin blender.

When designing a BINSERT® blender, it is desirable to be somewhat close to the limits for mass flow in order to obtain a high velocity differential between the center and outside of the unit. One of the unique advantages of the BINSERT® blender is that this velocity differential is propagated much higher into the bin than with a straight mass flow hopper.

A useful test to demonstrate the effectiveness of a BINSERT® blender consists in placing a thin layer of marker material of uniform thickness at the top surface of the blender and then discharging material while maintaining a constant level in the unit. This can be accomplished either by adding fresh, unmarked material or by recirculating the contents from the bottom of the blender back to its top surface.

The results of such a test in a 1/8 scale model using no recirculation are shown in Figure 1. Here we used a coarser marker material than the bulk which could be separated by sieving. The volume of the marker layer was 0.008 m^3 (0.3 $ft.^3$) which represents 6.6 percent of the volume of the blender. The material level was maintained at 0.5 diameter (0.5D) above the bottom of the cylindrical section. Material was removed from the blender in 0.003 m^3 (0.1 $ft.^3$) batches with each batch representing approximately 2.4 percent of the blender volume. Notice that the markers first start appearing when about one-half of the blender's contents have been withdrawn. The marker concentration in each batch goes up to a maximum of approximately 21 percent and then drops quickly down to approximately a 5 percent level from which it then trails off gradually to zero. It takes over two blender volumes for the concentration of the markers to diminish to zero.

The results of a similar test are shown in Figure 2 in which the blender level is maintained at a 1.5D level as opposed to 0.5D in Figure 1. Notice that the performance of the blender is not as good as when the lower level was maintained; however, the markers are still distributed over a significant percentage (approximately two-thirds) of the blender's volume.

The results of a single layer test without recirculation can be used to predict the performance of a BINSERT® blender with recirculation. Shown in dotted line in Figure 3 is the predicted result from a computer model of the blending process (described below). The solid line indicates the actual measured results using a constant level at 0.5D in the cylinder section.

The mathematical principles for a single layer used to predict the effect of recirculation are best illustrated by the following example:

For easy computational purposes consider a bin in which the material is divided into nine equal volume samples. Figure 4 illustrates the results of a single layer test in which a layer of one sample volume of 100 percent marker concentration is added to the top surface of a bin containing material with no markers. During this test a constant level of material in the bin is maintained by adding unmarked material to the top surface as material is discharged. As can be seen about 5 percent of the markers come out in the fourth sample, 30 percent in the fifth, 29 percent in the sixth, 21 percent in the seventh, 10 percent in the eighth, 5 percent the ninth and none later. If the original concentration of markers were only X percent in the top layer, then we would expect the material to be discharged in a similar sequence of samples except that the concentration of markers in the discharge samples would be X percent/100 percent times those measured in the single layer test when the concentration is 100 percent. For example if the concentration is only 50 percent, we would expect 2.5 percent, 15 percent, 14.5 percent, 10.5 percent, 5 percent and 2.5 percent for samples 4 through 9, respectively. For a constant level condition we expect that at any time we could add a layer of some different material and expect that it would come out in concentrations similar to the single layer test, but in a sequence starting from the time at which the layer was added to the top.

Recirculation is merely taking material withdrawn from the bottom and placing it on top. This new sample layer on top has a concentration of markers which is expected to be discharged in a similar fashion as a single layer, but merely a fraction thereof given by the concentrations of the discharge samples of the single layer test, and in a time sequence starting when it is added to the top. This idea is illustrated in Figure 5 which shows the concentrations of samples 4 through 9 in the original tests with the bars having the same fill patterns as those in Figure 4. When sample 4 (the solid black bar having a 5 percent marker concentration) is recirculated the markers of this sample will be discharged in quantities 5 percent of those in the single layer test starting 4 samples later as shown as the solid black portion of the bars of samples 8 through 13. Similarly when sample 5 (the darkest hatch filled bar having a 30 percent concentration of markers) is recirculated it will discharge the markers in quantities 30 percent of the original single layer test as illustrated by the darker hatched portion of samples 9 through 14. This process is repeated as each sample which is discharged is added to the top and discharged again. The marker concentration in the discharge sample is the sum of the concentrations of markers reaching the outlet from each layer added to the top. Figure 5 is only partially complete because sample 9, for instance, contains an amount of markers represented in solid black. Some of these will be discharged starting 4 samples later in sample 13 through 18. Figure 6 shows the cumulative effect after computations are complete for recirculating the material for two bin volumes or 18 samples.

This computational approach is easily carried out for smaller sample sizes relative to the bin volume and applied to one-pass-through blenders in addition to recirculating blenders. From these principles of computation, a good blender for recirculation is one which quickly discharges some of the incoming material and spreads the remainder over a period as close to one bin volume as possible. For one-pass-through blenders spreading the material over many bin volumes discharged is advantageous. If such a blender is operated in a recirculation mode the period of oscillations is long which requires a large number of recirculations before the material is uniformly blended.

The results of a single layer test can be scaled up to a full sized blender, and various conditions can be studied. For example, if the input stream is fluctuating in product quality or some other measurable variable, the effect on the output stream can be predicted. For example, consider a BINSERT® blender in which a constant level is maintained at 1.2D with a sinusoidal fluctuation of ±50 percent over a period of eight hours , i.e. the concentration of markers oscillates between 0 and 100 percent every 4 hours. Furthermore, assume that the blending bin is initially full of material containing no markers and operates at 180 kN/hr (20 tph). According to a computer prediction the markers should first start appearing after approximately two-thirds of the blender's volume have discharged and, by the time that 1 1/2 bin volumes have been discharged, the output should be steady at close to 50 percent, which is a perfect blend.

Through the use of multiple recirculations, blends can be achieved to any degree of uniformity required.

5 EXAMPLES OF BINSERT® BLENDER APPLICATIONS

5.1 Blending and degassing plastic pellets produces marketable product

Two of the first large-scale BINSERT® blenders have been operating successfully since 1981 at a polyethylene plant.

During reactor grade changes, several batches of polyethylene pellets are produced that are off-spec. These batches must be blended and degassed to produce a marketable, nonstandard product.

Two 4m diameter by 10m tall BINSERT® blenders are used to blend these batches of off-spec pellets to a uniform melt index. A purge gas distributor in the lower region of each BINSERT® is used to degas the pellets during blending. This removes the explosive gases (ethylene and acetate vapors) and makes the product safe.

During start-up the BINSERT® blenders were extensively tested to measure the blending performance and samples of material were taken during a long recirculation test to determine the time required for achieving a uniform batch. Performance tests and market acceptance of the product indicate that the desired quality is achieved after three recirculations of the product through the vessel.

5.2 Continuously blending alumina powder reduces rejection rate to zero

Alumina is a fine powder used in the manufacture of aluminum and various consumer products. Often delivered to plants in 900 kN (100 ton) railcars, alumina can vary in color within each railcar and between railcars.

Finished products can contain as much as 50 percent alumina, so the color of the alumina feedstock has an influence on the product color. Gradual changes in color can usually be detected and compensated for by the manufacturing process without any significant loss of product quality; however, rapid changes in color often result in significant rejection of finished product because of color variations.

A BINSERT® blender is being used successfully at one plant to store 1300 kN (150 ton) of alumina and provide continuous blending to smooth out color variations. With the BINSERT® blender, the rejection rate due to color variation in finished products has been reduced to zero.

This dramatic improvement was achieved at a cost of only the BINSERT® blender, which replaced an inadequate storage bin. There are no additional operating costs with this system. A similar BINSERT® blender has been installed at a new plant site utilizing the same process and has enabled this plant to start up without color quality problems.

5.3 Blending manganese dioxide dramatically improves product quality

During the production of manganese dioxide for use in dry cell batteries, various properties of the compound are monitored. These include particle size, pH, bulk density and electrical potential. Variations in these properties result in inferior products, product rejection and customer complaints.

In one situation it was shown that converting an existing bin having a capacity of eight, 90 kN (10 ton) batches into a BINSERT® blender would effectively allow the blending of one batch of material with thirteen other batches. Tests showed that with only a single pass through the converted bin (no recirculation), the standard deviation of the various parameters could be improved dramatically.

For example, if the standard deviation of bulk density of the batches before modifying the bin was 0.447, this could be reduced to 0.197 after only a single pass through the modified bin - a reduction of 56 percent! Similarly, the standard deviation of electrical potential could be reduced by 87 percent, pH by 52 percent, and particle size variation by 42 percent for the minus 74μ fraction and 66 percent for the minus 44μ fraction.

These dramatic improvements in product quality were possible with relatively inexpensive modifications to an existing bin and with no increase in plant operating costs.

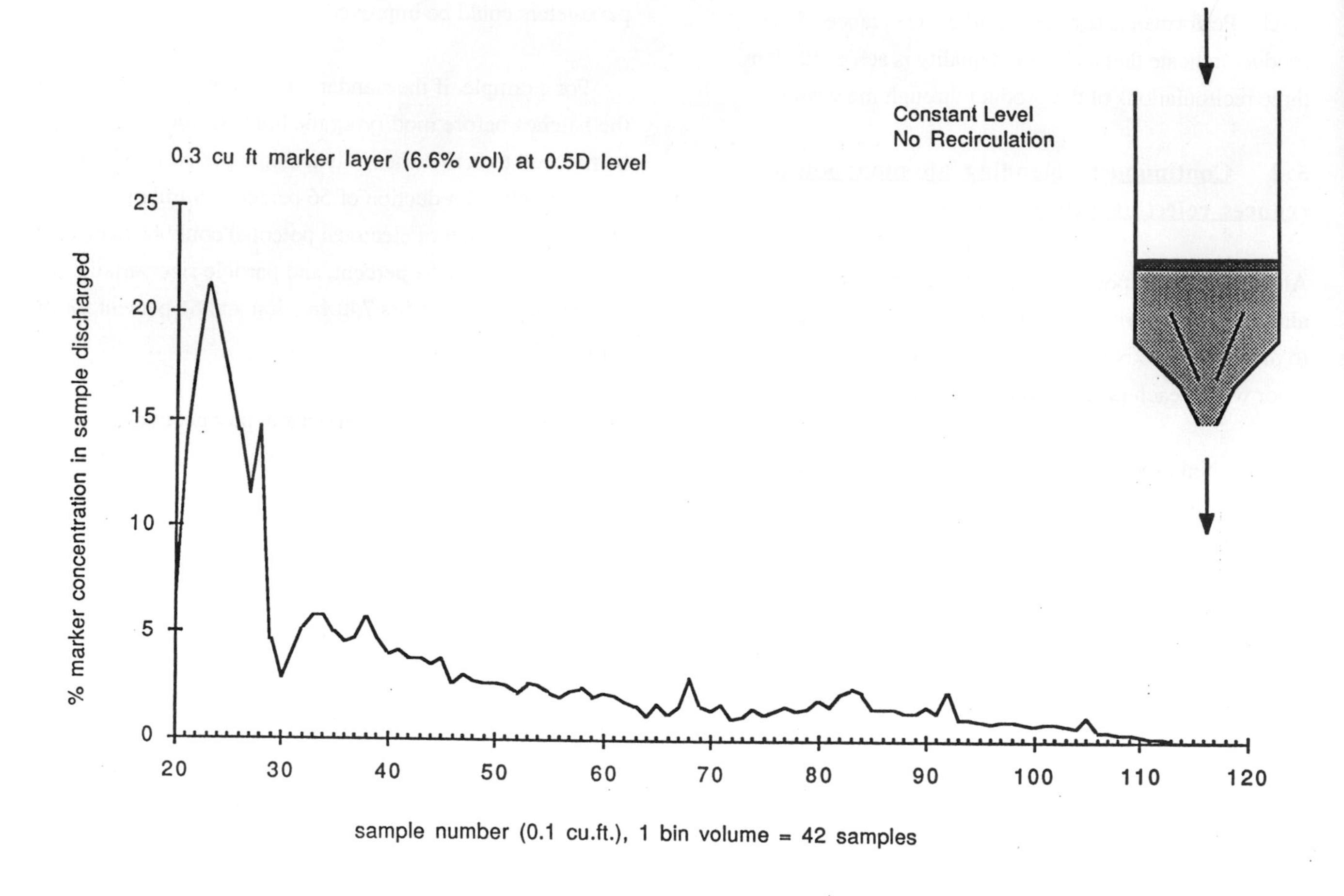

Fig 1 Single layer test at 0.5D level for one-pass-through blender

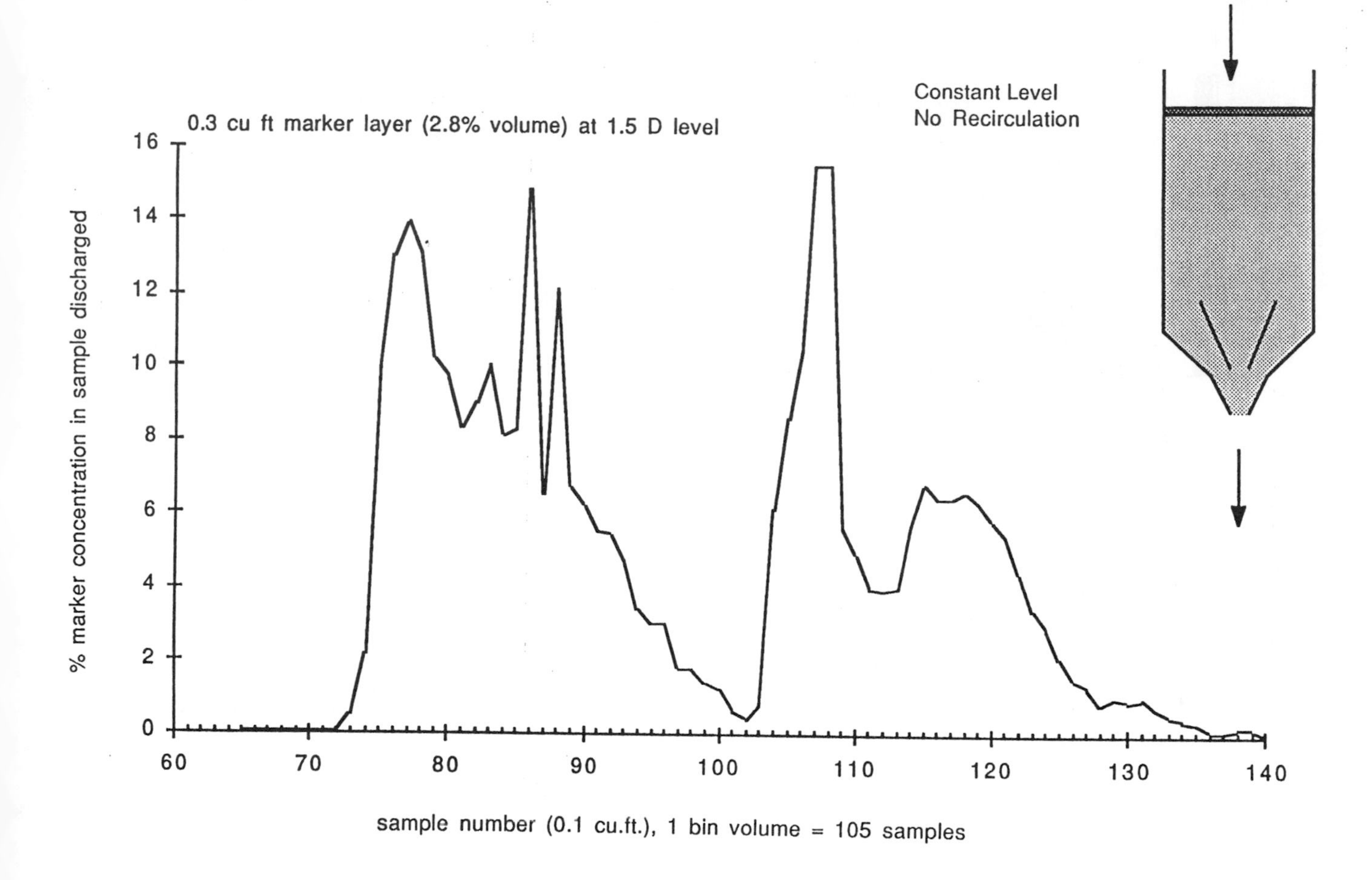

Fig 2 Single layer test at 1.5D for one-pass-through blender

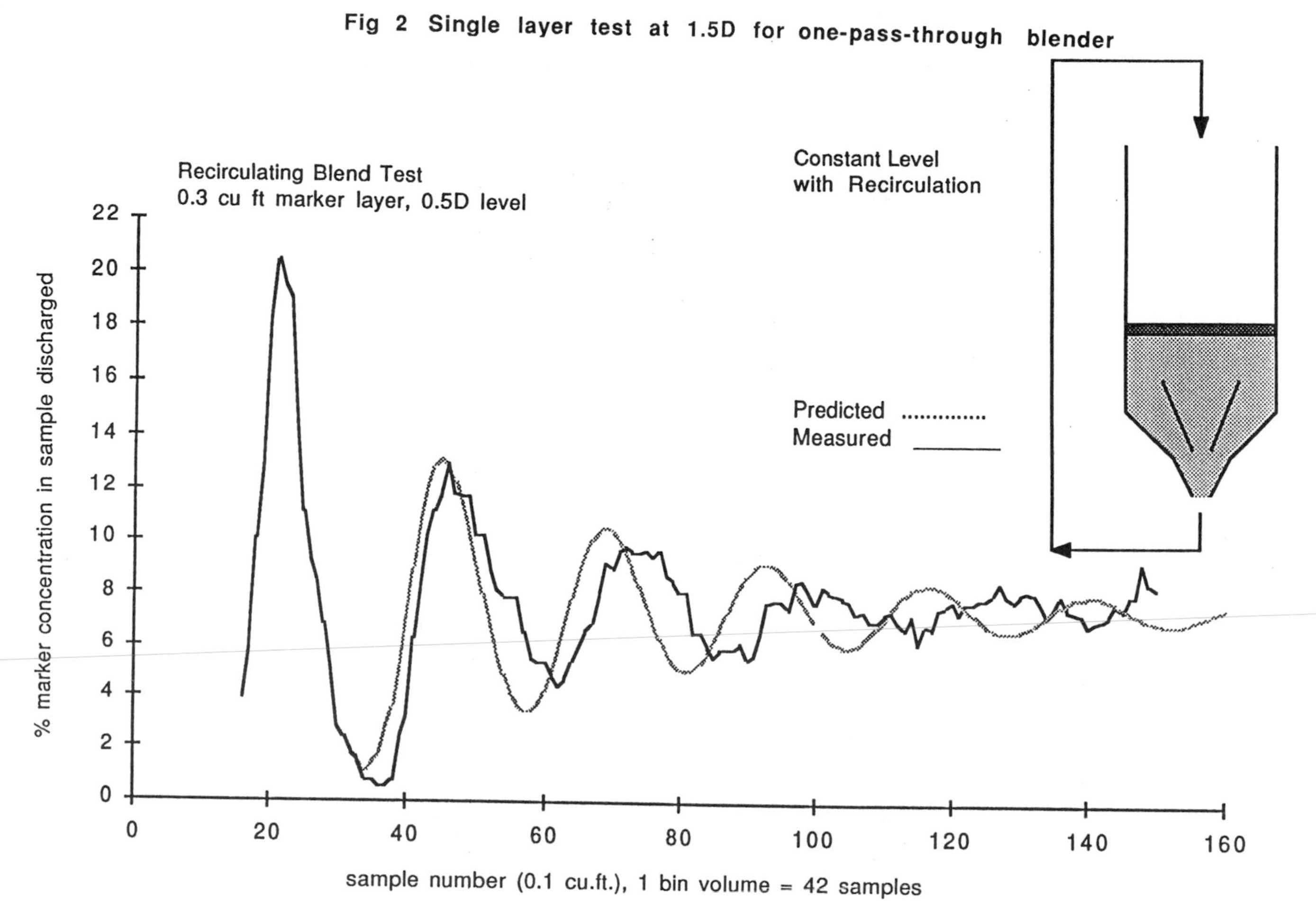

Fig 3 Comparison of calculated vs. measured values of discharge for multiple recirculations at 0.5D level

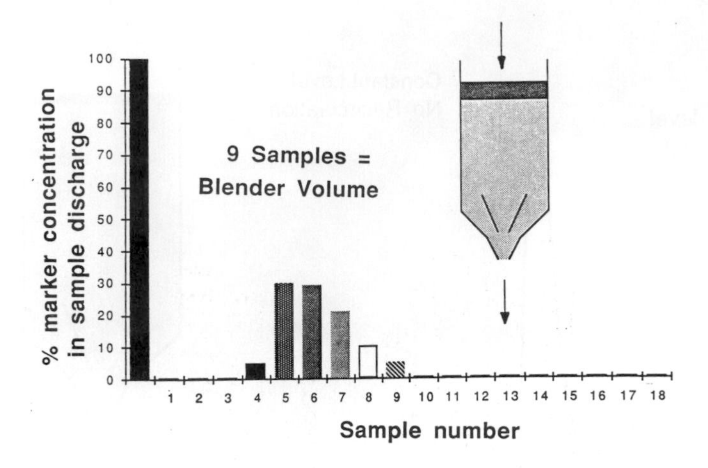

Fig 4 Illustration of constant level single layer test result

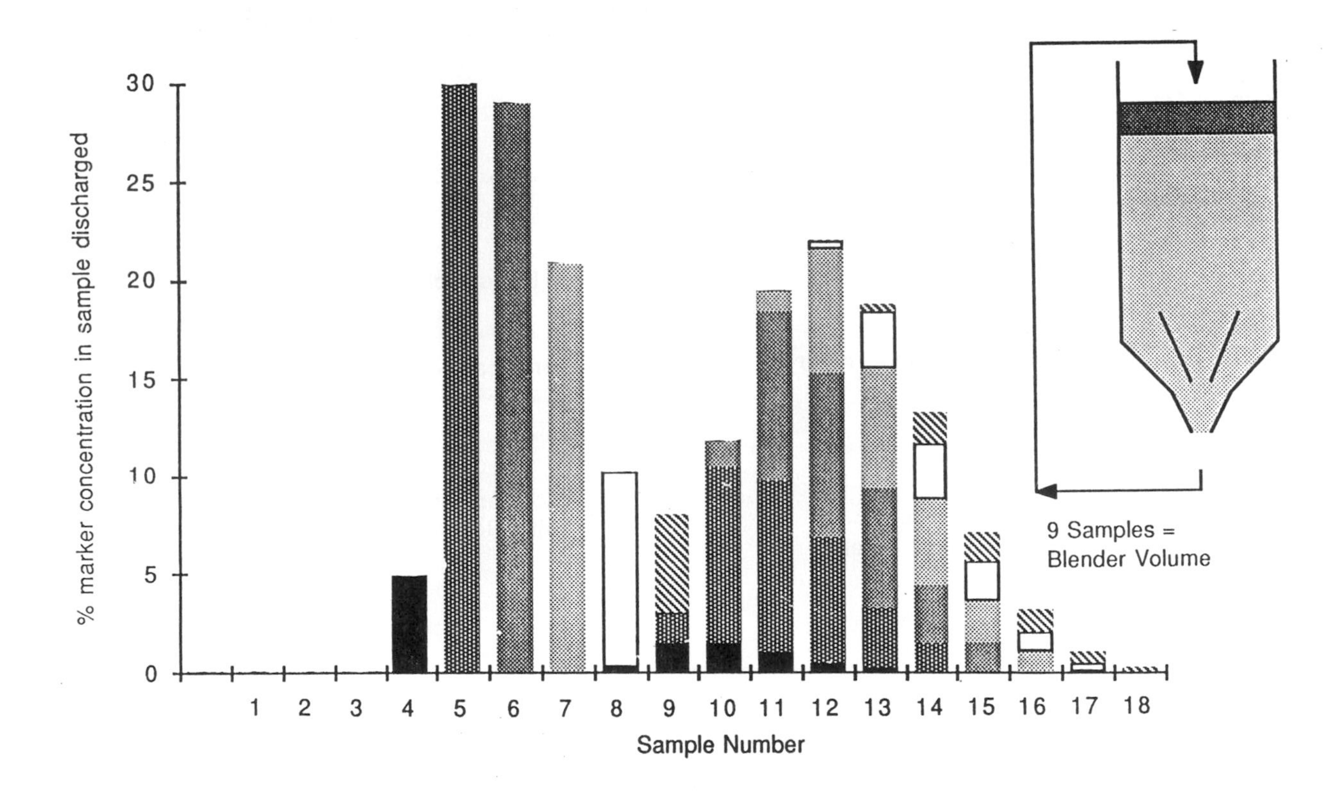

Fig 5 Partially developed computation of recirculation from single layer test of Fig. 4

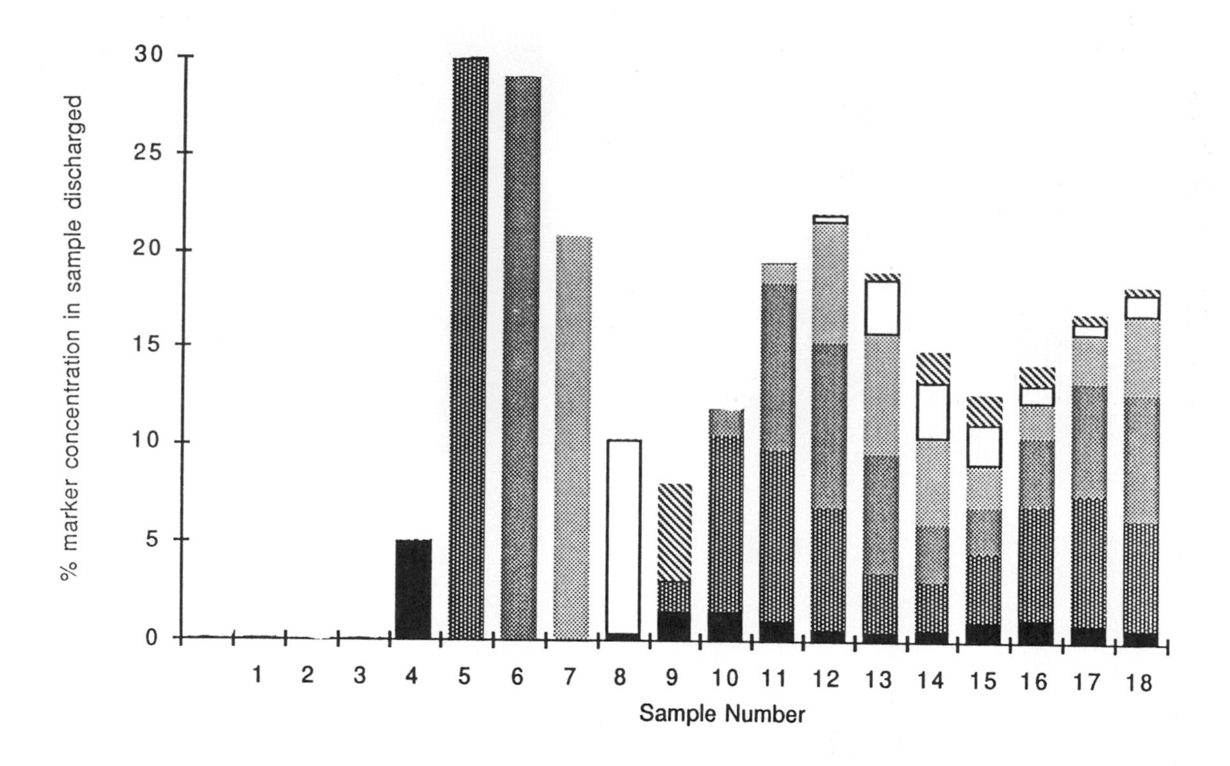

Fig 6 Final computational result for two bin volumes recirculated

C418/049

Design of bins and feeders for anti-segregation and blending

A W ROBERTS, BE, PhD, ASTC, CEng, MIMechE, FIEAust, FTS
Faculty of Engineering, University of Newcastle, New South Wales, Australia

SUMMARY This paper presents an overview of the basic principles of blending and mixing in bins. The problems of segregation during filling and discharge are briefly discussed and the salient characteristics of mass-flow as they relate to mixing and blending are indicated. The mechanics of mixing and blending in mass-flow bins using a re-cycling process are described, with an indication being given to the geometrical parameters of the bin that are required in order to achieve good performance. A review of the use of inserts to correct flow problems and to achieve mass-flow is given, with particular mention being made of the cone-in-cone insert developed by Johanson. The application of a hopper-in-hopper plane-flow bin with screw feeder to prevent segregation is described by reference to experiments performed on a pilot scale test rig. Criteria for the correct design of the bin and feeder are given.

1. INTRODUCTION

In process plants handling bulk solids or powdered materials, it is often necessary, as part of the plant quality control, to mix materials as well as prevent segregation during subsequent handling. While mechanical mixers are commonly employed, the use of well designed mass-flow bins can often achieve good blending performance, usually by a recycling process, provided the bin geometry is carefully chosen [1,2]. Even when mechanical mixers are employed, well designed blending bins and feeders can ensure that segregation will be avoided in the subsequent handling downstream of the mixer.

In-bin blending performance can be improved significantly through the use of inserts, such as the inverted cone [3] or, more particularly, the cone-in-cone hopper insert [4,5], both developed by Johanson . While the latter type of system is limited to axi-symmetric bins, the concept of a hopper-in-hopper insert may be readily extended to plane-flow bins.

The purpose of this paper is as follows:

(i) Review the basic concepts of bin flow and outline the mechanics of mixing during discharge in mass-flow bins indicating the required geometry for optimum blending performance.

(ii) Discuss the use of inserts in bins to improve the flow pattern in order to avoid segregation and achieve good mixing

(iii) Describe the application of a hopper-in-hopper plane-flow bin and screw feeder to prevent segregation and illustrate this concept by reference to experimental studies performed on a pilot scale test rig.

(iv) Summarise the objectives to be observed in feeder design to ensure correct interfacing with the feed hopper with a view to preventing segregation and enhancing blending.

2 SEGREGATION AND FLOW PATTERNS

2.1 Segregation in Bins and Chutes

Bulk solids of wide particle size range will segregate on being charged into a bin or during flow through a chute. If a bin is centre filled, then the fine particles will tend settle in the centre, while the coarse particles will congregate at the outside. If the bin is mass-flow, some re-mixing will occur during discharge, but the degree of re-mixing will depend on the bin geometry and the properties of the bulk solid. If the bin is funnel-flow, then the bulk solid will remain segregated during discharge.

In the case of fine powders, particularly if the bin is filled by means of a pneumatic conveying system, it is not uncommon for the fine particles, which remain aerated, to settle to the outside, while the coarser particles remain in the centre. To overcome this problem, a baffle plate or distribution device in the top of the bin may be used.

Segregation problems are more pronounced in the cases of off-centre loading and eccentric discharge. For this reason, eccentric loading and eccentric discharge should be avoided wherever possible.

The subject of segregation in bins is discussed in Refs.[6,7]

2.2 Importance of Flow Pattern

To avoid segregation during discharge from a bin, it is important that mass-flow be obtained. In the case of large storage capacities, where funnel-flow may be adopted in order to limit the bin height or where multiple outlets are employed, it is essential that mass-flow hoppers be incorporated. In this way the flow pattern is characterised by expanded-flow.

The characteristics of mass-flow and funnel-flow and the limits defining these two principal flow modes are well known. Since in the original work of Jenike [8,9], flow in a hopper is based on the radial stress theory, no account is taken of the influence of the surcharge head due to the cylinder on the flow pattern developed, particularly in the region of the transition. In Jenike's work, the limits for mass-flow are defined by

$$\alpha = f(\phi, \delta) \quad (1)$$

where α = Hopper half-angle
δ = Effective angle of internal friction
ϕ = Wall friction angle

It has been known for some time that complete mass-flow in a hopper is influenced by the cylinder surcharge head.

Referring to Figure 1, there is a minimum surcharge head H_{cr} which is required to enforce mass-flow in the hopper [7]; this height ranges from approximately 0.75 D to 1.0 D.

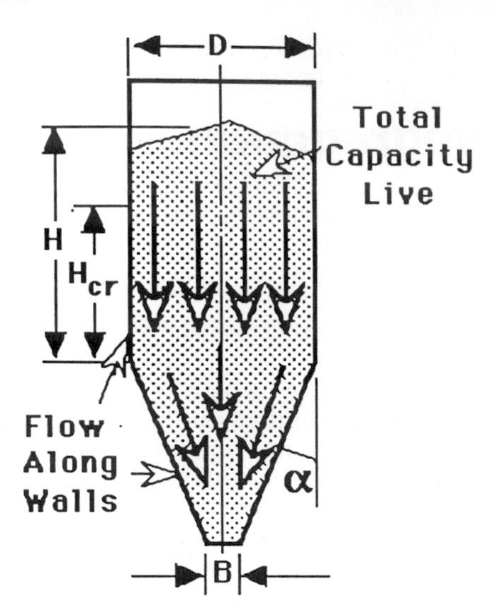

Figure 1 Mass-Flow

2.3 Mass and Funnel Flow Limits - More Recent Research

More recent research has shown that the mass-flow and funnel-flow limits require further explanation and refinement. For instance, Jenike [10] published a new theory to improve the prediction of funnel-flow; this led to new limits for funnel-flow which give rise to larger values of the hopper half-angle than previously predicted, particularly for high values of the wall friction angle. In the earlier theory, the boundary between mass-flow and funnel-flow was based on the condition that the stresses along the centre line of the hopper became zero. In the revised theory the flow boundary is based on the condition that the velocity becomes zero at the wall.

In a comprehensive study of flow in silos, Benink [11] has identified three flow regimes, mass-flow, funnel-flow and an intermediate flow as illustrated in Figure 2. Whereas the radial stress theory ignores the surcharge head, Benink has shown that the surcharge head has a significant influence on the flow pattern generated. He derived a fundamental relationship for H_{cr} in terms of the various bulk solid and hopper geometrical parameters, notably the H/D ratio of the cylinder and the effective angle of internal friction δ.

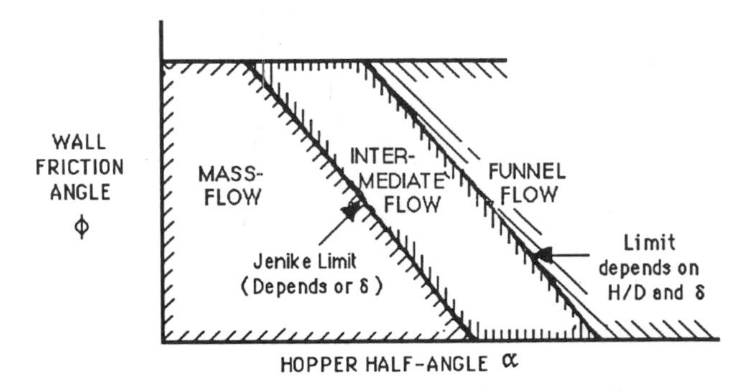

Figure 2. Flow Regimes for Plane-Flow Hopper defined by Benink [11]

Benink developed a new theory, namely the 'arc theory', to quantify the boundaries for the three flow regimes. This theory predicts the critical height H_{cr} at which the the flow changes. For surcharge heads H greater than H_{cr}, flow in the cylinder is uniform over the cross-section. For heads less than H_{cr}, flow in the cylinder is not uniform over the cross-section, with the bulk material in the centre moving faster than material at the walls due to the influence of the convergence at the cylinder/hopper transition.

2.4 Mass-Flow Hopper Flow Characteristics

In order that some appreciation of the blending and mixing capabilities of mass-flow hoppers be obtained, some salient aspects of the flow characteristics are briefly reviewed. A great many bulk solids are referred to as "coarse" in that their particle size range is such that the air permeability is sufficiently high to allow air to percolate through the stored solid with a minimum of resistance. The stress and velocity fields in converging channels have been studied in detail in the original work of Jenike and Johanson [8,12,13]. Consider the accelerated flow of a bulk solid in the region of the outlet of the hopper shown in Figure 3; in this case the air pressure gradient $\Delta p_r = 0$. Analysing the forces, it may be shown that

$$\bar{\sigma}_1 = \frac{\rho g B}{H(\alpha)} [1 - \frac{a}{g}] \qquad (2)$$

where
- ρ = Bulk density
- $\bar{\sigma}_1$ = Stress acting in arch at angle 45°
- a = Acceleration of discharging bulk solid
- B = Hopper opening
- H(α) = Factor to account for variation in arch thickness T, hopper half-angle α and hopper type, whether conical or plane-flow.

The minimum hopper opening to just prevent a cohesive arch from forming occurs when static equilibrium prevails; that is, the acceleration a = 0.

From (2) with a = 0,

$$B_{min} = \frac{\bar{\sigma}_1 H(\alpha)}{\rho g} \qquad (3)$$

The hopper half-angle is chosen from the mass-flow limits (Ref.[2,9,14]), while the condition for $\bar{\sigma}_1 = \sigma_c$ (Figure 3 (b)) is obtained from the intersection point of the Flow Factor ff line and Flow Function FF. The Flow Function is a bulk solid parameter and represents bulk strength while the Flow Factor is a flow channel parameter. The Flow Function is a plot of the unconfined yield strength σ_c as a function of the major consolidation pressure σ_1. The Flow Factor and function H(α) are given as design curves in Refs. [2,9].

The Flow Function is a plot of the unconfined yield strength σ_c as a function of the major consolidation pressure σ_1. The Flow Factor and function H(α) are given as design curves in Refs. [2,9].

Following the work of Johanson [15], it may be shown that the acceleration in equation (2) may be expressed as

$$a = g [1 - \frac{ff}{ff_a}] \qquad (4)$$

where
- ff = Critical Flow Factor based on the minimum arching dimension
- ff_a = Actual Flow Factor based on the actual opening dimension

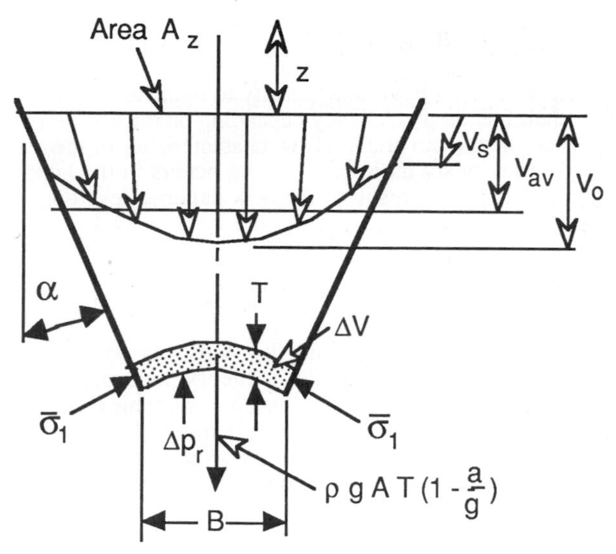

(a) Flow Channel

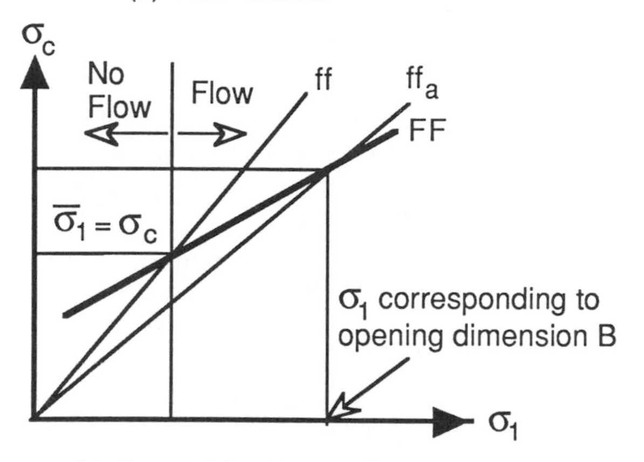

(b) Determining Hopper Geometry

Figure 3. Flow in a Hopper

$$ff_a = \frac{\sigma_1}{\sigma_c} \tag{5}$$

where σ_1 = Major consolidation pressure at outlet corresponding to dimension B

The acceleration has two components:

$$a = a_c + a_v \tag{6}$$

where a_c = Convergence component due to flow channel

a_v = Component due to velocity increase as flow is initiated

It may be shown that

$$a_v = g\left(1 - \frac{ff}{ff_a}\right) - \frac{2v^2(m+1)}{B}\tan\alpha \tag{7}$$

This shows that as the discharge velocity increases, $a_v \rightarrow 0$. Thus, an average terminal discharge velocity v_a is reached. With $a_v = 0$,

$$v_a = \sqrt{\frac{B\,g}{2(m+1)\tan\alpha}\left[1 - \frac{ff}{ff_a}\right]} \tag{8}$$

and the flowrate is

$$Q_0 = \rho\, B^{(1+m)}\, L^{(1-m)} \left[\frac{\pi}{4}\right]^m v_a \tag{9}$$

where m = 0 for plane-flow hopper
m = 1 for axi-symmetric or conical hopper
B = Width of slot or diameter of circular opening
L = Length of slot.

The flow rate given by (9) is the maximum possible. In the majority of cases, the discharge rate needs to be controlled and this is accomplished by means of a feeder.

2.5 Flow of Fine Bulk Solids

For fine powders discharge from a hopper will be impeded by the low permeability of the powder to air flow; that is $\Delta p_r \neq 0$. Flow rates very much lower than those computed using the method outlined in Section 2.4 above will occur. The analysis is much more complex and involves two-phase flow theory. The flow of fine powders has been studied by McLean [2,16], and more recently by Arnold and Gu [17,18]. Larger hopper openings are required than for the equivalent coarse bulk solids and sometimes air permeation is required to bias the pore pressure in the hopper in order to assist the discharge. Fine powders are prone to flooding and uncontrollable discharge if allowed to aerate so that extreme care must be exercised in designing and installing any air permeation system. Care must also be exercised in ensuring that the interface between the hopper and feeder are correctly designed to prevent problems due to flooding.

2.6 Velocity Characteristics in Hopper

The velocity distribution in a converging channel has been studied in some detail by Johanson [12,13]. A practical application of this earlier work of Johanson was presented in a more recent paper by Johanson and Royal [19], who were interested in calculating the sliding velocity at a hopper wall in relation to the wear of hopper linings. This concept has been followed up by Roberts et al [20-22] who have also examined hopper wall wear, as well as wear in chutes.

The average velocity along the converging hopper (Figure 3) is computed from the flow rate as follows:

$$v_{az} = \frac{Q_m}{\rho A_z} \tag{10}$$

Where Q_m = Mass flow rate at section
A_z = Cross-sectional area.

The velocity profile in the hopper is as depicted in Figure 4. The velocity of sliding at the wall, v_s, may be expressed in terms of the average velocity, v_{av}, by

$$v_s = K_v\, v_{av} \tag{11}$$

The parameter K_v for conical and plane-flow hoppers, computed in accordance with Ref.[19] is given in Ref.[21,22]. The maximum velocity, v_o, occurring at the centre of the flow channel may be computed as a function of the sliding velocity at the wall, v_s, as described in Ref.[19].

It is to be noted that velocity profile described by the velocity ratio v_s/v_o has a direct bearing on the blending or mixing characteristics of hoppers. An indication of the velocity variation at an appropriate hopper cross-section during mass-flow may be obtained by examining the velocity ratios v_s/v_o. By way of example, Figure 4 shows the variations of hopper half-angle α and velocity ratio v_s/v_o as a function of wall friction angle ϕ for conical and plane-flow hoppers. The curves correspond to an effective angle of internal friction $\delta = 50^o$ and to the mass-flow limits. In the case of the conical hoppers, the angle α is 3° less than the limiting angle, this being the normal design margin.

Figure 4(a) shows that for conical hoppers, the velocity ratio does not vary significantly, being substantially constant at $v_s/v_o = 0.3$. In the case of plane-flow hoppers, Figure 4(b), the ratio v_s/v_o increases with increase in hopper half-angle as indicated.

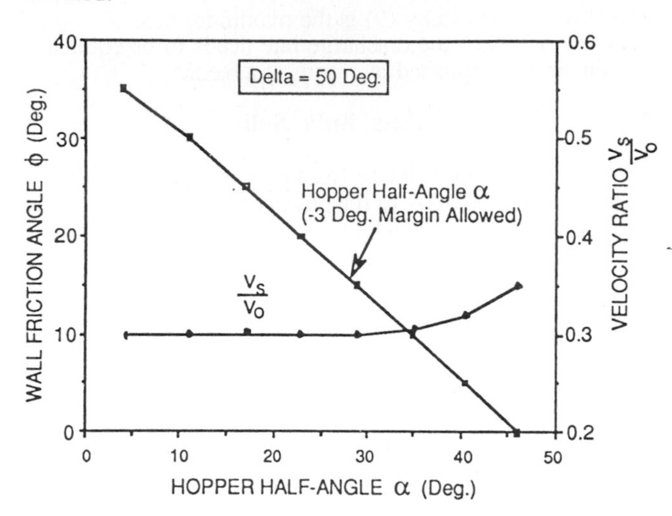

(a) Conical Hopper

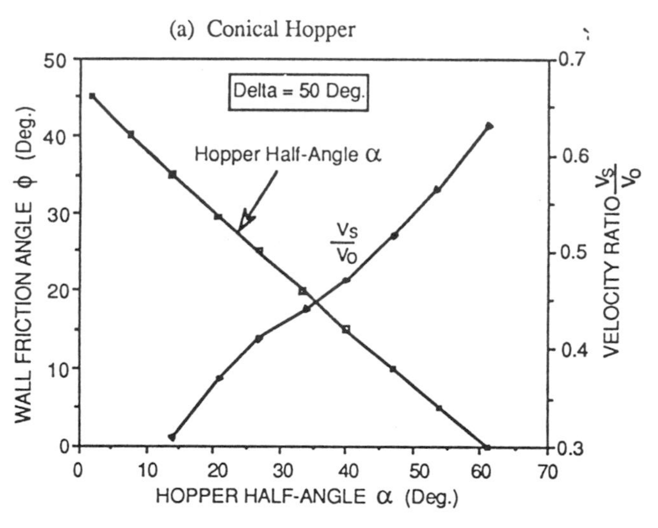

(b) Plane-Flow Hopper

Figure 4. ϕ and v_s/v_o as a Function of ϕ for Mass-Flow Limits - $\delta = 50^o$

The velocity ratios $K_v = v_s/v_{av}$ may be computed from the above v_s/v_o values in accordance with Ref.[19]. By way of example, for a plane-flow, mass-flow bin, with a wall friction angle of 20° and a corresponding hopper half-angle $\alpha = 35^o$ (Figure 3), $v_s/v_o = 0.44$; the velocity ratio $K_v = v_s/v_{av} = 0.57$. The wall friction angle of 20° would apply, for example, for coal on stainless steel type 304 with 2B finish. If structural steel plate is used, the wall friction angle is more likely to be 30°; on this basis, the hopper half-angle for mass-flow in a plane-flow hopper is $\alpha = 21^o$. The ratio $v_s/v_o = 0.37$ and the corresponding value of $K_v = 0.53$ which is not very different from the previous case for the stainless steel lining. The results for v_s/v_o and v_s/v_{av} in this example show the wide variation in velocity over a hopper cross-section, this being important with regard to blending or mixing during gravity flow.

3. IN-BIN BLENDING

Mass-flow hoppers have the capacity to blend and mix bulk solids, their blending efficiency depending on the geometrical proportions of the bin and the flow characteristics of the bulk solid. As previously indicated, mixing occurs in the hopper during discharge as a result of the flow patterns developed in the hopper.

3.1 Blending in Mass-Flow Bins

The concept of blending in mass-flow bins was explained by Johanson [1], who established a theory to examine the manner in which efficient blending may be achieved. This theory was modified by Roberts (Chapt. 8, Ref.[2]) who used the concept of a bin impulse response or 'weighting function' to model a mass-flow hopper as a blending system.

Figure 5 shows a mass-flow bin and its corresponding system representation. If a layer of marked particles is spread over the surface of the bin contents, as illustrated in Figure 5(a), this is equivalent to applying an impulse volumetric input to the bin considered as a system. In mathematical terms, the input is $U_o(V)$ N, where $U_o(V)$ is a unit volumetric impulse input and N is the number of marked particles. The system, in this case the bin, responds with its system impulse response which represents the volume concentration of particles discharging from the bin. The responding function is N h(V) where h(V) is known as the system impulse response or 'weighting function'. This function is triangular in form as shown in Figure 5(b); this characteristic has been verified by experiments. Assuming that the contents of the bin are re-cycled, the mixing efficiency may be assessed by examining the volume concentration of marked particles after each pass. Figure 6 illustrates the volume concentrations for the initial condition, first pass and second pass.

The explanation of the blending characteristics may be obtained by considering Figures 5 and 6 together. Referring to Figures 6, the blending characteristics of a mass-flow bin may be observed. During the first pass through the bin, the volume V_1 discharges before the first of the marked particles appear. In systems terminology, V_1 is volume transfer lag.The volume V_P corresponds to the discharge of the marked particles as the bin empties. The first pass distribution is triangular in shape and the total bin content is $V_T = V_1 + V_P$. If the contents of the bin are re-cycled a second time, the marked particles will discharge over the volume flow $2V_P$ and the distribution is as shown in Figure 6. For the third pass the marked particles discharge over the volume $3V_P$ and so on.

A general expression for the volume concentration for the 'n th' pass may be obtained from the convolution or superposition integral. The equation for the impulse response or weighting function is

$$h(V) = 0 \qquad \text{for } 0<V<V_1$$

$$h(V) = \frac{2}{V_P}\left[1 - \frac{V}{V_P}\right] \qquad (12)$$

$$\text{for } V_1<V<(V_1+V_P)$$

From linear systems theory,

$$C_n(V) = \int_{V_0}^{V_1} C_{(n-1)}(\varepsilon)\ h(V-\varepsilon)\ d\varepsilon \qquad (13)$$

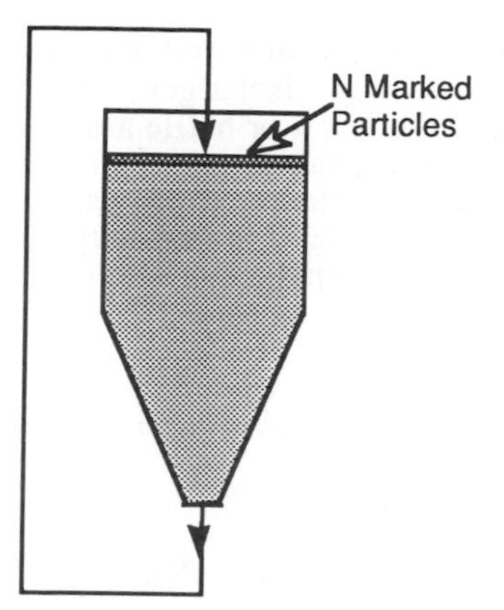

(a) Mass-Flow Blending Bin (Initial Condition)

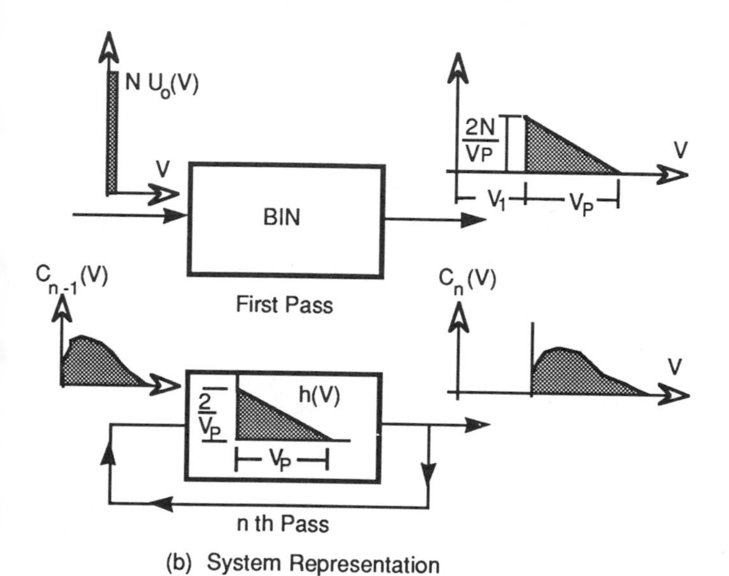

(b) System Representation

Figure 5. In-Bin Blending Model.

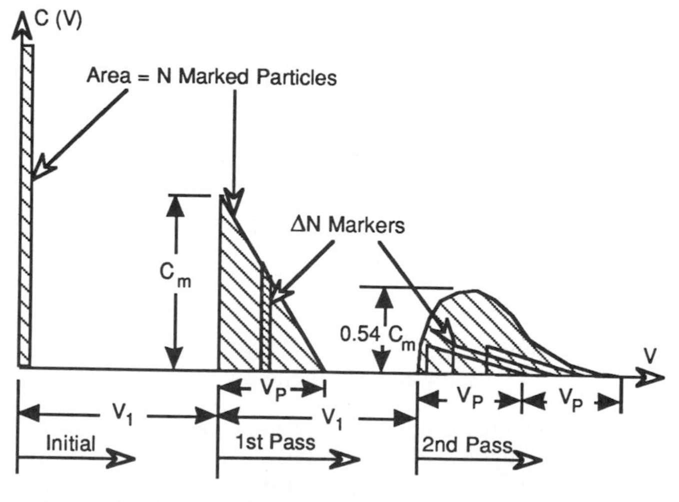

Figure 6. Volume Concentrations for In-Bin Blending.

or

$$C_n(V) = \int_{V_0}^{V_1} \frac{2\, C_{(n-1)}(\varepsilon)}{V_P} \left[\frac{V_P + \varepsilon - V}{V_P} \right] d\varepsilon \qquad (14)$$

where ε = dummy variable denoting volume concentration. Its introduction in the above equations follows from the derivation of the convolution integral.

The limits in equations (13) and (14) are as follows

$V_o = 0$ and $V_1 = V$ for $0 < V < V_P$
$V_o = (V - V_P)$ and $V_1 = V$ for $V_P < V < (n-1)V_P$
$V_o = (V - V_P)$ and $V_1 = (n-1)V_P$ for $(n-1)V_P < V < nV_P$

3.2 Requirements for Good Blending Performance

The blending efficiency may be gauged from the volume distributions. It is to be noted that for the 'n-th' pass the marked particles will discharge over the volume n V_P. However, it is apparent that the distribution is quite skewed, with the tail incrementing by V_P with each pass. For the assessment of the blending performance, the ratio of V_1/V_P is important. It has been shown that good blending can be obtained in two to three passes provided that a sufficient overlap is achieved. Overlap is illustrated in Figure 7.

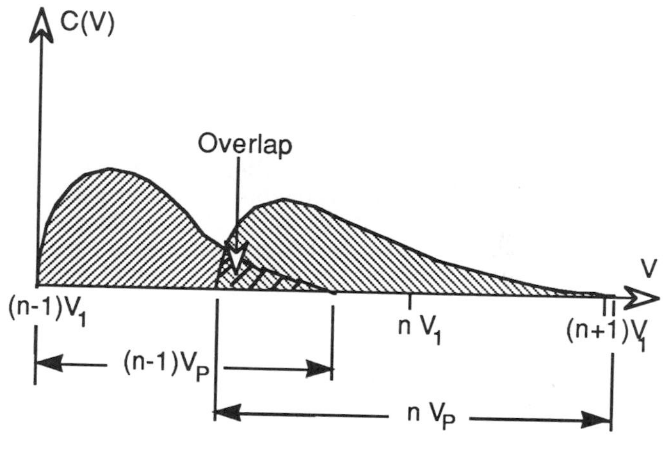

Figure 7. Overlap in Successive Passes

The overriding condition is for a uniform distribution to be achieved and this requires that

$$V_P/V_1 \geq 1.0 \qquad (15)$$

If this does not occur, then it does not matter how many passes are made, there will never be satisfactory blending or mixing. The condition described by (15) implies that the volume of first-pass mixed particles is at least half the total bin volume V_T (since $V_T = V_1 + V_P$).

Normally the condition defined by (15) can only be achieved in a bin with a large hopper half-angle and very little surcharge. Referring to Figure 8, the conditions for good blending or mixing performance are summarised as follows:

(i) The cylinder height should be kept as small as possible and certainly less than the critical height H_{cr} as discussed in Section 2.

(ii) Where possible side loading should be employed in preference to centre loading.

(iii) The hopper half-angle α should be kept as large as possible within the bounds of mass-flow. This is to promote a favourable flow pattern for good mixing to occur.

(iv) In the light of (iii) above, a plane-flow hopper which allows a larger hopper half-angle than that for a conical hopper has a distinct advantage. However, the disadvantage arises as a result of the long slotted hopper opening which makes feeding more difficult. The feeder must be designed for uniform draw of bulk solid.

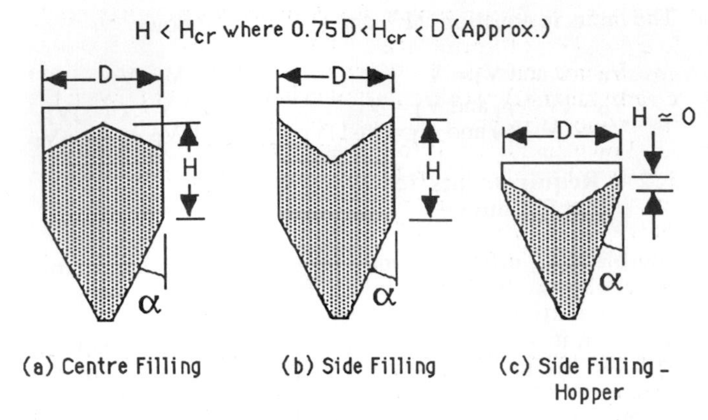

Figure 8. Examples Indicating Conditions for Good Blending Performance.

4. USE OF FLOW CORRECTING INSERTS.

Inserts may be used in bins to achieve a desirable flow pattern. Often they are used in the form of a 'retro-fit' device to correct an undesirable flow pattern. A knowledge of the the flow properties of the bulk solid, together with an understanding of the conditions controlling bin flow patterns will permit the application of innovative design in the use of inserts for a specific purpose. Inserts are classified as passive devices for flow promotion since they do not require energy for their operation.

Johanson [3] proposed the use of inverted conical-shaped inserts to correct the flow in axi-symmetric hoppers. In effect, the insert converts a conical hopper into a pseudo plane-flow hopper in which the slot is equivalent to a circular annulus. This follows since the hopper half-angle for a plane-flow hopper is about 8° to 10° larger than that of the equivalent conical or axi-symmetric hopper. Procedures for designing and correctly placing inserts are presented in Johanson's paper.

In some cases a second insert located higher up in the hopper is recommended. This arrangement has the advantage of minimising segregation as well as providing a solution to flow problems due to cohesive materials.

The application of inverted, conical-type inserts to in-bin blending has received limited attention and it is a subject requiring further research. The research of Tuzun and Nedderman [23] in which the flow round obstacles in relation to inserts, is worthy of note. From a practical point of view, a knowledge of the flow properties of the bulk solid and the boundary friction characteristics of the hopper and inserts can be used to advantage in correcting segregation problems. For instance, if segregation has occurred in a bin with fines congregating in the central region and coarse particles in the outer region, the selection of the appropriate lining surface for the insert as well as its slope angle is important. Since the boundary friction angle for fine particles is higher than for coarse particles due to the particle size and surface roughness interaction [22,24], the surface of the insert needs to be smoother than that of the hopper for similar slope angles so that the fines can flow more freely and mix with the coarser particles.

An overriding consideration in the use of inserts in bins for flow correction concerns the magnitude of the draw-down forces that can be generated. Since the inserts have to be supported from the hopper walls, it needs to be noted that for the design of the supports, loads greater than hydrostatic can occur; loads similar to those due to the piston effect (Chapter 1, Ref.[2]) may be experienced. For this reason, the placing of inserts in bins must be carefully considered in the light of alternative solutions.

The principle of the insert is also embodied in the vibrating 'live' bottom bin or pile discharger. In this case, the gyratory motion of the insert or baffle assists in promoting flow. It is important that the vibrator be cycled rather than run continuously in order to avoid compaction of the bulk solid in the bin (or stockpile) if the capacity of the conveyor transporting the material from the bin does not match the feed rate.

It is important to note that vibrations have a significant influence on the behaviour of bulk solid. This is discussed in Refs.[25-27].

5. THE CONE-IN-CONE INSERT

The foregoing problems may be solved by means of the cone-in-cone BINSERT™ invented by Johanson [4,5]. Alternative arrangements of this device are shown in Figure 9. Essentially the half-angle α of the inner cone is determined by mass-flow design requirement . The half-angle of the outer cone is then set at 2α, the hopper lining materials for the inner and outer cones being the same. Looked at another way, suppose a cylindrical bin with conical hopper has a hopper half angle 2α and no inner cone is fitted, then quite clearly the bin will operate under funnel-flow. By fitting the inner cone with half angle α, the flow changes to mass-flow. This illustrates the effectiveness of the Binsert in correcting problems due to funnel-flow as well as allowing large storage capacities to be achieved with reduced head room than is possible with a conventional mass-flow bin. The device is much more effective than the inverted cone insert described in Section 4.

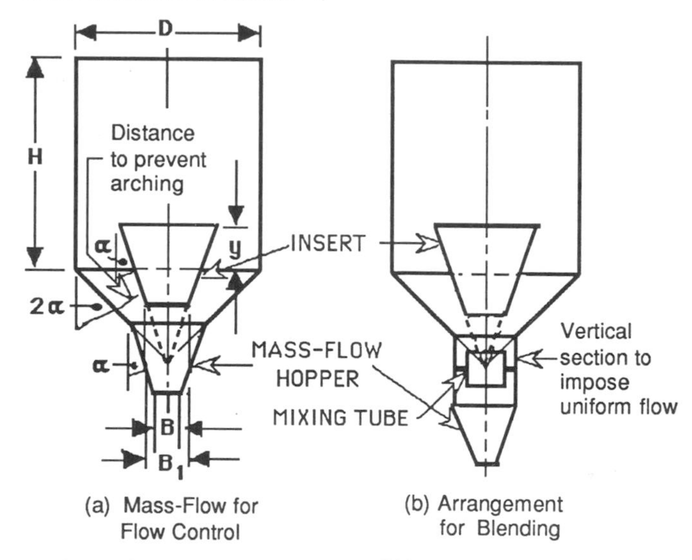

Figure 9. Johanson's BINSERT™ for Promoting Mass-Flow and Bin Blending.

The average velocities of the inner and outer flow channels are controlled by the velocity profile achieved at the bottom of the insert. For instance, the mass-flow cone of Figure 9(a) fitted to the bottom of the bin allows a higher velocity to occur in the inner channel than through the outer channel. This type of flow can be useful in blending. The vertical section shown in Figure 9(b) imposes a more uniform flow pattern throughout the hopper and can be of assistance in eliminating segregation problems that may have occurred when the bin is filled. The design of the two hoppers must include consideration of the bulk solid flow properties to ensure that not only is the hopper half-angle α correctly selected, but also the arching characteristics for the determination of the hopper openings and minimum spacing between the hoppers.

One important point concerns the application of the Binsert to bulk solids of wide particle size range, particularly if segregation has taken place with the larger particles located in the outer regions. This will mean that the larger particles will flow in the outer channel more readily than in the case of a bulk solid of uniform particle size. This must be taken into account in the design so that appropriate adjustments can be made to compensate for the variability in the flow. Care must be taken to ensure that the width of the outer annulus is not only large enough to prevent a cohesive arch from forming, but also is large enough to prevent mechanical arching due to the presence of larger particles. To avoid the latter condition, the opening should be at least 4 times the maximum particle size.

It is clear that the the cone-in-cone concept provides for more efficient blending than a single mass-flow hopper. The desired mixing characteristics enunciated in Section 3 are more readily achievable in view of the larger hopper half-angle for mass-flow and the fact that surcharge is possible.

Tests conducted on a small scale blending bin of the type shown in Figure 9(b) in which a thin layer of red coloured powder was spread over the surface of the contained white ascorbic acid powder showed that after two passes the coloured contents were well mixed and quite thoroughly blended after three passes.

It needs to be noted that, theoretically, the Binsert principle can be extended to three hoppers with hopper half-angles α, 2α and 3α for the inner, middle and outer hoppers respectively. However the complexity of this system would suggest that it would have limited application in practice.

6. PLANE-FLOW HOPPER-IN-HOPPER CONCEPT

The concept of a hopper-in-hopper approach to plane-flow bins has been successfully applied by the University of Newcastle Australia to the solution of an industrial handling problem in the food industry. The flow properties of the bulk solid were determined and these were used to select the appropriate half-angles and opening dimensions. The required half-angle was $\alpha = 22^o$ so this became the angle for the insert, while the hopper was designed with the half angle $2\alpha = 44^o$. A pilot scale bin and screw feeder was constructed and tested. Without the insert in place the bin clearly funnel-flowed as illustrated in Figure 10(a). With the inner hopper in place, the flow was mass-flow and uniform as shown in Figure 10(b). The final design of the bin is shown schematically in Figure 11. The importance of the mass-flow mixing hopper at the bottom of the bin must be stressed; the hopper is required to control the flow and effect good mixing. Particular attention must be given to the correct design of the feeder as discussed below.

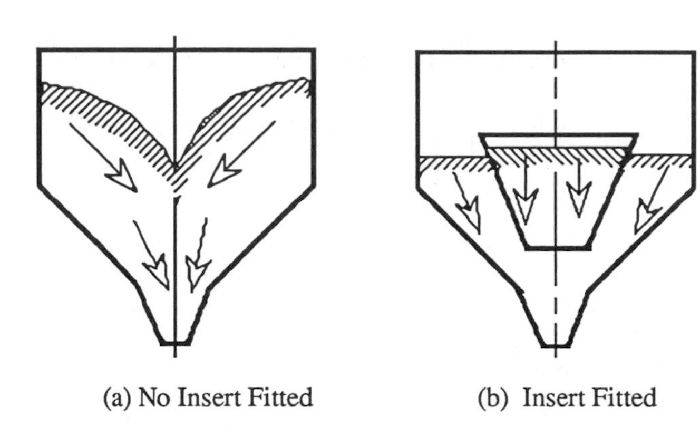

(a) No Insert Fitted (b) Insert Fitted

Figure 10. Flow Patterns in Plane Flow Bin

7. CONCLUDING REMARKS

In this paper an overview of some salient aspects of the flow of bulk solids in bins and hoppers has been given. The flow theories established by Jenike in the sixties have been briefly reviewed and current research which relates to the mass-flow and funnel-flow limits has been outlined.

Particular attention has been given to mixing and blending in bins and in the use of inserts to improve gravity flow and achieve better blending performance. Attention has been drawn to the cone-in-cone Binsert developed by Johanson which gives superior, in-bin blending performance over a

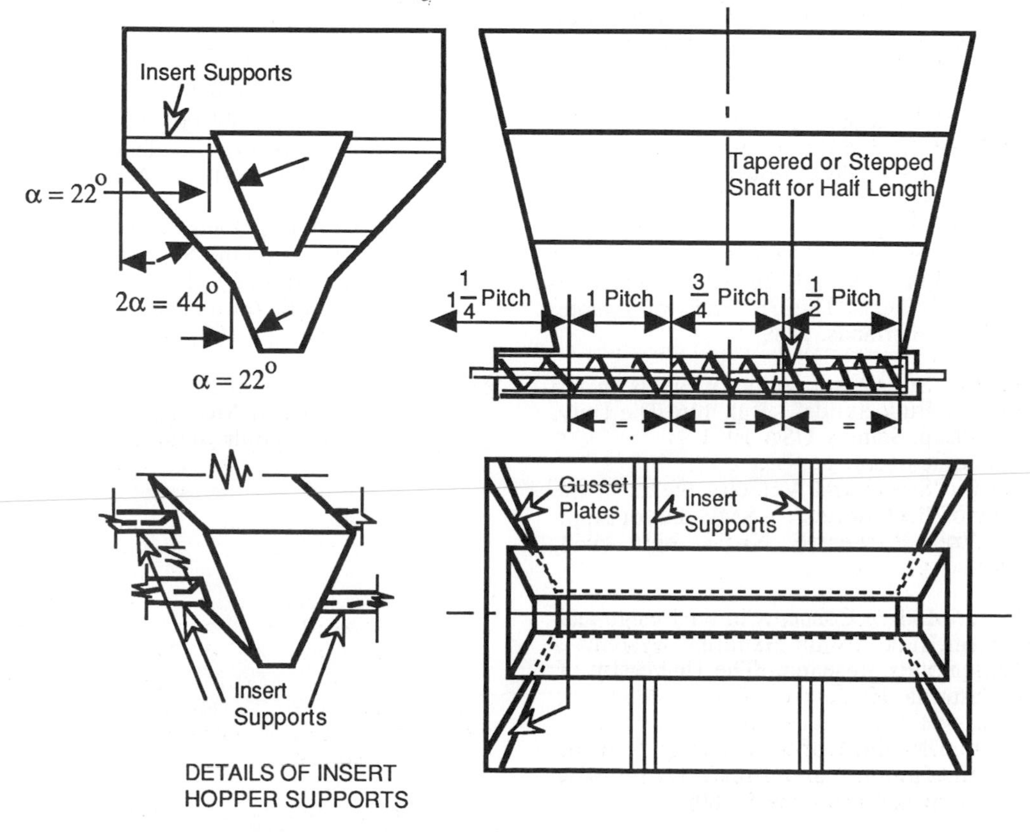

Figure 11. Bin and Feeder Design to Prevent Segregation

single, conical mass-flow hopper. Where head room is a problem, advantage may be made of the plane-flow bin using a hopper-in-hopper concept which allows larger hopper-half angles to be employed than is the case axi-symmetric bins with conical hoppers. Because the plane-flow hopper has a long outlet slot, it is most important that the feeder be correctly designed and interfaced with the hopper opening in order to achieve uniform draw-down in order to avoid the possibility of segregation occurring.

8. REFERENCES

1. Johanson, J.R. "In-Bin Blending". Chem. Engng. Progress, Vol. 66, No.6, 1970, (pp.50-55).

2. Arnold, P.C., McLean, A.G. and Roberts, A.W. "Bulk Solids: Storage, Flow and Handling". The University of Newcastle Research Associates (TUNRA), Australia, 1982.

3. Johanson, J.R. "The Use of Flow-Corrective Inserts in Bins". ASME, Jnl. for Engng. for Industry, Vol.88, Ser. B, No. 2, May 1966. (pp.224-230).

4. Johanson, J.R. "Controlling Flow Patterns in Bins by the Use of an Insert". Bulk Solids Handling, Vol. 2., No.3, September 1982. (pp.495-498).

5. Johanson, J.R. BINSERT™ Patent: U.S. Patent 44,286,883,

6. Johanson, J.R. "Particle Segregation and What To Do About It". Chemical Engineering, May 8, 1978, pp.183-188.

7. Thomson F.M. "Storage of Particulate Solids". Chapter 9, Handbook on Powder Science & Technology. (1984) Van Nostrand.

8. Jenike, A.W. "Gravity Flow of Bulk Solids". Bul. 108, The Univ. of Utah, Engn. Exp. Station, USA 1961.

9. Jenike, A.W. "Storage and Flow of Solids". Bul. 123, The Univ. of Utah, Engn Exp. Station, USA 1964.

10. Jenike, A.W. "A Theory of Flow of Particulate Solids in Converging and Diverging Channels Based on a Conical Yield Function". Powder Tech., Vol.50. (pp. 229-236).

11. Benink, E.J. "Flow and Stress Analysis of Cohesionless Bulk Materials in Silos Related to Codes". Doctoral Thesis, The University of Twente, Enschede, The Netherlands. 1989.

12. Johanson, J.R. "Stress and Velocity Fields in the Gravity Flow of Bulk Solids". Bul. 116, The Univ. of Utah, Engn Exp. Station, USA 1961.

13. Johanson, J.R. "Stress and Velocity Fields in the Gravity Flow of Bulk Solids". ASME, Jnl. of Appl. Mechanics, Vol. 31, Ser. E, No. 3, Sept. 1964. (pp.499-506).

14. Roberts, A.W. "Modern Concepts in the Design and Engineering of Bulk Solids Handling Systems". TUNRA Bulk Solids Research, The University of Newcastle, Australia, 1988.

15 Johanson, J.R. "Method of Calculating Rate of Discharge from Hoppers and Bins", Trans. Min. Engrs. AIME, Vol.232,1965. (pp.69-80).

16. McLean, A.G. "Flow Rates of Simple Bulk Solids from Mass-Flow Bins". PhD Thesis, The Univ. of Wollongong, Australia, 1979.

17. Arnold, P.C. and Gu, Z.H. "The Effect of Permeability on the Flowrate of Bulk Solids from Mass-Flow Bins". The Intl. Jnl. of Powder Handling and Processing, Vol. 2, No. 3, Sept. 1990, (pp.229-238).

18. Arnold, P.C., Gu, Z.H. and McLean, A.G., "On the Flowrate of Bulk Solids from Mass-Flow Bins". Proc. 2nd World Congress Particle Technology, Sept. 19-22, 1990, Kyoto, Japan, Vol. 2, (pp.2-9).

19. Johanson, J.R. and Royal, T.A. "Measuring and Use of Wear Properties for Predicting Life of Bulk Materials Handling Equipment". Bulk Solids Handling, Vol.2, No.3, Sept . 1982.(pp.517-523).

20. Roberts, A.W., Ooms, M. and Scott, O.J. "Surface Friction and Wear in the Storage, Gravity Flow and Handling of Bulk Solids". Proc. Conf. 'War on Wear', Wear in the Mining and Mineral Extraction Industry, Instn. of Mech. Engnrs, Nottingham U.K., 1984. (pp.123-134).

21. Roberts, A.W. "Friction, Adhesion and Wear in Bulk Materials Handling". Proc., AntiWear 88, The Royal Soc. London. 1988. Inst. of Metals, I.Mech.E.

22. Roberts, A.W., Ooms, M. and Wiche, S.J. "Concepts of Boundary Friction, Adhesion and Wear in Bulk Solids Handling Operations". Intl. Jnl. of Bulk Solids Handling, Vol.10, No. 2, May 1988.

23. Tuzun, U and Nedderman, R.M. "Gravity Flow of Granular Materials Round Obstacles - I. Investigation of the Effects of Inserts on the Flow Patterns Inside a Silo". Chem. Engng. Science, Vol 40, No.3, 1985. (pp.325-33).

24. Ooms, M. and Roberts, A.W. "Significant Influence of Wall Friction in the Gravity Flow of Bulk Solids". Intl. Jnl. of Bulk Solids Handling, Vol. 5, No. 6, 1985 (pp.1271-1277)

25. Roberts, A.W."Vibrations of Powders and Bulk Solids". Chapter 6, Handbook on Powder Science & Technology. (1984) Van Nostrand.

26. Roberts, A.W., Ooms, M. and Scott, O.J. "Influence of Vibrations on the Strength and Boundary Friction Characteristics of Bulk Solids and the Effect on Bin Design". Intl. Jnl. of Bulk Solids Handling, Vol.6, No.1. 1986. (pp.161-169).

27. Roberts, A.W. and Rademacher, F.J.C. "Induced Gravity Flow by Mechanical Vibrations". To appear in Intl. Jnl. of Bulk Solids Storage in Silos, UK.

C418/015

Microstructural effects on stress and flow fields of equal density granules in hopper

U TÜZÜN, BSc, PhD, MAIChemE and P ARTEAGA, BSc, MSc, MIChemE
Department of Chemical and Process Engineering, University of Surrey, Guildford, Surrey

SYNOPSIS A linear geometric model is used to describe the dependence of discharge rates, bulk densities and wall stresses measured during batch emptying of a model hopper on the microstuctural variations of powder beds obtained with binary and ternary mixtures of equal density granules. Experimental measurements are found to correlate well with the predicted phase transitions of the mixtures.

NOTATION

C'	discharge coefficient (= 0.58-0.59)
c	coarse particle
D	hopper diameter
D_o	orifice diameter
d	particle diameter
f	fine particle
g	gravitational acceleration
i	arbitrary integer number
K_{mix}	microstructural scaling factor for the empty annulus
m	medium particle size
N	number density
n	number of size classes in a discrete mixture
Q	volumetric flowrate
V_p	plug flow velocity
W	mass discharge rate
X	weight fraction
Z	empty annulus width

Greek

η_{mix}	fractional solids content of mixture
ρ	density
ρ_{mix}	bulk density of mixture
ρ_f	flowing bed density
σ_w	normal stress at wall
τ_w	shear stress at wall
ϕ_R	particle size ratio ($\phi_{R1} = d_c/d_m$; $\phi_{R2} = d_m/d_f$)
ϕ_w	wall friction angle

1. INTRODUCTION

Polydisperse powder mixtures are handled frequently in process industries such as chemical, pharmaceutical, and foodstuffs in operations involving mixing, blending, pasting, granulation, etc. Production of synthetic powders through processes like crystallisation and freeze drying often result in multimodal particle size and shape distributions. In mixing and blending operations, a common problem is to predict the required microstructural specification to achieve certain bulk properties such as maximum or minimum bulk density, uniform bulk concentrations of mixture components. Furthermore, during storage and transport, physical processes like segregation, agglomeration and attrition of particles are likely to alter the microstructure of the powder mixtures during flow with consequential variations in discharge rates and bulk stresses.

Traditionally, theoretical coupling of stress and flow fields of bulk solids is treated using continuum mechanics (1). Continuum approach often assumes constant or mean field values of the microstructural properties such as particle size , particle shape, and interstitial voidage to predict discharge rates and bulk stresses. This can perhaps explain the present scarcity of literature on the flow and stress fields of powder mixtures.

In this paper, we demonstrate the use of a linear geometric model (2), (3) to predict the fractional solids content of the powder bed as a function of particle size ratios and weight fractions of particles in binary and ternarymixtures. The model can be used to calculate the critical weight fractions inthe mixture for particles of a given size to constitute the "continuous" phase of the flowing bed microstructure while other particles appear as "dispersed" or "discrete" phases. Boundaries of phase transition between different size particles are shown to be a strong function of particle shape and particle size ratio. This in turn gives rise to microstructural criteria for feasibility of size segregation during the flow of equal density granular mixtures in hoppers (2), (3). Similar microstructural criteria are shown to apply here when predicting discharge rates and wall stresses in hoppers.

2. GRAVITY DISCHARGE EXPERIMENTS

2.1 Materials

Experimental materials with similar solid densities were used to prepare binary and ternary mixtures in 0.7 - 9.0 mm. size range. The particles are believed to be coarse enough for the flows not to be influenced by interstitial air drag effects (4); thus ensuring gravity- controlled flow fields in hopper discharge experiments. The choice of near spherical particle shape is also deliberate so as to provide quantitative comparison with the predictions of the simple microstructural model proposed. Table I shows a list of measured solid densities together with the results of the Quantimet analyses on particle sizes and shapes.

2.2 Preparation of Discrete Mixtures

The materials were mixed manually in small batches of some 0.5 - 1.0 kg. at a time and each batch was then placed into the experimental silo as successive layers. Great care was taken each time to lower the material batches on to the top surface of the material bed in the silo with the use of a short pipe section. It is believed that this procedure has minimised the possibility of size segregation due to particle impacts on the bed surface. Samples were also taken at random from the free surface of each material batch for size analysis. The results showed that the variations from the specified weight fractions were within 2 - 3 % .

2.3 Experimental Apparatus

Gravity flow experiments were performed in both bench top scale and pilot plant scale equipment. Fig.1(a) shows a photograph of the pilot plant scale equipment comprising several cylindrical bin sections of 0.4 m diameter flanged together to provide a tall silo (1 -2 m) with an interchangeable hopper section at the bottom to provide either mass or funnel flow. Each silo section was filled separately with the granular mixture before being reassembled prior to each flow experiment. The materials were discharged through a single circular orifice of 20 - 60 mm in diameter depending on the particle size range of the mixtures. Specially designed twin-axis contact load transducers were used to make simultaneous measurements of the normal and shear stress profiles along the silo walls. Fig.1(b,c) show photographs of a twin-axis, axially-symmetric load transducer and its mounting case used to monitor wall stresses during filling and batch emptying of the hopper contents. These transducers have a response time of less than a second and can be calibrated to detect loads down to 10 N m^{-2} with an accuracy of ± 2 %.

2.4 Discharge Rates

The accumulated mass discharge of the mixtures were measured continuously using a weighing platform supported on load cells whose digital output was connected to a data processor. The weight fractions of each size class present in the discharged material were determined by sieving samples from the silo discharge at frequent and regular time intervals.

2.5 Bulk Density Measurements

The experimental procedures for determining the static and the average flowing bulk densities in plug flow are discussed in detail in (2) and will not therefore be repeated here. Here the flowing bulk density ρ is given by ;-

$$\rho_f = \frac{W}{Q} = \frac{4\ W}{\pi\ D^2 V} \qquad (1)$$

where V_p is the plug flow velocity of the flowing bed in the cylindrical section of the silo which is determined by recording the rate of fall of the top material surface with the contact stress transducers placed at regular intervals along the silo wall; see Fig.1. Hence the values of ρ_f measured represent global averages at steady state conditions and not local values of the bulk density within the silo. The fractional solids content of the flow field is determined simply as $\eta_{mix} = \rho_f/\bar{\rho}_s$ where ρ_s is the mean solid density of the bed.

2.6 Measurement of Dynamic Wall Friction Angles

Simultaneous time traces of the normal and shear stresses obtained at various heights during filling and batch emptying were used to generate time traces of the corresponding insitu values of the static and dynamic wall friction angles. Here the wall friction angle is calculated simply from the measured shear to normal stress ratio at the wall; i.e $\phi_w = \tan^{-1} (\tau_w / \sigma_w)$. Fig.C shows some typical wall normal stress and corresponding wall friction angle traces obtained with binary mixtures which correlate significantly with the initial, pseudo-steady, and final transient stages of batch discharge.

2.7 Observations of Size Segregation during Gravity Discharge

Size segregation during gravity discharge was investigated by monitoring the change in the weight fraction of each particle size class present in the discharged samples collected at regular intervals during the three stages of batch emptying of a funnel-flow bin. The nature of the flow fields prevailing in the silo during the initial transient, pseudo-steady state and final transient stages are illustrated in the insert in Fig.2 and discussed in detail in (1) and (2).

Fig.2(a) - (c) show the variations with time in the weight fractions of the coarse, medium and fine size particles present in the material discharged from the experimental silo with three different ternary mixtures. Here the weight fractions of the different particle size classes are normalised by dividing the measured values X_i by the average concentration in the original fill X_o. Similarly, the cumulative mixture discharge was normalised by by expressing it in terms of the original mass filling the hopper, W / W_o. In these plots, values of $X_i / X_o > 1$ should indicate segregation of the particles of a given size class by means of "excess percolation" within the hopper. Correspondingly, values less than unity would suggest "selective retention" of particles of a given size. It therefore follows that when $X_i / X_o \cong 1$, there is negligible segregation of the particles of the corresponding size during discharge.

3. THE MICROSTRUCTURAL MODEL

3.1 Definition of the Bulk Microstructure

The microstructure of a polydisperse particle mixture can be quantified by three assembly parameters : i) The number density distribution of particles of different sizes and shapes ii) the size distribution of interstitial pores and iii) the geometry of the intersititial pores. To describe the microstructure, therefore, we need criteria to relate the number density of particles of different shapes and sizes to the size and geometry of the interstitial pores. The number density of the particles of a given size in the flow field is a function of their weight fraction in the mixture. With particles of different size but of equal density, it is possible to show that

$$N_{i+1}/N_i = \Phi_R^3 \cdot (X_{i+1} / X_i) \qquad (2)$$

where $\Phi_R = (d_i / d_{i+1}) > 1$.

The average pore size within the flow field will scale with the relative number ratios (N_{i+1}/N_i) of particle sizes making up the bulk. Visual observations [see (1)] indicate that the average pore size will be proportional to the particle size of the component i in the mixture when the number density of its particles exceeds a certain critical limit; i.e.

$$l_{pore} \propto d_i \quad \text{when} \quad N_i \geq N_{lim} \qquad (3)$$

When the condition in eqn (3) is met, the microstructure can be said to be continuous in the component i with the other components making up the dispersed discrete phases. We can then model the interactions between the continuous and dispersed particle phases within the bulk to establish microstructural feasibility limits for size segregation in mixtures of different size ratios. Here we propose a purely geometric model to relate the limiting number densities necessary for phase transformations to the weight fractions of the components in the discrete mixtures.

3.2 Binary Mixtures

We assume that the coarse particles will be embedded in a continuous bed of fines when the number of fine particles is sufficient to obstruct the entire surface of the coarse particles. With binary mixtures of spheres, this condition is given by :-

$$N_f / N_c = 4 \, d_c^2 / d_f^2 \qquad (4)$$

Substituting Eqn (4) into (2) results in:-

$$X_{f1} = \frac{4}{4 + \phi_R} \qquad (5)$$

where X_{f1} gives the limiting weight fraction of fines in the binary mixture when the microstructure is transformed from coarse continuous to fines continuous. It then follows that the coarse particles will have no relative mobility in a fines continuous bulk and the pore structure will remain unaffected by the coarse particle size.

Fig. **4**(b) shows a plot of the number ratio versus fines fraction for binary mixtures of different size ratios on which the limiting values X_{fL} are shown with dashed lines. As expected, as the particle size ratio is increased, the transformation to fines continuous flow is obtained at much smaller fines fractions in the mixture. More importantly, there is a remarkable match between the calculated values of X_{fL} in Fig. **4**(b) and the experimentally determined maxima of the mixture solids content for various size ratios seen in Fig. **4**(a). Clearly, with binary mixtures, the maximum mixture density is obtained when the coarse continuous bulk is transformed to fines continuous microstructure. Addition of more fines to a fines continuous bulk will only serve to reduce the bulk density as the solid volume of coarse particles is replaced by interstitial pores between the fines.

3.3 Ternary Mixtures

The physical mechanism proposed for size segregation of equal density particles is that the relative mobility of particles of a given size in the flowing mixture will be stopped when there are sufficient particles of other size classes to obstruct all the available surface of the said particles i.e. when these particles become embedded in a continuous phase. With ternary mixtures, three cases can be easily identified :

i) <u>Coarse Continuous Mixture</u> : This is obtained by the interaction between a hybrid continuous phase of coarse and medium sized particles and the coarse particles. To constitute a coarse continuous phase, we must have:-

$$N_c d_c^2 \geq \frac{1}{4} (N_f d_f^2 + N_m d_m^2) \qquad (6)$$

Hence in terms of weight fractions [see Eqn (2)], we have:-

$$X_c/d_c \geq \frac{1}{4}(X_f/d_f + X_m/d_m) \quad (7)$$

ii) Fines Continuous Mixture: Here we have to scale the interaction between the fine particles and a hybrid continuous phase of coarse and medium particles so that:-

$$\frac{X_f}{4\,d_f} \geq (X_c / d_c + X_m / d_m) \quad (8)$$

must be satisfied to constitute a fines continuous mixture.

iii) Medium Continuous Mixture : The condition for medium continuous mixture has two constraints. We must have :-

$$X_m/d_m \geq X_f/4d_f \text{ and } X_m/4d_m \geq X_c/d_c \quad (9)$$

A computer programme was written to calculate the phase transformation boundaries for ternary mixtures as a function of size ratios based on the equalities in Eqns.(6) - (8). The results obtained with three different ternary mixtures are presented in the form of triangular diagrams of weight fractions as seen in Fig.5. The boundaries of the coarse, medium and fine continuous zones in these diagrams are separated by a region in which all three size classes are found to coexist as three dispersed phases.

4. WALL STRESSES

In a series of experiments using radish seeds and acrylic beads (see Table I), the profiles of normal and shear stresses at the wall of the model hopper were monitored using the twin-axis load transducers described above. As seen in the insert in Fig.6, the transducers were placed at three strategic positions within the silo ; i) in the cylindrical section immediately above the plane of transition to the conical hopper section to provide comparison with the"greatdepth" values of wall stresses given by the Janssen-Walker analysis (4), ii) immediately below the transition plane to facilitate comparison with Switch Stresses predicted by Walters (4), and iii) in the radial flow section immediately above the hopper outlet (4). Measurements to date include mono-sized beds of the two materials as well as the binary mixtures of 30% and 80% by weight of acrylic beads respectively representing the "coarse continuous" and "fine continuous" microstructures discussed above; refer to Fig. 4(a).

Further experiments are underway with binary mixtures of different size ratios and with ternary mixtures. Fig.6 compares the normal stresses and insitu wall friction angles measured with binary mixtures of 30% and 80% fines by weight.

Normal stress traces seen in Fig.6 confirm the "great depth" limit, switch stress below the transition and subsequent reduction towards the outlet. Furthermore, these time traces correlate well with the initial, pseudo- steady, and final stages of batch discharge; refer to Fig.2. The wall normal stresses measured with the two binary mixtures are quite similar as would be expected from the solids fraction profiles presented in Fig.4(a). Clearly, at a size ratio of 3.5 ÷ 1 , the bulk densities of the mixtures are almost identical at 30% and 80% by weight.

The insitu wall friction angles shown in Fig. show that the values obtained with the 80% fines by weight mixture are consistently lower at all heights than the corresponding values for the 30% fines by weight mixture. This is explained by comparing the surface roughness of radish seeds with acrylic beads. An SEM investigation has revealed that acrylic beads have very smooth surface whereas radish seeds are rough. Clearly, these results indicate the wall friction to scale with the properties of the particles making up the "continuous phase" of the binary mixture. Hence a 30% mixture shows frictional properties of the coarse particles (i.e. radish seeds) while the 80% by weight mixture has frictional properties similar to a bed of fine particles (i.e. acrylic beads).

A tribological model has been proposed (5) to predict the static and dynamic wall friction angles measured in silos based on a consideration of the contact frictional properties of single particles. This model is now being adopted to predict the wall friction angles of binary mixtures by considering the relative number density of particles of different sizes in the bed.

5. DISCHARGE RATES

The materials were discharged through a 50 mm diameter circular orifice which is deliberately chosen small enough to highlight the "empty annulus" effect;but large enough not to cause geometric arching during discharge; see (1).

The cumulative discharge of different particle size classes in the mixtures was recorded during the initial transient, pseudo-steady and final transient stages of the batch emptying of the hopper. As discussed in detail in (2) and (3), size segregation is most noticeable in with a coarse continuous mixture, whereas a fine continuous mixture results in negligible segregation during discharge; refer to Fig.2 above. These results are in complete agreement with the predictions of the linear geometric microstructural model described above and discussed in (1) and (2).

For discharge from a flat bottomed cylindrical silo; the gravity discharge rate at steady state could best be described by a modified form of the Beverloo etal's (6) correlation ;-

$$W_{mix} = C'\rho_{mix}(X_i, \phi_{Ri})g^{1/2}.(D_0 - Z(X_i, \phi_{Ri}))^{5/2} \quad (10)$$

where i = 1, n-1 and n is the number of particle size classes in the discrete mixture. Experimental results obtained with binary and ternary mixtures with different finite size ratios indicate that the width of the empty annulus at the orifice scales linearly with the mean diameter of the "continuous phase" of the discrete mixtures; i.e.

$$Z\ (X_i\ ,\ \Phi_{Ri}) = K_{mix}\ d_{con} \qquad (11)$$

The experimental values of the mass discharge rates obtained with binary and ternary mixtures have been used to determine values for K_{mix}. All values were found to fall in the range of 1.0 - 2.0 $\bar{d}_{con}$; see Fig.3 for some typical results obtained with ternary mixtures.

6. CONCLUSIONS

The work presented here is an attempt in microstructural modelling of the gravity induced flow in hoppers of free flowing particle mixtures. The interactions between particles of different sizes but of near equal solid density are considered to be purely geometric in nature in the absence of any significant interstitial fluid pressures and interparticle cohesive forces affecting the flow fields. The proposed linear geometric model allows for the boundaries of the continuous and dispersed phase transformations to be calculated as a function of particle size ratios and relative weight fractions of particles in the mixtures.

Extensive experimental evidence presented verifies that the flowing bulk densities, mass discharge rates and size segregation patterns as well as the dynamic wall friction angles all correlate consistently with the predicted microstructural transformations of the flow fields.

7. ACKNOWLEDGEMENT

The authors are grateful for the financial support provided by SERC under the Specially Promoted Programme in Particle Technology.

REFERENCES

(1) TüZüN U., *Tribology in Particulate Technology*, Edt. by Briscoe B.J. and Adams M.J., 1988, Adam Hilger Publ., Bristol, 38.

(2) ARTEAGA, P. and TüZüN, U., *Chem. Engng. Sci.*, 1990, 45, 205.

(3) TüZüN, U. and ARTEAGA, P., *Chem. Engng. Sci.*, 1991, in press.

(4) NEDDERMAN, R. M., *Trans. Inst. Chem. Engrs.*, 1982, 60 , 259.

(5) TüZüN, U., ADAMS, M.J., BRISCOE B.J., *Chem. Engng. Sci.*, 1988, 43, 1083.

(6) NEDDERMAN, R. M., TüZüN, U., SAVAGE, S. B., and HOULSBY, G.T., *Chem. Engng.Sci.*, 1982, 37, 1597.

TABLE I MATERIAL PROPERTIES

material	solid density (Kg m^{-3})	weight mean diameter (mm)	number mean diameter (mm) (*)
maple pea	1.22	7.23 ± 0.90	7.85 ± 0.60
radish seed	1.09	2.41 ± 0.30	2.65 ± 0.20
acrylic bead	1.11	0.75 ± 0.08	0.82 ± 0.08

(*) Projected area equivalent sphere diameter determined by Quantimet .

Sphericity : $\%\ S = \left(4\ \pi \times \frac{\text{surface area}}{(\text{perimeter})^2} \right) \times 100.$

	maple pea	radish seed	acrylic bead
% S	90.6 ± 6.7	89.7 ± 3.5	88.5 ± 5.2

a

FIG. 1

PILOT PLANT RIG

(a) Model Silo;

(b) Twin-axis Load Transducer;

(c) Transducer Mounting Case.

b

c

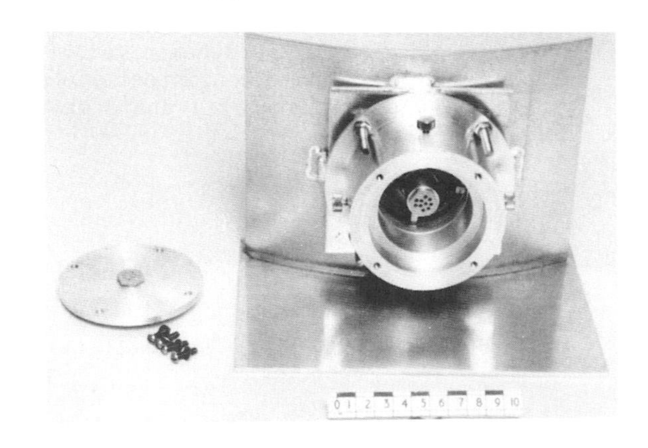

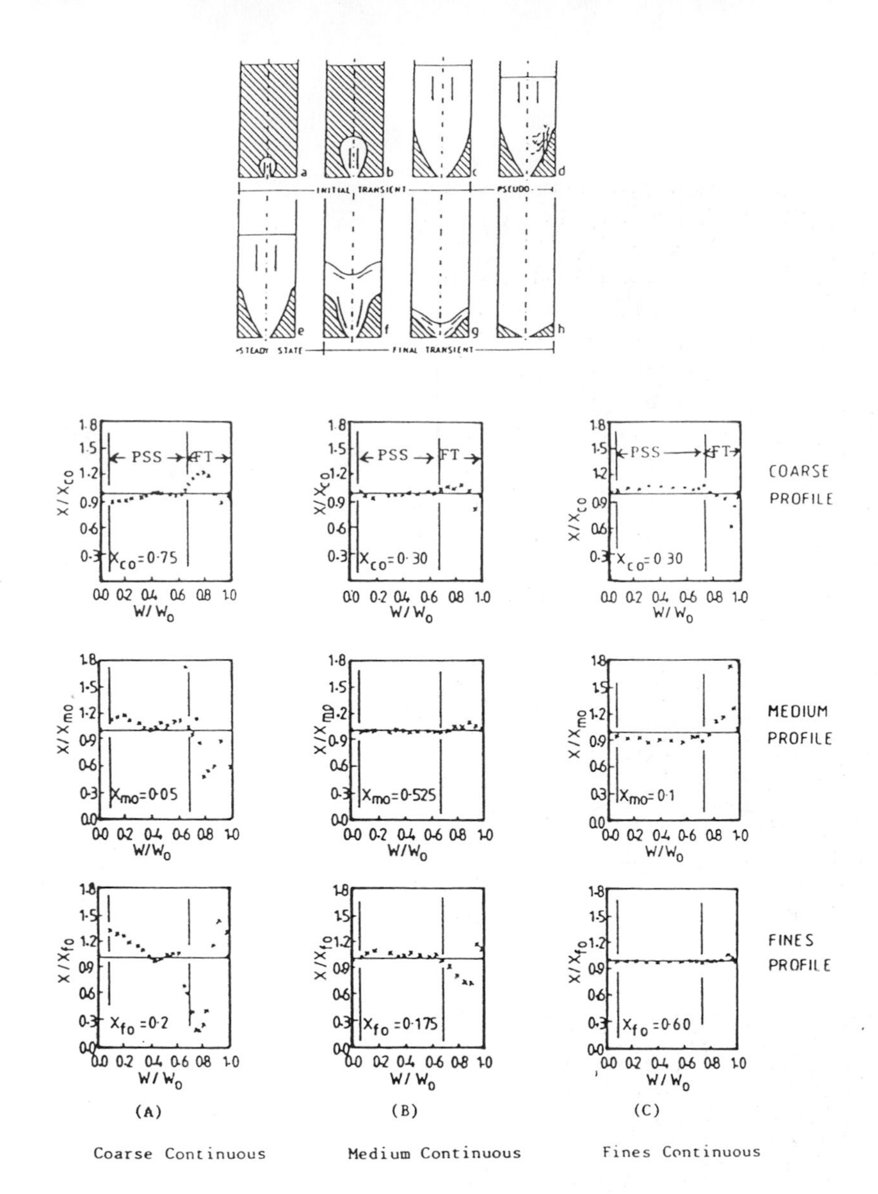

FIG. 2 SIZE SEGREGATION OF TERNARY MIXTURES DURING BATCH DISCHARGE FROM A FUNNEL-FLOW HOPPER

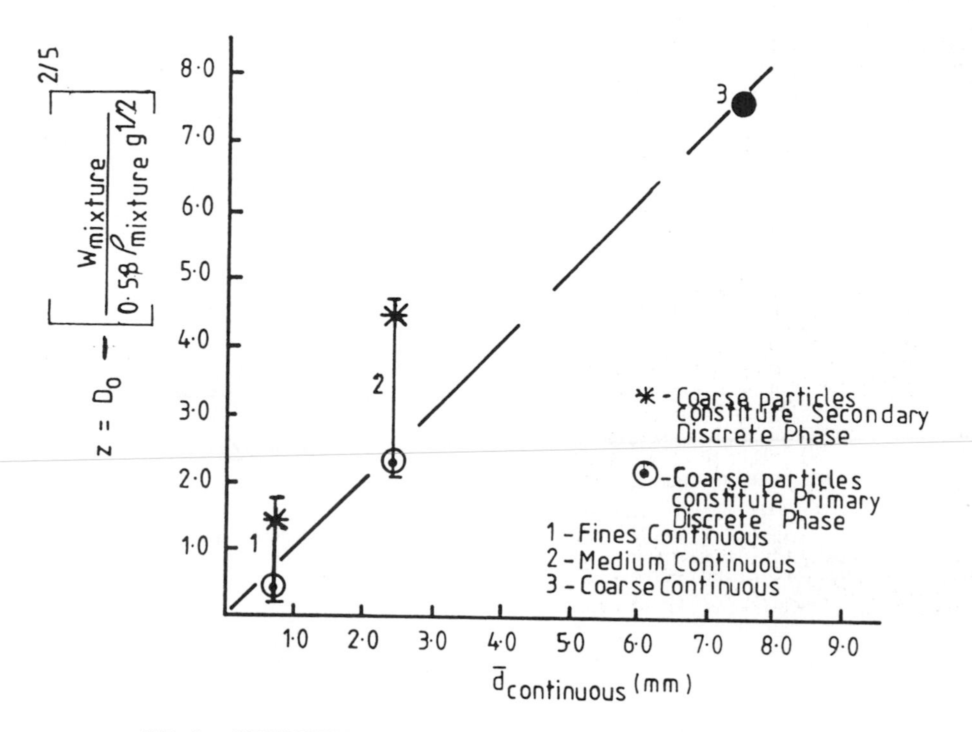

FIG. 3 EXPERIMENTAL VARIATION OF THE EMPTY ANNULUS WIDTH WITH THE MEAN PARTICLE DIAMETER OF THE CONTINUOUS PHASE OF THE TERNARY MIXTURE

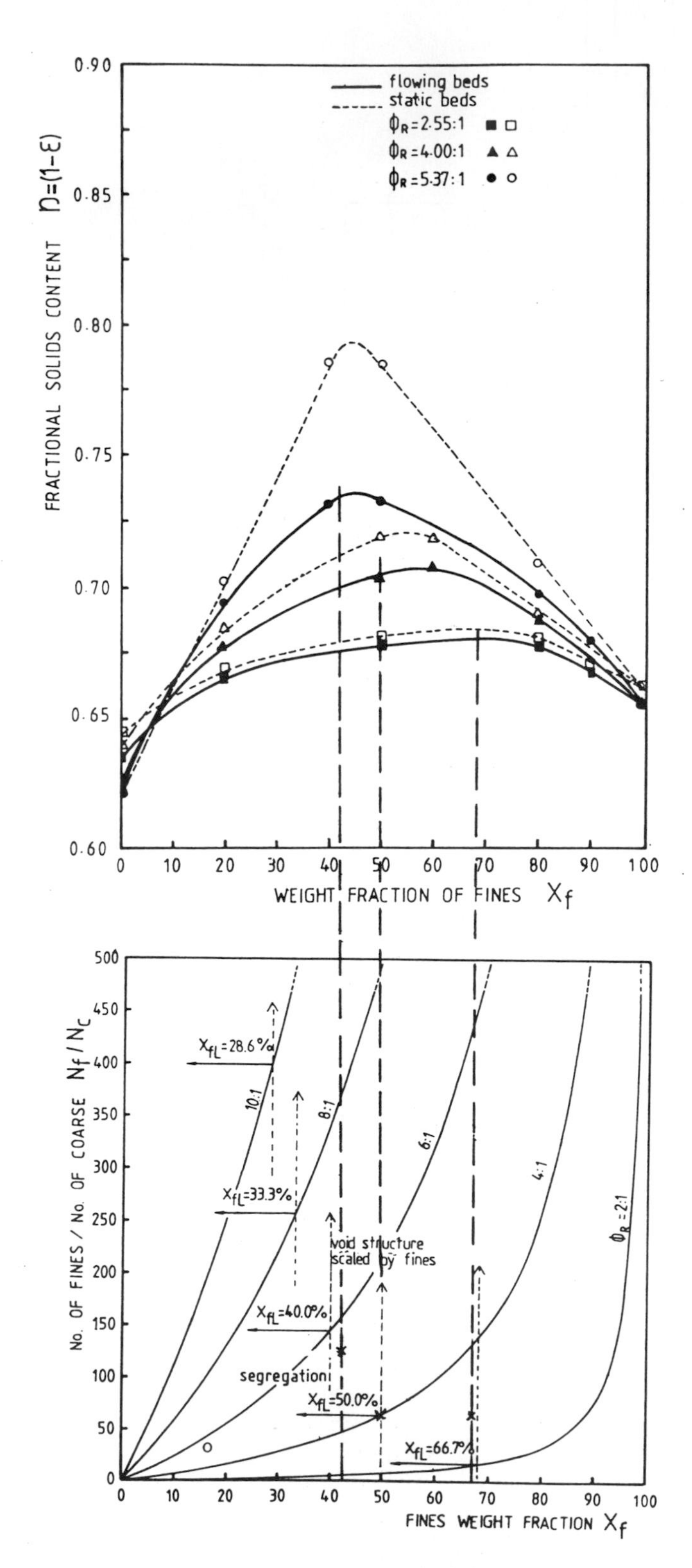

FIG. 4 VARIATION IN BINARY MIXTURE SOLIDS FRACTIONS AND CONTINUOUS PHASE TRANSITION BOUNDARIES WITH SPHERES OF FINITE SIZE RATIOS

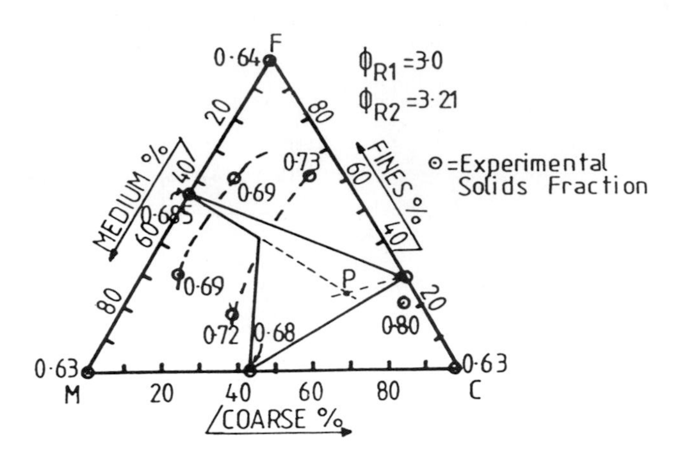

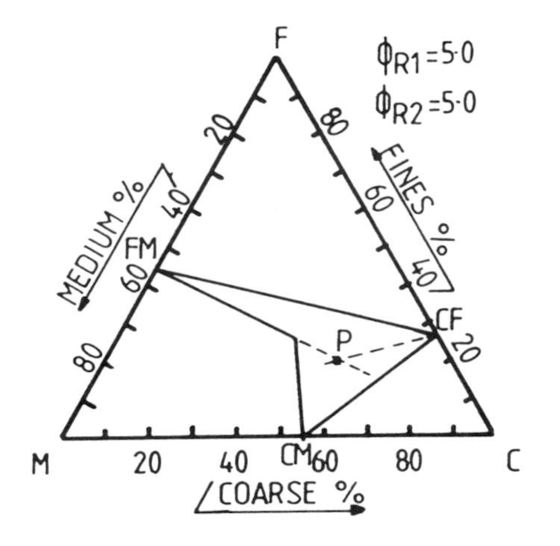

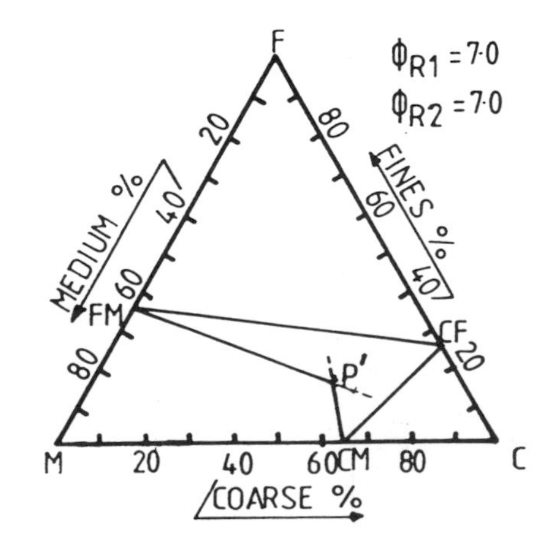

FIG. 5 THEORETICAL MICROSTRUCTURAL PHASE TRANSITION BOUNDARIES OF TERNARY MIXTURES OF SPHERES OF DIFFERENT FINITE SIZE RATIOS

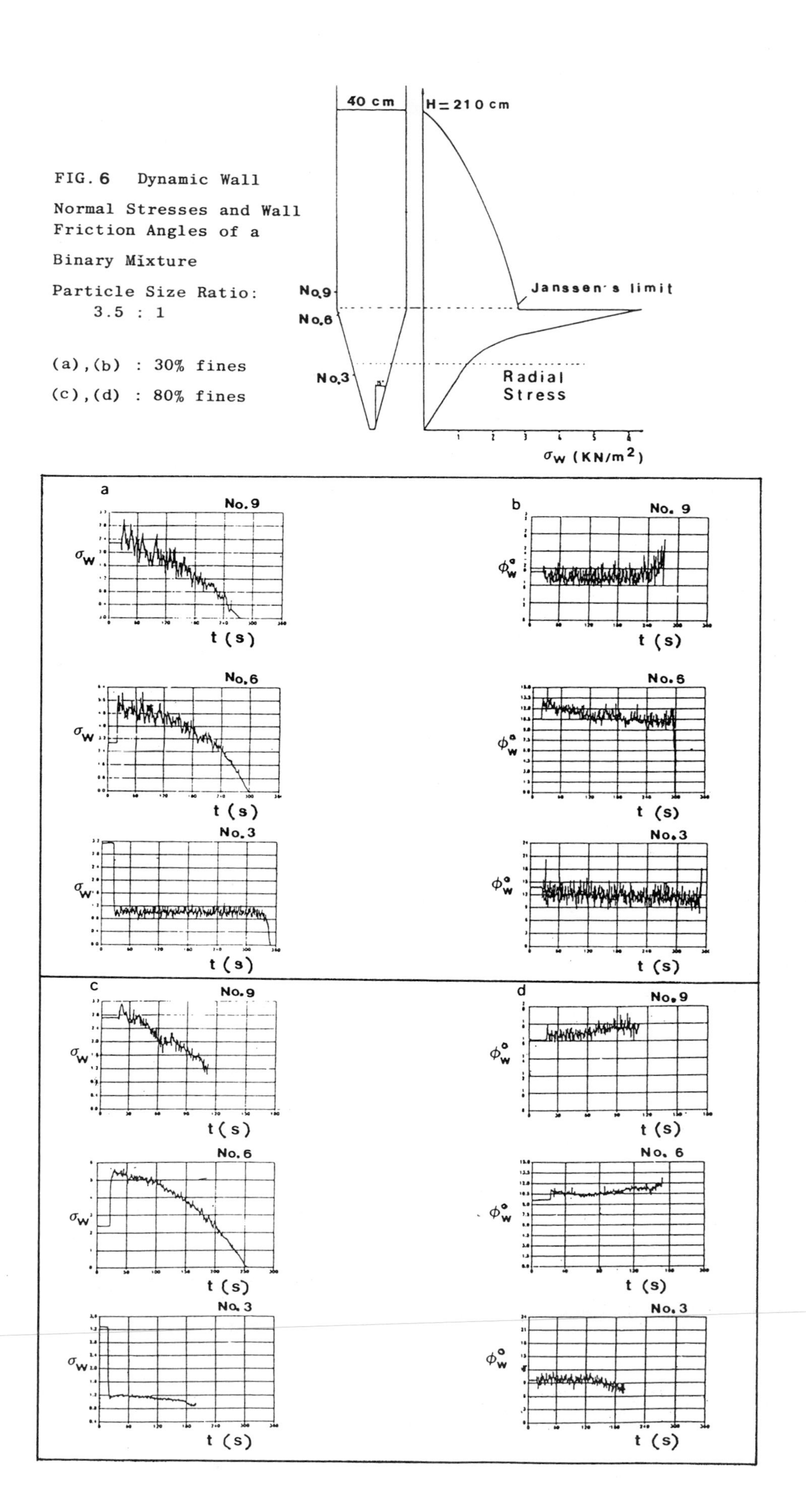

FIG. 6 Dynamic Wall Normal Stresses and Wall Friction Angles of a Binary Mixture

Particle Size Ratio: 3.5 : 1

(a),(b) : 30% fines

(c),(d) : 80% fines

C418/054

Stepped pipelines – a new lease of life for lean phase pneumatic conveyors

M S A BRADLEY, BSc, PhD and A R REED, BSc, PhD
The Wolfson Centre for Bulk Solids Handling Technology, Thames Polytechnic, London

SYNOPSIS

Lean phase pneumatic conveyors have three main advantages over dense phase conveyors, namely lower capital cost and greater tolerance of design inaccuracies and changes in product type. The criticisms usually levelled again lean phase conveyors are that they cost more to run, damage the product more, and suffer from wear more than do dense phase conveyors. However, these difficulties can often be overcome provided that air velocities in the pipeline are kept down as low as possible consistent with reliable operation; this is helped by enlarging the bore of the pipeline at one or more points along its length.

'Stepping' the bore of the pipeline at one or more points along its length serves to compensate for the expansion of the air as pressure decreases along the line, thus keeping the air velocities down to reasonable levels. Using this technique it is often possible to overcome the shortcomings which can be experienced with lean phase conveyors of a single bore size, thus avoiding the need to turn to dense phase transport with its associated higher capital costs.

In this paper the advantages of lean phase conveyors are explored in more detail, and the techniques available for designing pipelines with increases in bore size are examined. It is argued that for many applications, even where it is desired to minimise air velocity in order to keep product degradation and/or wear down, a well-designed lean phase system with a stepped pipeline can be a better choice than a commercial dense phase system.

1. INTRODUCTION

Lean phase pneumatic conveyors (wherein the product flows along in suspension in the air) have several advantages over dense phase conveyors; notably

(i) They do not require such close design, being more tolerant of the inevitable inaccuracy in design

(ii) they are more tolerant of change in product type, or uncertainty in the exact quality of the product to be conveyed

(iii) the capital cost of the system is generally lower

However, there are a number of criticisms very often levelled against lean phase pneumatic conveyors. These are that they tend to cause product attrition, they suffer from wear of bends and other components and they consume more energy, in comparison with dense phase conveyors. These are the arguments usually put forward to sell dense phase systems.

All of these problems are related to air velocity. Where they arise with lean phase conveyors, it is often found that the air velocities being used are higher than they need be; this leads to more pressure drop, and hence higher running costs, as well as creating unnecessarily high levels of damage to the product and wear of system components caused by the high speed collisions between solid particles and pipe walls.

These difficulties need not arise. Lean phase conveyors are frequently capable of performing acceptably well provided the air velocities are kept down to a safe minimum. This can be facilitated by increasing the bore of the pipe at successive points along the line, to compensate for the natural expansion of the air which occurs as pressure drops along the pipeline.

One or two additional benefits arise with lean phase conveyors because they do not normally use pressure vessel blow tank feeders. No pressure vessel inspections or special insurance are required. There is not the potential for damage to be done by the tank full of high pressure air at the end of the batch cycle expanding down the line at high velocity taking the last of the product with it. Also, depending on the type of feeder used, flow is normally continuous rather than in batches. These difficulties with dense phase conveyors, all of which can be overcome with care, do not figure with lean phase conveyors.

2. LEAN AND DENSE PHASE

The difference between lean and dense phase conveying is not always well understood, so at this point a few words of explanation will be in order; a brief outline is given below whilst a fuller discussion appears in Appendix 1.

Lean phase flow is universally understood to mean a condition of flow in which the particles travel along in suspension in the air, dispersed across the pipe cross section. This requires the air velocity to exceed a certain minimum value, typically of about 15 to 18m/s although it varies a little with the product. Most products will convey in lean phase.

Dense phase flow is taken to mean a condition which occurs at air velocities insufficient to suspend the particles fully, and in which at least some of the particles move along the bottom of the pipe in contact with each other, either as a bed or as waves, dunes or plugs. Not all products will convey in this mode, many simply block the pipe if the air velocity is insufficient to suspend the particles fully.

3. THE EFFECT OF AIR VELOCITY; MINIMUM CONVEYING VELOCITY

As pointed out above, it is the air velocity which differentiates between lean phase and dense phase flow. The figures of 15 to 18 m/s plus, quoted above for lean phase conveying, apply to the vast majority of products, and all will convey in lean phase. Of those products which will convey in dense phase, i.e. those which will flow satisfactorily at lower air velocities, most tend to convey down to about 3 to 5 m/s. Few products have minimum conveying velocities in between these two ranges.

4. THE EFFECT OF EXPANSION; ADVANTAGES OF A STEPPED BORE LINE

The natural expansion of air as pressure falls means that in a constant bore pipeline, air velocity will increase towards the end of the line. If the air velocity is set to the minimum at inlet, then inevitably it will be more and more in excess of this minimum further along the pipeline.

4.1 Examples of excessive air velocity arising from expansion in a single bore size

4.1.1 In a lean phase system; Taking a typical lean phase conveyor with an inlet pressure of 1 bar gauge (2 bar absolute), the air velocity at outlet will be twice that at inlet. For a product which will convey at say 15 m/s and upwards, this means that the air velocity at the end of the pipeline will be 30 m/s, i.e. twice that required for satisfactory conveying.

4.1.2 In a dense phase system; Taking an alternative system with a higher pressure at inlet, say a dense phase system operating at 4 bar gauge (5 bar absolute) with a product which will convey at 5m/s and upwards, then the outlet velocity will be at least 25 m/s even if the air velocity at inlet is kept right down; if a small 'safety margin' of say just 2 m/s is used so that the inlet air velocity is 7 m/s, then the outlet velocity will be 35 m/s, i.e. SEVEN TIMES as high as need be, and higher even than in the lean phase system mentioned above.

Thus it is inevitable that if a pipeline of constant bore size is used, and the pressure drop is anything above about 0.3 bar, the air velocity will be excessive towards the end of the pipe. Consequently any excessive product attrition, bend wear or pressure drop will occur towards the end of the line. All of these rise disproportionately with air velocity so the effect is very marked.

It will also be clear that dense phase systems, operating at higher pressures, suffer from this even more than do lean phase systems. Many commercial systems described as 'dense phase' actually operate in lean phase towards the end of the pipeline with air velocities in excess of those required for lean phase conveying, because they do not employ a stepped bore line.

4.2 Using a stepped pipeline for the lean phase conveyor

Taking example 1 above, of a lean phase system operating at 1 bar gauge, what would be the advantage of putting a single step into the line? Suppose that the pipe bore was 104mm at the inlet. To achieve the necessary velocity of 15m/s, an air volume flow rate of 7.6m^3/min actual would be necessary. The expansion of this to 15.2m^3/min at the end of the line would give 30m/s in the 104mm bore pipe, but if the pipe was enlarged to 130mm bore towards the end, then the velocity in this would only be 19.2m/s, much more acceptable. The position of the change in size must of course be chosen so that the air velocity at the start of the 130mm section does not fall below 15m/s, requiring an air volume flow rate of 11.9m^3/min which would occur at a pressure of 0.6 bar gauge; so the step should be positioned at the point where the air pressure is down to this value. The air velocity in the 104mm line just prior to the expansion would be 23.4m/s, somewhat higher than the 19.2m/s at the end of the pipe but still much more acceptable than the 30m/s in the single bore line.

So it can be seen that in this simple case, the maximum air velocity in the pipeline has been reduced from 30m/s to 23.4m/s. The beneficial effect of this will be apparent when considering that the rate of wear of system components is generally agreed to rise as a fourth or fifth power of velocity, so even taking a conservative view this is likely to result in a 63% reduction of wear.

5. CONSIDERATIONS WITH STEPPED BORE PIPELINES

From the foregoing it has been seen that the increase in air velocity along the pipeline, arising because of the increasing volume of the air as pressure reduces, can be partially overcome by enlarging the pipe bore along the line. Ideally one would specify a pipe with a gentle taper from inlet to outlet but of course this is not practical, so we must make do with available pipe sizes. The question which arises is 'where to put the steps?'.

5.1 Criterion for placing the step(s)

The step to the next larger pipe size must not come too soon along the pipe or air velocity will fall too low and the pipe will block. On the other hand, if the step is too far towards the end of the pipe then the air velocity will still be excessive even after the step and the full benefit of the stepping will not be realised.

Two approaches have gained popularity for determining the best positions for the steps; in Australia and continental Europe a dimensionless number often referred to as a 'Froude number' (containing the same types of quantities, but certainly not the same thing, as the original Froude number which was developed for ship resistance) is used, ensuring that this dimensionless quantity is kept above a certain minimum value which relates to the product in question.

Recent work at The Wolfson Centre, however, suggests that it is better to rely on the minimum conveying air velocity for the product in question, simply designing so that the air velocity at inlet and after each step is equal to the minimum air velocity for reliable conveying, determined from the conveying trials.

This matter is addressed in a little more detail in Ref. 2. It will no doubt remain contentious for some time to come, however it is recommended here that the step positions are based purely on air velocity. It is believed that design using the 'Froude number' method will result in unnecessary conservatism so the full benefit of stepping will not be gained.

5.2 Design of Stepped Bore Pipelines

Accuracy of design is more important with a stepped pipeline than with one of constant bore size, because if the pressure profile along the plant line does not turn out as expected then the air velocities at the steps will be either lower or higher than intended, which may lead to either pipe blockages or excessive degradation and/or wear. In this respect the fact that lean phase systems are easier to design than dense phase systems, and more tolerant of change in product type, means that it is easier to get an accurate, reliable design with the lean phase system, thus maximising the benefits of stepping.

5.3 Conveying trials

To design a line accurately, it is necessary to predict the pressure profile along a pipeline so that the air velocities can in turn be predicted. In order to make the prediction of the pressure profile, it is necessary to use data on pressure drop in bends and straight lengths which has been obtained from tests conveying the product in question in a specially instrumented test rig. There is no way around this, unfortunately; in spite of the vast amount of literature on the subject, no theory or mathematical models, or computer programs, have been found which can predict this sufficiently accurately.

It is therefore necessary to undertake conveying trials and use data for design. The approach which has been popular in the past has been to build a pilot pipeline and measure the total pressure drop along this for a suitable range of conveying conditions, then to scale the results to predict the total pressure drop expected in the plant pipeline. The scaling must take account of differences in bore and length, and also number and positions of bends, between pilot and plant lines. Unfortunately this method is not suitable for the design of stepped bore pipelines because (a) there is no way of scaling to allow for a step, and (b) even if this could be achieved it will not give the information on air velocities part way along the pipeline, to allow proper positioning of the step. It may be thought that the line could be designed as two pipelines of different bore sizes, the smaller feeding into the larger, but in practice this will give a misleading result because the effects of the differences in air density ranges between the two pipelines cannot be scaled for.

5.4 Recommended design method

The method of design which has been developed at The Wolfson Centre, Thames Polytechnic, is to convey the product through a pilot pipeline which is instrumented with a number of pressure transducers along two long straight sections with a bend in between. This enables the pressure drop per metre in the straight lengths, and the pressure loss caused by a bend, to be measured accurately over the range of conveying conditions used for the tests. Using the data resulting from this, it is then possible to model the plant pipeline, predicting the pressure profile and air velocity from inlet to outlet. Steps can be put in at any position in the model of the proposed line and the air velocity at all points re-examined.

The design method is enlarged upon in another paper in this volume (Ref. 2) and is described in detail in Ref. 1 so the reader is referred to those for more detail.

5.5 Available pipe sizes

If it is desired to stay with the nominal bore piping often used for the smaller sizes of conveying pipelines (say up to 8in. or 200mm), for which fittings are readily available, then the ratio of velocity before a step to that after is fixed by the ratios of areas from one pipe size to the next. In this range the area ratios from one size to the next lay between 1.4 and 1.7, so it should never be necessary to have air velocities anywhere in the pipe more than 1.7 times the value which is taken as the minimum safe value. For example, if the minimum safe velocity for lean phase flow for a given product is taken as 14 m/s, it will not be necessary to have more than 24 m/s anywhere in a well designed stepped pipeline.

For larger bores, scheduled tubing is frequently specified, with bends fabricated to suit. The available sizes are then closer than for the nominal bore pipes so the situation is further improved. It is of course true that scheduled tubing is available in the smaller sizes as well, but the experience of The Wolfson Cemtre is that most clients prefer to stay with nominal bore sizes where they are available, to ease maintenance.

5.6 Design of transitions

No firm evidence is known to exist to show the best way to make the transition between the pipes of different sizes. Both plain steps and slow tapers (included angle about 1 in 20) have been used at The Wolfson Centre and no operational differences have come to light; nevertheless some designers hold that the use of plain steps increases product degradation, so until proper test work is undertaken on this it is recommended that slow tapers are employed.

6. CONCLUSIONS

Where the adverse effects of high air velocity are a significant factor in the design of a conveyor, specifically if product degradation or pipeline wear are important considerations, it is not always necessary nor even beneficial to turn to dense phase transport with its attendant cost and potential problems.

To summarise the advantages of lean phase conveyors, compared with dense phase, they are:-

* More tolerance of the inevitable slight inaccuracies in design
* Not normally affected very much by variation in product quality
* Lower capital cost, with much less complexity of equipment
* Normally no pressure vessels to inspect and insure
* No damaging blow-down of the high pressure air in the tank at the end of the cycle
* Continuous operation rather than batch operation (may or may not be significant)

A lean phase conveyor, designed with a little care and using a stepped pipeline, is quite capable of operating with air velocities which are not excessive, and indeed are often less than the air velocities found at the end of the pipeline in many dense phase systems which have been seen.

It should of course be noted that dense phase systems can also benefit from the use of stepped pipelines, but the point being made here is that by using a stepped line, the advantages of lean phase conveying can be gained whilst reducing many of the disadvantages often levelled against lean phase conveyors - frequently by the vendors of dense phase systems.

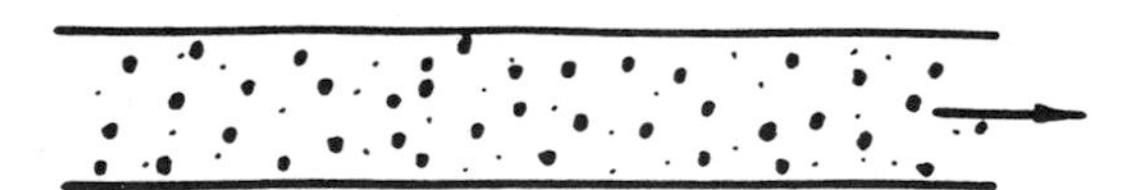

(a) Suspension ('lean phase') Flow

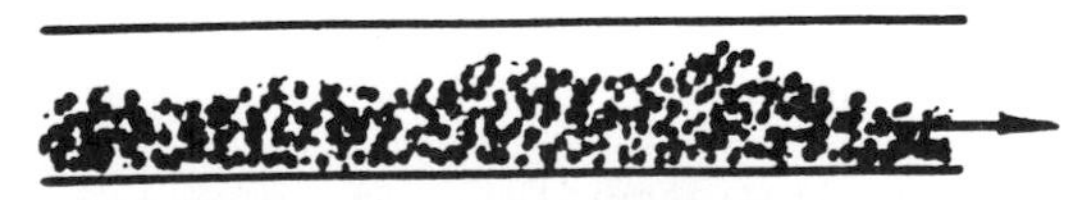

(b)(i) Moving Bed Flow

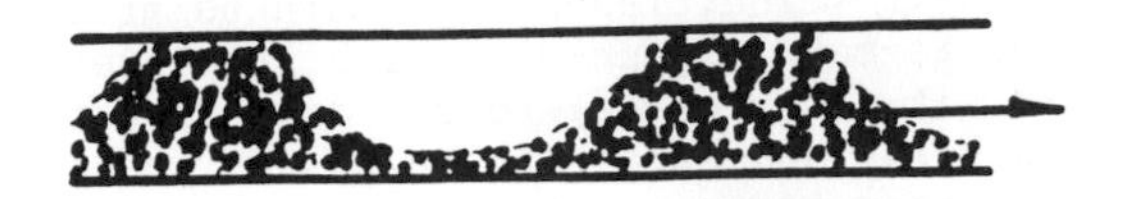

(b)(ii) Plug (or Slug) Flow

(b) Two Modes of Non-Suspension Flow

Fig. 1
Modes of Flow in Pneumatic Conveyors

APPENDIX 1

DIFFERENCES IN MECHANISMS AND CHARACTERISTICS BETWEEN LEAN AND DENSE PHASE SYSTEMS

The differences arise because of different air velocities in the pipelines. Air velocity is the controlling factor, and the effect of the solids loading ratio (ratio of mass flow rate of solids to mass flow rate of air) is minimal. The term 'Phase Density' is frequently used to mean the solids loading ratio; this is a misleading term which should be avoided, because it has nothing to do with the phase of the flow.

A1.1 Lean Phase (or Dilute Phase) conveying

If the air velocity is sufficiently high, the natural turbulence in the air is high enough to pick up the particles in the pipe so that they flow along in a suspension, well dispersed across the pipe section. This is lean phase flow and usually it requires an air velocity above about 15 to 18 m/s, though with some heavy materials a higher air velocity may be required. The scene which an observer looking into the pipe through a sight glass might see, is represented in fig. 1. Lean phase conveying is thus quite well defined, as suspension flow; all products will convey in lean phase.

Lean phase conveying systems often operate at relatively low pressures, the most common example being the rotary valve and Roots blower combination often seen in food and mineral plants, operating at up to 1 bar gauge. Capital cost is relatively low.

A1.2 Dense phase flow

If air velocity is reduced, so the turbulence in the air reduces to the point where it is no longer sufficient to keep the particles in suspension. Some particles begin to settle towards the bottom of the pipe where they sweep along as a moving bed. With further reductions in air velocity, the majority of the particles come into contact in the lower part of the pipe cross section, where they may still sweep along as a moving bed or the bed may be static with dunes or slugs of product moving along in the top part of the pipe. Thus there is a variety of flow types which might be called "dense phase"; if the phrase was not so well established it would be better known as "non-suspension flow".

Many products will not convey in dense phase, and simply block the pipe if the air velocity is reduced much below about 12 m/s. Work done at Thames Polytechnic by Jones (Ref. 3) suggests that only products which are sufficiently permeable to air, or tend naturally to retain air, are suitable for this type of flow. For example, cement powder will convey in dense phase, and so will some coarse sands; but fine sand will not. Changes in such simple things as supplier of a product, or the process used to produce it, can cause problems; cases are known of where a dense phase system has been designed and installed to convey a particular product, and the supplier has made a minor modification to his process which, although not altering the chemical composition, makes a slight difference to the particle size range or shape, and this has led to unreliable operation of the conveyor.

Dense phase conveying systems operate at relatively high pressures, usually about 2 to 4 bar gauge, using blow tank feeders and high pressure air from screw or reciprocating compressors; the capital cost is comparatively high, not only for the air mover but for the blow tank and also the control gear which must be of some sophistication. They operate in a batch mode, by filling the tank, pressurising, discharging, depressurising and starting again. The flow even during the conveying part of the cycle is of course pulsing, rather than smooth as in a lean phase system. Additionally, a blow tank operating over 1 bar g is a pressure vessel; as such it needs regular examination by a licenced inspector, and also special insurance.

APPENDIX 2

REFERENCES

(1) BRADLEY, MSA and REED, AR. An Improved Method of Predicting Pressure Drop Along Pneumatic Conveying Pipelines. Powder Handling and Processing, 1990 Vol. 2 No.3.

(2) BRADLEY, MSA and REED, AR. Latest Techniques for the Design of Reliable and Energy Efficient Pneumatic Conveyors. Proc. IMechE Conf. BULK 2000, London, Oct. 1991.

(3) JONES, MG. The Influence of Bulk Particulate Properties on Pneumatic Conveying Performance, PhD Thesis, Thames Polytechnic, 1988.

C418/063

Towards improved design for bulk solids handling plant

L BATES, CEng, MIMechE
Ajax Equipment (Bolton) Limited, Bolton, Lancashire

Towards improved design for Bulk Solids Handling Plant.

Synopsis.

Poor performance of bulk handling plant often arises because account has not been taken of some relevant properties of the bulk materials. An I.Mech.E. document is introduced which systematically directs attention to behavioural related features of bulk solids to give a basis for improved design. The need for wider education in the technology and better test instruments are also highlighted.

The demands placed by the manufacturing and process industries upon designers of bulk solids storage and handling equipment in the latter part of the twentieth century is characterised by the increase in both size and number of installations, growing automation and instrumentation, tighter health and safety regulations, stricter environmental control, the need for unattended reliability, higher quality control and the ever expanding range of bulk materials and their condition which have to be handled and processed.

All these pressures occur within a framework of more comprehensive international competition

These factors emphasis the need for improved equipment design.

Yet, it is common experience, confirmed by a detailed study (1), to find that plants processing bulk solids perform far worse than plants that process only liquids and gasses. This report also showed that recently built plants perform little or no better than those built in the mid to late sixties.

Within these type of plants the key performance problems are invariably associated with mechanical and physical difficulties, particularly those related to the storage and handling of the bulk materials, rather than features of process chemistry. (2),(3).

These findings, which are supported by the D T I Neale Report (4), suggests that the fundamental problems underlying poor performance in these areas are not being addressed or resolved by the industry.

The business system that is so effective in translating advances in process chemistry into commercial technology appears to break down almost immediately, and at all levels, when the problems are outside process considerations.

The failure to anticipate and accommodate the behavioural nature of the bulk solids handled is the root cause of most problems, which are aggravated because the effects only arise at plant commissioning stage.

These range from catastrophic failures to persistent operating difficulties and performance shortfalls. Whilst the former receive rapid and extensive attention to account for, correct and ensure the circumstances do not repeat, the more mundane, but widespread effects of performance deficiencies do not command a similar approach to corrective action and the prevention of recurrence.

Three associated areas of weakness lay at the root of this situation.

1. The most fundamental reason is that there has been no commonly accepted style of embracing format for defining the relevant

physical properties of bulk solids for handling purposes.

The erratic, inconsistent, inadequate, and sometimes downright misleading manner in which such performance sensitive information is presented, even for major contracts, erodes the foundation for a fair and equitable contract between and equipment supplier and a user.

No one would expect a bridge or engine manufacturer to produce designs without well defined information regarding the properties of the constituent materials. In solids handling it is not uncommon for the assumption to be made that a bulk material will flow in all conditions in a manner similar to the behaviour of a small,loosely poured, fresh sample of the product. The paucity of data upon which the designer is expected to derive an efficient equipment specification often appals.

For these reasons the Bulk Materials Handling Committee of the Institution of Mechanical Engineer prepared a document, relating to improved design for solids handling plant, which emphasises the need for properly quantified data, and proposes a basis of specification of bulk solids for materials handling applications". (5).

2. There is no commonly accepted procedural methodology for relating the behaviour related properties of the bulk materials to be handled to the characteristics of any specific form of equipment under consideration.

This is a question of industrial education in the technology. In recent years a start has been made by the formation of various specialised centres. (6) and (7). and the inclusion of particulate solids technology as a subject in many university courses.

Seminars organised by the I.Mech.E. and others play a part and the quality of specialised technical journals in this field is also making a contribution. (8), (9).

However, it is considered that more specific guidance is required for users with regard to machine selection, in a manner similar to that of a recent publication by B.M.H.B. (10) "Users guide to particle attrition in Materials Handling equipment", which outlines the sensitivity of particles to breakdown in differing types of handling equipment and a WSI report on the selection of equipment for paste handling (11).

The I.Mech.E. Bulk Materials Handling Committee are therefore preparing a document indicating "The range of usefulness of conveyors for Solids Handling" (12).

3. A further feature which determines the state of the art in practice is that there is very limited industrial testing equipment available to derive quantitative values of relevant Bulk Solids properties.

Such equipment as exists is of limited value for the assessment of overall handling equipment performance. This position is exacerbated by a lack of knowledge as to what are the features which characterises the behaviour of a bulk material for a particular handling situation.

This remains a neglected area of attention and no obvious steps are in hand to redress the situation for new forms of test equipment.

Research in U.K. has tended to be dominated by extended study based on the Jenike type shear cell approach, with a view to securing consistent data for the application of this specific design procedure for hoppers.

This testing device is not readily available, its operations and interpretation demands skilled operatives and its use is limited to a narrow range of storage applications. The net result is that this technique is normally commissioned only by large companies for large scale storage installations, such as British Steel coal and ore silos.

For such applications it has made a significant contribution, but for the vast number of smaller scale and differing handling applications there is a dearth of factual design data.

Fundamental to this problem is that there are no absolute physical properties of bulk solids suitable for "once only" measurements and use in data bank references. The interaction of many variables, such as environmental, operational, time and differing contact surfaces, would lead to the use of misleading information taken from a rigid data base.

There is a clear need for the development of a family of powder testing instruments suitable for industrial use in the field, in order to secure measured values of bulk solid properties in situations which reflect the circumstances of product use.

Until progress is made in this direction there will be no firm foundation upon which improved designs may be solidly based.

Much remains to be done. However means are now available to satisfy stage one and define the bulk material and its properties relevant to a materials handling application.

In the document put forward there is a review of the background, a form of specification for the physical properties relating to material flow and behaviour, safety of personal, value and use of the product, the local and broad environment and to the durability and mechanical features of the equipment used. Notes are appended explaining the background and features of powder testing.

For the purposes of establishing the criteria which effect the performance of the equipment the term term "hazard" is applied to any feature of the bulk material which may give rise to a potential impediment to the smooth and satisfactory operation of the overall installation.

The definition of the form of such hazards, as with the assessment of the seriousness with which such "hazards" influence the equipment design inevitably have specialised and subjective elements which relate to the specific application under consideration. Standards acceptable in a steel works will not apply in a pharmaceutical or food factory, and visa versa.

It must be emphasised that the data is application related.

Where there may be some phenomena which is unacceptable to the installation, such as "dead" regions of storage or flow, permanent residue, segregation, excessive particle attrition, or "poor" product condition in a manner described, then the onus is upon the specifier of the relevant properties to draw attention the those specific characteristics of the material which would make these aspects of the materials behaviour unsuitable in the case under consideration.

The final arbitrator of equipment suitability must be the user, who not only determines the physical task to be accomplished, but must also approve of all aspects of the operating conditions and of the resulting condition of the product as delivered from the system.

The compilation of the data must therefore be undertaken and authorised by one who is aware of the full requirements of the duty and of the implications of these bearing upon the equipment to be supplied. When powder testing is to be undertaken to secure particular quantified values, agreement must be reached upon the method to be employed and the manner in which the information is to used.

All of the information must be available to the equipment designer. The registration of this date may then be incorporated in any contractual conditions relating to performance guarantees. Without such data guarantees are often meaningless and merely provide fertile ground for disputes and litigation.

For an initial evaluation of the general nature of the bulk material it may be assessed by means of a "personality" classification. fig. 1.

Apart from characterising the material this will draw attention to the factors which require more detailed consideration in relation to the equipment to be used and sensitive aspects of operation.

The features outlined in the form of specification are intended to be embracing for general consideration. Various sections may not be relevant to a particular case, The form is intended to be used such that features may be marked as "not relevant" rather than be omitted from consideration by default.

Additional information, pertaining to the characteristics of specialised forms of solids storage, handling or processing equipment, may be added as required.

The following sections of the I.Mech.E. document lists the contents, Fig 2, and section 6 and 7, which relate to the specification of the bulk material and the hazards to fully satisfactory operation which may be relevant to a particular application.

'PERSONALITY' classification for Bulk Solids.

For a preliminary assessment as to the form of 'handling treatment' required.

'NORMAL' Materials	Stable and consistent in behaviour
Like people, very few materials can be classed as normal.	Very predictable Homogeneous nature Uniform measured properties. Insensitive to environment and handling
'NEUROTIC' Materials	Awkward to handle, varies in nature
changes in nature according to conditions. May be stable in a controlled environment.	Not stable in state or condition Prone to flush, run and leak everywhere or settles to an intractable stubborn mass clings or sticks,very sensitive to changes
'SCHIZOPHRENIC' Materials	Significantly change in their behaviour pattern or nature
has radically different behaviour pattern,often irreversibly.	Behaviour alters with liquor content. Changes condition with time or temperature Segregates to two or more bulk conditions with differing properties. Deteriorates or reacts to environment.
'PARANOID' Materials	Desperate concern for particular property
hypersensitive about particular aspects, such as contact with foreign bodies.	To maintain purity, hygiene, texture, homogeneity, or other specific property.
'MASOCHISTIC' Materials	Fragile or easily damaged.
self destructs at the slightest opportunity	Easily abraded or spoilt. vulnerable to environmental changes or local hazards value easily effected by changes. very sensitive to conditions and effects.
'SADISTIC' Materials	Aggressive to their surroundings.
Dangerous to people and equipment, ranging from difficult or unpleasant to horrible and positively evil.	Abrasive, sharp, extreme temperature. Corrosive, explosive, inflammable, irritant, fouls smells or contaminates. Attacks on contact, carcinogenic or plain nasty.

Fig 1

Bulk Solids Handling Committee - Institution of Mechanical Engineers

Bulk Storage and Handling Equipment - Design for improved performance.

C O N T E N T S.

Note: Sections 6 & 7 only are reproduced on following pages.

Fig 2

6.0.- Recommended form of Specification of a Bulk Solid for the Design of Bulk Storage and Handling Equipment.

6.1. Mono Product, Compound, Blend.

6.2 Bulk Material - Generic Name.
- Trade Name.
- Chemical Formula.

Source of Sample.
Size of Sample.
Date of Sample.

6.3 Descriptive Form - Appearance - Colour.
- Texture.
- Condition - Coarse, fine, etc..

6.4 Uniformity - Homogenous.
- Consistent.
- Stability.

6.5 Physical Composition of Bulk Material.
Particle Size Distribution.
Particle shape.

6.6 Moisture Content - Total - Free Moisture.
- "Inherent" Moisture - Bound Moisture.

6.7 Density - Loose Poured - Lightly Tapped
- Is material compatible? - Areated.

6.8 Flow Condition - Free flowing - Cohesive.
- Interlocking.
- Loose poured Repose Angle.

6.9 Slip Properties.

	Contact Material.	Surface Finish.	Sliding Friction Value.
(1)			
(2)			
(3)			

6.10 Shear strength

Shear Cell date (if applicable)
Instantaneous Unconfined Yield Strength.
Effective Internal Angle of Friction.
Effect of Time Consolidation.

Operational Features. - See following pages for details.

6.11 Friability/susceptibility to particle damage

6.12 Tendency to segregate.

6.13. Hygroscopic, Deliquescing.

6.14 Hardness. Mohr Number.

Potential Hazards. - See following pages for details.

6.15. Corrosive, Caustic.

6.16. Sharp, penetrating

6.17. Subject to extreme temperatures.

6.18. Any other relevant information.

7.0. - Hazards associated with the bulk material.

For systematic consideration, these hazards are grouped according to how the consequences may fall. These consequences then require to be graded in terms of seriousness. For this a form of grading system is required.

7.1. Degree of hazard,

In order to register the degree to a hazard may give rise to harm, injury or unsatisfactory operation, the following grading system directs attention to the potential magnitude of problems which may arise when storing or handling a specific bulk material.

Design attention must then be directed to provide adequate protection against the nature and scale of the hazards indicated.

Grade Seriousness of the hazard

Grade 1 - Negligible, or of such minor effects that no special care is required by personnel or equipment design.

Grade 2 - Significant effects, requiring preventative, protective or precautionary measures to be taken. The containment of dust, fumes, the enclosure of hot or sharp products, pigments or of mild irritants are typical.

Grade 3 - Serious effects, requiring specific or additional degrees of attention, e.g. Corrosive products, active pharmaceuticals. The containment or extraction of harmful dusts, protection of operatives from heat, cold or injurious materials.

Grade 4 - Extreme effects, presenting severe risks of harm, injury or catastrophic failure, e.g. Explosion Hazard, radioactivity, excessive temperatures, toxic or aggressive chemicals.

These gradings must be defined and authorised by a named, responsible person who is identified for further reference. Appropriate C O S H H documentation must be available to the designer.

Hazards which are graded 2 or higher are further defined in the following sheets.

7.2. Forms of hazards

Grading

(i) Hazards - Effecting the Health, Safety or comfort of employees and persons who may be placed in the vicinity of the plant. Dust, temperature, noise.

(ii) Hazards - Effecting the Environment, local to the equipment or broader, relating to the plant, dust or effluent emissions. Mess and spillage, site contamination.

(iii) Hazards - Effecting the efficient and reliable performance of the operation of the equipment. Capacity, condition of the product, Starting, stopping and standing conditions. Residue, cross contamination.

(iv) Hazards - Effecting the Integrity, durability and service needs of the equipment. Wear and tear, exceptional loadings, impact, pressures, corrosion.

(v) Hazards - Effecting the condition of the product for handling or subsequent process operations.

(vi) Hazards - Effecting the ultimate use or value of product for sale.

It should be noted that the statutory requirements of the Control of Substances Hazardous to Health (C.O.S.H.H) apply equally to the examination and handling of samples for any investigatory or testing work as to the final application of the equipment under consideration.

Some of the properties of the material may have potential effects on more than one of the above groups. However, it is good practice to consider each section independently, as the significance of each hazard may then be individually considered.

7.2. (i) Hazards which may affect the health, safety or comfort of the equipment operator, service personnel or others associated with the location or operation of the equipment.

	Hazard Type.		Effect	Grade
(a)	Toxic material.		Ingestion or absorption.	
(b)	Carcinogenic.		Minor to fatal, short or long term.	
(c)	Radioactive.			
(d)	Harmful to : -			
		Lungs	Pungent odour. Causes respiratory, bronchial, asthmatic or silicosis problems. Powder ingestion or deposition.	
		Eyes	Gas, liquid or dust or reactant.	
		Nose	Pungent, irritant, causes sneezing.	
		Skin	Corrosive, acidic or alkali. Provokes allergies or dermatitis. Sensitises or irritates the skin. Absorbed through the skin. Sharp, penetrating, barbed, excessively hot or cold.	
(e)	Gases/Fumes		Explosive, poisonous, flammable, or suffocating. Dirty, smelly or dusty.	
(f)	Environment		Dangerously or uncomfortably high or low temperatures.	
(g)	Other.			
(h)	Special Precautions.			
(j)	Safety Instructions.			

7.2. (ii) Hazards which affect the environment, either local to the equipment used or causing broader atmosphere contamination or effluent pollution.

	Hazard Type.	Effect	Grade
(a)	Gases/Fumes Vapours/Grit	Dirty, smelly or dusty vapour, fumes or smoke.	
(b)	Noise/Vibration	Excessive or unsocial noise and/or vibration.	
(c)	Infestation	Raw material, process or end product supports bacterial action of attracts vermin or pests.	
(d)	Spillage/Mess	Unsightly, dangerous, slippery, and/or dirty	
(e)	Atmospheric	Airborne dust and odours.	

(f) Pollution — Via drains, percolation/leaching through soil, absorption, etc..

(g) Other.

(h) Special Precautions or environmental conditions.

7.2. (iii) Hazards which may affect the efficient and reliable performance of the equipment, including starting and running effects, cleaning and reliability, including provisions for exceptional circumstances of operation.

	Hazard Type.	Effect	Grade
(a)	Poor Flow condition	High shear strength, compacts to a firm poor flowing condition. Interlocks, mats, Sticks to contact surfaces. Melts or fuses, crystalises, bonds.	
(b)	Moisture migration	Preferential condensation causes build up on walls.	
(c)	Fumes/Gases/Dust	Contamination of material. Effect on contact surface of material. Explosion/fire hazard.	
(d)	Noise/Vibration	Effect on operator. Causes damage to, or shaking loose of, equipment and/or cause compaction of the material to a poor flow condition.	
(e)	Moisture absorption	hygroscopic, deliquescent, hydrates, decomposes, etc.	
(f)	Cakes/Sets	Material forms hard mass.. Resists shear in equipment.	

(g) Any other impediment to smooth and reliable operation

7.2 (iv) Hazards which may affect the integrity or durability of the equipment.

	Hazard Type.	Effect	Grade
(a)	Material	Corrosive, adhesive, abrasive or explosive.	
(b)	Site	Vibrations, poor access, isolated wide temperature variations, extreme conditions, moisture,	
(c)	Servicing	Limited, inadequate or irregular maintenance.	
(d)	Environmental	Flooding, earthquake or tremors, natural hazards..	
(e)	Operation	Continuous, Frequent use, Surges, overloads.	
(f)	Operators	Unattended, Poor quality labour,	
(g)	Importance	Need for dependable performance. Rapid restoration of duty, Strategic spares.	

(h) Any other features relating to the durability of the equipment.

7.2. (v) Hazards which may cause the bulk material to be adversely affected for subsequent processes or production operations.

Hazard Type.	Effect	Grade
(a) Heat/Cold	Ambient or Process operating at too high or too low a temperature.	
(b) Moisture	Variations in moisture content, too much/too little moisture in the material.	
(c) Infestation	Rodents, insects, bacteria.	
(d) Texture of material	Process causes material to coagulate, cake, set, compress, melt, fuse, soften, cool, heat up, aerate, matt, agglomerate.	
(e) Quality of material	Process causes the material to decompose, express oils, moisture, etc., degrade, cross-contaminate, age. decay.	
(f) Friability	Susceptible to the generation of fines. undersized, excessive surface area,	
(g) Particle Size variation	Liable to segregate in storage and flow. packaging problems,	

(h) Other factors relating to further operations and processes.

7.2. (vi) Hazards which affect the value, quality or suitability of the bulk material for its ultimate use or sale.

Hazard Type.	Effect	Grade
(a) Infestation spoilage.	Rodents, pests, infestation or waste products in process	
(b) Texture of material	process causes material to coagulate, cake, set, compress, melt, fuse, soften, cool, heat up, aerate, matt, etc.. Poor "feel" or taste.	
(c) Quality of material	Process causes material to decompose, expresses oils, moisture, etc., degrade, cross contaminate, putrefy, set, age.	
(d) Appearance	Damage to surface condition or particles may adversely effect the appearance of the bulk or the particles on which the quality is judged.	
(e) Homogeneity	Segregation which results in differing qualities appearing at the discharge point of the system may influence the product consistency, quality or value.	

(f) Other qualities or requirements influencing the value or utility of the product.

Many parts of the document are elementary, but in practice omission to take query these frequently give rise to major misunderstandings and problems. For example the first sections of the specification are directed to establish the identity of the material and whether it is a uniform material of "simple" composition.

Compounds, composites, blends and mixtures may be subject to changes in formulation or structure, whilst having the same notional description.

Products such as food and mineral recipes, intermediates in various industries and particularly semi processed goods, carrying the same descriptive reference, may vary in composition and thereby have differing handling characteristics.

In terms of material uniformity is necessary to establish whether the material is universally similar in composition and nature through the bulk and if it consistent on a batch to batch, day to day and year round basis, i.e. fluctuations in supply conditions, alternative sourcing, seasonal or natural variations are to be indicated

It must then be determined whether any characteristic of the material will lead to changes taking place throughout the products life in the plant, which will effect any performance feature of the operation.

Uniformity - Is the material homogeneous in nature, isotropic, even throughout the bulk in such features are particle or granule size, moisture content, and all features which may effect behaviour?

If not there may be features which require the plant to handle varied bulk materials, such as fines or large lumps, in isolation as well as there being part of the general mass.

Consistency - Is the material always of the same nature or in a condition depending upon its source, process history or other pre-delivery variable ?

Natural products, such as ores and organic product, will vary according to conditions of origin. Processed material may vary in condition, temperature or circumstances of production. Subtle changes of particle composition, of what is ostensibly the "same material" by a differing manufacturer may lead to variations in the way that the materials handles or reacts to handling.

Stability - Is the material inert and unchanging, invariant to time effects handling, processing and ambient conditions. Not subject to influence by degradation, attrition, adsorption of moisture, thermal, chemical or electrostatic effects, decomposition?

Where the cause of the variations lie in the conditions of the application rather than those of material supply it is necessary to identify the intrinsic properties of the solid which make it venerable to such changes.

Supplementary to the form of specification are notes setting out the purpose of the various sections, definitions and method of establishing the required measurements.

In conclusion the steps taken to introduce quantified values to bulk solids properties relevant to a specific material handling application is but a start to the process of achieving improved performance in the operation of industrial operations involving loose solids.

The dissemination of the existing technology requires to be continued at all levels, from students to senior management, in order that the requirements from equipment, in all its aspects, can be appreciated.

As the technology develops further the integration of disciplines and interaction of ideas will lead to a better understanding of this most complex section of rheology. One of the features which distinguish this field is the vast permutation of variables which apply, not only to the material itself, but to the ambient conditions and the scale and nature of the processes involved.

The work of Jenike introduce the critical state concept to powder technology and set the technology on a scientific footing with the provision of a measuring tool and design method for storage silos. It also had the effect of drawing together the study of bulk solids handling as a specific discipline. Progress since has tended to be slow because of the great difficulty in drawing together a comprehensive framework embracing all potential variables.

The provision of a formal means of defining a bulk solid for storage and handling applications, whilst essentially a document of low technology, will hopefully play a part in the educational programme, as well as fulfilling a need for setting out design information and giving a sound basis for establishing contractual obligations.

R E F E R E N C E S

(1) Merrow E. W. "Linking R & D to problems experienced in Solids Processing" Chem.Eng.Processing. May 85, p 14 -22

(2) Merrow E. W.,Phillips K. E. & Myers C. W. "Understanding cost growth and performance shortfalls in pioneer process plants". Rand Corporation Report. Section V. (1981)

(3) Arnold Prof. P. Editorial, Powder Handling and Processing, Vol 1 No 3 Sept. 1989

(4) Neale Associates. "A survey of Mechanical Handling Equipment to discover new product opportunities and relevant subjects for basic research work" D.T.I Report No. S R S 368. 1988

(5) Bulk Materials Handling Committee of I.Mech.E. "A Procedure for improved design of Solids Handling Plants". 1991.

(6) The Wolfson Centre for Bulk Solids Handling. Woolwich (Based on Thames Polytechnic)

(7) The Industrial Centre for Bulk Solids Handling. Glasgow. (Based on Glasgow College).

(8) Transtech Publications. Germany. " Bulk Solids Handling "

(9) Transtech Publications. Germany. " Bulk Solids Processing "

(10) B.M.H.B. "User Guide to Particle Attrition in Materials Handling Equipment" 1990.

(11) Warren Spring Laboratory. Wet Solids Handling Project. Club Project.

C418/013

Handling plant for bulk solids – improved productivity through a systems approach to design

P C ARNOLD, BE, PhD, CEng, MIMechE, FIEAust
Department of Mechanical Engineering, University of Wollongong,
New South Wales, Australia

SYNOPSIS A systems approach to design of bulk solids handling plant is required if the plant is to be cost effective. The design philosophy must also take account of the many developments in technology which have occurred in recent years. The consequences of failing to heed this advice will be illustrated by several case studies covering a wide range of storage and conveying situations.

1. INTRODUCTION

The necessity of taking a systems approach to design of bulk solids handling plant cannot be over-emphasised nor is it new [1]. Developments in technology have lead to significant advances being made in the processes involving the production and utilisation of bulk solids. Generally such processes involve significant elements of bulk solids storage and handling; these storage and handling elements usually form series links with the processing elements in the total system chain. If the handling elements are the weak links in the chain then the performance of the total system suffers when those links fail to perform. The resulting problems of, for example, reduced production, lack of quality control, mess and spillage, increased manpower requirements can make significant inroads into otherwise healthy profit margins. Such disasters are generally avoidable by implementing sound system design practices when the overall process is being designed and detailed.

These assertions will be illustrated graphically by several case studies covering:

- storage bins,
- feeders,
- stockpiles,
- segregation,
- pneumatic conveying,
- chutes,
- handling of fine powders.

2. CASE STUDIES

(a) Bottom reclaim stockpile system for loading coal trains.

A coal stockpile system had a bottom reclaim system consisting of forty-two windows serviced by two rotary plough feeders. The load-out system was required to load 8000 tonne trains in four hours via a 300 tonne surge bin. The system gave inconsistent reclaim rates resulting in extended train loading times which reduced potential shipping rates and incurred railway penalty payments and scheduling difficulties. Initially six of the windows were modified as shown in Fig 1; later a further six windows were modified. The modifications enabled the achievement of reliable train loading times of two and one half to three hours, Fig 2. Full details of this case study have been reported [2].

(b) Aeration of a bin leads to an air shortage at a coal mine.

The three fine feed bins to a coal preparation plant were fed by a common rotary plough feeder. To overcome arching problems at the outlet a sophisticated aeration system was fitted to the outlet slot. As the coal was minus 12 mm the end result was the usage of huge quantities of air with little effect on the flow instability problems. The mine was being starved of air and plans were being made to augment the compressed air capacity. The problem was a simple mismatch of bin outlet and feeder caused by the presence of a number of structural columns protruding into the outlet area, Figure 3. By a simple modification similar to the one outlined in (a) above the flow was stabilised and the requirement for air to the bins was removed.

(c) A limestone hopper extension overloads the feeder.

The limestone dump hopper at a cement plant was duplicated and uprated. As Figure 4 indicates the hopper was wedge shaped and fed by an apron feeder. The hopper fed from the rear and required the feed to be dragged under dead material thereby over loading the feeder. Higher capacity motors and gearboxes were tried with the result that the apron feeder itself was overstressed. The solution finally adopted was a redesign of the hopper/feeder interface to achieve a fully active hopper outlet, Figure 5. The feeder power required after the modification was less than originally specified.

(d) A power station only achieves 70% availability due to coal handling problems.

Coal fired power stations depend on a reliable flow of coal throughout the entire coal handling plant. A modern 4 x 660 MW plant had its availability reduced to 70% through difficulties experienced with coal bins, transfer chutes and undersized coal valves.

(e) A 1200 tonne coal bin delivers 300 tonnes.

The coal storage bins at a number of minesites consist of multiple outlet bins with significant flat areas between outlets. The end result is normally a reduced live capacity over time as outlets block or stable ratholes develop around outlets preventing feed to the outlet(s). The elimination of these flat areas by properly designed hoppers can usually restore full live capacity provided the structure is designed to take the loads imposed on the walls, Figure 6.

(f) A pulverised coal bin has continual flooding problems.

A classic paper which illustrates this problem was written some years ago by Bruff and Jenike [3]. Figure 7 depicts the original system while Figure 8 shows the modified system. Yet despite the lessons learnt by that experience we are still being faced with trouble-shooting similar problems even in newly installed plants. Feeding of fine powders onto open feeders under controlled conditions is one application where a mass-flow bin giving the powder sufficient residence time to deaerate is mandatory.

(g) A nickel ore bin floods when the material flow properties are changed.

Care must be taken not to change adversely the flow properties of a bulk solid by changing process or operational parameters. In this case study the nickel ore was changed from free-flowing to floodable when precipitator dust was added to the material flow stream.

(h) Grain silos fail when changed from barley to wheat.

Seemingly subtle changes like storing a different grain can have a catastrophic effect. As the illustrating slides will show a change from barley to wheat resulted in severe structural failure of a corrugated steel silo.

(i) A bin filled by pneumatic conveying has flow problems due to inverse segregation.

Segregation problems can affect the performance of a bin as well as the quality of the bulk solid, Figure 9. Particular care must be taken in charging bins with pneumatic conveyors [4].

(i) A feed silo develops severe ratholing problems.

Jenike and Johanson reported [5] four stable ratholes 6.7 m (22 ft) in diameter forming in a 18.3 m (60 ft) diameter silo containing rice hulls. The ratholes resulted from the presence of 840 mm (2 ft 9 in) flat area between hoppers. In the original design wedge-shaped tents were recommended between the hoppers to connect the flow channels. Unfortunately they were omitted in the final design stage resulting in the stable ratholes depicted in Figure 10.

(k) A pneumatic conveying system causes 30% of grain to be rejected due to breakage.

Traditionally pneumatic conveying has the reputation of using high energies and causing major problems with plant erosion and product degradation. With modern developments in the technology many more materials can be handled with minimal degradation and with significantly less energy requirements. In a recent consultancy for industry conducted in the Bulk Solids Handling Laboratory at the University of Wollongong it was demonstrated to a grain processing firm that their grain, in various processed states could be handled without and discernible breakage using low velocity pneumatic conveying technology.

(l) A sugar silo collapses on its initial filling/ discharge.

A 3000 tonne sugar silo which had a cylindrical vertical section and a bifurcated hopper with eight outlets collapsed on its initial usage; as the slides will indicate the collapse was catastrophic! The bin was designed for a symmetric wall load profile and collapsed when one of the peripheral outlets was opened. Wall loads in silos are directly influenced by the flow pattern that develops as the bin is operated. Variables that affect the flow pattern such as hopper geometry, multiple outlets, segregation, feeder/hopper interfacing must be taken into account when predicting the wall loads.

(m) Design of transfer chutes must take account of material trajectories, segregation and flow properties.

The weak links in many conveyor systems are the transfer chutes. These are often designed without any consideration being given the accurately predicting and controlling the material trajectory through the chute, taking account of the segregation of coarse and fine particles and the different flow properties that each of these components are likely to exhibit. Little attention is given to such details as ledges, valley angles, chute slopes, impact points, loading profiles on receival belts, Figures 11 and 12. Considerable information on the design of transfer chutes can be found in Reference [6].

3. CONCLUDING REMARKS

The concept of systems design is not new. For example, Sutton and Scofield writing on the topic many years ago [1] maintained that problems and inefficiencies that occur in bulk handling usually arise from a failure to consider the system as a whole. Considerable attention and a large component of the budget is expended on ensuring that the processing units perform their proper function. Budget overruns on the processing units often means that materials handling systems which link the total system together are the targets for cutting expenditure. This cost cutting usually results in inferior materials handling plant being designed and installed. As the total system is normally a series linked system with the processing units linked by materials handling components then the end result is that the total system has severe weak links. As these weak links begin to fail and cause serious and costly loss of productivity. Under such circumstances it becomes obvious, even to the 'bean counters' that the money saved on the inferior materials handling plant was a false economy.

4. REFERENCES

1. Sutton, H.M. and Scofield, C., **A Systems Approach to In-Plant Bulk Handling.** Bulk Vol. 1, 1974.

2. Arnold, P.C., Fiedler, V.A., Corr, N.G. and Clarkson, C.J., **Improving the Performance of a Gravity-Reclaim Stockpile System.** Bulk Solids Handling, Vol. 3, No. 4, 1983, pp 727 - 734.

3. Bruff, W. and Jenike, A.W., **A Silo for Ground Anthracite.** Powder Technol. Vol 1, 1967/68 pp 252 - 256.

4. Arnold, P.C., **On the Influence of Segregation on the Flow Pattern in Silos.** Paper presented at CHISA 90, Prague, Czechoslovakia, August, 1990. Bulk Solids Handling Vol. 11 No. 2, 1991.

5. Jenike and Johanson Inc., **Twenty-two Foot Diameter Ratholes can be Stable.** Flow of Solids Newsletter Vol. 3, No. 1, 1983.

6. Arnold, P.C. and Hill, G.L. , **Design of Conveyor Chutes with Special Attention to Blockage, Wear and Conveyor Direction Change.** End of Contract Report, NERDDP Project No. 1188, 1991.

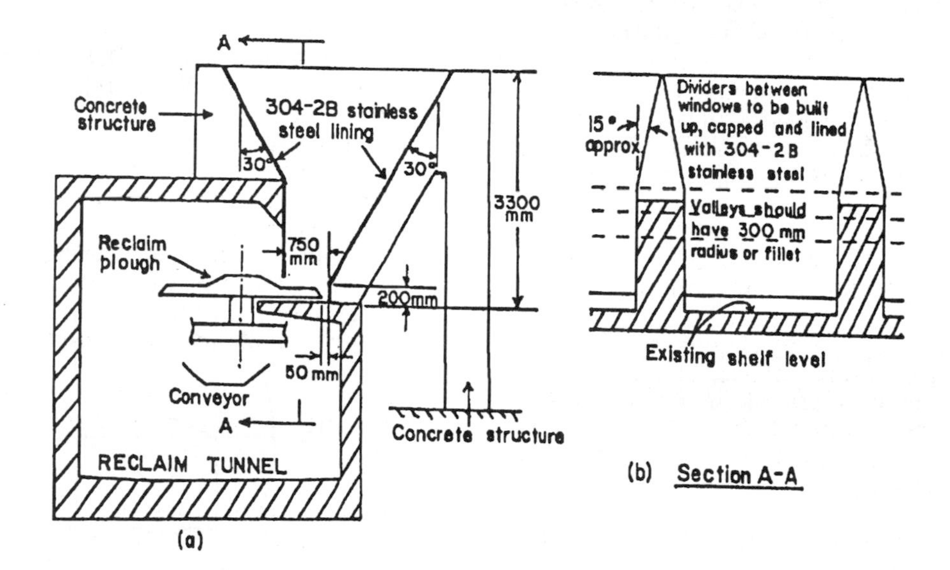

FIG. 1 MODIFICATIONS TO EXISTING RECLAIM TUNNEL WINDOWS

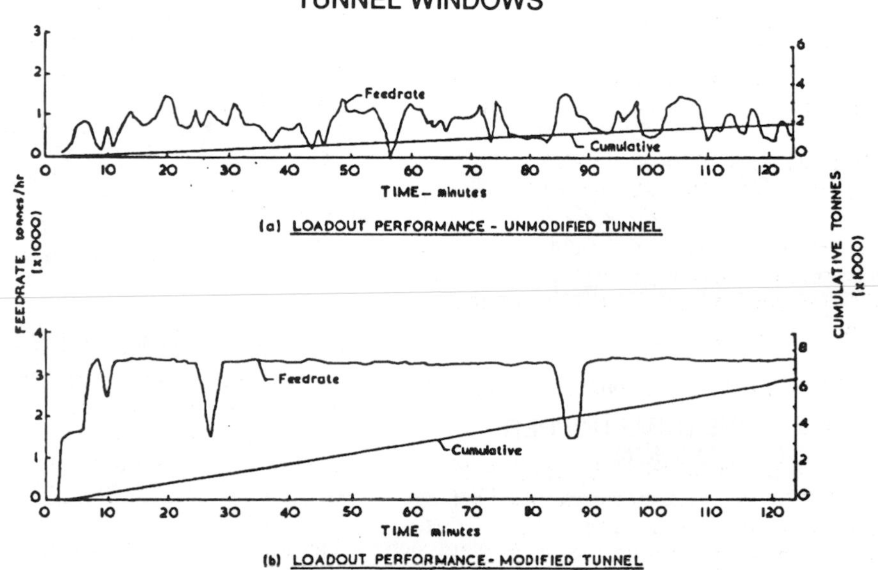

FIG. 2 COMPARATIVE LOADOUT DATA BEFORE AND AFTER MODIFICATIONS

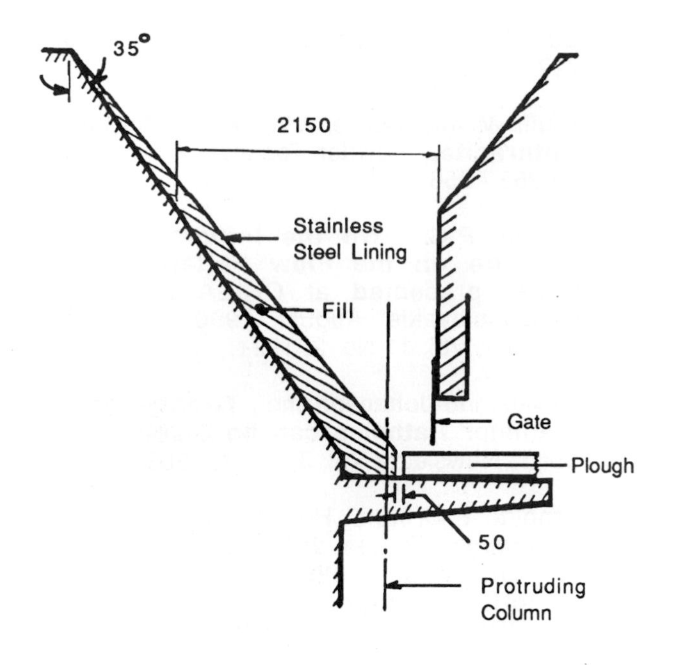

FIG. 3 PROFILE THROUGH SLOT OUTLET SHOWING OFFENDING COLUMNS AND MODIFICATIONS TO GEOMETRY

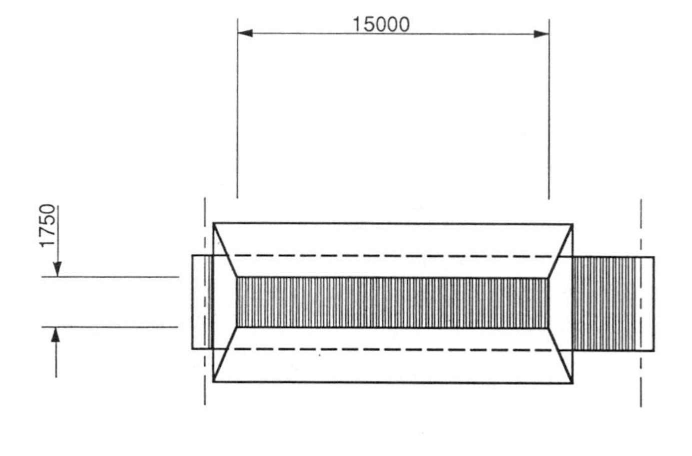

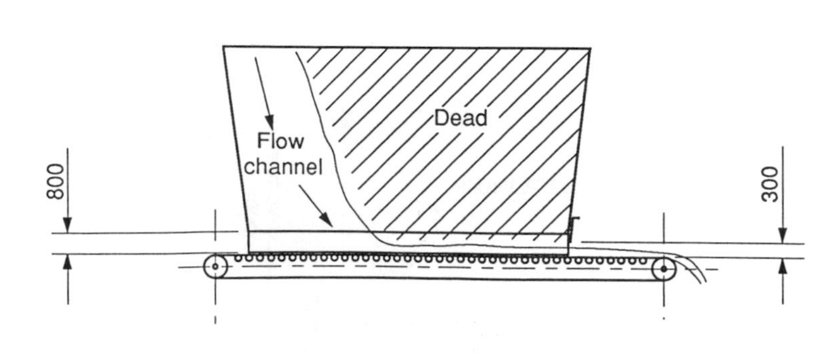

FIG. 4 LIMESTONE DUMP HOPPER INITIAL DESIGN

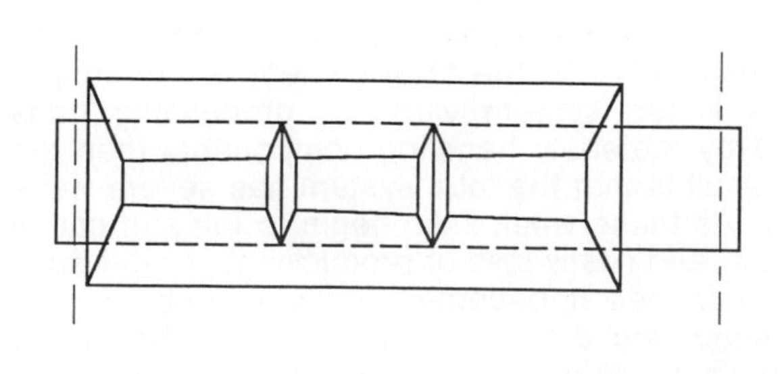

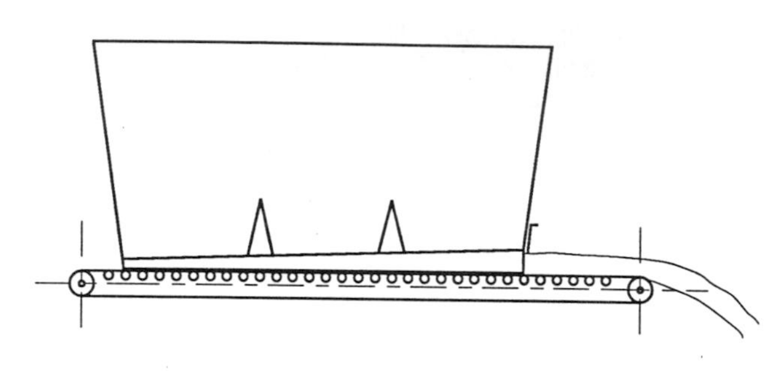

FIG. 5 LIMESTONE DUMP HOPPER MODIFICATIONS

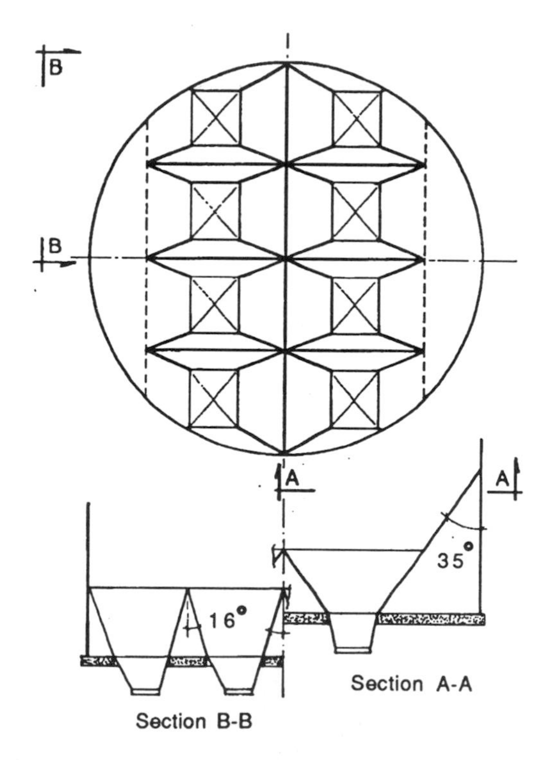

FIG. 6 MODIFICATIONS TO MULTI-OUTLET COAL BIN

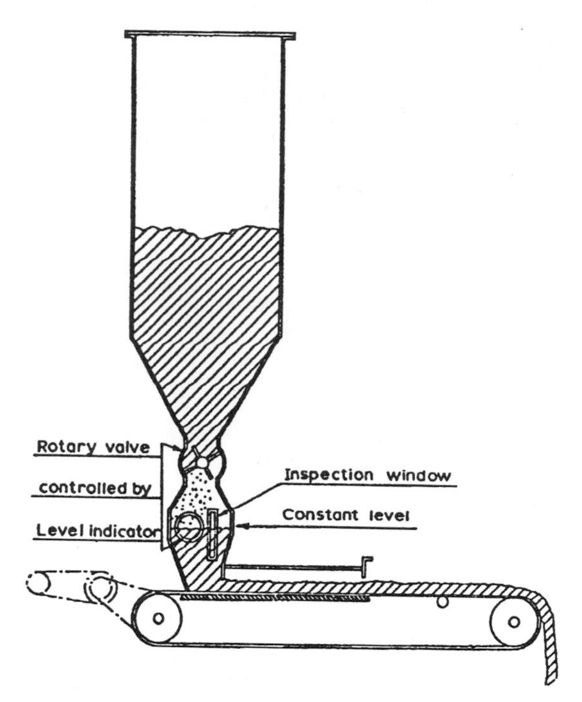

FIG 7 PULVERISED COAL BIN ORIGINAL SYSTEM
(from Ref. 3)

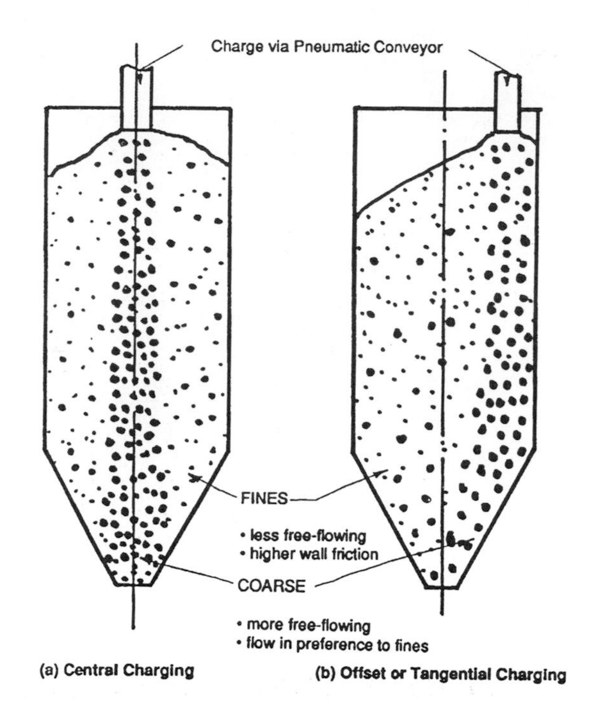

FIG. 9 SEGREGATION PROBLEMS WITH PNEUMATIC CONVEYING PROMOTES FUNNEL-FLOW

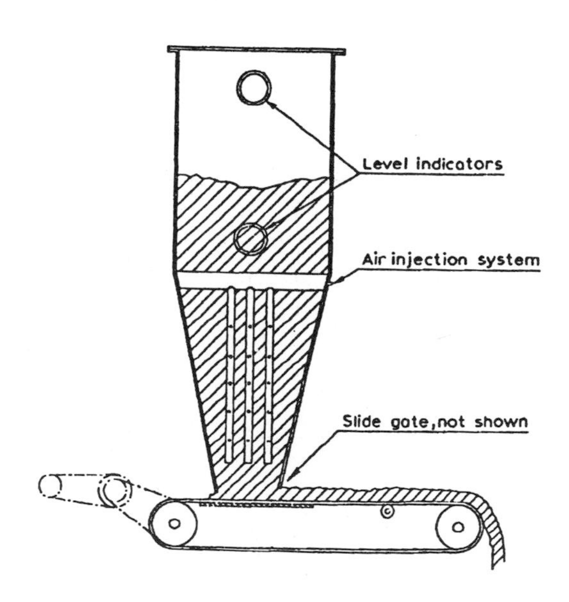

FIG 8 PULVERISED COAL BIN MODIFIED SYSTEM
(from Ref. 3)

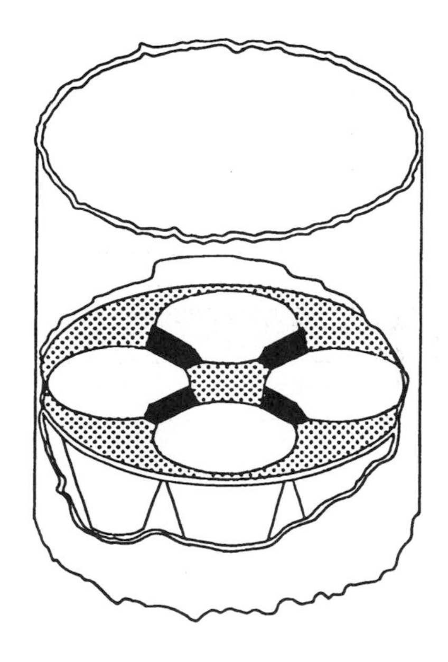

FIG. 10 SEVERE RATHOLING IN FEED SILO
(from Ref. 5)

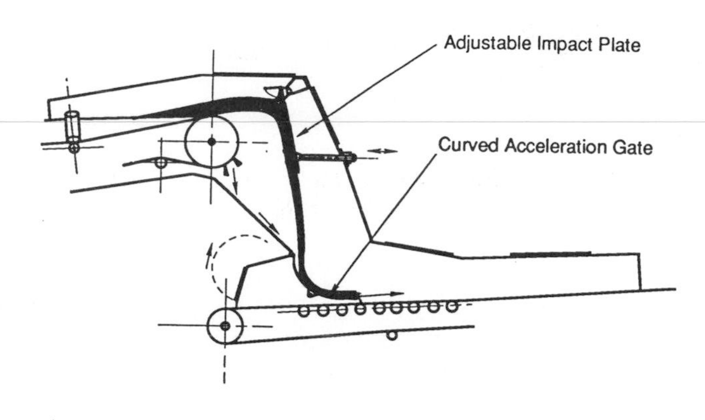

FIG. 11 TRANSFER CHUTE WITH IMPACT PLATE AND ACCELERATION GATE

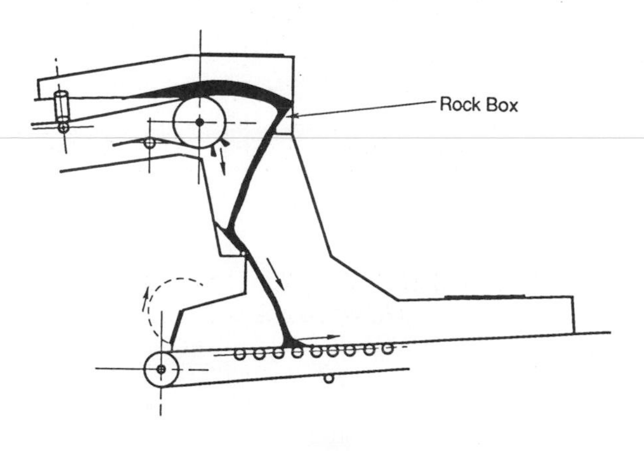

FIG. 12 TRANSFER CHUTE WITH ROCK BOX DEFLECTORS

C418/005

Pneumatic conveying: system types and components

B VELAN, BE, MIE
Scorpio Engineering Pvt Limited, Bangalore, India

SYNOPSIS Pneumatic Conveying can be a viable alternate to conventional conveying systems for bulk solids especially for in-plant applications. System types and commonly specified components are described. System and component reliability is discussed so that the end user can specify, evaluate and operate Pneumatic Conveying Systems satisfactorily.

1.0 INTRODUCTION

With continuous improvement in technology over the past 20 years, pneumatic conveying now competes favourably with conventional bulk material conveying systems both in terms of operational cost and conveying reliability. In addition to improvement in basic system design, a variety of reliable control components and equipments are available that enable both vendors and users to exploit the flexibility and relative ease of operation of such systems.

2.0 SYSTEM TYPES

Engineers planning to install a pneumatic conveying system will need to decide whether to use "dilute" or "dense" phase systems or commercial variations of these basic types. A distinct technical demarcation exists between these two systems. Dilute phase conveying will mean suspension flow where the conveying gas, usually air, is pumped through the conveying pipe at relatively low pressures and high velocities. The material to be conveyed is introduced into the pipeline in a controlled manner allowing the particles to be carried in suspension to the destination point. The material loading factor, the weight ratio of material to air, is generally in the range of 10-15 at the higher ends (i.e. 10 kg. to 15 kg. of material per kg. of air).

In dense phase conveying however the material is moved in the conveying pipe to the destination in non-suspension flow, in collapsing/reforming dunes or in plugs of full pipe cross section. The pressures required are higher than those required for dilute phase conveying and the material loading considerably greater, even going upto 200 depending on the ability of the material to be conveyed in this mode. Dense phase conveying is essentially a batch operation as compared to dilute phase conveying which is a continuous operation. The higher pressures involved in dense phase conveying and the higher materrial loading require robust equipment and the specification of valves, actuators etc. should consider the cyclic nature of operation.

3.0 KEY SYSTEM COMPONENTS

Components are generally common for both types of systems though equipments for use with dense phase systems would be designed to handle the higher pressures in system operation.

3.1 Dilute phase system components and equipments

3.1.1. Centrifugal Blowers

The high velocity air or gas stream that is required to suspend and pneumatically transport material particles is generated commonly by either centrifugal blowers or by twin lobe positive displacement blowers. Small systems like hopper loading units for plastic extruders use regenerative blowers. These equipments can be installed upstream or downstream of the pneumatic conveying sytem depending on whether system is of pressure type or vaccuum type. Centrifugal blowers would be useful for small conveying duties with low pressure drops and where the material is generally known not to be troublesome in storage or handling. Line chokes in systems having centrifugal blowers will create major problems because the blower cannot generate positive pressures to clear chokes and will operate inefficiently at shut off head when such chokes occur.

3.1.2. Twin Lobe Roots Blowers

Roots blowers are generally more prevalent than centrifugal blowers or fans for inplant material transfer. These blowers are the industry workhorses in positive and negative pressure dilute phase conveying systems. They consist of twin lobes in a cast iron housing, machined to close tolerances. At every half revolution a small quantity of air is trapped between the lobes and is discharged positively. At high speeds and capacities, pulsations arising out of this action are characteristic of this equipment and may require acoustic hoods to comply with noise pollution laws. The characteristic curve of this blower lends itself to considerable power saving in the partial load range. There is a very small change in volume for relatively large changes in discharge pressure as compared to centrifugal blowers. The capacity of these blowers cannot be controlled by dampers as they are positive displacement equipment.

In pneumatic conveying systems where system resistance due to material flow in the pipeline cannot be accurately calculated, these blowers tend to oversized. Fine tuning during plant commissioning is done by by-passing some of the air from a tee connection on the discharge line or by changing blower speed.

3.1.3 Pipeline

Pipelines for dilute phase system are generally steel, ERW pipes. Where the conveyed material is not compatible with steel, stainless steel and aluminium are generally used. Stainless steel pipes can be thin walled for economy. In the petrochemicals industry special internal treatment to pipelines is sometimes required to reduce the frictional heat and consequent pellet melting when the pellets hit the pipe insides during pneumatic transport. This leaves a characteristic streamer/angel hair which is the result of solidification of pellet. Many proprietary internal treatments are available.

Long radius bends have radii generally in the range of 12/15 times pipe diameter. These bends should be crinkle free on the insides and require special techniques and equipments for bending. Pipe couplings can be conventional flanged type or can have sleeve type couplings or quick release cam-groove couplings. In the food industry quick release couplings are essential for cleaning in place (CIP).

The need for quick coupling removal is a maintenance requirement. In case of choked pipelines, use of separate compressed air inlets at critical sections of conveying pipeline especially at bends is also resorted to. In long conveying lines, the pipelines should be stepped to higher diameters towards the discharge end to reduce conveying velocities. This is specially required where delicate materials are handled and require controlled conveying velocities.

3.1.4 Material Pick-up/Feeding Devices

In dilute phase conveying, which are material into air systems, there should be means to load/feed the air/gas stream with controlled amounts of material such that the solids to air ratio by weight is compatible with characteristics of the material being conveyed. These units are generally rotary airlocks and are also called rotary feeders or rotary valves. When called rotary feeders, they generally discharge to atmosphere. These units allow material to fall into the pressurised air stream below while preventing the pressurised air to escape through them. These are critical components of dilute phase systems and require to be specified, selected and sized carefully. Bulk densities of materials vary considerably in stored, poured, tapped and loose forms and should be considered in the form in which it enters the rotary valve for sizing the valve. The air retention characteristics of the material will also drastically change the bulk densities which lead to capacity and operational problems if not taken into account when sizing the rotary airlock. The unit consists of multi-vaned rotors (normally eight) in a rigid housing with inlet and outlet openings which can be directly below each other or tangentially offset. The offset type prevents total fill of the pockets and thus eliminates shearing of particles between the stator and moving rotor vanes. Such units are generally specified in the plastics industry to avoid pellet shearing.Blow through types are used where sticky or cohesive materials do not empty from the rotor pockets. The conveying air stream is passed axially through arriving pockets to enable them to be blown through. Fig. 4 shows some of these types.

Another method of feeding materials is with a venturi eductor. This device which uses the venturi principle to pull vaccum and thereby material from the overhead hopper is generally used when blow back of extremely fine powders becomes problematic with the use of rotary valves which need necessarily to have a clearance between rotor vane tips and the housing inner diameter. However, their use is restricted generally to tonnages upto 5T/hr., bulk densities upto 700 kg./m3 and distances upto 100/150m. Their main disadvantage is the high pressure drop across them when compared to the actual conveying line pressure drop downstream. This can be as high as 5/6 times the downstream pressure drop. This depends of course on the throughput and conveying distance but would be a general figure. They are however useful in short distance transfer and the lack of any moving parts is a distinct maintenance advantage.

3.1.5 Air/Material Separating Equipments

In pneumatic conveying systems the ultimate destination of the material is always a storage silo or storage hopper. At this point, equipment that will effectively separate material from the conveying air/gas with minimum carryover of the fine material particles to the atmospheric exit is required. Depending on the particle size and its distribution in the mass, these units can be any one of the following:

Gravity Receivers
Cyclone Separators
Bag Filters

The gravity receiver is only a simple receiving vessel with suitable baffles at the point of material/air stream inlet (tangential to assist separation). Because of the relative large difference between the incoming conveying pipeline size and receiver diameter, there is a sudden deceleration of the air/material stream and the material drops out of suspension.

In many cases where the material particles are relatively large (greater than 2mm) with no fines generated during conveying, this gravity receiver alone is sufficient to separate almost all of the material. Any carryover can be arrested by providing a fine mesh filter at the outlet of the gravity receiver.

Cyclone separators are well known separating devices and use centrifugal force for their action. High efficiency and standard units are available and it would be advisable to have cyclone separators preceding a bag filter when conveying fine powder (<100 μm) so that the latter can be economically sized. Cyclone separators are available in high throughputs and high efficiency types. The latter are narrower and taller compared to the former and have a higher pressure drop across them.

Bag filters provide a physical barrier (the filter cloth) for the separation of material and air and are essential where fine particles and being conveyed. Bag filters can efficiently separate fine dusts. The bags need to be cleaned continuously to provide continued efficiency of separation. Reverse pulse jets of compressed air perform this function and are timed to release into the bags in bursts of 100 to 300 milliseconds at frequencies between 1 and 30 seconds.

3.1.6 Material Diverters

When a pneumatic conveying system feeds to many receiving points, it is necessary that the material is properly diverted to the destination points. Diverters performs this function and can be used for both dilute and dense phase systems. Depending on the material size and shape, diverters are available in the following types:

Flap type
Plug type

All diverters are built for diversion generally when conveying has temporarily stopped. Onstream diversion is generally not performed. For multiple destinations, multi-way pipe switches are available that can be indexed to any one of 8-16 outlets or more from a single inlet line either manually or mechanically (motor driven). Fig. 5 shows some of these types.

3.2 Dense phase system components and equipments

3.2.1 Pressure Tanks

Pressure tanks or blow tanks are the primary feeding devices in dense phase pneumatic conveying systems. Blow tanks shapes vary but generally have a conical portion with or without an upper cylindrical portion. They are vessels with dished ends and require to be manufactured to boiler/pressure vessel codes. The internals of these tanks may contain fluidizing devices, material pick-up hoods or may be plain depending on the application, the material characteristics and the discharge type (whether top or bottom discharge). Portable vessels are available which can be used as independent pneumatic loading devices for storage tanks, silos etc.

3.2.2 Valves

In dense phase conveying, the material inlet and material discharge valves are critical components and should be carefully manufactured/selected to ensure reliable operation. Common problems involve rapid wear of valve seats and seals and incomplete sealing. The proper selection of these valves are extremely critical to reliable dense phase conveying operations. Conventionally, the transporter has two inlets,one below the other. The top valve is generally a slide valve and shuts against bulk. The lower valve is a butterfly valve and shuts against vessel pressure. The top valves closes before the bottom one when the vessel is full.

The other valves for pressure vessel operation include on/off duty valves for air pressurisation, air pulsing, vent, scavenging, probe and line conditioning. With cycle times ranging from 10-20 cycles per hour depending on batch size, the duty requirements of these valves is relatively high and require wear resistant and low friction contact parts.

3.2.3 Pipeline Air Injection Systems

These have special applications and are used for difficult materials like carbon black, titanium dioxide etc. The pipelines can be made flexible or can have air injection points along the line to reduce pressure build-up and to condition the material/air ratio as the conveying proceeds. A critical requirement for these devices is their ability to inject air into line without malfunctioning due to material blow back.

3.2.4 Air/Material Separation Devices

Because of the low conveying velocities and high material to air ratios, air requirements for conveying are generally much lower in dense phase conveying systems. A direct economic advantage of this is the fact that cyclone separators and large bag filters are not required. In many cases where the material particle sizes are large (greater than 3mm) with no fines, the conveying pipeline can be terminated at the receiving hopper with just an oversized vent for exit of the conveying air. Where the exit air quality is critical, bag filters can be used but are quite small in size (and cost). Provision must however be made in the selection of the bag filter for the increase in conveying air volume as it expands to atmosphere at the end of the conveying line.

4.0 SYSTEM AND COMPONENT RELIABILITY

The successful and efficient operation of a complete pneumatic conveying system will depend primarily on proper system design and correct specifications of the components which should be properly matched to the system requirement specially with relation to the product being handled. To begin with, the layout should be finalised before any design work is undertaken. This is because seemingly small changes in layouts may require longer pipe runs and additional bends which could contribute to major pressure drops not considered during initial system design. Pipe size would normally not exceed 250-300mm diameter at the upper end for dilute phase systems. For dense phase systems a maximum could be 100-150mm diameter. A minimum number of bends and horizontal and ver.tical runs would be ideal. Inclined runs for dilute phase systems are to be avoided as power requirement increases for these lines. For dilute phase systems the pipelines can be taken up vertically about 30-50 pipeline diameters after the material feed point. This is because of the need to accelerate material at feed point to the pick-up velocity: a bend immediately at the feed point would impose additional acceleration load and would therefore mean higher pressure drops. In dense phase systems this does not prevail and the pipeline can be taken vertically up within 10-20 diameters of the blow tank discharge point. Elevated horizontal runs can be taken on existing plant pipe racks. If these are not available, the runs can be adjacent to plant walls or columns from where cantilever supports can hold the pipe.

4.1 System Reliability

A reliable pneumatic conveying system will deliver the required capacity with no associated problems. However, this is easier said and than done because of the large number of variables involved. The designer should consider each section of the pneumatic conveying system independently while at the same time should be able to match interfaces between units of each section. For example: when considering the design of a storage silo feeding to a pneumatic conveying system, the silo volume should be calculated based on the poured bulk density of the material since this silo will receive material from a pneumatic or mechanical device which carries slightly aerated material. When designing the pneumatic conveying system downstream of this silo, one should consider the heaviest (tapped) bulk density for pneumatic transport as insurance against unforeseen energy demands from the actual system after installation. (Unexpectedly heavy air leakage from rotary valves for example)

Some of the common system problems are:

a. System Chokes
b. Excessive dust and blow back
c. Exessive power consumption
d. Low throughput (in dense phase systems)

System chokes: In dilute phase systems, material will block the pipeline if there is insufficient velocity of the conveying air/gas. This could be due to an undersized air mover or very high material to air loadings. The latter can be reduced by reducing the rotary valve speed. The system will operate but at reduced throughput.

Inadequate pressure ratings of the air movers can also significantly reduce velocities. A few more bends in the line than originally configured can lead to high pressure drop conditions and therefore blockage. Increasing the same blower speed (if possible) can give increased air flow but at the cost of pressure. Increase in line size will need greater air flows to maintain the same velocity. For a system which has been installed and which has either low throughput or choking problems, the above trials can be made to attempt rectification. If a new air mover with increased flow rate is required, the pressure rating has to be re-specified for the same line diameter as increased air flow and therefore increased velocity will give increased pressure drops. Rectifying a system requires that all parameters be evaluated singly as well as with respect to their effect on other parameters.

Line blocks in dense phase systems would mean improper air distribution between the pressure vessel and line fluidization. It could also mean inadequate air. This can be rectified by altering timer sequences and periods on the vessel pressurization and line air injection timers for such systems. Throughput is dependent on batch size and number of batches per hour. Batch size by weight can significantly alter with materials that retain air. Vessel volumes have to be calculated allowing for this. The maximum number of batch despatches per hour should be between 10 to 15 to allow for vessel pressurization times and valve operating times.

Excessive dust and blow back: In dilute phase pressure systems fed by a rotary airlocks and handling fine powders (less than 50 microns), there is likely to be blow back of the fine powder through the clearances between rotor vane and stator. In such situations it is preferable to locate the blower at the downstream end and convert the system to a vacuum system. Alternately a bag dump hopper can be located above the rotary valve but this would be a more expensive proposition. Instead of one air mover in a pressure system, two air movers can be used, one each at the upstream and downstream ends of the system.

The system pressure drop can be distributed between these blowers such that the rotary airlock near the pressure blower sees lesser pressure than when using a single pressure blower. This will also help to eliminate or reduce fines blow back. Whenever fabric filters are used in a dilute phase system handling fine powders, it would be advisable to size the filter for low air to cloth ratios,typically between 1 and 2 m3/min/m2. While this increases cloth area, it reduces individual bag loading and prevents high pressure drops across bags thereby reducing dust bleeding through the fabric. Selection of a less porous fabric is also one solution but there would be a higher pressure drop across bags and need for more frequent bag cleaning. Filter bags automatically cleaned with reverse jets of compressed air are subject to mechanical strain due to the frequent ballooning and collapsing of bags. Bag life can be extended by a proper cleaning cycle which is best field-set after the system has been started and running for about a month or two. Small increases in offtimes of the bag cleaning cycle when maintaining operating pressure differential can lead to significant savings in compressed air consumption.

Excessive Power Consumption: This is a problem occuring only with dilute phase systems using centrifugal blowers. Blowers overdesigned in anticipation of unforeseen pressure drops will draw excessive current when actual plant conditions do not have the designed pressure drops. The remedy will only be to introduce resistance into the line by using a damper but this would lead to lower velocities downstream of the damper and possible choking. A balance will therefore have to be obtained and power consumption reduced but with a corresponding reduction in throughput.

Low Throughput (in dense phase systems): Dense phase systems are batch systems and achieve hourly throughputs by despatching a number of batches of material. A change in the number of batches despatched per hour will change the rated throughputs. 10-15 batches per hour are normal batch transfer rates in in-plant applications. This could reduce if vessel pressurisation times, valve opening/closing times and actual conveying times are incorrectly calculated. If this happens, there is no way in which throughput can be improved except by changing the dense phase tranporter vessel to a larger size. Another significant factor contributing to reducing the number of batch transfers per hour is the gravity filling time of powder into the transporter vessel. It is always a good decision to have large size inlet valves (200-300m dia.) to reduce the filling time. In cases of powders with difficult flow properties, flow aid devices must be used to increase the rate of discharge from the hopper or silo above the dense phase tranporter.

When manually feeding material through a bag dump station into a dense phase transporter, the time required for cutting open and discharging bags is to be taken into account for calculating the filling time in the overall cycle time.

Air requirements for dense phase transportation depend on the material to air ratio for the particular material and on the troughput requirement. Calculation of this air quantity should be done on the steady state throughput at each cycle rather than on the average throughput from a number of cycles. This will mean a design figure for throughput larger than the rated throughput requirement.

4.2 Component Reliability:

Failures of operating components of pneumatic conveying systems like slide gates, diverter valves, instrumentation, filters etc. are largely due to the all pervasive nature of dust. The problems aggravate especially when handling fine powders less than 50 microns size. Particular features to look for in individual components of pneumatic conveying systems are effective sealing, low friction sliding materials and methods that prevent powders from remaining and collecting in the units

4.2.1. Rotary Airlocks

Rotary valves are generally used in dilute phase pneumatic conveying systems and should have minimum air leakages at the maximum differentials (generally 1 bar). This would require extremely fine machining tolerances and relatively elaborate seals between rotor vanes and body. Valves handling fine powders have continuouus compressed air purge lines that keep powders passing through the equipment away from the critical moving parts. The housing and shaft should be of very rigid construction to absorb vibration due to particles "rolling" between the clearance of vane tips and housing. Rotary valves speeds are generally limited to 30 rpm for fine powders and can be much lesser depending on the material flowability. Higher speeds do not allow sufficient time for material to fill rotor pockets as they move past the feed opening. Where the differential is high between the feed and the discharge openings, a larger number of rotor vanes will reduce leakage with the vanes acting as a "labyrinth" seal. At the same time too many rotor vanes will form narrow wedges in which material will neither fill nor discharge. Normal vane quantities are eight.

4.2.2. Diverters

These equipments are prone to jamming if fine powder collects in the minute clearance between housing and plug (see fig. 5 - plug diverter).

The key here is again effective sealing to prevent material from reaching the areas which need to be free of any deposition. A compressed air purge line can often solve the jamming problem in existing installations. Although this means a loss of air, the diverters generally perform satisfactorily afterwards. However the construction of the unit should allow for the purge air to escape suitably.

4.2.3. Filters

Automatically cleaned or manually shaken fabric filters are an essential part of any pneumatic conveying system except where large (3 to 8mm) dust free particles are being conveyed.

Fabric filters need to be sized carefully and it would be prudent to be conservative in both selection of fabric material and in the area required for filtration. The mechanical strength of the fabric is important in automatically cleaned units where there is a constant flexing of the fabric during the life of the unit. This fatigues the fabric and could lead to premature fabric rupture. Almost all filter units manufactured today use needled felt with various properties which depend on the operating environment. The compressed air required for cleaning such units should be dry and should have sufficient pressure to create the necessary rapid ballooning/collapsing of the bags which dislodge entrapped material. The air receiver which is part of the unit should be adequately sized to hold enough air and the incoming compressed air line should be large enough to instantly replenish the consumed compressed air.

5.0 PRACTICAL TROUBLESHOOTING TIPS

In an installed plant that is not performing satisfactorily, a systematic troubleshooting procedure can often yield positive results provided the suspension velocities for dilute phase systems and the phase density for dense phase systems has been properly specified for the system by the vendor. This is generally not a problem except when one is conveying a material which has not been pneumatically conveyed before. In such a case it is imperative that the above parameters are established beforehand in a test facility.

5.1. Dilute Phase Systems

5.1.1. Confirm Correct Conveying Velocity:

The normal range of bulk densities of commonly used industrial powders is between 500 kg./m3 to 1000 kg./m3. The corresponding conveying velocities would approx. be 25m/sec. to about 40m/sec. For bulk densities between these values, the relationship is not linear but can be verified from any well known mechanical or chemical engineering reference volume. The air velocity should be established at a point near the entry point of the material into the system. 50mm NB threaded welding sockets can be welded to the conveying pipe at this point and a pitot tube velocty measuring kit preferably graduated in direct reading velocity units can then be used to establish the conveying velocity. A few readings can be taken at the same point to confirm consistency in values. A pressure gauge at the blower outlet or near the vaccuum generator inlet will give the empty line pressure drop. This gauge should have a full scale reading of not more than 1 kg.cm2 since almost all dilute phase systems operate below this limit. If the velocities are found lower than that required, the blower curves can be consulted to check if speed could be increased and the value of the corresponding pressure developed. An increase in blower speed results in greater air volume through the system and therefore higher velocities. This will result in pressure drops and hence the pressure capability of the blower must be checked. Pressure drops can be estimated from emperical formulae in any of the well known reference works on the subject.

5.1.2. Confirm Feed Rate into the System:

The material will be conveyed into the system through a rotary valve. Throughput should now be practically established since volumes are sized based on an assumed value of filling efficiency which could be different in actual operating conditions. For this the conveying pipes connections can be removed below the rotary valve and the valve run for a minute or so with material passing through it from the feed hopper or silo above the rotary valve. The quantity collected in a given time will then be a rough approximation of the rotary valve throughput rate.

5.1.3. Confirm Phase Density:

The material to air ratio can now be established by computing the weight ratio of material to air by converting the computed air flow from 5.1.1., converting it to weight units and dividing the material discharge rate established in 5.1.2. by this figure. If this figure is in the range of 2 to 8, the system still has a possibility of successful operation. (Always check the vendor figures for material to air ratio at the time of order finalisation).

5.1.4. Check Adequacy of Filter Area:

Normal air to cloth ratios are in the range of 1 to 2 m3/min./m2 of cloth area. If this is available there is adequate filter area. However, for fine powders less than 10 microns size, the air to cloth ratio should be in the region of 1 or less for effective filtration. Filter bags must be checked for glazing (clogging of pores) which can result if the cleaning mechanism has become ineffective. If cloth area is insufficient, a new filter unit has to be put in.

5.2 Dense Phase Systems:

Malfunctioning or totally inoperable dense phase systems would require more expert troubleshooting than dilute phase systems. It is always advisable to get the vendor to put in sufficient "fine tuning" capability in the dense phase vessel in order to be able to experiment at site in case of difficulty in operation. This involves methods of introducing additional air into the vessel as well as into the conveying line in various combinations of frequency and duration which are enabled by timers and solenoid valves. The most common fault in such systems is that conveying begins, pressure begins to drop in the dense phase transporter indicating that material has moved into the line and is moving and there is a sudden arrest of pressure drop with the result that the pressure gauge stays put with pressure in the vessel. This indicates line blockage and could result primarily from inadequate air but also from improperly designed geometry of the dense phase transporting and conveying pipe interface. In such a situation provision must be available for venting the captive pressure in the vessel, isolating the vessel from the choked pipeline and clearance of choke by a separate compressed air jet into the line at the start of the conveying pipe.

Other failures in dense phase systems will involve instrumentation. This needs be specified with IP55 or higher protection grades especially when the plant uses fine powders.

The key to most dense phase systms is the ability to fludize and keep air in the system during the entire transportation phase. For materials which naturally retain air, this is not a problem and dense phase conveying is a simple and straightforward affair. For materials which have difficulty in retaining air, air must be forced into the system judiciouly and the efficacy of such systems is in direct proportion to vendor experience. There are other classes of materials like geometric pellets (prevalent in the petrochemical industry) which though not retaining air, are most easily transported in dense phase although at lower phase densities than obtaining for fine fluidizable powders.

As explained elsewhere,throughput in dense phase systems is a function of batch transfer per hour and low throughputs could result because of improperly sized equipment and incorrectly specified air requirements. The non-suspension flow in dense phase systems do not depend on high velocities in particle suspension but require air for "pushing" the material through the pipe in a combination of flow work and air expansion. An interesting point is that whereas in dilute phase systems if the minimum conveying velocity is not available the system cannot be rectified, in dense phase systems the system can begin to convey with the same air quantity but at much lesser throughputs.

Dense phase conveying is more energy efficient than dilute phase conveying and is generally less prone to problems faced in dilute phase conveying. A choice between the two systems would depend on a large number of factors but if the powder is uniform, fine, fluidizable and conveyable in dense phase, the choice is clear. However, for irregularly shaped and non-uniform particles and where the material cannot transport in dense phase, there is no option to dilute phase transport.

****

References:

1. Mills, David. Troubleshooting Pneumatic Conveying Systems. Chem Engg. June 1990.

2. Kraus, Milton. Pneumatic Conveying Systems. Chem Engg. Oct. 1986.

3. Marcus RD, et al. Pneumatic Conveying of Solids.

Pneumatic Conveying System Types	
Suspension flow (Dilute Phase)	Pressure Vacuum Combination pressure-vacuum
Non-suspension flow (Dense Phase)	Single Plug Multiple Plugs (timed pulse) External Air bypass Internal Air bypass

Table 1 Types of Pneumatic Conveying Systems

Pneumatic Conveying System Components					
Air or gas Movers	Line Feeding Equipment	Diverting Equipment	Air-material Separation Equipment	Conveying Piping & Bends	Controls & Instrumentation
Centrifugal fan					
Twin Lobe Blower (Suspension flow-Dilute Phase)	Rotary Air-lock Venturi Eductors	Two way plug Two way flap	Gravity Separators Cyclone Separators	Pipes Bends -sharp radius	Pressure Indication & control Temp. Indication & control
Reciprocating and screw compressors (Non-suspension flow-Dense Phase)	Suction Nozzles	Multiway Diverters Slide Plate Valves Unit Valves	Fabric Filters	-long radius	Level indication & control

Table 2 System Components

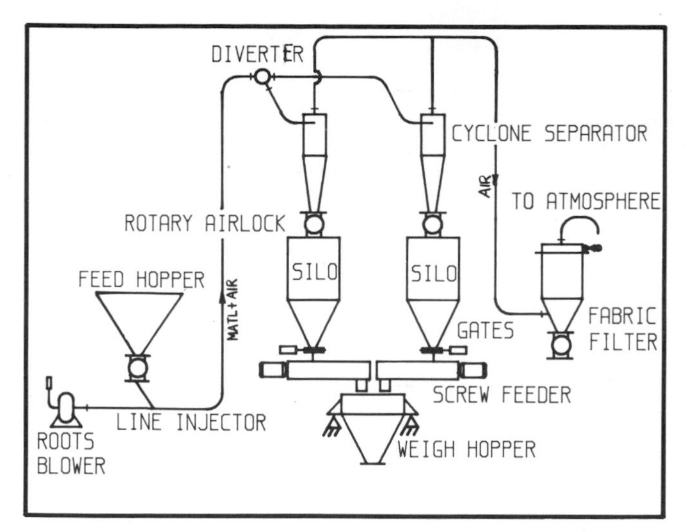

Fig 1 Dilute Phase Pneumatic Conveying System

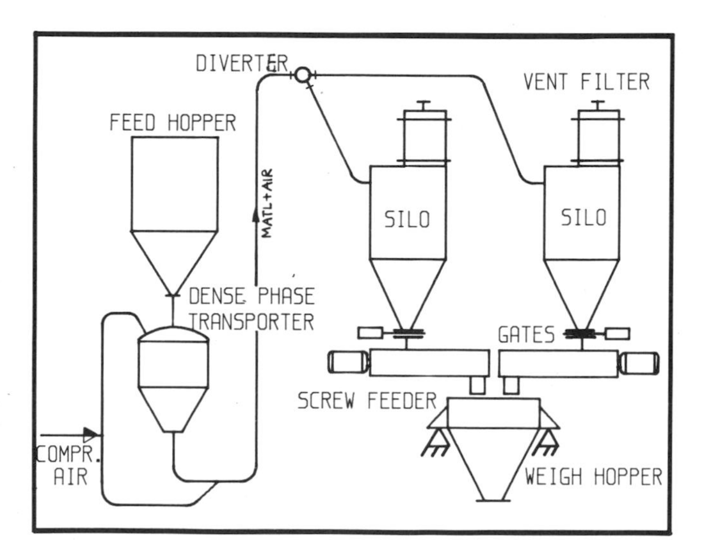

Fig 2 Dense Phase Pneumatic Conveying System

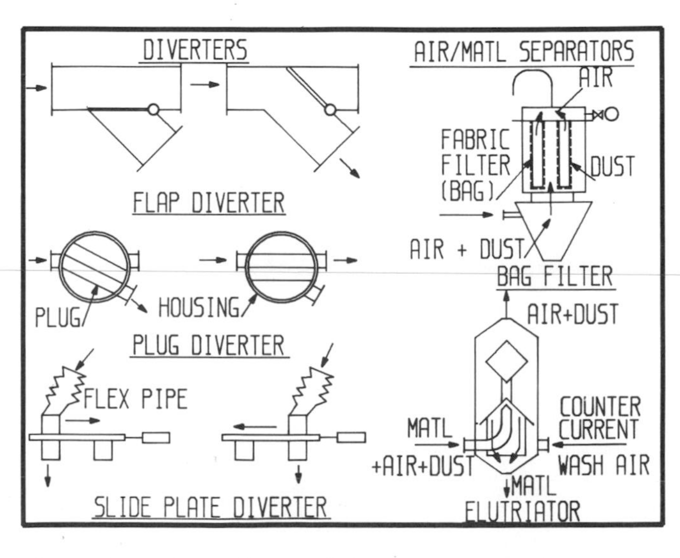

Fig 5 Diverting and Separating Equipment

MATERIAL PROPERTIES CHART

1.1.0 MATERIAL NAME________

1.2.0 BULK DENSITY (Kg/m3)
- a POURED ______________
- b TAPPED ______________
- c STORED ______________
- d AERATED ______________

1.3.0 MOISTURE CONTENT

1.4.0 ANGLE OF SLIDE (COMML. STEEL PLATE)

1.5.0 SIEVE ANALYSIS

SIEVE SIZE	RETAIN (%)	THROUGH (%)
8________	____________	____________
14________	____________	____________
20________	____________	____________
35________	____________	____________
48________	____________	____________
65________	____________	____________
100________	____________	____________
200________	____________	____________
325________	____________	____________

AVERAGE PARTICLE)
SIZE SMALLER THAN)
325 MESH. ______________

1.6.0 PROPERTIES

[ ] FREE FLOWING	[ ] SLUGGISH	[ ] ABRASIVE
[ ] CORROSIVE	[ ] EXPLOSIVE	[ ] FLUIDISABLE
[ ] HYGROSCOPIC	[ ] DELIQUESCENT	[ ] PACKS UNDER PRESSURE
[ ] INTERLOCKS	[ ] FRAGILE	[ ] FRIABLE

1.7.0 PARTICLE SIZE
- a MAX ______________
- b MIN ______________

1.8.0 ANGLE OF REPOSE (DEG)
- a POURED ______________
- b DRAINED ______________

1.9.0 TEMPERATURE AT ENTRY TO PNEU. CONV. SYSTEM (DEG. CELSIUS) ________

Fig 3 Material Properties Chart

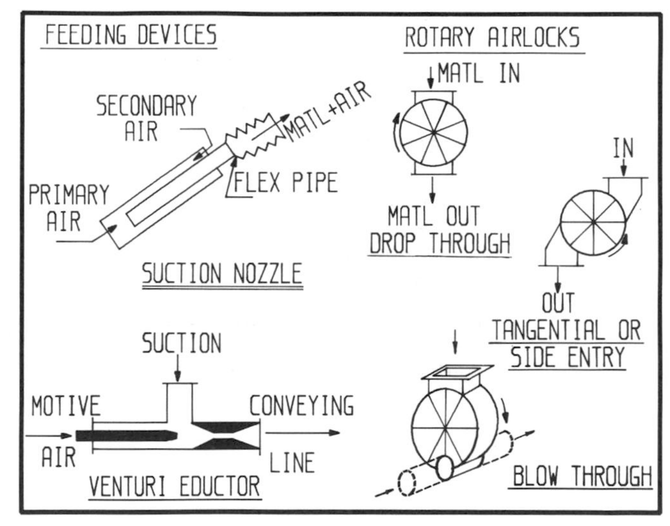

Fig 4 Material Feeding Equipment